New Insights into Milk and Dairy Products: Quality and Sustainability

New Insights into Milk and Dairy Products: Quality and Sustainability

Guest Editors

Piero Franceschi
Paolo Formaggioni

Basel • Beijing • Wuhan • Barcelona • Belgrade • Novi Sad • Cluj • Manchester

Guest Editors

Piero Franceschi
University of Parma
Parma
Italy

Paolo Formaggioni
University of Parma
Parma
Italy

Editorial Office
MDPI AG
Grosspeteranlage 5
4052 Basel, Switzerland

This is a reprint of the Special Issue, published open access by the journal *Foods* (ISSN 2304-8158), freely accessible at: https://www.mdpi.com/journal/foods/special_issues/milk_dairy_quality_sustainability.

For citation purposes, cite each article independently as indicated on the article page online and as indicated below:

Lastname, A.A.; Lastname, B.B. Article Title. *Journal Name* **Year**, *Volume Number*, Page Range.

ISBN 978-3-7258-2801-2 (Hbk)
ISBN 978-3-7258-2802-9 (PDF)
https://doi.org/10.3390/books978-3-7258-2802-9

© 2024 by the authors. Articles in this book are Open Access and distributed under the Creative Commons Attribution (CC BY) license. The book as a whole is distributed by MDPI under the terms and conditions of the Creative Commons Attribution-NonCommercial-NoDerivs (CC BY-NC-ND) license (https://creativecommons.org/licenses/by-nc-nd/4.0/).

Contents

About the Editors vii

Preface ix

Paolo Formaggioni and Piero Franceschi
New Insights into Milk and Dairy Products: Quality and Sustainability
Reprinted from: *Foods* 2024, 13, 1969, https://doi.org/10.3390/foods13131969 1

Tawfiq Alsulami, Mohamed G. Shehata, Hatem S. Ali, Abdulhakeem A. Alzahrani, Mohamed A. Fadol and Ahmed Noah Badr
Prevalence of Aflatoxins in Camel Milk from the Arabian Peninsula and North Africa: A Reduction Approach Using Probiotic Strains
Reprinted from: *Foods* 2023, 12, 1666, https://doi.org/10.3390/foods12081666 9

Lene Idland, Erik G. Bø-Granquist, Marina Aspholm and Toril Lindbäck
The Ability of Shiga Toxin-Producing *Escherichia coli* to Grow in Raw Cow's Milk Stored at Low Temperatures
Reprinted from: *Foods* 2022, 11, 3411, https://doi.org/10.3390/foods11213411 24

Konstantinos Papadimitriou, Marina Georgalaki, Rania Anastasiou, Athanasia-Maria Alexandropoulou, Eugenia Manolopoulou, Georgia Zoumpopoulou and Effie Tsakalidou
Study of the Microbiome of the Cretan Sour Cream Staka Using Amplicon Sequencing and Shotgun Metagenomics and Isolation of Novel Strains with an Important Antimicrobial Potential
Reprinted from: *Foods* 2024, 13, 1129, https://doi.org/10.3390/foods13071129 36

Natalia Tsouggou, Aleksandra Slavko, Olympia Tsipidou, Anastasios Georgoulis, Svetoslav G. Dimov, Jia Yin, et al.
Investigation of the Microbiome of Industrial PDO Sfela Cheese and Its Artisanal Variants Using 16S rDNA Amplicon Sequencing and Shotgun Metagenomics
Reprinted from: *Foods* 2024, 13, 1023, https://doi.org/10.3390/foods13071023 57

Li Sun, Annika Höjer, Monika Johansson, Karin Hallin Saedén, Gun Bernes, Mårten Hetta, et al.
Associations between the Bacterial Composition of Farm Bulk Milk and the Microbiota in the Resulting Swedish Long-Ripened Cheese
Reprinted from: *Foods* 2023, 12, 3796, https://doi.org/10.3390/foods12203796 76

Beatriz Nunes Silva, José António Teixeira, Vasco Cadavez and Ursula Gonzales-Barron
Mild Heat Treatment and Biopreservatives for Artisanal Raw Milk Cheeses: Reducing Microbial Spoilage and Extending Shelf-Life through Thermisation, Plant Extracts and Lactic Acid Bacteria
Reprinted from: *Foods* 2023, 12, 3206, https://doi.org/10.3390/foods12173206 93

Shu Huey Lim, Nyuk Ling Chin, Alifdalino Sulaiman, Cheow Hwang Tay and Tak Hiong Wong
Microbiological, Physicochemical and Nutritional Properties of Fresh Cow Milk Treated with Industrial High-Pressure Processing (HPP) during Storage
Reprinted from: *Foods* 2023, 12, 592, https://doi.org/10.3390/foods12030592 111

Fernanda Machado, Ricardo V. Duarte, Carlos A. Pinto, Susana Casal, José A. Lopes-da-Silva and Jorge A. Saraiva
High Pressure and Pasteurization Effects on Dairy Cream
Reprinted from: *Foods* 2023, 12, 3640, https://doi.org/10.3390/foods12193640 125

Cindy Bande-De León, Laura Buendía-Moreno, Adela Abellán, Pamela Manzi, Bouthaina Al Mohandes Dridi, Ismahen Essaidi, et al.
Clotting and Proteolytic Activity of Freeze-Dried Crude Extracts Obtained from Wild Thistles *Cynara humilis* L. and *Onopordum platylepis* Murb.
Reprinted from: *Foods* 2023, 12, 2325, https://doi.org/10.3390/foods12122325 138

Fabrizio Domenico Nicosia, Ivana Puglisi, Alessandra Pino, Andrea Baglieri, Rosita La Cava, Cinzia Caggia, et al.
An Easy and Cheap Kiwi-Based Preparation as Vegetable Milk Coagulant: Preliminary Study at the Laboratory Scale
Reprinted from: *Foods* 2022, 11, 2255, https://doi.org/10.3390/foods11152255 152

Cristina Anamaria Semeniuc, Mara Mandrioli, Matilde Tura, Beatrice Sabrina Socaci, Maria-Ioana Socaciu, Melinda Fogarasi, et al.
Impact of Lavender Flower Powder as a Flavoring Ingredient on Volatile Composition and Quality Characteristics of Gouda-Type Cheese during Ripening
Reprinted from: *Foods* 2023, 12, 1703, https://doi.org/10.3390/foods12081703 165

Luis Gustavo Lima Nascimento, Davide Odelli, Antônio Fernandes de Carvalho, Evandro Martins, Guillaume Delaplace, Paulo Peres de sá Peixoto Júnior, et al.
Combination of Milk and Plant Proteins to Develop Novel Food Systems: What Are the Limits?
Reprinted from: *Foods* 2023, 12, 2385, https://doi.org/10.3390/foods12122385 183

Li Chen, Hiroaki Taniguchi and Emilia Bagnicka
Microproteomic-Based Analysis of the Goat Milk Protein Synthesis Network and Casein Production Evaluation
Reprinted from: *Foods* 2024, 13, 619, https://doi.org/10.3390/foods13040619 206

Guowu Yang, Juanxiang Zhang, Xiaoyong Ma, Rong Ma, Jinwei Shen, Modian Liu, et al.
Polymorphisms of *CCSER1* Gene and Their Correlation with Milk Quality Traits in Gannan Yak (*Bos grunniens*)
Reprinted from: *Foods* 2023, 12, 4318, https://doi.org/10.3390/foods12234318 219

Xiaoyong Ma, Guowu Yang, Juanxiang Zhang, Rong Ma, Jinwei Shen, Fen Feng, et al.
Association between Single Nucleotide Polymorphisms of *PRKD1* and *KCNQ3* Gene and Milk Quality Traits in Gannan Yak (*Bos grunniens*)
Reprinted from: *Foods* 2024, 13, 718, https://doi.org/10.3390/foods13050781 230

Piero Franceschi, Wancheng Sun, Massimo Malacarne, Yihao Luo, Paolo Formaggioni, Francesca Martuzzi and Andrea Summer
Distribution of Calcium, Phosphorus and Magnesium in Yak (*Bos grunniens*) Milk from the Qinghai Plateau in China
Reprinted from: *Foods* 2023, 12, 1413, https://doi.org/10.3390/foods12071413 239

Greta Castellini, Serena Barello and Albino Claudio Bosio
Milk Quality Conceptualization: A Systematic Review of Consumers', Farmers', and Processing Experts' Views
Reprinted from: *Foods* 2023, 12, 3215, https://doi.org/10.3390/foods12173215 249

Francesca Martuzzi, Piero Franceschi and Paolo Formaggioni
Fermented Mare Milk and Its Microorganisms for Human Consumption and Health
Reprinted from: *Foods* 2024, 13, 493, https://doi.org/10.3390/foods13030493 268

Kevin Linehan, Dhrati V. Patangia, Reynolds Paul Ross and Catherine Stanton
Production, Composition and Nutritional Properties of Organic Milk: A Critical Review
Reprinted from: *Foods* 2024, 13, 550, https://doi.org/10.3390/foods13040550 287

About the Editors

Piero Franceschi

Piero Franceschi works at the Department of Food and Drug Science of Parma University. Member of the Editorial Board of Foods since year 2021, he graduated in Molecular Biology in the 2001 and earned a Ph.D. degree in Animal Productions in 2005. Since the year 2000, he has investigated the themes of milk production, cheese-making processes, and sustainability in the dairy sector, before working for the Research Centre on Animal Production of Reggio Emilia and, after, since 2012, has worked for the University of Parma. In 2021 he was Co-Editor of the Special Issue "New Insight into the Milk" of the scientific journal *Animals*. Co-Author of over 200 publications, in the year 2021 he was Contract Professor for the Course of "Innovative and Sustainable Technologies for Dairy and Meat Products" of the degree course in "Innovative and Sustainable Animal Production".

Paolo Formaggioni

Paolo Formaggioni works at the Department of Veterinary Science of Parma University. He graduated in Chemistry in 1998 and in Food Science and Technology in 2013, and completed his Ph.D. degree in Animal Production in 2005. From 2016, he has been a teacher of the "Parmigiano Reggiano Cheese" course at the Department of Food and Drug Science of Parma University. In 2021 he was Co-Editor of the Special Issue "New Insight into the Milk" of the scientific journal *Animals*. His general interests are in food science and technology. His main research fields are milk and cheese quality; cheese yield and cheesemaking efficiency; cheese proteolysis and lipolysis; effect of animal breed, of milk cooling, and of milk somatic cells on milk quality for cheesemaking; milk of different species. He is Co-Author of more than 200 publications.

Preface

Currently, there is a need to consider the increasingly substantial issues related to sustainability, whether they are economic, environmental, or social. In the dairy sector, these topics are strongly felt, and, even if in recent years, considerable progress has been made in their improvement, much work remains to be performed. For this reason, the Editors of this Special Issue considered it important to collect some research contributions on dairy quality and sustainability. This Special Issue is thus addressed to all readers and researchers interested in research in the dairy field, those who have an overview of some of the current trends in research in this sector. These innovative studies can serve as a valuable resource for generating new research ideas and developing novel and more sustainable dairy products.

With this SI, we have taken the opportunity to contribute to the push towards a production system that is more respectful of future generations.

Piero Franceschi and Paolo Formaggioni
Guest Editors

Editorial

New Insights into Milk and Dairy Products: Quality and Sustainability

Paolo Formaggioni [1,*] and Piero Franceschi [2,*]

1 Department of Veterinary Science, University of Parma, Via del Taglio 10, 43126 Parma, Italy
2 Department of Food and Drug Science, University of Parma, Via delle Scienze 27/A, 43124 Parma, Italy
* Correspondence: paolo.formaggioni@unipr.it (P.F.); piero.franceschi@unipr.it (P.F.)

Citation: Formaggioni, P.; Franceschi, P. New Insights into Milk and Dairy Products: Quality and Sustainability. *Foods* **2024**, *13*, 1969. https://doi.org/10.3390/foods13131969

Received: 10 June 2024
Accepted: 19 June 2024
Published: 21 June 2024

Copyright: © 2024 by the authors. Licensee MDPI, Basel, Switzerland. This article is an open access article distributed under the terms and conditions of the Creative Commons Attribution (CC BY) license (https://creativecommons.org/licenses/by/4.0/).

1. Introduction

The dairy industry is confronting a major challenge that could profoundly affect its future and its long-standing role as a cornerstone of human nutrition [1,2].

FAO, in its last report on the global trends, estimated an increase in the world's population to almost 10 billion people by 2050 and thus, by 2050, agriculture must produce about 50% more food and feed than it did in 2013 [3]. Meeting the increased demand for food will not be the only challenge; indeed, beyond producing food and feed, the agriculture and food industries will have to produce it in a sustainable way [4].

Therefore, the primary mission will lie in providing adequate solutions to meet the demand for nutritionally balanced and environmentally, economically and socially sustainable products [5]. To this aim, generally, ensuring the safety of dairy products is the most important requirement. However, the quality of milk and its derived products is crucial. For instance, the capacity of milk to coagulate with rennet is essential for cheese production, as it impacts both the yield and the quality of the cheese [6]. In addition to this, rheological and microbiological properties are significant for obtaining dairy products that serve various purposes [6]. Therefore, gaining new insights into how genetic, physiological, pathological, environmental, and technological factors influence the quality of milk and dairy products will contribute to the progress of the sector.

Furthermore, milk and dairy products are vital sources of nutrients for humans, providing proteins, fats, calcium, and vitamins. The dairy industry also produces several by-products, such as whey and buttermilk, which are valuable, due to their high nutritional content and can be repurposed for other uses [7]. This repurposing also helps in reducing the environmental pollution caused by the industry [7].

In addition, milk from certain species has not been extensively studied or characterised; this untapped potential could be used to create new dairy products.

2. An Overview on the Published Articles

This Special Issue, titled "New Insights into Milk and Dairy Products: Quality and Sustainability", has collected 19 articles (14 original research studies and 5 reviews) that make significant contributions to the field. These articles can be categorised into four main groups: the first category includes studies that address the microbiological aspects of milk, cheeses, and other dairy products; the second category has a focus on product innovation, featuring seven studies that discuss technological advancements and compositional aspects related to process improvements, new milk coagulants, and novel dairy foods; the third category regards papers in which genetics are applied to milk quality, by means of identification of functional genes or markers; the fourth category includes three reviews and an original study addressing various and base aspects of dairy science.

Microbiological aspects of milk, cheeses, and dairy products. Alsulami et al. [contribution 1] focused their study on the sustainability of camel milk, a nutritious dairy product widely consumed in the Middle East, particularly in Saudi Arabia. The study addressed the major concern of aflatoxin contamination in camel milk. Samples from the Arabian Peninsula and North Africa were analysed for aflatoxins B1 and M1 by means of ELISA and fluorometer methods. The results showed significant aflatoxin M1 contamination and cross-contamination with aflatoxin B1. This study also demonstrated the effectiveness of two probiotic bacteria strains, *Lactobacillus plantarum* NRC21 and *Lactobacillus acidophilus* NRC06 (both categorised as probiotic strains), in inhibiting toxigenic fungi and reducing aflatoxin levels in both liquid media and spiked samples of camel milk. The findings suggest that fermenting camel milk with these probiotic strains could be an effective method for reducing aflatoxin contamination.

Idland et al. [contribution 2] also address hygienic–sanitary issues and their impact on human health. In particular, they investigated the presence and proliferation of Shiga toxin-producing *Escherichia coli* (STEC) in raw milk, considering its association with human illnesses like diarrhoea, haemorrhagic colitis, and haemolytic uremic syndrome. The pathogenicity of STEC is linked to its production of Shiga toxin (Stx) and its capability to attach to the intestinal epithelium through the adhesion protein intimin, which is encoded by the eae gene (intimin-encoding gene). The authors studied four eae-positive STEC isolates from Norwegian dairy herds, analysing their ability to grow in unpasteurised milk (UPM) under various temperature conditions. Genome analysis indicated that three of the isolates were clonal, suggesting transmission between farms. All isolates produced Shiga toxin, with one capable of growing at 8 °C, posing a significant health risk. This study suggests the need for better consumer awareness and proper refrigeration practices; moreover, even properly stored unpasteurised milk poses an increased risk of illness, especially for vulnerable populations like young children, the elderly, and immunocompromised individuals.

Papadimitriou et al. [contribution 3] investigated the microbiome of Staka (a Cretan sour cream) utilising various methods, including culture-based techniques, amplicon sequencing, and shotgun metagenomics. Staka is a traditional Greek cultured cream made from spontaneously fermented sheep's milk or a mixture of sheep's and goat's milk. The study revealed that the samples were predominantly composed of *Lactococcus* and *Leuconostoc* species, along with other bacteria such as *Streptococcus* and *Enterococcus*, and Gram-negative genera like *Enterobacter*, *Pseudomonas*, *Buttiauxella*, *Escherichia-Shigella*, and *Hafnia*. They also found common genera of yeasts and moulds. Through shotgun metagenomics, specific species were identified. This study also isolated novel strains from Staka with antibacterial potential. Furthermore, several LABs, *Hafnia paralvei* and *Pseudomonas* spp., have antimicrobial activity against pathogenic bacteria, indicating their potential use in food safety and biomedical applications.

Tsouggou et al. [contribution 4] examined the microbiome of industrial PDO Sfela cheese and its artisanal variants utilising 16S rDNA amplicon sequencing and shotgun metagenomics, to better understand its complex microbial ecosystem, which is crucial for sustainably enhancing production and safety measures. Sfela, a white, salted PDO Greek cheese from the Peloponnese region, is made from sheep's milk, goat's milk, or a mixture of both. The study examined two PDO industrial Sfela samples and two non-PDO variants (Xerosfeli and Sfela touloumotiri) using MALDI-TOF MS, 16S rDNA amplicon sequencing, and shotgun metagenomics. Analysis with culture media revealed the presence of *Lactiplantibacillus plantarum*, *Enterococcus faecium*, *Pediococcus pentosaceus*, *Levilactobacillus brevis*, and *Streptococcus thermophilus* as the most common bacterial species. Shotgun analysis revealed *Streptococcus thermophilus* dominance in industrial Sfela 1 and high levels of *Lactococcus lactis* in industrial Sfela 2. The samples of artisanal Xerosfeli and Sfela touloumotiri were mainly composed of *Streptococcus thermophilus* and *Tetragenococcus halophilus*, respectively. Additionally, *Debaryomyces hansenii* was the only yeast present in quantities exceeding 1%, and was detected only in Sfela touloumotiri samples.

Sun et al. [contribution 5] investigated the relationship between the microbiota present in raw milk and that found in the resulting cheese, specifically a Swedish long-ripened cheese, with the aim of understanding why cheeses generally take longer to mature now than in the past. Three commercial farm clusters were created to introduce variability in the microbiota of dairy silo milk utilised for the production of cheese. This latter process took place in three different periods throughout the year, with milk from each farm cluster being collected separately, pasteurised, and processed into cheese. Samples were collected at different stages, including herd bulk milk and dairy silo milk, starting cultures, early processing samples and samples of cheese at various maturation stages (7–20 months) and analysed by means of 16S rRNA amplicon sequencing. The microbiota in herd bulk milk varied significantly among periods and among clusters, while the microbiota in dairy silo milk showed differences only across periods. The microbiota in cheese differed by periods and groups, with *Leuconostoc* and *Lactococcus* emerging as the predominant genera in both early processing and samples of cheese. Surprisingly, the abundance of *Lactobacillus* was very low during cheese ripening, and even at later stages starter lactic acid bacteria were dominant.

The review by Silva et al. [contribution 6] discusses the microbiological quality of cheese made from raw milk. This review focuses specifically on alternative systems for milk sanitisation other than pasteurisation. Plant extracts and lactic acid bacteria seem to offer promising methods for minimising microbial contamination in artisanal raw milk cheese. This is attributed to their natural components, like phenolic compounds in plant extracts, and the capacity of lactic acid bacteria to produce antimicrobial substances such as organic acids and bacteriocins. Furthermore, thermisation is considered an alternative strategy to pasteurisation. It aims to preserve the sensory qualities of artisanal cheeses while also effectively inactivating microorganisms by disrupting their cellular structure and functions. This review explores the antimicrobial mechanisms, benefits, drawbacks, and practical applications of all three strategies.

Process technological improvements, new milk coagulants, and novel foods. The papers in this section can in turn be grouped into three subsections, each constituted by two studies: high pressure as a sanitising process for milk or cream; new vegetable coagulants for milk; the addition of vegetable to milk to obtain novel foods. The six papers are presented in this order.

Lim et al. [contribution 7] investigated the impact of high-pressure processing (HPP) for 10 min at 600 MPa on the quality, safety, and shelf life of raw milk stored at 6 °C for 60 days. This method provides an alternative to thermal processing and preserves the nutritional integrity of raw milk. HPP-treated milk satisfied all microbiological safety standards and demonstrated a shelf life exceeding 60 days, even under hot and humid conditions. Additionally, it effectively preserved nearly all vitamins and minerals, including phosphorus (99.4%), calcium (99.3%), and magnesium (99.1%). However, over the 60-day storage period, there was partial degradation of vitamins A (25%), B3 (91%), B5 (35%), B6 (80%), and C (85%), as well as minerals like zinc (18%) and potassium (5%), compared to fresh milk. This study highlights the significant benefits of adopting advanced HPP processing technology in the dairy industry for maintaining the physico-chemical and nutritional properties of milk and extending its shelf life beyond 60 days.

The study by Machado et al. [contribution 8] also evaluated the effects of high-pressure processing (HPP) at 450 and 600 MPa for 5 and 15 min at 7 °C, compared to thermal pasteurisation at 75 °C for 15 s, on the microbiological and physico-chemical quality of dairy cream. The assessments were made both immediately after processing and during refrigerated storage at 4 °C. HPP-treated samples were still microbiologically safe even after 51 days of storage, unlike samples treated with thermal pasteurisation. HPP effectively reduced populations of *Escherichia coli* and *Listeria innocua* by more than 6 log units to undetectable levels (1.00 log CFU/mL). The pH, colour, and fatty acid profiles of the cream remained stable under various processing conditions and throughout storage. Furthermore, there was a tendency for volatile compounds (VOCs) to increase in all treated samples during storage, especially acids and aliphatic hydrocarbons. These findings suggest that HPP can significantly extend the shelf life of highly perishable dairy cream by at least 15 days compared to thermal pasteurisation.

Bande-De-Léon et al. [contribution 9] studied the effect of two novel plant milk coagulants (freeze-dried extracts from *Cynara humilis* L. (CH) and *Onopordum platylepis* Murb. (OP)) and compared their coagulation and proteolytic activities with those of *Cynara cardunculus* L. (CC). They examined the impact of extract concentration (5–40 mg/mL), pH (5–8), temperature (20–85 °C), and $CaCl_2$ concentration (5–70 mM) on the milk coagulation activity (MCA) of CC, CH, and OP extracts. At the same concentration, CC exhibited significantly higher MCA values. Among these various extracts, the clotting activity of OP increased with increasing temperature, reaching a maximum value at 70 °C. Adding $CaCl_2$ improved the coagulation ability of the extracts, particularly for OP and CH. Additionally, the proteolytic activity and the hydrolysis rate increased over time and with higher enzyme concentrations, with CC showing the highest caseinolytic activity among all the extracts.

Nicosia et al. [contribution 10] studied the use of kiwi fruit aqueous extract as a cheese coagulant. In particular, they used SDS-PAGE to detect actinidin, the kiwi enzyme responsible for the hydrolysis of casein, in various parts (pulp, peel, and whole fruit) of both ripe and unripe kiwi fruits. Actinidin was present in both the peel and the pulp. While the peel extract could partially hydrolyse skimmed milk, it could not degrade casein. On the contrary, the pulp extract demonstrated hydrolytic activity against both the casein and the skimmed milk. Ripe kiwi extracts showed higher hydrolytic activity than unripe ones. Using a 3% (v/v) extract from ripe kiwi pulp resulted in a curd yield of 20.27%, similar to that achieved with chymosin. In summary, the extraction method for the aqueous kiwi extract is fast, economical, chemical-free, and environmentally friendly, making it an effective vegetable coagulant for cheese production.

Semeniuc et al. [contribution 11] developed a Gouda-type cheese from cow milk, flavoured with pollen of lavender (0.5 g/L of mature milk), and aged for 30 days at 14 °C, with 85% of relative humidity. They evaluated the physico-chemical, microbiological, and textural properties of the lavender cheese and control cheese at 10-day intervals. Lavender pollen significantly influenced the sensory and microbiological properties, as well as the volatile compounds of the cheese, but had a modest impact on its physico-chemical and textural properties. During the ripening process, the moisture, carbohydrate contents, pH, viscosity, elasticity, and chewiness of both cheeses decreased, while increases were observed for protein content, titratable acidity, ash, sodium chloride content, hardness, streptococci, lactobacilli, and volatile compounds. Despite lavender's known antibacterial effects against *Escherichia coli* and *Clostridium butyricum*, it did not inhibit the growth of microorganisms of the starter culture; instead, it promoted the growth of lactic acid bacteria. Sensory evaluations revealed that the overall score for the lavender cheese was slightly lower but comparable to the control cheese, with which consumers were more familiar.

In their review, Lima Nascimento et al. [contribution 12] examine the integration of milk proteins with plant proteins, as a strategy to address the functional and sensory limitations of plant proteins. This combination results in various colloidal systems, such as suspensions, emulsions, gels, and foams, commonly found in food products. For example, adding plant protein can solve technical problems, like preventing gel formation in high-protein milk drinks during packaging. Dairy proteins can also improve the solubility of certain plant proteins in dispersions of mixed protein. By understanding the technical and functional properties of these proteins and their responses to environmental conditions, such as temperature and pH, new products like yogurt, analogues of cheese, and beverages with desired textures and sensory properties can be developed. To further optimise these systems and their applications in the food industry, innovative techniques to modify the technical and functional characteristics of both plant and dairy proteins, like pulsed electric fields, precision fermentation, and high hydrostatic pressure, are being evaluated.

Genetics applied to milk quality: identification of functional genes or markers. The synthesis of milk protein is regulated by a complex network involving numerous signalling pathways. Chen et al.'s [contribution 13] study aimed to elucidate these pathways in goat mammary epithelial cells (GMECs) utilising microproteomic techniques, and to identify the key genes involved. Their analysis identified over 2253 proteins and annotated 323 pathways.

This study highlighted the significant role of the IRS1 (insulin receptor substrate 1) gene in influencing casein content in goat milk and the pathways that are involved in milk protein synthesis in GMECs. Altering IRS1 expression increased the amounts of β-casein and κ-casein but decreased α-casein levels. By identifying proteins that, in response to IRS1 silencing, were differentially expressed, the researchers gained new insights into the network and signalling pathways associated with goat milk protein synthesis. These findings could lead to new strategies for modifying the content of casein in the dairy goat sector and for developing milk products.

Yang et al. [contribution 14] studied the polymorphisms in the CCSER1 gene of Gannan yaks, identifying three SNP loci and analysing their relationship with milk quality. The study found that all three SNPs showed moderate polymorphism. The analysis of correlations revealed that the mutant genotype at the CCSER1 g.183,843A>G locus significantly increased milk fat content ($p < 0.05$). Additionally, mutant genotypes at the CCSER1 g.222,717C>G and g.388,723G>T loci significantly increased both casein and protein contents in milk from Gannan yak ($p < 0.05$). Thus, mutations at these loci (g.183,843A>G, g.222,717C>G, and g.388,723G>T) notably improved the quality of milk from Gannan yak. The identification of these SNPs can allow the development of further research and application in selecting Gannan yaks, helping in the improvement and optimisation of their milk quality.

The study by Ma et al. [contribution 15] also investigated the polymorphism of Gannan yak genes. In particular, they specifically focused on the PRKD1 and KCNQ3 genes. This was a pioneering study examining the connection between these gene polymorphisms and dairy traits; in particular, it evaluated the relationship between the single nucleotide polymorphisms (SNPs) of these two genes and the milk quality of Gannan yaks, aiming to identify potential molecular markers for breeding. Using a technology Illumina yak cGPS 7K liquid-phase microarray, they detected, in 172 lactating Gannan yaks, three new SNPs: two in the PRKD1 gene (g.283,619T>C, g.283,659C>A) and one in the KCNQ3 gene (g.133,741T>C). Association analysis revealed significant correlations between these gene polymorphisms and milk quality traits. Specifically, mutations at these loci were found to significantly improve the levels of fat, lactose, protein, casein, non-fat milk solids (SNF), and acidity in Gannan yak milk. Therefore, genotyping the PRKD1 and KCNQ3 genes can effectively improve milk quality in Gannan yaks.

Reviews and experimental studies on general aspects of dairy. Franceschi et al. [contribution 16] studied the distribution between the colloidal and soluble phases of calcium, phosphorus and magnesium, and their level within yak milk casein micelles. They compared nine samples of yak milk from Qinghai, China, with nine bulk cow milk samples. The authors found similar levels of colloidal calcium, higher levels of magnesium, and lower levels of colloidal phosphorus per casein unit in yak milk. Yak milk was characterised by high casein and mineral content, and was particularly rich in colloidal forms of Ca, P, and Mg, which could enhance the bioavailability of calcium and phosphorus during digestion. Moreover, yak milk had lower prosthetic phosphorus per casein unit, suggesting less phosphorylated amino acids, which may affect the micelle structure, as well as the processability and the digestion of yak milk casein in comparison to cow milk casein.

The aim of the systematic review of Castellini et al. [contribution 17] was to examine and categorise the conceptual attributes of milk quality, as perceived by citizen consumers, farmers, and processing experts. Using PRISMA guidelines, they screened 409 papers and assessed 20 full-text papers. This review identified 12 main attributes defining milk quality, the most prominent of which were nutritional quality/healthiness, hygienic quality, and knowledge and attitudes of workers. Farmers and processing experts had similar technically focused perceptions of milk quality, emphasising expertise and knowledge. On the contrary, citizen consumers gave a representation of milk quality in more simplistic and subjective terms. This study underlined the need to create a common platform for communication and exchange of knowledge, to align the different perceptions and expectations of milk quality.

The review by Martuzzi et al. [contribution 18] summarises recent research on microorganisms in fermented products from mare milk and their potential functional properties. Mare milk is consumed by approximately 30 million people worldwide and is primarily used in Asia and Eastern Europe to produce fermented and alcoholic beverages like koumiss, airag, or chigee, made using bacteria and lactose-fermenting yeast cultures. The review details the main bacterial and yeast species found in these products, highlighting a complex population that includes lactic acid bacteria, yeasts, acetic acid bacteria, and white moulds. The focus of this review is on the potential health benefits and functional properties of these mare milk products, making them highly nutraceutical foods, with the aim of optimising their use in industrial production, particularly for koumiss.

The review by Linehan et al. [contribution 19] compared milk production, nutritional, and compositional properties between conventional and organic dairy systems, highlighting the health benefits of organic milk and the global landscape of organic dairy production. During the past two decades, organic dairy has increased in importance, due to concerns over the use of antibiotics, fertilisers, and pesticides, as well as animal health, and increasing environmental and self-health awareness. Most reports suggest that milk generally has beneficial effects on human health, with few, if any, negative effects. Organic milk offers some further benefits, due to a lower omega-6 to omega-3 ratio, attributed to pasture-based practices of feeding. However, organic milk production can be difficult in some regions, due to high costs and geographical conditions. The review also highlighted future perspectives and identified knowledge gaps in organic dairy management.

3. Further Remarks and Conclusions

An interesting aspect of this Special Issue is that contributions related to milks other than cow's milk have also been included: camel milk [contribution 1], mare milk [contribution 18], yak milk [contributions 14–16], goat milk [contribution 13], goat milk, sheep milk, or a mixture of goat and sheep milk [contributions 3,4].

Overall, this Special Issue "New Insights into Milk and Dairy Products: Quality and Sustainability" presents readers with a wealth of innovative information that can serve as a valuable resource for sustainability, for generating new research ideas and developing novel dairy products.

To conclude, we want to express our gratitude to the research teams mentioned earlier for their valuable contributions to this collection of articles present in this Special Issue. Their studies demonstrate the diverse and interdisciplinary nature of cheese and dairy research.

Author Contributions: Conceptualisation, P.F. (Piero Franceschi) and P.F. (Paolo Formaggioni); methodology, P.F. (Piero Franceschi) and P.F. (Paolo Formaggioni); software, P.F. (Piero Franceschi) and P.F. (Paolo Formaggioni); formal analysis, P.F. (Piero Franceschi) and P.F. (Paolo Formaggioni); investigation, P.F. (Piero Franceschi) and P.F. (Paolo Formaggioni); resources, P.F. (Piero Franceschi) and P.F. (Paolo Formaggioni); data curation, P.F. (Piero Franceschi) and P.F. (Paolo Formaggioni); writing—original draft preparation, P.F. (Piero Franceschi) and P.F. (Paolo Formaggioni); writing—review and editing, P.F. (Piero Franceschi) and P.F. (Paolo Formaggioni); visualisation, P.F. (Piero Franceschi) and P.F. (Paolo Formaggioni); supervision, P.F. (Piero Franceschi) and P.F. (Paolo Formaggioni); project administration, P.F. (Piero Franceschi) and P.F. (Paolo Formaggioni). All authors have read and agreed to the published version of the manuscript.

Data Availability Statement: The data presented in this study are available on request from the corresponding author.

Acknowledgments: This work has been carried out in the frame of the ALIFAR project, funded by the Italian Ministry of University through the program "Dipartimenti di Eccellenza 2023-2027".

Conflicts of Interest: The authors declare that there are no conflicts of interest in this research article.

List of Contributions:

1. Alsulami, T.; Shehata, M.G.; Ali, H.S.; Alzahrani, A.A.; Fadol, M.A.; Badr, A.N. Prevalence of aflatoxins in camel milk from the Arabian peninsula and North Africa: a reduction approach using probiotic strains. *Foods* **2023**, *12*, 1666. https://doi.org/10.3390/foods12081666.
2. Idland, L.; Bø-Granquist, E.G.; Aspholm, M.; Lindbäck, T. The ability of shiga toxin-producing *Escherichia coli* to grow in raw cow's milk stored at low temperatures. *Foods* **2022**, *11*, 3411. https://doi.org/10.3390/foods11213411.
3. Papadimitriou, K.; Georgalaki, M.; Anastasiou, R.; Alexandropoulou, A.M.; Manolopoulou, E.; Zoumpopoulou, G.; Tsakalidou, E. Study of the microbiome of the Cretan sour cream Staka using amplicon sequencing and shotgun metagenomics and isolation of novel strains with an important antimicrobial potential. *Foods* **2024**, *13*, 1129. https://doi.org/10.3390/foods13071129.
4. Tsouggou, N.; Slavko, A.; Tsipidou, O.; Georgoulis, A.; Dimov, S.G.; Yin, J.; Vorgias, C.E.; Kapolos, J.; Papadelli, M.; Papadimitriou, K. Investigation of the microbiome of industrial PDO Sfela cheese and its artisanal variants using 16S rDNA amplicon sequencing and shotgun metagenomics. *Foods* **2024**, *13*, 1023. https://doi.org/10.3390/foods13071023.
5. Sun, L.; Höjer, A.; Johansson, M.; Saedén, K.H.; Bernes, G.; Hetta, M.; Gustafsson, A.H.; Dicksved, J.; Lundh, Å. Associations between the bacterial composition of farm bulk milk and the microbiota in the resulting Swedish long-ripened cheese. *Foods* **2023**, *12*, 3796. https://doi.org/10.3390/foods12203796.
6. Silva, B.N.; Teixeira, J.A.; Cadavez, V.; Gonzales-Barron, U. Mild heat treatment and biopreservatives for artisanal raw milk cheeses: reducing microbial spoilage and extending shelf-life through thermisation, plant extracts and lactic acid bacteria. *Foods* **2023**, *12*, 3206. https://doi.org/10.3390/foods12173206.
7. Lim, S.H.; Chin, N.L.; Sulaiman, A.; Tay, C.H.; Wong, T.H. Microbiological, physicochemical and nutritional properties of fresh cow milk treated with industrial High-Pressure Processing (HPP) during storage. *Foods* **2023**, *12*, 592. https://doi.org/10.3390/foods12030592.
8. Machado, F.; Duarte, R.V.; Pinto, C.A.; Casal, S.; Lopes-da-Silva, J.A.; Saraiva, J.A. High pressure and pasteurization effects on dairy cream. *Foods* **2023**, *12*, 3640. https://doi.org/10.3390/foods12193640.
9. Bande-De León, C.; Buendía-Moreno, L.; Abellán, A.; Manzi, P.; Al Mohandes Dridi, B.; Essaidi, I.; Aquilanti, L.; Tejada, L. Clotting and proteolytic activity of freeze-dried crude extracts obtained from wild thistles *Cynara humilis* L. and *Onopordum platylepis* Murb. *Foods* **2023**, *12*, 2325. https://doi.org/10.3390/foods12122325.
10. Nicosia, F.D.; Puglisi, I.; Pino, A.; Baglieri, A.; La Cava, R.; Caggia, C.; Fernandes de Carvalho, A.; Randazzo, C.L. An easy and cheap kiwi-based preparation as vegetable milk coagulant: preliminary study at the laboratory scale. *Foods* **2022**, *11*, 2255. https://doi.org/10.3390/foods11152255.
11. Semeniuc, C.A.; Mandrioli, M.; Tura, M.; Socaci, B.S.; Socaciu, M.I.; Fogarasi, M.; Michiu, D.; Jimborean, A.M.; Mureșan, V.; Ionescu, S.R.; Rotar, M.A.; Gallina Toschi, T. Impact of lavender flower powder as a flavoring ingredient on volatile composition and quality characteristics of Gouda-type cheese during ripening. *Foods* **2023**, *12*, 1703. https://doi.org/10.3390/foods12081703.
12. Lima Nascimento, L.G.; Odelli, D.; Fernandes de Carvalho, A.; Martins, E.; Delaplace, G.; Peres de sá Peixoto Júnior, P.; Nogueira Silva, N.F.; Casanova, F. Combination of milk and plant proteins to develop novel food systems: what are the limits? *Foods* **2023**, *12*, 2385. https://doi.org/10.3390/foods12122385.
13. Chen, L.; Taniguchi, H.; Bagnicka, E. Microproteomic-based analysis of the goat milk protein synthesis network and casein production evaluation. *Foods* **2024**, *13*, 619. https://doi.org/10.3390/foods13040619.
14. Yang, G.; Zhang, J.; Ma, X.; Ma, R.; Shen, J.; Liu, M.; Yu, D.; Feng, F.; Huang, C.; Ma, X.; La, Y.; Guo, X.; Yan, P.; Liang, C. Polymorphisms of CCSER1 gene and their correlation with milk quality traits in Gannan yak (*Bos grunniens*). *Foods* **2023**, *12*, 4318. https://doi.org/10.3390/foods12234318.
15. Ma, X.; Yang, G.; Zhang, J.; Ma, R.; Shen, J.; Feng, F.; Yu, D.; Huang, C.; Ma, X.; La, Y.; Wu, X.; Guo, X.; Chu, M.; Yan, P.; Liang, C. Association between single nucleotide polymorphisms of PRKD1 and KCNQ3 gene and milk quality traits in Gannan yak (*Bos grunniens*). *Foods* **2024**, *13*, 781. https://doi.org/10.3390/foods13050781.

16. Franceschi, P.; Sun, W.; Malacarne, M.; Luo, Y.; Formaggioni, P.; Martuzzi, F.; Summer, A. Distribution of calcium, phosphorus and magnesium in yak (*Bos grunniens*) milk from the Qinghai plateau in China. *Foods* **2023**, *12*, 1413. https://doi.org/10.3390/foods12071413.
17. Castellini, G.; Barello, S.; Bosio, A.C. Milk quality conceptualization: A systematic review of consumers', farmers', and processing experts' views. *Foods* **2023**, *12*, 3215. https://doi.org/10.3390/foods12173215.
18. Martuzzi, F.; Franceschi, P.; Formaggioni, P. Fermented mare milk and its microorganisms for human consumption and health. *Foods* **2024**, *13*, 493. https://doi.org/10.3390/foods13030493.
19. Linehan, K.; Patangia, D.V.; Ross, R.P.; Stanton, C. Production, composition and nutritional properties of organic milk: A critical review. *Foods* **2024**, *13*, 550. https://doi.org/10.3390/foods13040550.

References

1. Augustin, M.A.; Udabage, P.; Juliano, P.; Clark, P.T. Towards a More Sustainable Dairy Industry: Integration Across the Farm–factory Interface and the Dairy Factory of the Future. *Int. Dairy J.* **2013**, *31*, 2–11. [CrossRef]
2. Malcata, X.F. Critical Issues Affecting the Future of Dairy Industry: Individual Contributions in the Scope of a Global Approach. *J. Dairy Sci.* **1999**, *82*, 1595–1611. [CrossRef] [PubMed]
3. Food and Agriculture Organization of the United Nations (FAO). *The Future of Food and Agriculture–Trends and Challenges*; FAO: Rome, Italy, 2017; pp. 1–151, ISBN 978-92-5-109551-5. Available online: https://reliefweb.int/report/world/future-food-and-agriculture-trends-and-challenges?gad_source=1&gclid=CjwKCAjwvIWzBhAlEiwAHHWgvZvwji90_QWPTHNVzv5Lj19KCw0sdFzRvUS-u_suhOqzPIq6SNsz5hoCPSAQAvD_BwE (accessed on 6 June 2024).
4. Food and Agriculture Organization of the United Nations (FAO). *The Future of Food and Agriculture–Alternative Pathways to 2050*; FAO: Rome, Italy, 2018; pp. 1–224, ISBN 978-92-5-130158-6. Available online: https://openknowledge.fao.org/server/api/core/bitstreams/2c6bd7b4-181e-4117-a90d-32a1bda8b27c/content (accessed on 6 June 2024).
5. Faccia, M. Chemical and Technological Characterization of Dairy Products. *Foods* **2020**, *9*, 1475. [CrossRef] [PubMed]
6. Auldist, M.J.; Johnston, K.A.; White, N.J.; Fitzsimons, W.P.; Boland, M.J. A Comparison of the Composition, Coagulation Characteristics and Cheesemaking Capacity of Milk from Friesian and Jersey Dairy Cows. *J. Dairy Res.* **2004**, *71*, 51–57. [CrossRef] [PubMed]
7. Franceschi, P.; Martuzzi, F.; Formaggioni, P.; Malacarne, M.; Summer, A. Seasonal Variations of the Protein Fractions and the Mineral Contents of the Cheese Whey in the Parmigiano Reggiano Cheese Manufacture. *Agriculture* **2023**, *13*, 165. [CrossRef]

Disclaimer/Publisher's Note: The statements, opinions and data contained in all publications are solely those of the individual author(s) and contributor(s) and not of MDPI and/or the editor(s). MDPI and/or the editor(s) disclaim responsibility for any injury to people or property resulting from any ideas, methods, instructions or products referred to in the content.

Article

Prevalence of Aflatoxins in Camel Milk from the Arabian Peninsula and North Africa: A Reduction Approach Using Probiotic Strains

Tawfiq Alsulami [1,*], Mohamed G. Shehata [2,3], Hatem S. Ali [4], Abdulhakeem A. Alzahrani [1], Mohamed A. Fadol [1] and Ahmed Noah Badr [5,*]

1. Food Science & Nutrition Department, College of Food and Agricultural Sciences, King Saud University, Riyadh 11451, Saudi Arabia
2. Food Technology Department, Arid Lands Cultivation Research Institute, City of Scientific Research and Technological Applications(SRTA-City), Borg El Arab 21934, Egypt; gamalsng@gmail.com
3. Food Research Section, R&D Division, Abu Dhabi Agriculture and Food Safety Authority (ADAFSA), Abu Dhabi P.O. Box 52150, United Arab Emirates
4. Food Technology Department, National Research Centre, Dokki, Cairo 12622, Egypt; hatem.owyean1@gmail.com
5. Food Toxicology and Contaminants Department, National Research Centre, Dokki, Cairo 12622, Egypt
* Correspondence: talsulami@ksu.edu.sa (T.A.); noohbadr@gmail.com (A.N.B.); Tel.: +20-1000327540 (A.N.B.)

Citation: Alsulami, T.; Shehata, M.G.; Ali, H.S.; Alzahrani, A.A.; Fadol, M.A.; Badr, A.N. Prevalence of Aflatoxins in Camel Milk from the Arabian Peninsula and North Africa: A Reduction Approach Using Probiotic Strains. *Foods* **2023**, *12*, 1666. https://doi.org/10.3390/foods12081666

Academic Editors: Piero Franceschi and Paolo Formaggioni

Received: 7 March 2023
Revised: 7 April 2023
Accepted: 12 April 2023
Published: 17 April 2023

Copyright: © 2023 by the authors. Licensee MDPI, Basel, Switzerland. This article is an open access article distributed under the terms and conditions of the Creative Commons Attribution (CC BY) license (https://creativecommons.org/licenses/by/4.0/).

Abstract: Camel milk is known as a source of nutritional and health supplements. It is known to be rich in peptides and functional proteins. One main issue facing it is related to its contamination, mainly with aflatoxins. The present study aimed to evaluate camel milk samples from different regions while trying to reduce its toxicity using safe approaches based on probiotic bacteria. Collected samples of camel milk were sourced from two main regions: the Arabic peninsula and North Africa. Samples were tested for their contents of aflatoxins (B_1 and M_1) using two techniques to ensure desired contamination levels. Additionally, feed materials used in camel foods were evaluated. Applied techniques were also tested for their validation. The antioxidant activity of camel milk samples was determined through total phenolic content and antioxidant activity assays. Two strains of probiotic bacteria (*Lactobacillus acidophilus* NRC06 and *Lactobacillus plantarum* NRC21) were investigated for their activity against toxigenic fungi. The result revealed high contamination of aflatoxin M_1 for all samples investigated. Furthermore, cross-contamination with aflatoxin B_1 was recorded. Investigated bacteria were recorded according to their significant inhibition zones against fungal growth (11 to 40 mm). The antagonistic impacts were between 40% and 70% against toxigenic fungi. Anti-aflatoxigenic properties of bacterial strains in liquid media were recorded according to mycelia inhibition levels between 41 to 52.83% against *Aspergillus parasiticus* ITEM11 with an ability to reduce aflatoxin production between 84.39% ± 2.59 and 90.4% ± 1.32 from media. Bacteria removed aflatoxins from the spiked camel milk in cases involving individual toxin contamination.

Keywords: aflatoxin M_1; aflatoxin B_1; aflatoxin removal; antioxidant activity; camel milk; ELISA technique; feed contamination; probiotic bacteria

1. Introduction

Milk is a nutrient-rich beverage that possesses health benefits. Milk contains essential nutrients, minerals, and vitamins and is also considered an excellent source of protein. Generally, it is recognized as a nutrient-rich fluid produced by female mammals to feed their offspring. The most commonly consumed types of milk are buffalo, sheep, goat, and cow's milk, with cow's milk being favorable in Western countries [1]. Dairy consumption is sometimes a controversial issue, raising the critical question of whether milk consumption is healthy or a source of harm. Camel milk forms the dietary habits of global nomads and desert populations with all the nutrients represented in other milk varieties [2]. Both fresh and fermented, camel milk has been consumed for human nutrition and for illnesses treated in traditional medicine [3]. Evidence suggests that camel milk has immunomodulatory effects and is readily absorbed by the body. Children who lack the enzyme lactase and are allergic to cow's milk do well on a diet of camel milk. There is evidence that drinking camel milk may help protect the body from the harmful effects of toxins and microbial infection [4].

Like other types of milk, camel milk is a metabolite secreted by the mammalian gland, which is affected by feed ingredients and any potential contamination. Feed contaminants that can pass into the excreted milk include heavy metals, pesticide residues, hormones, and mycotoxins. These contaminants indirectly threaten public health due to their accumulation in small quantities via regular consumption of milk and dairy products. Mycotoxins represent the most significant danger among these contaminants due to their classification by the International Agency for Cancer Research as pre-carcinogens. Mycotoxins are classified into 400 types, of which the most serious to public health are aflatoxins (AFs). Due to cross-contamination, milk can be infected with aflatoxin (B or G) types. It also may be contaminated with metabolic products from feed contamination, as in aflatoxin M types.

Recently, probiotics and lactic acid bacteria have been used as influential factors in reducing contamination in dairy products [5]. Previous studies indicated the role played by these strains due to their activity through antifungal action or their role in reducing the secretion of mycotoxins [6,7]. Two probiotic strains were recorded with antifungal activity via a simulated in vivo investigation [8]. The previous investigation reveals the functionality of some strains of probiotic bacteria (*lactobacillus acidophilus* and *bifidobacteria*) in the reduction of AFM_1 contamination [9]. It should be noted that three different mechanisms can explain the in vivo action of bacterial strains against mycotoxins. Bacterial cell walls can chelate mycotoxins and generate a complex that facilitates removal throughout the biological system [10]. In this method, mycotoxins can leave the body securely, preventing them from causing tissue injury. Other mechanisms are linked to bacterial metabolites [11].

Camel milk is consumed in the middle East and Arab regions in considerable quantities. It is handled in markets and sold for local consumption in some areas such as Saudi Arabia. A few investigations discuss this point, but none recommend a solution. The study investigated aflatoxin contamination in camel milk, which is known to be used for nutritional and immunological consumption. The research was targeted to explore the current situation of aflatoxin (AF) contamination in commercial samples of camel milk. Also attempted to find the link between the source of feeding and the contamination levels and reduce these contamination levels using a safe approach. Additionally, provides solutions to consuming camel milk with a low hazard of mycotoxin contamination. The fermentation using two probiotic strains was applied as a part of the strategy to enhance product safety.

2. Materials and Methods

2.1. Sample Collection

Camel milk samples were collected as commercial samples from markets in the Arabian Peninsula (Saudi Arabia and United Arab Emirates) and North Africa (Egypt and Libya). Samples were collected in 1 kg quantities each (5 samples/region), and each country was classified into three areas.

Using the same manner described above, we also collected samples of feed materials to evaluate contamination levels of camel milk. Two feed materials were collected: ready-to-use (imported manufactured) feed and wild, green-feed materials. The aflatoxin content of evaluated feed materials was utilized to further recommend healthy camel milk consumption.

2.2. Materials and Chemicals

Microbiological media, including potato dextrose agar (PDA), De-man Rogosa and Sharpe (MRS), yeast extract agar (YES), potato dextrose broth (PDB), and Czapek-Dox agar

(CDA) were BD Difco analytical media acquired from Fisher Scientific, Guldensporenpark, Merelbeke, Belgium. Methanol, Ethanol, Di-methyl-sulfoxide (DMSO), Trolox (6-hydroxy-2,5,7,8-tetramethyl chroman-2-carboxylic acid), ABTS+ (2,2′-azino-bis-3-thylbensothiazoline-6-sulfonic acid), DPPH (2,2-diphenyl-1-picryl-hydrazine-hydrate), and other solvents applied were of analytical grade, Merck Co., Ltd., Burlington, MA 1803, USA.

Two ELISA kits (an aflatoxin M_1 Kit and a total aflatoxin kit) were applied to determine AF content. The provided materials inside the Elabscience test kit® (14780 Memorial Drive, Houston, TX 77079, USA) included a Microtiter plate pre-coated with linked antigen, Horseradish peroxidase conjugate (HRP), AF standard solutions required to generate a calibration curve, chromogen (tetra-methyl-benzidine), and a stop reagent.

2.3. Sample Preparation for the Analysis

Before the AF determination, collected samples were prepared according to the methodology described by the manual of the applied technique of the ELISA kits. The milk sample was centrifuged ($5000 \times g$/10 min/4 °C) for the cream separation, the formed cream layer was discarded, and 40 µL of milk was taken for analysis using the ELISA technique. Feed samples (1 g) were ground with aqueous methanol (10 mL; 80%) and 0.1 g of NaCl. The slurry was filtered using filter paper (Whatman no.1), followed by the 0.45 µm filter, where the filtrate cleanup was completed using an AflaTest® column. The column was washed twice before aflatoxin was eluted with 1 mL methanol (HPLC grade). A quantity of 40 µL was applied in the same way as it was for the milk-analysis step.

2.4. Determination of Aflatoxin Content Using the ELISA Technique

The aflatoxin content for the collected samples was determined according to the methodology described in the kits' manuals. The samples and standard solutions were injected into prepared plate wells. A total of 80 µL of the HRP solution was added to wells that were immediately sealed and oscillated (10 s) before undergoing shaded incubation (40 min). When the incubation ended, wells were washed using 260 µL of washing buffer (4 replicates, intervals of 30 s) and inverted for the drying step. Reagent A (50 µL) and reagent B (50 µL) were added to each well, and the plate was re-incubated (15 min; 25 °C) for the reaction performance. The stop reaction solution (50 µL) was added to each well when the reaction time ended, and the optical density was immediately measured for the wells at 450 nm. A calibration curve was performed using the standard concentration of kits to calculate aflatoxin concentrations.

2.5. Determination of Aflatoxins Using the VICAM Technique

The aflatoxin content was determined following the methodology described previously [12]. In summary, 100 g (100 mL) of representative samples was blended with 10 g of NaCl and 200 mL of 80% aqueous methanol. The slurry was homogenized for one minute using a high-speed blender and then filtered using Whatman paper (No. 1). Before re-filtering, the filtrate (5 mL) was diluted with Milli-Q water (20 mL). Ten milliliters of the filtrate was purified on a VICAM immunoaffinity column (Aflatest, VICAM, Milford, MA, USA). The column was washed with 10 mL of Milli-Q water before the aflatoxin was eluted with 1 mL of methanol. The eluted fraction was measured with the VICAM fluorometer after diluting twice with Milli-Q water (VICAM Series 4EX Fluorometer). All operations were carried out following the manufacturer's instructions.

2.6. Validation of the Applied Methodologies

Before analyzing the samples, the ELISA and VICAM techniques were tested to guarantee the validity of the findings. Validation of ELISA was accomplished by calculating recoveries. The mean coefficient of variation for fresh milk spiked with varying concentrations of AFs (10, 20, 40, 80, and 160 ng/L), and results are summarized below in Section 3 of the results.

2.7. Determination of Antioxidant Activity in Camel Milk

Total phenolic content, DPPH (2, 2-diphenyl-1-picryl-hydrazyl-hydrate free radical method), and ABTS+ scavenging (2, 2′-Azinobis [3-ethylbenzothiazoline-6-sulfonic acid]-diammonium salt) were determined to indicate camel milk's antioxidant activity. The previous methodology (with modifications) was followed to evaluate the antioxidant activity of camel milk [13]. Collected samples were first centrifuged ($5000 \times g/4\,°C/10$ min) to separate the cream layer. Briefly, phenolic content was measured in milk before and after bacterial fermentation. Creamless milk (1 mL) was blended with ethanol (1.0 mL, 95% v/v) and deionized water (5 mL). The Folin–Ciocalteu reagent (0.5 mL; 50% v/v) was added to each sample, and after vigorous mixing, the solutions were let to stand ($25\,°C/5$ min). Sodium carbonate solution (1.0 mL, 5% g/100 mL) was added, and then the absorbance was measured after an hour of incubation (at 725 nm). The total phenolic content was measured as a microgram equivalent of Gallic acid (μg GAE/mL milk).

DPPH inhibition was determined by mixing 250 μL of milk with DPPH (3 mL of 60 mmol/L in ethanol) [14]. The mixture was shaken thoroughly and stood ($25\,°C/20$ min). The absorbance readings were measured (at 517 nm), and the DPPH inhibition (%) was calculated as follows:

$$\%DPPH = (Ac - As/Ac) \times 100, \quad (1)$$

where Ac represents absorbance of the control, and
As represents absorbance of the sample.

The same manner was applied for the ABTS+ scavenging determination with the required suitable solutions described previously [15], and the absorbance was measured at 734 nm. The inhibition was expressed according to the following equation:

$$\%ABTS+ = (Ac - As/Ac) \times 100 \quad (2)$$

where Ac represents absorbance of the control, and
As represents absorbance of a sample.

2.8. Activation of Bacterial Strains

Two strains of probiotic bacteria, *Lactobacillus acidophilus* NRC 06 and *Lactobacillus plantarum* NRC 21, were gifted from the Dairy Department, National Research Centre, Cairo, Egypt. The strains were reactivated once in sterile skimmed milk media (11%) and twice in MRS media before the application. The bacterial concentration was adjusted using a hemocytometer chamber at $1.3–1.7 \times 10^9$ CFU/mL.

2.9. Preparation of Bacterial Supernatant

Bacterial supernatant was prepared with the bacterial growth in 1 L of the MRS media [16]; the bioactive components were regained using a dichloromethane and media broth mixture at a ratio of 3:1. The supernatant was recovered using a rotary evaporator (Heidolph, HeiVAP, GmbH, Landsberger, Germany). It was kept in an amber vial until further applications.

2.10. Determination of the Antifungal Effect

The antifungal effect of bacterial strains and their supernatants was evaluated against fungal strains of toxigenic fungi [17]; these strains are known to contaminate camel feed material. The toxigenic fungal strains were *Aspergillus flavus*, *A. parasiticus*, *A. niger*, *A. fumigatus*, *Penicillium oxysporium*, *P. notatum*, *Fusarium graminaerum*, and *Alternaria alternate*. These strains were isolated from feed materials and identified by the Food Toxicology and Contaminant Department, NRC, Egypt.

The ability of bacterial strains to suppress isolated toxigenic fungal growth was investigated [18]. The bacterial antagonism was performed in vitro using PDA media on Petri dishes. On a Petri plate, a disc of fungi was inoculated on one side, whereas a bacterial disc was inoculated on the other. Suitable distances were left between each bacterial culture site and the Agar discs of the examined fungus. Negative control of fungal agar discs without bacterial culture spots was performed. The Petri plates were then incubated (5 days/30 °C). The percentage of fungal growth inhibition was estimated using the following formula:

$$\%A = \left(1 - \left(\frac{X}{Y}\right)\right) \times 100 \qquad (3)$$

where %A: represents antagonism ratio,

X: represents the distance between the fungal edge and bacterial edge, and

Y: represents the distance between the treated fungi's upper edge and the control's upper edge.

The well-diffusion assay was applied to evaluate the antifungal activity of the bacterial supernatant; each well was loaded with 100 μL of bacterial supernatant, as described previously [19]. The results were expressed as millimeter diameters (mm) of the inhibition zone achieved around the well; the greater the inhibition diameter, the more efficient the supernatant.

2.11. Determination of the Anti-Aflatoxigenic Effects

The anti-aflatoxigenic effects of bacterial strains were evaluated using the YES media containing a productive fungal strain of *A. parasiticus* ITEM 11, as described previously [20]. The experiment was divided into two major groups of flasks. The first group used the fungal growth in the presence of a bacterial strain using potato dextrose broth (PDB) media to suit the two microorganisms. This group comprised two flasks infected with fungus (1.37×10^3 CFU/mL) and bacteria (1.71×10^9 CFU/mL), whereas the control flask was inoculated with fungi. Flasks were incubated (30 °C/5 days), and mycelial reduction was expressed as dry weight and a ratio of inhibition against the control.

The second group was tested after the fungus was grown and mycelia were discarded. *A. parasiticus* fungal spores inoculated the flasks containing YES broth. The flasks were incubated (30 °C/9 days) to enable aflatoxin production [19]. By the end of incubation, the media were filtered using Whatman (No. 1) filter paper followed by a micro syringe filter (Millipore, 45 μm). Bacterial strains were enriched on MRS media (24 h) and centrifuged to collect the bacterial pellets that inoculated to the filtrate of fungal media. The flasks were incubated (37 °C/2 h) before measurements of the aflatoxin content were taken. Aflatoxin concentrations in media were measured before and after bacterial pellet treatment.

2.12. Application of Bacteria for Milk Fermentation

Bacterial strains of *Lactobacillus acidophilus* NRC 06 and *Lactobacillus plantarum* NRC 21 were utilized in camel-milk fermentation. Samples of camel milk were spiked with aflatoxin M_1 (220 ng/mL) and Aflatoxin B_1 (400 ng/mL). Camel milk was packed in sterile bottles, inoculated with bacteria at 1.7×10^9 CFU/mL, incubated (37 °C/2 h), and then cooled overnight. Camel milk was inoculated with bacteria strains as individuals and as a mix. Aflatoxin concentrations were measured in samples after 24 h of treatment.

2.13. Statistical Analysis

At least three results were given as means ± standard deviation (SD). ANOVA was used to determine if there was a significant difference between the means, and Duncan's multiple range test ($p = 0.05$) was performed. The SPSS V.22.0 and Graph Pad Prism V.7.0 statistical programs were used to analyze the data expressed as means ± SD.

3. Results

Collected samples from the four countries were inspected concerning the presence of AFs for knowledge of potential contamination in camel milk. The results also illustrate the variation in aflatoxin content in camel milk of the Arabian Peninsula and North African regions. Moreover, two feed material sources, including dry imported and wild plant feeds, were analyzed to detect potential contamination hazards. To our knowledge, wild plants are the primary feed material consumed in North Africa, and imported dry feed is the primary feed material consumed in the Arabian Peninsula.

3.1. Aflatoxin Determination

3.1.1. AFM_1 Evaluation in Camel Milk

The AF content of camel milk was determined to identify natural contamination caused by the AFM_1 toxin and to check for the occurrence of cross-contamination with the AFB_1 toxin. Table 1 shows the AF contamination for the collected samples determined using two techniques (ELISA and VICAM). A high presence of aflatoxin contamination was demonstrated in collected samples from the coastal region (Region 1). The farthest area of the coast seemed to have the lowest contamination level (Region 3). For the samples collected in the United Arab Emirates, there were no significant differences between the region samples concerning aflatoxin $_{M1}$ content. This result could be explained by the fact that these samples were taken from the most extended coastal areas occupying a narrow geographical region.

Table 1. Determination of aflatoxin M_1 in camel milk collected from two regions, the Arabian Peninsula and North Africa, evaluated using ELISA and VICAM techniques.

	AFM_1 Detected via ELISA (ng/L)			AFM_1 Detected via VICAM (ng/L)		
	Region 1	Region 2	Region 3	Region 1	Region 2	Region 3
Saudi	205.8 ± 69.93 [b]	166.6 ± 23.56 [a]	150.8 ± 31.23 [a]	206.6 ± 26.44 [a]	168.2 ± 22.52 [b]	152.8 ± 33.56 [b]
Emirates	291.1 ± 73.13 [a]	225.8 ± 65.53 [a]	256.1 ± 40.4 [a]	293.2 ± 74.86 [a]	223.6 ± 39.84 [a]	256.4 ± 62.24 [a]
Egypt	312.2 ± 21.45 [a]	177.4 ± 16.31 [b]	124.4 ± 25.43 [c]	314.8 ± 22.84 [a]	178.0 ± 16.81 [b]	124.6 ± 27.87 [c]
Libya	124.4 ± 15.63 [a,b]	66.2 ± 19.42 [b]	99.4 ± 17.46 [a,c]	128.8 ± 16.31 [a,b]	70.6 ± 18.61 [b]	100.0 ± 20.86 [a,c]

The data were expressed as means ± SD ($n = 5$; $p < 0.05$). For each technique, the data with the same superscript letter in the same rows show no significant differences.

The AFM_1 contamination levels recorded in the United Arab Emirates and Egypt samples seemed similar in Region 1. We noticed that the primary type of feed in these areas was dried-manufactured feed without any natural feed from wild plants [21]. Camel milk samples from Region 3 in Egypt, which mainly utilized wild plants in camel feeding with little dry-feed material, recorded lower AFM_1 contamination. Bedouin pastoralists in these areas referred to their dried feed as partial consumption due to the dried climate seasons and rarely found wild plants. In Libya, wild pastoralism was found to be the primary type. This behavior may explain the lowest contamination levels of the AFM_1 in camel milk samples from this area.

3.1.2. AFB_1 Evaluation in Camel Milk

The main cause of the AFM_1 contamination was AFB_1 as it transformed metabolically from contaminated feed consumed by the mammalian, resulting in AFM entering the animals' bodily fluids. Furthermore, AFB_1 could have been present through cross-contamination in milk samples during handling, transportation, or storage. Collected samples were inspected for AFB_1 cross-contamination, and the results reflect its occurrence in all camel milk samples (Table 2).

The cross-contamination levels with the AFB_1 in the investigated camel milk samples were similar. This result reflects the need to pay attention to hygiene practices during the production and product-handling stages. The contamination levels were remarkable and exceeded the permissible limits in the collected samples. This indicates the need to review the stages of production and storage well to preserve the therapeutic properties of this type of milk. It is known that camel milk is healthy and can be relied upon to boost immunity levels as it is rich in vital peptides and functional proteins. However, the accidental or direct contamination of these kinds of milk may make it a source of hazard to public health. The risk of this contamination is related to mycotoxins as they are invisible and require specialized approaches for detection. Therefore, the best practice is to check and adequately review the stages of production and the quality of feeding to reduce contamination incidence caused by mycotoxins.

Table 2. Determination of aflatoxin B_1 in camel milk collected from two areas, the Arabian Peninsula and North Africa, evaluated via ELISA and VICAM techniques.

	AFB_1 Detected via *ELISA* (ng/L)			AFB_1 Detected via *VICAM* (ng/L)		
	Region 1	Region 2	Region 3	Region 1	Region 2	Region 3
Saudi	80.8 ± 12.07 [a]	112.6 ± 8.45 [b]	108.0 ± 13.46 [c]	82.2 ± 13.67 [a]	113.2 ± 9.88 [b]	108.1 ± 14.11 [c]
Emirates	152.8 ± 54.28 [a]	180.2 ± 18.56 [a]	75.4 ± 13.69 [b]	156.4 ± 54.55 [a]	180.6 ± 38.35 [a]	76.0 ± 13.56 [b]
Egypt	57.8 ± 8.24 [b]	101.6 ± 13.33 [a,b]	108.8 ± 21.66 [a]	58.8 ± 6.79 [a]	101.2 ± 13.91 [b]	106.0 ± 21.56 [b]
Libya	61.2 ± 15.52 [a,b]	80.8 ± 6.7 [a]	42.4 ± 8.86 [b]	64.6 ± 16.9 [a,b]	80.4 ± 5.57 [a]	41.8 ± 8.69 [b]

The data were expressed as means ± SD ($n = 5$; $p < 0.05$). For each technique, the data with the same superscript letter in the same rows show no significant differences.

3.2. Method Validation of Aflatoxin Determination

The validity of the method was first evaluated using spiked aflatoxin concentrations for aflatoxin M_1 (AFM_1) and aflatoxin B_1 (AFB_1). The determination results are recorded in Table 3, wherein the recovery seems acceptable for accurately evaluating aflatoxin content.

Table 3. Data validation for the samples using VICAM and ELISA techniques to determine AFM_1 and AFB_1 recovery.

Spiked	ELISA Technique		Coefficient Variation (%)	VICAM Technique		Coefficient Variation (%)	Samples (n)
	AF Determined (pg/L)	AF Recovered (%)		AF Determined (pg/L)	AF Recovered (%)		
			AFM_1				
10	10.01 ± 0.01	100 ± 0.01	0	10.01 ± 0.02	100 ± 0.02	0	7
20	19.98 ± 0.03	99.9 ± 0.03	0.1	19.96 ± 0.02	99.8 ± 0.03	0.2	7
40	39.87 ± 0.05	99.67 ± 0.02	0.33	39.89 ± 0.08	99.73 ± 0.02	0.27	7
80	79.74 ± 0.11	99.67 ± 0.05	0.33	79.79 ± 0.14	99.74 ± 0.11	0.26	7
160	159.52 ± 0.28	99.7 ± 0.14	0.3	159.64 ± 0.11	99.78 ± 0.14	0.22	7
			AFB_1				
10	10.01 ± 0.01	100 ± 0.01	0	10.01 ± 0.01	100 ± 0.01	0	7
20	19.99 ± 0.01	99.95 ± 0.02	0.05	19.99 ± 0.01	99.95 ± 0.02	0.05	7
40	39.89 ± 0.12	99.73 ± 0.03	0.27	39.93 ± 0.06	99.83 ± 0.09	0.17	7
80	79.65 ± 0.23	99.56 ± 0.18	0.44	79.77 ± 0.1	99.71 ± 0.22	0.29	7
160	159.21 ± 0.47	99.5 ± 0.21	0.5	159.54 ± 0.34	99.71 ± 0.16	0.29	7

The data were expressed as means ± SD ($n = 7$; $p < 0.05$). AF: aflatoxin; AFM_1: aflatoxin M_1; AFB_1: aflatoxin B_1.

It was noticed that the recovery at different concentrations showed acceptable levels, and few changes were recorded regarding the factor influencing coefficient variation. The results at this stage provide clarity regarding the aflatoxin evaluation.

3.3. Aflatoxin Determination in Plant Feeds

Feed samples were investigated for sources of risks that may be linked to AFM_1 in camel milk. First, wild plant samples consumed in natural pastoralism contexts were analyzed, and the results are shown in Table 4. AFB_1 was present in collected plant material during the investigation; however, AFB_1 was present in samples at low levels. Determination of the changes in AFB_1 using the two techniques of ELISA or VICAM revealed that the presence of AFB_1 was limited, showing small values. These results indicate that natural pastoralism was not the main cause behind AFM_1 contamination of camel milk samples.

Table 4. Determination of Aflatoxin B_1 in plant feed materials collected from camel pasture areas of the Arabian Peninsula and North Africa.

	AFB_1 Detected via *ELISA* (ng/kg)			AFB_1 Detected via *VICAM* (ng/kg)		
	Region 1	Region 2	Region 3	Region 1	Region 2	Region 3
Saudi	5.2 + 1.18 [a]	9.4 + 1.71 [b]	7.2 + 1.74 [a,b]	10.2 + 2.49 [a]	6.4 + 1.95 [b]	9.6 + 2.36 [a]
Emirates	12.2 + 3.05 [a]	5.2 + 2.55 [c]	7.8 + 1.80 [b]	13.8 + 2.59 [a]	6.2 + 6.95 [b]	5.2 + 1.77 [b]
Egypt	8.8 + 3.28 [a]	8.0 + 3.76 [a]	8.0 + 2.06 [a]	10.8 + 1.48 [a]	10.4 + 3.51 [a]	9.4 + 2.83 [a]
Libya	5.0 + 3.10 [a]	4.0 + 2.05 [a]	6.6 + 2.07 [a]	7.8 + 0.84 [a]	6.0 + 1.81 [a]	8.0 + 2.34 [a]

The data were expressed as means ± SD ($n = 5$; $p < 0.05$). For each technique, the data with the same superscript letter in the same rows show no significant differences.

3.4. Aflatoxin Determination in Manufactured Feeds

The next step involved the investigation of manufactured dried feed material imported for use as camel feed. The manufactured dried feed materials consumed as camel feed were analyzed, and the results are shown in Table 5. AFB_1 was present in the investigated samples; however, AFB_1 was present in dry feed samples at high contamination levels. Changes in AFB_1 determination using the two techniques of ELISA or VICAM were recorded as limited and fluctuated only slightly. These results may reveal that the consumption of manufactured feed was the source of the AFM_1 contamination of camel milk samples.

Table 5. Determination of Aflatoxin B_1 in manufactured feed collected from camel pasture areas of the Arabian Peninsula and North Africa.

	AFB_1 Detected via *ELISA* (ng/kg)			AFB_1 Detected via *VICAM* (ng/kg)		
	Region 1	Region 2	Region 3	Region 1	Region 2	Region 3
Saudi	376.6 ± 73.96 [a]	461.2 ± 75.49 [a]	377.6 ± 48.86 [a]	378.2 + 73.19 [a]	463.0 + 72.42 [a]	381.0 + 109.08 [a]
Emirates	732.4 ± 159.51 [a]	646.2 ± 81.38 [a]	719.0 ± 165.6 [a]	731.2 + 157.28 [a]	645.2 + 81.10 [a]	719.2 + 161.27 [a]
Egypt	437.8 ± 49.70 [a]	360.4 ± 70.44 [a]	416.2 ± 93.03 [a]	438.0 + 50.19 [a]	363.6 + 69.79 [a]	417.6 + 93.07 [a]
Libya	365.6 ± 64.22 [a]	367.6 ± 58.04 [a]	321.4 ± 68.83 [a]	365.0 + 67.26 [a]	368.0 + 59.01 [a]	321.2 + 69.81 [a]

The data were expressed as means ± SD ($n = 5$; $p < 0.05$). For each technique, the data with the same superscript letter in the same rows show no significant differences.

3.5. Determination of Antioxidant Activity

The total phenolic content and antioxidant activity of camel milk were part of our bioactivity investigation of the camel milk. The collected samples of camel milk were investigated for their antioxidant activity using two assays (DPPH and $ABTS^+$). Furthermore, the total phenolic content of camel milk samples was determined to reflect their partial activity as antioxidants. The results (Figure 1) showed that camel milk samples collected from North Africa were distinct in their total phenols and antioxidant activity content. Additionally, the samples collected from Libya for contained more antioxidants than those collected from Egypt.

The samples collected from the Arabian Peninsula were lower in their levels of antioxidants compared to North Africa. Camel milk samples from the Arabian Peninsula were collected from the Kingdom of Saudi Arabia and the United Arab Emirates. The low content of antioxidants in these samples may be due to the consumption of these components to maintain the product's quality and safety against microbial contamination during production or handling; it is not caused by any inherent lack of essential elements in the camel milk of these regions. The antioxidant activity of the food product is known to play a vital function in delaying microbial spoilage. Again, the primary type of feeding, such as using wild plants, might contribute to these results due to their bioactive components.

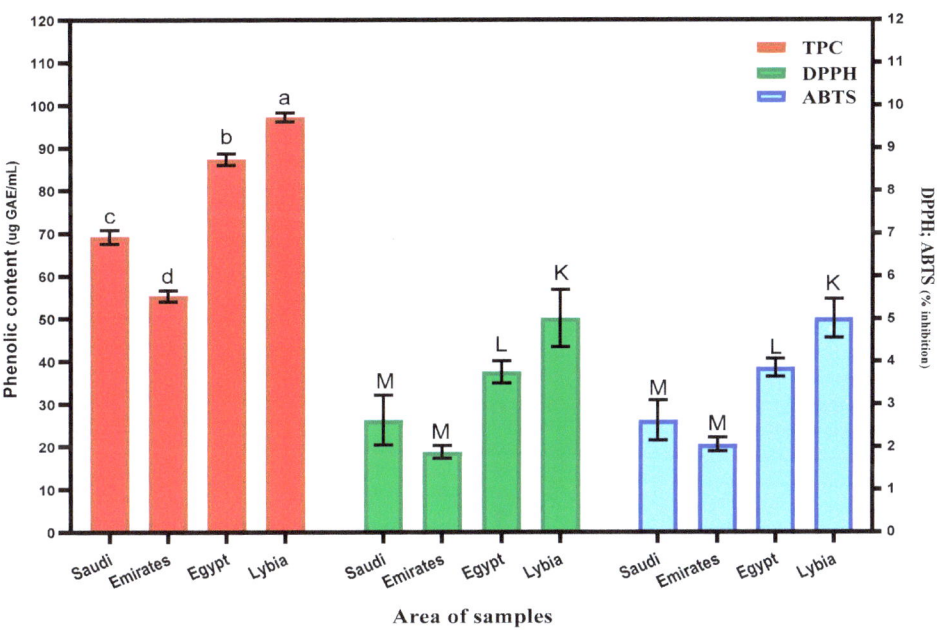

Figure 1. Total phenolic content and antioxidant activity for the collected camel milk from the Arabian Peninsula and North African regions. The columns with the same superscript letters show non-significant differences ($p = 0.05$). TPC: total phenolic compound contents determined as microgram Gallic acid equivalents per milliliter of milk sample. DPPH: DPPH (2, 2-diphenyl-1-picryl-hydrazyl-hydrate free radical solution), ABTS+: ABTS+ scavenging (2, 2′-Azinobis [3-ethylbenzothiazoline-6-sulfonic acid]-diammonium salt).

3.6. Antifungal Activity of Applied Probiotic Strains

The antifungal activities of the two applied strains (*L. acidophilus* NRC06 and *L. plantarum* NRC21) were evaluated using two assays, and two representative methods expressed the obtained results. The supernatant collected from the bacterial growth was applied using a well-diffusion assay. The activity in this method is described as inhibition zone diameter, which is recorded in Figure 2A. The results show that the strains possessed high inhibition zone diameters, particularly for *Fusarium* and *Alternaria*: two toxigenic fungi strains under investigation. Other fungi growth was recorded as being inhibited by lower levels, but they were still significant compared to the control (complete fungal growth).

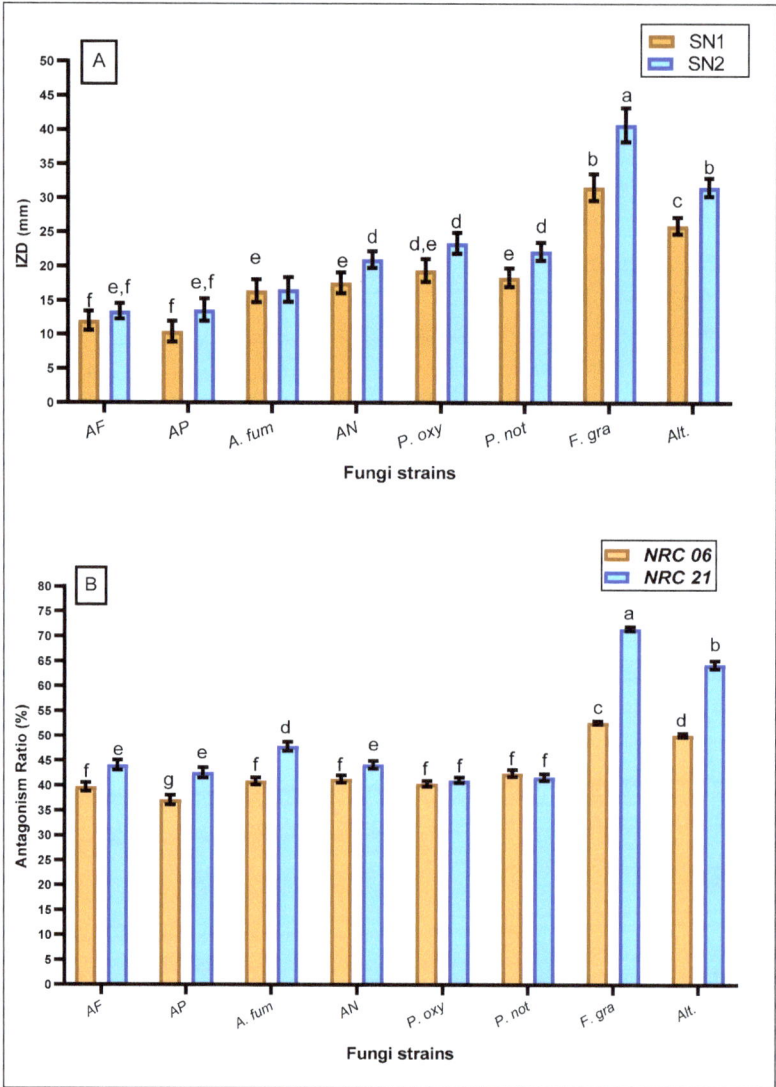

Figure 2. Antifungal activity of applied probiotic strains against toxigenic fungal strains. (**A**) Antifungal activity evaluated with bacterial supernatants determined according to zone inhibition diameter. (**B**) Antifungal activity of bacterial cells determined according to antagonistic ratio (%). For each of the (**A**,**B**), the columns with different superscript letters show significantly differences. SN1: supernatant of *L. acidophilus* NRC06; SN2: supernatant of *L. plantarum* NRC21; NRC06: bacterial cells of *L. acidophilus* NRC06; NRC 21: bacterial cells of *L. plantarum* NRC21. AF: *Aspergillus flavus*; AP: *A. parasiticus*; A. fum: *A. fumigatus*; AN: *A. niger*; P. oxy: *Penicillium oxysporium*; P. not: *P. notatum*; F. gra: *Fusarium graminarum*; Alt: *Alternaria alternata*.

Bacterial cells used antagonistically to stop the growth of toxic fungi were successful according to our results (Figure 2B). For the two strains, the effect of bacterial cells as inhibitors of *Aspergillus* and *Penicillium* fungi ranged from 40 to 50%. This ratio, however, has been documented to be up to 70% or more for some fungi, such as for the genus *Fusarium*. It was noticed that the inhibition influence was efficiently detected by utilizing two bacterial strains against eight strains of toxigenic fungi.

3.7. Anti-Aflatoxigenic Effects of Bacterial Strains

The results in Table 6 show the extent to which the strain of fungus (*A. parasiticus ITEM 11*), which is known to highly produce aflatoxins, was affected by the presence of probiotic bacteria in the fungal growth media. The effect on the fungus strain, associated with the presence of bacteria, was shown to exhibit a reduction in fungi mycelial growth and in aflatoxin secretion levels in the growth media compared to the control growth media.

The data reflected a high ratio of mycelial growth reduction at 41.003% ± 0.013 using the bacterial strain NRC 06. This inhibition ratio increased to 52.83% ± 0.07 by applying the NRC 21 bacterial strain. The reduction in aflatoxin concentration in the fungal growth media ranged between 84.39% ± 2.59 and 90.4% ± 1.32 for the utilization of bacterial treatment.

Table 6. Anti-aflatoxigenic effects of bacterial strains against the fungal growth of *A. parasiticus* and toxin production reduction in growth media.

	Flasks Containing NRC 06—Strain	Flasks Containing NRC 21—Strain	Control Flasks
Mycelia weight (g)	3.1117 ± 0.144 [b]	2.4876 ± 0.208 [c]	5.2741 ± 0.131 [a]
MIR (%)	41.003 ± 0.013 [b]	52.83 ± 0.07 [a]	–
AFB_1 (ng/mL)	76.11 ± 14.37 [b]	46.77 ± 9.81 [c]	487.6 ± 12.48 [a]
RR—AFB_1 (%)	84.39 ± 2.59 [b]	90.40 ± 1.32 [a]	–

The data were expressed as means ± SD ($n = 5$; $p < 0.05$). The data with the same superscript letter in the same rows show no significant differences. NRC06: bacterial cells of *L. acidophilus* NRC06; NRC 21: bacterial cells of *L. plantarum* NRC21. MIR: mycelial inhibition reduction; RR—AFB_1: reduction ratio recorded for aflatoxin B_1 concentration.

3.8. Aflatoxin Reduction in Spiked Camel Milk Inoculated by Bacteria

The camel milk samples collected from Egypt were chosen as median samples for the present part of the study. The samples utilized for the subsequent steps were collected from pastoral nomads of the northwestern desert area (Matruh to Siwa).

The previous strains of bacteria, which were recorded to have antifungal and anti-aflatoxigenic impacts, were tested in spiked samples of camel milk. Table 7 shows the applied strains' capability to remove the aflatoxin content from camel milk. Moreover, the aflatoxin removal results from samples indicated that the approach efficiency is high. Aflatoxin removal using bacterial strains was recorded for aflatoxin B_1 and aflatoxin M_1. The efficiency of the NRC 21 bacterial strain for Aflatoxin removal was 100% as the treated sample recorded detected no Aflatoxins. The results reveal that there was more bacterial efficiency in removing toxins from individual spiked samples than in the mixed spiked samples. However, the gap between removing toxins from individual samples or mixed-toxin samples was still limited.

Table 7. Aflatoxin reduction using spiked camel milk fermented using bacterial strains of NRC 06 and NRC 21.

	AFB_1—Spike CM (ng/mL)	AFM_1—Spiked CM (ng/mL)	CM Containing AF Mixture	
			AFB_1 (ng/mL)	AFM_1 (ng/mL)
Spiked control	482.8 ± 5.24 [a]	299.2 ± 6.13 [a]	492.5 ± 3.71 [a]	316.9 ± 2.14 [a]
NRC 06	103.61 ± 7.64 [b]	54.71 ± 6.84 [b]	91.34 ± 4.55 [b]	76.18 ± 6.24 [b]
NRC 21	ND	ND	23.66 ± 4.89 [c]	5.94 ± 3.17 [c]

The data were expressed as means ± SD ($n = 5$; $p < 0.05$). The data with the same superscript letter in the same columns show no significant differences. For each column, the result with different superscript letters were significantly different.

4. Discussion

Camel milk is one source of biologically active substances because it may contain functional proteins and peptides that have activity against several pathogens [22]. Camel milk and its production areas are often associated with Bedouins and desert regions, as it is known for its widespread use in those areas [23]. Camel milk peptides are linked to nutraceutical impacts when consumed [21,24]. Camel milk could be contaminated by food hazards, like other milk types, and these contaminants may turn it from a source of health benefits to a source of health issues. Camel milk could be contaminated due to production, handling, transportation, storage, or marketing conditions. While good hygiene practices are required for safe production, this may be difficult to apply in some production areas, which affects the safety of camel milk.

Aflatoxins are a significant hazard, considering their classification as pre-carcinogens [25,26]. These chemical hazards may contaminate camel milk directly (AFM secreted from AFB biotransformation) or indirectly through cross-contamination with AFB toxins [27]. Collected samples of camel milk were investigated for both contamination types (direct or indirect), and the results showed evidence of both types (Tables 1 and 2). Two geographical areas were included in the present study: the Arabian Peninsula and North Africa. Two countries represented each area for the collected samples along with six regions (three regions for each country). The results reflected complete contamination of the collected camel milk samples (with AFB_1 and AFM_1). The presence of aflatoxins in milk samples in ascending order according to concentration was as follows: Libya < Egypt < Saudi Arabia < United Arab Emirates. This result shows that the Arabian Peninsula was a more hazardous production area compared to North Africa. The aflatoxin concentrations in tested samples were evaluated using a VICAM fluorometer and ELISA reader. The method validation of each approach was evaluated to ensure result accuracy (Table 3). Feed materials were investigated, including wild plant feed and manufactured feed imported from outside the country.

Aflatoxin contamination in camel milk has been previously tested in camel milk samples by other researchers [28–30]. Still, these studies are few in number, and none of them studied the relationship between the contamination of feed in the places of production and the levels of aflatoxins in the milk produced in the same areas.

The current study evaluated the two types of feed used for camels. The results indicated high levels of aflatoxin B_1 contamination in the manufactured feed abundant in the Arabian Gulf region. Additionally, manufactured fodder is the alternative source in case of scarcity of wild plants and during drought periods throughout the year. The results highlight the high level of AFB_1 detected in the manufactured feed type (Table 5). From these results, it could be concluded that the primary source of hazard for aflatoxin contamination is manufactured feed. The low level of aflatoxin contamination in wild plant feed may be linked to their bioactive components. These components play defensive roles on behalf of the plants and preserve their spoilage.

It is essential to search for a method that aids in removing aflatoxin contamination from camel milk and maintains its nutrition and health benefits. Several strategies have been applied to detoxify aflatoxin in dairy products, such as using non-traditional oils [31]. Probiotic strains could play this function and support the milk's beneficial properties. Two isolates of local strains, *L. acidophilus* NRC 06 and *L. plantarum* NRC 21, were investigated to evaluate their antifungal activity. The strains showed a high inhibition impact against eight strains of toxigenic fungi (Figure 2). The application of bacterial supernatant using an agar-well diffusion assay showed an inhibition zone diameter range between 11–40 mm (Figure 2A). Using the bacterial cell antagonistic impact, the ratio of inhibition shown reached up to 70% (Figure 2B).

A bacterial genus of *Lactobacilli* previously known for presenting probiotic properties has been known to bind into pathogens and limit their growth [32,33]. Lactobacillus strains also produce several secondary metabolites, including bacteriocins, active peptides, hydrogen peroxide, and organic acids. Bacteriocins of *lactobacilli* strains, such as *L. plantarum*, have been previously characterized. Pure substances were tested for their antifungal and anti-aflatoxigenic impacts [34,35]. The bioactivity characteristics of *L. plantarum* make it a distinguishable application for food preservation [36]. Incorporating probiotic bacteria as an antifungal agent in food may minimize the incidence of fungal spoilage and toxicity, and it may also extend shelf life and reduce mycotoxin concentrations [37]. The presence of probiotics in food might change its physiochemical and organoleptic features. These changes may be linked to the impacts referred to previously. In contrast, the predominant population of fungi infecting a typical meal should be considered when choosing the most effective probiotics/combinations of probiotic bacteria to prevent fungal development [7,10]. The reason for this is that antifungal activities of probiotics are fungal-strain-specific, which means that a probiotic strain may be very active against one fungus strain while having no impact on the growth of another [32,36].

Previous results have shown that strain differences in AFB_1 removal are unequal as bacterial strains are differentiated in their activity [6]. Contrary to Gram-negative bacteria, Gram-positive bacteria removed aflatoxin more efficiently [6,38]. It is also worth noting that a study conducted by Line and Brackett [39] pointed out that the concentration and the growth stage of the cells applied, besides the incubation time, possessed a function in the elimination rates of mycotoxins as well as in the efficiency of their removal from the growth media.

It is clearly shown from this study that both NRC 06 and NRC 21 have significant effects in inhibiting toxigenic fungal contamination in growth media (Table 6). Additionally, these strains could reduce aflatoxin levels of AFB_1 and AFM_1 in liquid media (Table 6) and in spiked samples of camel milk (Table 7). Both NRC 06 and NRC 21 are classified as probiotic strains. Although these strains were recorded to detoxify aflatoxins, they can also remove other mycotoxins. The present bacterial strains are a potential approach for reducing aflatoxin during the food pathway metabolism in the gastrointestinal tract. The application of investigated bacteria to remove aflatoxin as in individual or in combination contexts (AFB_1 + AFM_1) in spiked samples of camel milk could recommend their utilization as a fast treatment for camel milk before consumption. This study also recommends the fermented consumption of camel milk instead of fresh consumption due to the high contamination recorded in the collected samples.

5. Conclusions

Camel milk is a beneficial dairy product consumed widely for its nutritional and health benefits. Recently, aflatoxin contamination has been known to threaten several food products, including dairy food materials. Camel milk samples were collected from the Arabian Peninsula and North Africa and contaminated. Samples analyses using two validated techniques (ELISA and Fluorometer) indicated the presence of AFB_1 and AFM_1. The AFM_1 in camel milk was high in the Arabian Peninsula region. Cross-contamination with AFB_1 was also recorded. However, feed material was recorded as positively contaminated. Two probiotic strains of NRC 06 and NRC 21 showed distinguished antifungal activity. These strains were able to inhibit the growth of eight toxigenic fungi strains. They also removed aflatoxin from the simulated media. Finally, the NRC 06 and NRC 21 bacterial strains effectively reduced aflatoxin content whether applied individually or in mixtures to spiked camel milk after incubation treatment. Based on these results, we recommend the fermentation of camel milk using probiotic strains as an approach to limit aflatoxin contamination in camel milk. Further studies are also recommended to find suitable solutions to aflatoxin contamination in dairy products.

Author Contributions: Conceptualization, M.G.S. and A.N.B.; data curation, T.A. and M.A.F.; formal analysis, M.G.S., A.A.A. and A.N.B.; funding acquisition, H.S.A.; investigation, T.A., H.S.A. and A.N.B.; methodology, M.G.S. and A.N.B.; project administration, A.N.B.; resources, T.A., A.A.A., H.S.A. and M.A.F.; software, T.A. and A.A.A.; supervision, M.G.S.; validation, H.S.A. and M.A.F.; writing—review & editing, A.N.B. All authors have read and agreed to the published version of the manuscript.

Funding: This research was supported by the Ministry of Education in Saudi Arabia for funding this research work through project no. (IFKSURG-2-937).

Data Availability Statement: The data used to support the findings of this study are included in the article.

Acknowledgments: The authors extend their appreciation to the Deputyship for Research & Innovation, Ministry of Education in Saudi Arabia for funding this research work through the project no. (IFKSURG-2-937).

Conflicts of Interest: The authors declare no conflict of interest for the present manuscript data.

References

1. El-Salam, M.H.A.; El-Shibiny, S. Bioactive Peptides of Buffalo, Camel, Goat, Sheep, Mare, and Yak Milks and Milk Products. *Food Rev. Int.* **2013**, *29*, 1–23. [CrossRef]
2. Kaskous, S. Importance of camel milk for Human Health. *Emir. J. Food Agric.* **2017**, *28*, 158–163. [CrossRef]
3. Kumar, D.; Verma, A.K.; Chatli, M.K.; Singh, R.; Kumar, P.; Mehta, N.; Malav, O.P. Camel milk: Alternative milk for human consumption and its health benefits. *Nutr. Food Sci.* **2016**, *46*, 217–227. [CrossRef]
4. Ahamad, S.R.; Raish, M.; Ahmad, A.; Shakeel, F. Potential Health Benefits and Metabolomics of Camel Milk by GC-MS and ICP-MS. *Biol. Trace Elem. Res.* **2017**, *175*, 322–330. [CrossRef] [PubMed]
5. Petrova, P.; Arsov, A.; Tsvetanova, F.; Parvanova-Mancheva, T.; Vasileva, E.; Tsigoriyna, L.; Petrov, K. The Complex Role of Lactic Acid Bacteria in Food Detoxification. *Nutrients* **2022**, *14*, 2038. [CrossRef] [PubMed]
6. El-Nezami, H.; Kankaanpaa, P.; Salminen, S.; Ahokas, J. Ability of dairy strains of lactic acid bacteria to bind a common food carcinogen, aflatoxin B1. *Food Chem. Toxicol.* **1998**, *36*, 321–326. [CrossRef] [PubMed]
7. Pierides, M.; El-Nezami, H.; Peltonen, K.; Salminen, S.; Ahokas, J. Ability of Dairy Strains of Lactic Acid Bacteria to Bind Aflatoxin M1 in a Food Model. *J. Food Prot.* **2000**, *63*, 645–650. [CrossRef] [PubMed]
8. Mahjoory, Y.; Mohammadi, R.; Hejazi, M.A.; Nami, Y. Antifungal activity of potential probiotic Limosilactobacillus fermentum strains and their role against toxigenic aflatoxin-producing aspergilli. *Sci. Rep.* **2023**, *13*, 388. [CrossRef]
9. Sarlak, Z.; Rouhi, M.; Mohammadi, R.; Khaksar, R.; Mortazavian, A.M.; Sohrabvandi, S.; Garavand, F. Probiotic biological strategies to decontaminate aflatoxin M1 in a traditional Iranian fermented milk drink (Doogh). *Food Control* **2017**, *71*, 152–159. [CrossRef]
10. El-Nezami, H.; Polychronaki, N.; Salminen, S.; Mykkänen, H. Binding rather than metabolism may explain the interaction of two food-grade Lactobacillus strains with zearalenone and its derivative ά-zearalenol. *Appl. Environ. Microbiol.* **2002**, *68*, 3545–3549. [CrossRef]
11. Mogahed Fahim, K.; Noah Badr, A.; Gamal Shehata, M.; Ibrahim Hassanen, E.; Ibrahim Ahmed, L. Innovative application of postbiotics, parabiotics and encapsulated Lactobacillus plantarum RM1 and Lactobacillus paracasei KC39 for detoxification of aflatoxin M1 in milk powder. *J. Dairy Res.* **2021**, *88*, 429–435. [CrossRef] [PubMed]
12. Hafez, E.; Abd El-Aziz, N.M.; Darwish, A.M.G.; Shehata, M.G.; Ibrahim, A.A.; Elframawy, A.M.; Badr, A.N. Validation of New ELISA Technique for Detection of Aflatoxin B1 Contamination in Food Products versus HPLC and VICAM. *Toxins* **2021**, *13*, 747. [CrossRef] [PubMed]
13. Shori, A.B.; Baba, A.S. Antioxidant activity and inhibition of key enzymes linked to type-2 diabetes and hypertension by Azadirachta indica-yogurt. *J. Saudi Chem. Soc.* **2013**, *17*, 295–301. [CrossRef]
14. Abu-Sree, Y.H.; Abdel-Fattah, S.M.; Abdel-Razek, A.G.; Badr, A.N. Neoteric approach for peanuts biofilm using the merits of Moringa extracts to control aflatoxin contamination. *Toxicol. Rep.* **2021**, *8*, 1685–1692. [CrossRef] [PubMed]
15. Soleymanzadeh, N.; Mirdamadi, S.; Kianirad, M. Antioxidant activity of camel and bovine milk fermented by lactic acid bacteria isolated from traditional fermented camel milk (Chal). *Dairy Sci. Technol.* **2016**, *96*, 443–457. [CrossRef]
16. Abdel-Nasser, A.; Fathy, H.M.; Badr, A.; Hathout, A.; Barakat, O.S. Prevalence of Aflatoxigenic Fungi in Cereal Grains And Their Related Chemical Metabolites. *Egypt. J. Chem.* **2022**, *65*, 455–470. [CrossRef]
17. Abu-Seif, F.A.; Badr, A.N. Anti-aflatoxigenic of Agave Extracts to Increase Their Food Safety Applications. *Egypt. J. Chem.* **2022**, *65*, 407–418. [CrossRef]
18. Slama, H.B.; Cherif-Silini, H.; Chenari Bouket, A.; Qader, M.; Silini, A.; Yahiaoui, B.; Alenezi, F.N.; Luptakova, L.; Triki, M.A.; Vallat, A.; et al. Screening for *Fusarium* Antagonistic Bacteria From Contrasting Niches Designated the Endophyte *Bacillus halotolerans* as Plant Warden Against Fusarium. *Front. Microbiol.* **2019**, *9*, 3236. [CrossRef]
19. Abdel-Razek, A.G.; Badr, A.N.; Alharthi, S.S.; Selim, K.A. Efficacy of Bottle Gourd Seeds' Extracts in Chemical Hazard Reduction Secreted as Toxigenic Fungi Metabolites. *Toxins* **2021**, *13*, 789. [CrossRef]
20. Badr, A.N.; Ali, H.S.; Abdel-Razek, A.G.; Shehata, M.G.; Albaridi, N.A. Bioactive Components of Pomegranate Oil and Their Influence on Mycotoxin Secretion. *Toxins* **2020**, *12*, 748. [CrossRef]
21. Alavi, F.; Salami, M.; Emam-Djomeh, Z.; Mohammadian, M. Chapter 36—Nutraceutical Properties of Camel Milk. In *Nutrients in Dairy and Their Implications on Health and Disease*; Watson, R.R., Collier, R.J., Preedy, V.R., Eds.; Academic Press: Cambridge, MA, USA, 2017; pp. 451–468. [CrossRef]

22. Swelum, A.A.; El-Saadony, M.T.; Abdo, M.; Ombarak, R.A.; Hussein, E.O.S.; Suliman, G.; Alhimaidi, A.R.; Ammari, A.A.; Ba-Awadh, H.; Taha, A.E.; et al. Nutritional, antimicrobial and medicinal properties of Camel's milk: A review. *Saudi J. Biol. Sci.* **2021**, *28*, 3126–3136. [CrossRef] [PubMed]
23. Degen, A.A.; El-Meccawi, S.; Kam, M. The Changing Role of Camels among the Bedouin of the Negev. *Hum. Ecol.* **2019**, *47*, 193–204. [CrossRef]
24. Ali Redha, A.; Valizadenia, H.; Siddiqui, S.A.; Maqsood, S. A state-of-art review on camel milk proteins as an emerging source of bioactive peptides with diverse nutraceutical properties. *Food Chem.* **2022**, *373*, 131444. [CrossRef] [PubMed]
25. Badr, A.N.; Naeem, M.A. Protective efficacy using Cape-golden berry against pre-carcinogenic aflatoxins induced in rats. *Toxicol. Rep.* **2019**, *6*, 607–615. [CrossRef] [PubMed]
26. Mukhtar, F. Implication of aflatoxins as potent carcinogens. *Bayero J. Pure Appl. Sci.* **2019**, *12*, 39–45. [CrossRef]
27. Nazhand, A.; Durazzo, A.; Lucarini, M.; Souto, E.B.; Santini, A. Characteristics, Occurrence, Detection and Detoxification of Aflatoxins in Foods and Feeds. *Foods* **2020**, *9*, 644. [CrossRef]
28. Yousof, S.S.M.; El Zubeir, I.E.M. Chemical composition and detection of Aflatoxin M1 in camels and cows milk in Sudan. *Food Addit. Contam. Part B* **2020**, *13*, 298–304. [CrossRef]
29. Shokri, H.; Torabi, S. The effect of milk composition, yeast-mould numbers and seasons on aflatoxin M1 amounts in camel milk. *J. Food Saf.* **2017**, *37*, e12300. [CrossRef]
30. Fallah, A.A.; Fazlollahi, R.; Emami, A. Seasonal study of aflatoxin M1 contamination in milk of four dairy species in Yazd, Iran. *Food Control* **2016**, *68*, 77–82. [CrossRef]
31. Badr, A.N.; El-Said, M.M.; El-Messery, T.M.; Abdel-Razek, A.G. Non-traditional oils encapsulation as novel food additive enhanced yogurt safety against aflatoxins. *Pak. J. Biol. Sci.* **2019**, *22*, 51–58. [CrossRef]
32. Gerbaldo, G.A.; Barberis, C.; Pascual, L.; Dalcero, A.; Barberis, L. Antifungal activity of two Lactobacillus strains with potential probiotic properties. *FEMS Microbiol. Lett.* **2012**, *332*, 27–33. [CrossRef] [PubMed]
33. Ruiz, F.O.; Gerbaldo, G.; Asurmendi, P.; Pascual, L.M.; Giordano, W.; Barberis, I.L. Antimicrobial Activity, Inhibition of Urogenital Pathogens, and Synergistic Interactions Between Lactobacillus Strains. *Curr. Microbiol.* **2009**, *59*, 497–501. [CrossRef]
34. Shehata, M.G.; Badr, A.N.; El Sohaimy, S.A. Novel antifungal bacteriocin from *lactobacillus paracasei* KC39 with anti-mycotoxigenic properties. *Biosci. Res.* **2018**, *15*, 4171–4183.
35. Shehata, M.G.; Badr, A.N.; El Sohaimy, S.A.; Asker, D.; Awad, T.S. Characterization of antifungal metabolites produced by novel lactic acid bacterium and their potential application as food biopreservatives. *Ann. Agric. Sci.* **2019**, *64*, 71–78. [CrossRef]
36. Nasrollahzadeh, A.; Mokhtari, S.; Khomeiri, M.; Saris, P.E.J. Antifungal Preservation of Food by Lactic Acid Bacteria. *Foods* **2022**, *11*, 395. [CrossRef]
37. Afshar, P.; Shokrzadeh, M.; Raeisi, S.N.; Ghorbani-HasanSaraei, A.; Nasiraii, L.R. Aflatoxins biodetoxification strategies based on probiotic bacteria. *Toxicon* **2020**, *178*, 50–58. [CrossRef] [PubMed]
38. Peltonen, K.D.; El-Nezami, H.S.; Salminen, S.J.; Ahokas, J.T. Binding of aflatoxin B1 by probiotic bacteria. *J. Sci. Food Agric.* **2000**, *80*, 1942–1945. [CrossRef]
39. Line, J.E.; Brackett, R.E. Factors Affecting Aflatoxin B_1 Removal by Flavobacterium aurantiacum. *J. Food Prot.* **1995**, *58*, 91–94. [CrossRef]

Disclaimer/Publisher's Note: The statements, opinions and data contained in all publications are solely those of the individual author(s) and contributor(s) and not of MDPI and/or the editor(s). MDPI and/or the editor(s) disclaim responsibility for any injury to people or property resulting from any ideas, methods, instructions or products referred to in the content.

Article

The Ability of Shiga Toxin-Producing *Escherichia coli* to Grow in Raw Cow's Milk Stored at Low Temperatures

Lene Idland [1], Erik G. Bø-Granquist [2], Marina Aspholm [1] and Toril Lindbäck [1,*]

[1] Department of Paraclinical Sciences, Faculty of Veterinary Medicine, Norwegian University of Life Sciences, 1432 Ås, Norway
[2] Department of Production Animal Clinical Sciences, Faculty of Veterinary Medicine, Norwegian University of Life Sciences, 1432 Ås, Norway
* Correspondence: toril.lindback@nmbu.no; Tel.: +47-97982726

Abstract: Despite the lack of scientific evidence, some consumers assert that raw milk is a natural food with nutritional and immunological properties superior to pasteurized milk. This has led to the increased popularity of unpasteurized cow milk (UPM) and disregard for the risks of being exposed to zoonotic infections. Dairy cattle are healthy carriers of Shiga toxin (Stx)-producing *E. coli* (STEC), and contaminated UPM has caused STEC outbreaks worldwide. The association between STEC, carrying the *eae* (*E. coli* attachment effacement) gene, and severe diseases is well-established. We have previously isolated four *eae* positive STEC isolates from two neighboring dairy farms in the Southeast of Norway. A whole genome analysis revealed that isolates from different farms exhibited nearly identical genetic profiles. To explore the risks associated with drinking UPM, we examined the ability of the isolates to produce Stx and their growth in UPM at different temperatures. All the isolates produced Stx and one of the isolates was able to propagate in UPM at 8 °C ($p < 0.02$). Altogether, these results highlight the risk for STEC infections associated with the consumption of UPM.

Keywords: Shiga toxin-producing *Escherichia coli*; raw cow's milk; unpasteurized; storage; temperature; food safety; Shiga-toxin; bacteriophage

1. Introduction

Enterohemorrhagic *Escherichia coli* (EHEC) is a globally distributed intestinal pathogen associated with human diarrhea, hemorrhagic colitis, and hemolytic uremic syndrome (HUS) [1]. The term "EHEC" is restricted to Shiga toxin-producing *E. coli* (STEC) associated with human disease. The main reservoir of STEC is the ruminant digestive tract and undercooked beef and unpasteurized milk are considered high-risk foods for STEC infections [1,2]. In 2020, 4446 cases of EHEC disease and 13 deaths were reported in the EU [3]. The first large outbreak of EHEC occurred in the USA in 1982 and was caused by a strain of serotype O157:H7 [4]. Since then, other serotypes have also been associated with outbreaks of EHEC disease [5–7]. The most known non-O157:H7 strain is O104:H4, which caused 855 cases of HUS and 50 fatalities during a large European outbreak in 2011 [2]. EHEC has a low infectious dose of 10–100 colony-forming units [8,9], and insufficient food decontamination practices increases the risk for EHEC infections.

STEC can produce two different types of Shiga toxin, Stx1 and Stx2, both comprising several subtypes. Stx2 is more often associated with HUS than Stx1, and Stx2a is considered as the most potent subtype of the toxin [2]. The Stx-encoding genes are carried by temperate bacteriophages [2], and the pathogenic potential of STEC has been suggested to be influenced by the "EHEC phage replication unit" (Eru) located in the phage genome [10,11]. The life cycle of temperate phages is regulated by the CI repressor protein, which represses the transcription of the replication proteins during the lysogenic state of the phage [12,13]. The de-repression of CI results in the production of Stx and new phage particles [14]. Based on similarities in its amino acid sequence, the CI protein of Stx phages has been grouped into eight major clades (I–VIII) [11]. Exactly how the variability in the CI sequence influences its regulatory properties and potentially the virulence properties of its host STEC strain have not been explored so far.

Stx production combined with the ability to adhere to the intestinal epithelium via the adhesion protein intimin are believed to be necessary for STEC to cause severe disease. The intimin-encoding gene *(eae)* is part of the locus of the enterocyte effacement pathogenicity island (LEE-PAI), which encodes proteins responsible for introducing attaching and effacing (A/E) lesions to the epithelial cells [15]. Similar to CI, intimins display a structural diversity that potentially reflects differences in host cell tropism. The most common types of intimin are α, β, γ, ε, ζ, and η [15]. The β-type has been shown to predominate in non-O157 STEC strains from diarrheal patients, while cattle isolates more often carry the ζ-type [16]. The presence of *eae* is associated with a higher risk of developing HUS [17].

EHEC is regarded as an emerging public health challenge as new pathotypes and serotypes constantly appear [18–20]. Milk contaminated with pathogens causes foodborne disease worldwide, and 33% of all reported milk-borne disease outbreaks in England and Wales between 1992–2000 were caused by EHEC [21]. Previous studies have shown that 27, 13, and 5% of cattle from Portugal [22], US [23], and the EU [3] carry STEC, respectively. A study from Finland showed that 2% of on-farm, in-line milk filters were positive for STEC of the serotype O157:H7 [24], while in Norway, STEC has been detected in 7% of milk filters [25]. As STECs are carried by asymptomatic cows and frequently occur in dairy farm environments [26], the milk from these sources can easily be contaminated during the milking process. The lack of effective preventive measures in the primary production of milk makes pasteurization necessary to ensure food safety. Pasteurization at 72 °C for 15 s has shown to be very effective for the inactivation of STEC [27].

Low-temperature storage is important for preventing microbial growth in milk [28]. Previous studies have shown that STEC is not able to grow at 4 °C, but proliferation has been observed at inadequate refrigeration temperatures [29,30]. It has been shown that *E. coli* of the serotype O157:H7 grows in unpasteurized and pasteurized milk with a 2- to 3-log CFU/mL increase at 8 °C within a time period of seven days [31]. The European Food Safety Authority (EFSA) recommend that certain unpasteurized and low-pasteurized dairy products should be stored below 5 °C to minimize microbial growth [32]. However, the temperature in domestic refrigerators has been shown to vary between 7.0 ± 2.7 °C and 6.1 ± 2.8 °C for southern and northern European countries, respectively [33]. In addition, short breaks in the cold chain, for example, during meals, represent an additional but unexplored factor that may add to the risk of consuming UPM.

To further assess the food safety risk associated with the consumption of UPM, we need to gain more knowledge on the genetic- and growth characteristics of the STECs isolated from raw cows' milk. In the present study, we have compared the genome of four STECs isolated from milk [25] with a focus on their content of virulence-associated genes and Stx phages. The isolates were tested for their survival and growth in UPM milk, incubated at recommended and abused storage temperatures, and for the production of Stx at the body temperature of a human host. Altogether, the results highlight the risk for EHEC infections associated with the consumption of UPM, particularly if the milk has been stored at an abused temperature.

2. Materials and Methods

2.1. Culturing Conditions

This study comprises four *stx-* and *eae*-positive *E. coli* isolates from Norwegian dairy farms [25]. Three of the isolates were from the same farm, two from fecal samples (S2 and S4) and one from an in-line milk filter sample (S3), while the fourth isolate was isolated from a fecal sample (S1) at a nearby farm. The isolates were collected at two different sampling occasions separated by five months (Table 1). Raw milk, from the dairy cattle breed Norwegian Red, was collected from a bulk tank at the Center for Livestock Experiments at the Norwegian University of Life Sciences and used as cultivation medium in the growth experiments. The milk was collected in batches of approximately 2 L at two different occasions (September 2021 and April 2022) and aliquoted in 40 mL batches in Falcon tubes and frozen at −20 °C until use.

Table 1. Characteristics of the Shiga toxin-producing *E. coli* isolated from dairy farms located in the southeast of Norway [25].

	S1	S2	S3	S4
Source	Cattle feces (Farm B)	Cattle feces (Farm A)	Milk filter (Farm A)	Cattle feces (Farm A)
Year of isolation	2019 (November)	2020 (January)	2020 (June)	2020 (June)
Country	Norway	Norway	Norway	Norway
Pathotype	STEC	STEC	STEC	STEC
Serotype	ONT:H28	O108:H25	ONT:H28	ONT:H28
NCBI accession no	JANWGF000000000	JANWGE000000000	JANWGD000000000	JANWGC000000000
LEE operons	five	five	five	five
Intimin type	gamma	alpha	gamma	gamma
ehxA	yes	yes	yes	yes
astA ST toxin	yes	yes (2)	yes	yes
Stx type	Stx1a	Stx2a	Stx1a	Stx1a
Eru type	lambdoid	Eru1	lambdoid	lambdoid
Stx phage CI clade	V	II	V	V

To explore the ability of the STEC isolates to grow at different temperatures, over-night cultures of the respective isolates grown in Lysogeny broth (LB) were diluted to OD_{600} = 0.3, whereof 0.5 µL were transferred to 40 mL of thawed raw milk. Immediately after inoculation, 10 µL of the milk samples was plated on CHROMagar™ STEC (Kanto Chemical Co., Tokyo, Japan) and incubated at 37 °C for 24 h to enumerate the start concentration of STEC. The inoculated raw milk samples were then incubated at five different temperature settings: optimal refrigerator temperature (4 °C), abused refrigerator temperatures (6 °C and 8 °C), room temperature (20 °C), and a temperature setting mimicking the situation when milk is kept at room temperature during meals (4 °C except for 1.5 h daily at 20 °C). To determine the temperature fluctuation of the samples incubated this way, the temperature was recorded in an uninoculated 40 mL raw milk sample every 15 min during the 20 °C incubation and until the milk temperature had returned to 4 °C, which encompassed a total time of 4.5 h. For enumeration of STEC in the raw milk samples incubated at different temperatures, dilutions of the samples were plated on CHROMagar™ STEC agar after 24, 48, and 72 h of incubation. The growth ratio, used as indicator of growth, was calculated by dividing the number of STEC colonies appearing on the plates after 24, 48, and 72 h by the number of the STEC colonies present in the cultures at time zero.

To determine growth of the STEC isolates in laboratory media without the impact of competing bacteria, each isolate was inoculated into 40 mL LB and incubated at 20 °C. For enumeration, appropriate dilutions of the cultures were plated on LB agar after 0 and 24 h. All experiments were performed in three biological replicates, except for STEC incubated in raw milk at 20 °C, which was only performed with two replicates. To exclude the presence of STEC in the two raw milk batches used, 6 × 100 µL raw milk samples from each batch were plated on CHROMagar™ STEC agar and incubated according to manufacturer's instructions.

2.2. Stx Production

A volume of 100 µL overnight LB-cultures was transferred to 5 mL fresh LB and incubated at 37 °C with agitation at 250 rpm until the optical density reached 0.5 at 600 nm (OD_{600}). Half of these cultures were induced by addition of 0.5 µg/mL of Mitomycin C (MMC). Both induced and uninduced cultures were incubated further for 3 h. Six samples, three induced and thee uninduced, were processed and analyzed with respect to Stx content for each STEC isolate. The Stx content was measured in 1:20 dilutions of the cultures using the semi-quantitative enzyme immunoassay RIDASCREEN® Verotoxin kit (R-biopharm, Darmstadt, Germany) according to the manufacturer.

2.3. Genome Sequence Analyses

DNA for long-read sequencing was extracted using Nanobind CBB Big DNA Kit (NB-900-001-01, Circulomics, Baltimore, MD, USA), according to the manufacturer's instructions (Nanobind HMW DNA Extraction protocol for Gram-Negative Bacteria, 2021). Oxford Nanopore Technologies' "Ligation Sequencing kit" (SQK-LSK109, Oxford Nanopore Technologies Plc., Oxford, UK) was used for library preparation and "Native Barcoading Expansions" 1–12 (EXP-NBD104, Oxford Nanopore Technologies Plc., Oxford, UK) for barcoding the libraries. Nanopore sequencing was performed on a FLO-Min106 (R9.4.1, Oxford Nanopore Technologies Plc., Oxford, UK) flow cell. Recovered reads were assembled using the Flye assembler implemented in the "Dragonflye"-pipeline (https://github.com/rpetit3/dragonflye, v.1.0.12 (accessed on 25 March 2022)), which also performs adapter removal and assembly polishing. Virulence and antimicrobial resistance genes, core genome MLST type, and serotype were identified using the following tools on the CGE website: VirulenceFinder 2.0 [34,35], ResFinder 4.1 [36–38], cgMLSTFinder 1.1 [39,40], and SerotypeFinder 2.0 [41]. Prophage sequences were identified and annotated using the Phaster web software [42]. Isolate diversities were examined by SNP using Snippy v. 4.6.01 (https://github.com/tseemann/snippy (accessed on 20 May 2022)) and Mauve v2.4.0 (https://darlinglab.org/mauve/mauve.html (accessed on 5 May 2022)) were used to align the genomes (default parameters). This Whole Genome Shotgun project has been deposited at DDBJ/ENA/GenBank under the accession JANWGC000000000 to JANWGF000000000 (Table 1).

2.4. Statistics

For all growth experiments, a two-tailed paired Student's *t*-test, performed via Microsoft Office Excel, was used to test for statistically significant differences between average CFU determined at two different time points. *p*-values equal to or below 0.05 were considered significant. Standard deviation was calculated using Excel.

3. Results

3.1. Genetic Characterization of STEC Isolates from Raw Milk

Three of the four STEC isolates included in this study originated from the same farm (S2, S3, and S4); two were collected from fecal samples (S2 and S4) and one from an in-line milk filter sample (S3). Isolate S2 was collected five months prior to S3 and S4. The fourth isolate (S1) originates from a fecal sample from a second farm located within 10 km from farm one. The characteristics of the four STECs are listed in Table 1.

A genome sequence analysis revealed that isolates S1, S3, and S4 are highly similar and differ by only 19–23 SNPs, suggesting that these isolates are clonal (Figure 1).

S1, S3, and S4 exhibit 5.2 Mb chromosomes and the sequence analysis shows that they are of the serotype ONT:H28 and that they belong to the core genome multi-locus sequence type (cgMLST) 7679. Their genomes harbor the LEE-PAI-encoding intimin gamma (*eae*) and the gene encoding the translocated intimin receptor (Tir). The LEE-PAI is 99% identical over 33.3 kbp to the *E. coli* O157:H7 strain EDL933 (NCBI accession number NZ_CP008957) from the US outbreak in 1982 [4]. The lambdoid Stx1 phage of isolates S1, S3, and S4 is 99% identical over 22.8 kbp to Phage BP-4795 (*E. coli,* strain 4795/97, serotype O84:H4 human, Germany 1997) [15,43]. The CI repressor of this phage belongs to Clade V [11]. All three isolates carry a circular plasmid of 55 kbp encoding a heat-stable toxin (*astA*) and enterohaemolysin (*ehxA*) [44,45]. The heat-stable toxin is known to cause sporadic diarrhea in humans and animals [46], while enterohaemolysin is associated with bloody diarrhea and HUS [47]. Furthermore, in the genome of each isolate, a total of 18 prophages of varying completeness were identified by Phaster [42,48]. The Stx phage harbored by these stains is of the lambdoid type and encodes Stx1a [4,49].

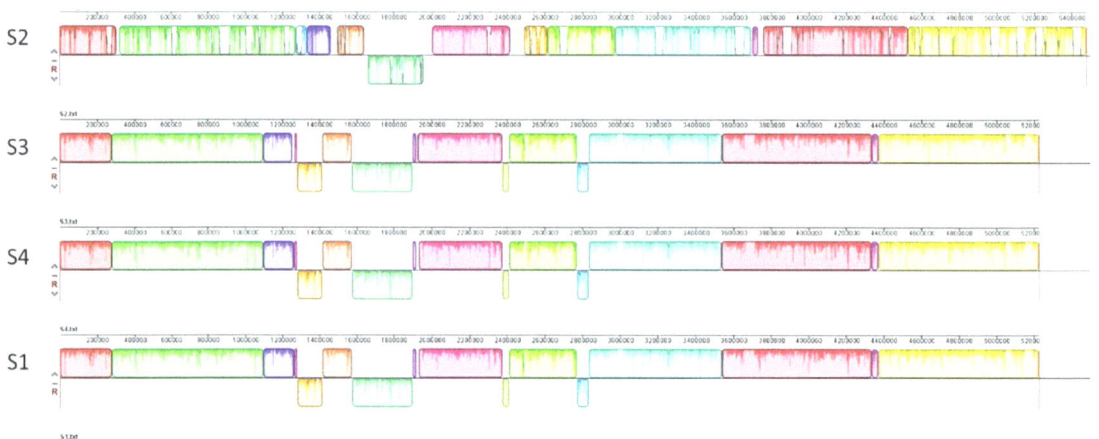

Figure 1. Multiple genome alignment was performed using the Mauve software. Each sequence is represented by a horizontal panel of blocks. The colored blocks indicate homologous sequence regions between the genomes. Blocks below the center line in each genome are inverted sequences with respect to the other genomes.

The genome of isolate S2 is highly different from those of S1, S3, and S4 (Figure 1). It comprises two circular contigs including a chromosome of 5.4 Mbp and a plasmid of 80 kbp. A DNA-typing analysis revealed that the isolate belongs to serotypes O108:H25 and cgMLST 141324. S2 carries a bacteriophage of Eru type 1 and a CI repressor belonging to Clade II [10,11]. The phage encodes the Stx2a type of Stx [50] and shares 99% identity over 22.2 kbp covering the replication region of the Stx2 phage TL-2011c (NCBI accession number NC_019442), which was carried by a highly virulent EHEC strain that caused an outbreak in Norway in 2006 [51].

Similar to the other three isolates, S2 harbors LEE-PAI including both *eae* and *tir*. The DNA sequence of the five LEE operons shows 87% identity over 30 kbp to the corresponding sequence of *E. coli* O157:H7 strain EDL933 (NCBI accession number NZ_CP008957). The 80 kbp plasmid of isolate S2 contains both *astA* and *ehxA*. Phaster identified 30 prophage regions on the chromosome and one prophage on the plasmid in isolate S2. ResFinder 4.1 did not detect antimicrobial resistance genes in any of the four isolates.

3.2. Stx Production

To explore the virulence potential of the STEC isolates, the Stx production was examined during growth in LB at 37 °C, with and without induction by MMC. All four isolates produced Stx and the levels were higher three hours post-induction with MMC compared to the uninduced samples (Figure 2).

3.3. Growth Characteristics of STEC Isolates in Raw Milk at Different Storage Temperatures

To examine the ability of the four STEC isolates to survive and grow in UPM, 40 mL raw milk samples were inoculated with approximately 3000–5000 CFU/mL of STEC culture. The samples were then incubated at 4 °C, 6 °C, 8 °C, and 20 °C for 72 h. After 0, 24, and 72 h, the samples were plated on Chromagar™ STEC for enumeration. The growth ratios were calculated by dividing the number of STEC at 24 and 72 h by the number of bacteria inoculated into the milk.

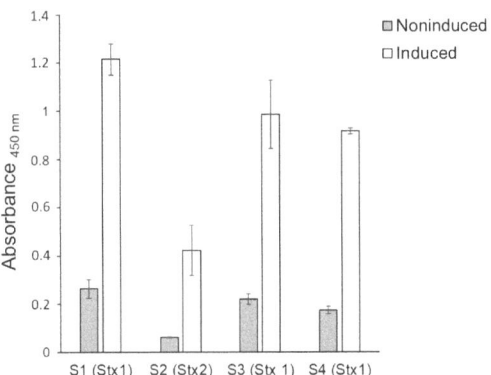

Figure 2. Semi-quantitative determination of Stx production of STEC isolates S1, S2, S3, and S4 after three hours induction with Mitomycin C (0.5 µg/mL). Error bars represent standard deviation.

At 4 °C, an average reduction in CFU (growth ratio below 1) was observed for all four isolates after storage for 24 h. The reduction was, however, not significant for any of the four isolates (Figure 3a). For isolates S1, S2, and S3, the number of CFU was further reduced over the next 48 h, while the level of isolate S4 remained unchanged (Figure 3a). The reduction in bacterial levels seen after 72 h, compared to the levels at the start of cultivation, was only significant for isolate S3 ($p < 0.01$). At 6 °C, a decrease in CFU/mL was observed during the first 24 h ($p \leq 0.05$ for isolate S1 and S4) but the cell death stopped after 24 h (Figure 3b). At 8 °C, S1, S3, and S4 multiplied over the first 24 h of storage (growth ratio above 1), and all strains showed increased CFU counts after 72 h (Figure 3c). The increase in CFU/mL after 72 h of storage, compared to the CFU at the start of cultivation, was significant only for isolate S1 ($p < 0.02$). There was a large difference in growth between isolate S2 and the three other isolates at 20 °C (Figure 3d).

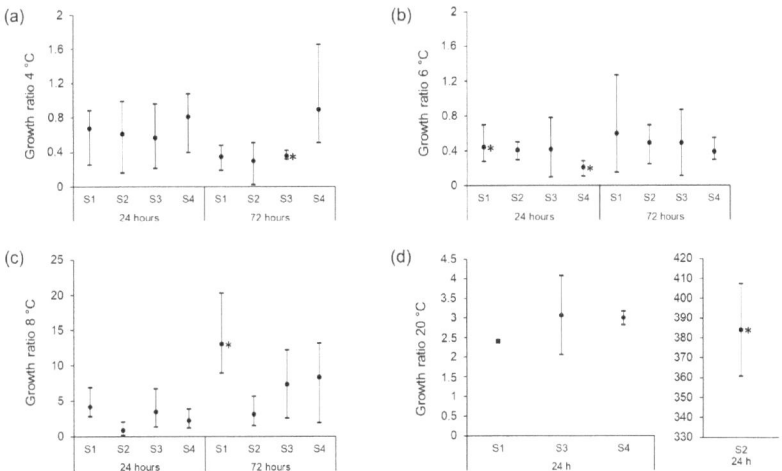

Figure 3. Chart showing the minimum, maximum, and average of growth ratios for STEC isolates S1, S2, S3, and S4 in unpasteurized milk at 4 °C (**a**), 6 °C (**b**), 8 °C (**c**), and 20 °C (**d**). Growth ratios below 1 indicate cell death while a growth ratio above 1 indicates growth. Asterisks represent statistical differences from pairwise comparisons between inoculation point and 24 or 72 h using two-tailed paired Student's t tests (* $p \leq 0.05$).

Under abused conditions, wherein the inoculated milk samples were kept at 4 °C but exposed to 20 °C for 90 min every 24 h, a trend of positive growth ratios was observed after 72 h of storage. However, only the increase in CFU/mL between 24 h and 48 h ($p < 0.01$) and between 24 h and 72 h ($p < 0.05$) for isolate S2 were significant (Figure 4a). The average growth ratios were lower than those observed at 8 °C (Figure 3c). The growth ratios of the four isolates inoculated into LB and incubated for 24 h indicate that the ability to grow in LB at 20 °C is similar for the four isolates (Figure 4b), and that they multiply faster in LB compared to unpasteurized milk at 20 °C.

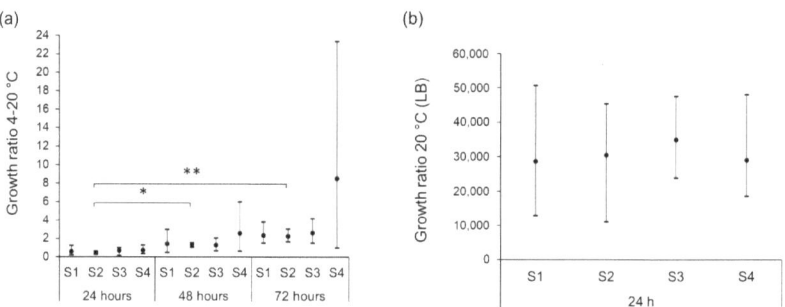

Figure 4. Chart showing the minimum, maximum, and average of growth ratios for STEC isolates S1, S2, S3, and S4 in unpasteurized milk at 4 °C under a temperature abuse scheme of 90 min at 20 °C every 24 h (**a**) and at 20 °C in LB-broth (**b**). Asterisks represent statistical differences from pairwise comparisons determined using two-tailed paired Student's t tests (* $p < 0.01$; ** $p < 0.05$).

Recordings of the temperature fluctuation in the 40 mL raw milk showed that after reaching 20 °C, it took >3 h for the milk to reach below 5 °C (Figure 5).

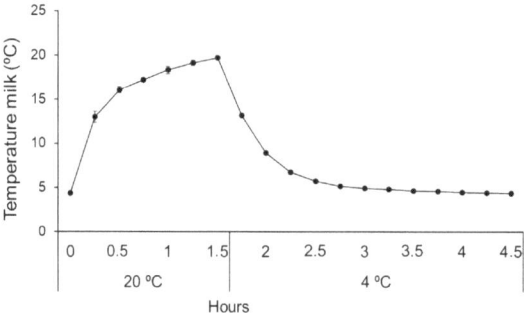

Figure 5. Temperature fluctuation in 40 mL UPM incubated at 4 °C, interrupted with incubation at 20 °C for 90 min. Error bars represent standard deviation.

4. Discussion

Cattle represent a reservoir of STEC, and the consumption of unpasteurized milk is, therefore, considered an important risk factor for contracting milk-borne STEC infections [1,2]. Herein, we explore the pathogenic potential of four *eae*-positive STECs (S1–S4) isolated from Norwegian dairy herds and their ability to grow in UPM stored under optimal and abused temperature conditions.

The genome analysis showed that isolates S1, S3, and S4 are clonal even though they were isolated from two different farms and S1 was isolated seven months prior to S3 and S4. This indicates that STEC has been transmitted between the two farms and persisted in the farm environment over time. Previous studies have shown that *E. coli* O157:H7 can survive for 99 days in soil [52] and 13 weeks in lake water at 15 °C [53]. The clonal isolates

S4 from feces and S3 from a milk filter were isolated the same day from the same farm, which strongly suggests that STEC can be transmitted from feces to the raw milk.

To explore the potential of the four isolates to cause disease, the genomes of the isolates were examined with respect to known virulence-associated genes. Isolates S1, S3, and S4 carry genes encoding Stx1a, while isolate S2 carries genes encoding Stx2a. Stx2a is considered the most potent Stx subtype and is associated with high virulence and HUS [50,54,55]. As isolate S2 has the potential to produce the more potent Stx2a form of Stx, it is likely to be more virulent than the other three isolates described in this study. All four isolates produced Stx, and the production was increased in the presence of MMC. In a study by Muniesa et al. (2004), 18% of 168 *stx2*-carrying STEC strains, isolated from cattle, were MMC-inducible [56]. Our results indicate a higher production of Stx1 by isolates S1, S2, and S4 compared to the degree of Stx2 production by isolate S2. The kit used for the detection of Stx, the enzyme immunoassay RIDASCREEN® Verotoxin kit (R-biopharm, Darmstadt, Germany), detects all known Stx-types [57]; however, a direct comparison between the amount of Stx1 and Stx2 produced is not applicable as the RIDASCREEN® Verotoxin kit has a lower detection limit for Stx1 (12.5 pg/mL) than for Stx2 (25 pg/mL). The degree of Stx production was examined at 37 °C, as this is the temperature in the human gut where the toxin's production occurs.

Stx-encoding prophages are very diverse and recent studies have suggested that their pathogenic potential is determined by the phage replication region, encoding the phage repressor protein CI and the phage replication proteins [10,11]. The EHEC phage replication unit Eru1, which is carried by the highly pathogenic EHEC strains that caused the Norwegian O103:H25 outbreak in 2006 and the large O104:H4 outbreak in Europe in 2011, is also carried by the S2 isolate described in this study [10]. Eru1 is often carried together with a Clade II CI repressor, as is the case for the S2 isolate, and may also indicate a high pathogenic potential [11]. It has previously been suggested that phage production is not induced by MMC in the Eru1 type of Stx-phages [10]. Contrary to this suggestion, we show herein that Stx production is induced by MMC in isolate S2, which suggests that Stx production and the production of new phage particles are regulated differently even in phages belonging to the same Eru type.

All four STEC isolates characterized in this study carry the gene encoding intimin, which has been associated with an increased ability to cause severe disease [55]. They also carry the large O157 plasmid harboring the virulence gene *ehxA*, encoding enterohaemolysin, which is present in most isolates from clinical STEC-infections [55]. The gene *astA*, encoding the heat-stable EAST1 toxin, which is present in several human diarrheagenic *E. coli* pathotypes was also found in in the genomes of the four isolates [46]. An EAST1-positive *E. coli* strain has been suggested to be the culprit of a large diarrhea outbreak in Japan that affected 2697 children [58]. The *astA* gene is, however, also commonly found among *E. coli* isolates collected from the environment [59]. The presence of genes encoding Stx, intimin, and enterohaemolysin as well as the EAST1 toxin in *E. coli* isolates from Norwegian dairy farms strongly indicate that Norwegian raw milk may contain highly pathogenic *E. coli*.

As raw milk may contain highly pathogenic bacteria such as STEC, *Listeria monocytogenes*, *Campylobacter*, and *Salmonella*, the temperature used for its storage is critical. In this study, we observed that at 4 °C the STEC levels slightly decreased over 72 h; however, only the reduction of S3 was significant ($p < 0.01$). At 6 °C, there was a trend towards decreased STEC levels over the first 24 h of storage, whereafter the levels were constant over the next 48 h. At 8 °C, there was an increasing trend in the STEC levels. Due to the large variation between the three biological replicates in the growth experiments, the results are not conclusive. However, at each temperature, at least one isolate showed a clear increase or decrease in CFU ($p \leq 0.05$), indicating that temperatures between 6 and 8 °C for more than 24 h may allow STEC to multiply in raw milk. These results are comparable to previous studies that have shown that *E. coli* O157:H7 is capable of growing in raw milk at 7 and 15 °C [60], but not at 5 °C [31]. Another study showed that *E. coli* O157:H7 did not decrease during storage at 4 °C for five days. However, the study used streptomycin-resistant strains and raw milk supplemented with streptomycin, which may have influenced the natural microbiota of the raw milk [28]. The large growth variations between replicates of the same isolate in our study indicate that even though the growth is not statistically significant, sudden multiplications of STEC can occur in individual milk samples. The experimental conditions in the present study are not directly comparable to natural conditions since the raw milk was inoculated with 3000–5000 CFU/mL and such a high number of STEC is not likely to be present in fresh bulk tank milk. The transition from LB media at 37 °C—used for pre-culturing the isolates—to raw milk at low temperatures may also have influenced the survival of the isolates.

To mimic a real-life scenario of temperature abuse during meals, the milk was stored at 4 °C interrupted by exposure to room temperature (20 °C) for 1.5 h per day. Under these conditions, a general increase in CFU/mL milk was observed after 72 h; however, the increase was only significant for isolate S2 ($p < 0.05$). The recordings of the temperature in 40 mL of raw milk moved from 4 °C to 20 °C showed that the sample reached room temperature after 1.5 h. In a real-life situation, we assume that the volumes of raw milk stored for consumption are larger than 40 mL and the temperature fluctuation in the stored milk will be less pronounced.

Isolate S2 showed rapid growth during the storage of UPM at 20 °C, while the growth rates of the clonal isolates S1, S3, and S4 were slower. However, in LB media, all isolates showed similar growth rates and reached higher concentrations than they did in UPM, stored for the same time. The growth inhibition of the three clonal isolates may be due to the presence of milk-borne antimicrobial components such as lactoperoxidase, lysozyme, xanthine, oxidase, lactoferrin, immunoglobulins, and bacteriocins, or by competing microorganisms [61]. Previous studies have shown a better survival of *E. coli* inoculated in pasteurized milk compared to *E. coli* inoculated in UPM [31]. This is not surprising, since UPM contains an indigenous microbiota that can influence the growth of STEC. Notably, *E. coli* O157:H7 has been shown to be unresponsive to the antimicrobial activity of the lactoperoxidase–thiocyanate–hydrogen peroxide system (LPS) in milk, and this may also be the case for isolate S2 [60].

The survival and growth levels were only examined over a period of 72 h post-inoculation, as raw milk is not recommended to be stored for a very long time before consumption [62]. However, temperature abuse in consumers' handling practices is common both during transport and storage. Most consumers are unaware of their refrigerator's temperature [63], and studies show that household refrigerators often hold higher temperatures than recommended. Furthermore, milk is often kept at locations in the refrigerator where the temperature varies, for example, in refrigerator door racks [63–66]. This is particularly important to consider regarding the risk of disease from low-dose pathogens such as EHEC [67].

5. Conclusions

STEC isolates harboring genes associated with pathogenicity such as *stx1/2*, *eae*, *ehxA*, and *astA* are present in Norwegian dairy farms, and potentially pathogenic STEC isolates are able to can grow in raw milk stored at temperatures above 6 °C. As previous studies show that domestic refrigerators often hold higher temperatures than recommended, the growth of STEC in stored raw milk is a likely scenario. Altogether, the results suggest that UPM from Norwegian dairy farms may contain highly pathogenic STEC strains, and that the storage of UPM under suboptimal refrigeration conditions increases the risk for hemorrhagic colitis and HUS. To reduce the risk associated with the consumption of UPM, consumers need more knowledge regarding the importance of keeping the milk sufficiently chilled to prevent the growth and survival of STEC and other pathogenic bacteria. They should also be made aware of that even correctly stored UPM is associated with an increased risk for illness and that young children, elderly, and immunocompromised individuals are particularly vulnerable.

Author Contributions: Conceptualization, L.I., E.G.B.-G., M.A. and T.L.; methodology, L.I., E.G.B.-G., M.A. and T.L.; supervision, M.A., E.G.B.-G. and T.L.; writing—original draft preparation, L.I.; writing—review and editing, L.I., T.L., M.A. and E.G.B.-G. All authors have read and agreed to the published version of the manuscript.

Funding: This research received no external funding.

Institutional Review Board Statement: Not applicable.

Informed Consent Statement: Not applicable.

Data Availability Statement: The data presented in this study are available on request from the corresponding author.

Acknowledgments: We thank Nofima and Helga Ness for the use of the Biosafety lab for performing the STEC storage experiment, and Anette Wold and Tove Maugesten for helpful instructions.

Conflicts of Interest: The authors declare no conflict of interest.

References

1. Karmali, M.A. Factors in the emergence of serious human infections associated with highly pathogenic strains of shiga toxin-producing *Escherichia coli*. *Int. J. Med. Microbiol.* **2018**, *308*, 1067–1072. [CrossRef] [PubMed]
2. Joseph, A.; Cointe, A.; Mariani-Kurkdjian, P.; Rafat, C.; Hertig, A. Shiga toxin-associated hemolytic uremic syndrome: A narrative review. *Toxins* **2020**, *12*, 67. [CrossRef] [PubMed]
3. European Food Safety Authority & European Centre for Disease Prevention and Control. The European Union One Health 2020 Zoonoses Report. *EFSA J.* **2021**, *19*, e06971. [CrossRef]
4. Riley, L.W.; Remis, R.S.; Helgerson, S.D.; McGee, H.B.; Wells, J.G.; Davis, B.R.; Hebert, R.J.; Olcott, E.S.; Johnson, L.M.; Hargrett, N.T. Hemorrhagic colitis associated with a rare *Escherichia coli* serotype. *N. Engl. J. Med.* **1983**, *308*, 681–685. [CrossRef]
5. Gould, L.H.; Mody, R.K.; Ong, K.L.; Clogher, P.; Cronquist, A.B.; Garman, K.N.; Latrop, S.; Medus, C.; Spina, N.L.; Webb, T.H. Increased recognition of non-O157 Shiga toxin–producing *Escherichia coli* infections in the United States during 2000–2010: Epidemiologic features and comparison with *E. coli* O157 infections. *Foodborne Pathog. Dis.* **2013**, *10*, 453–460. [CrossRef]
6. Hedican, E.B.; Medus, C.; Besser, J.M.; Juni, B.A.; Koziol, B.; Taylor, C.; Smith, K.E. Characteristics of O157 versus non-O157 Shiga toxin-producing *Escherichia coli* infections in Minnesota, 2000–2006. *Clin. Infect. Dis.* **2009**, *49*, 358–364. [CrossRef]
7. Hadler, J.L.; Clogher, P.; Hurd, S.; Phan, Q.; Mandour, M.; Bemis, K.; Marcus, R. Ten-year trends and risk factors for non-O157 Shiga toxin–producing *Escherichia coli* found through Shiga toxin testing, Connecticut, 2000–2009. *Clin. Infect. Dis.* **2011**, *53*, 269–276. [CrossRef]
8. Tuttle, J.; Gomez, T.; Doyle, M.; Wells, J.; Zhao, T.; Tauxe, R.; Griffin, P.M. Lessons from a large outbreak of *Escherichia coli* O157:H7 infections: Insights into the infectious dose and method of widespread contamination of hamburger patties. *Epidemiol. Infect.* **1999**, *122*, 185–192. [CrossRef]
9. Sperandio, V.; Nguyen, Y. Enterohemorrhagic *E. coli* (EHEC) pathogenesis. *Front. Cell. Infect. Microbiol.* **2012**, *2*, 90. [CrossRef]
10. Llarena, A.-K.; Aspholm, M.; O'Sullivan, K.; Węgrzyn, G.; Lindbäck, T. Replication region analysis reveals non-lambdoid shiga toxin converting bacteriophages. *Front. Microbiol.* **2021**, *12*, 640945. [CrossRef]
11. Fagerlund, A.; Aspholm, M.; Węgrzyn, G.; Lindbäck, T. High diversity in the regulatory region of Shiga toxin encoding bacteriophages. *BMC Genom.* **2022**, *23*, 230. [CrossRef] [PubMed]
12. Bednarz, M.; Halliday, J.A.; Herman, C.; Golding, I. Revisiting bistability in the lysis/lysogeny circuit of bacteriophage lambda. *PLoS ONE* **2014**, *9*, e100876. [CrossRef] [PubMed]
13. Casjens, S.R.; Hendrix, R.W. Bacteriophage lambda: Early pioneer and still relevant. *Virology* **2015**, *479*, 310–330. [CrossRef] [PubMed]
14. Zeng, L.; Skinner, S.O.; Zong, C.; Sippy, J.; Feiss, M.; Golding, I. Decision making at a subcellular level determines the outcome of bacteriophage infection. *Cell* **2010**, *141*, 682–691. [CrossRef] [PubMed]
15. Zhang, W.; Kohler, B.; Oswald, E.; Beutin, L.; Karch, H.; Morabito, S.; Caprioli, A.; Suerbaum, S.; Schmidt, H. Genetic diversity of intimin genes of attaching and effacing *Escherichia coli* strains. *J. Clin. Microbiol.* **2002**, *40*, 4486–4492. [CrossRef]
16. Yang, X.; Sun, H.; Fan, R.; Fu, S.; Zhang, J.; Matussek, A.; Xiong, Y.; Bai, X. Genetic diversity of the intimin gene (*eae*) in non-O157 Shiga toxin-producing *Escherichia coli* strains in China. *Sci. Rep.* **2020**, *10*, 3275. [CrossRef]
17. De Rauw, K.; Buyl, R.; Jacquinet, S.; Piérard, D. Risk determinants for the development of typical haemolytic uremic syndrome in Belgium and proposition of a new virulence typing algorithm for Shiga toxin-producing *Escherichia coli*. *Epidemiol. Infect.* **2019**, *147*, E6. [CrossRef]
18. European Food Safety Authority & European Centre for Disease Prevention Control. The European Union summary report on trends and sources of zoonoses, zoonotic agents and food-borne outbreaks in 2009. *EFSA J.* **2011**, *9*, 2090. [CrossRef]
19. Bosilevac, J.M.; Koohmaraie, M. Prevalence and characterization of non-O157 Shiga toxin-producing *Escherichia coli* isolates from commercial ground beef in the United States. *Appl. Environ. Microbiol.* **2011**, *77*, 2103–2112. [CrossRef]

20. European Food Safety Authority. Scientific Opinion of the Panel on Biological Hazards (BIOHAZ)-Monitoring of verotoxigenic *Escherichia coli* (VTEC) and identification of human pathogenic VTEC types. *EFSA J.* **2007**, *5*, 579. [CrossRef]
21. Gillespie, I.; Adak, G.; O'brien, S.; Bolton, F. Milkborne general outbreaks of infectious intestinal disease, England and Wales, 1992–2000. *Epidemiol. Infect.* **2003**, *130*, 461–468. [CrossRef] [PubMed]
22. Ballem, A.; Gonçalves, S.; Garcia-Meniño, I.; Flament-Simon, S.C.; Blanco, J.E.; Fernandes, C.; Saavedra, M.J.; Pinto, C.; Oliveira, H.; Blanco, J.; et al. Prevalence and serotypes of Shiga toxin-producing *Escherichia coli* (STEC) in dairy cattle from Northern Portugal. *PLoS ONE* **2021**, *15*, e0244713. [CrossRef] [PubMed]
23. Venegas-Vargas, C.; Henderson, S.; Khare, A.; Mosci, R.E.; Lehnert, J.D.; Singh, P.; Ouellette, L.M.; Norby, B.; Funk, J.A.; Rust, S. Factors associated with Shiga toxin-producing *Escherichia coli* shedding by dairy and beef cattle. *Appl. Environ. Microbiol.* **2016**, *82*, 5049–5056. [CrossRef] [PubMed]
24. Jaakkonen, A.; Castro, H.; Hallanvuo, S.; Ranta, J.; Rossi, M.; Isidro, J.; Lindström, M.; Hakkinen, M. Longitudinal Study of Shiga Toxin-Producing *Escherichia coli* and *Campylobacter jejuni* on Finnish Dairy Farms and in Raw Milk. *Appl. Environ. Microbiol.* **2019**, *85*, e02910-18. [CrossRef]
25. Idland, L.; Granquist, E.G.; Aspholm, M.; Lindbäck, T. The prevalence of *Campylobacter* spp., *Listeria monocytogenes* and Shiga toxin-producing *Escherichia coli* in Norwegian dairy cattle farms: A comparison between free stall and tie stall housing systems. *J. Appl. Microbiol.* **2022**, *132*, 3959–3972. [CrossRef]
26. Geue, L.; Segura-Alvarez, M.; Conraths, F.; Kuczius, T.; Bockemühl, J.; Karch, H.; Gallien, P. A long-term study on the prevalence of Shiga toxin-producing *Escherichia coli* (STEC) on four German cattle farms. *Epidemiol. Infect.* **2002**, *129*, 173–185. [CrossRef]
27. D'aoust, J.-Y.; Park, C.; Szabo, R.; Todd, E.; Emmons, D.; McKellar, R. Thermal inactivation of *Campylobacter* species, *Yersinia enterocolitica*, and hemorrhagic *Escherichia coli* O157: H7 in fluid milk. *J. Dairy Sci.* **1988**, *71*, 3230–3236. [CrossRef]
28. Leclair, R.M.; McLean, S.K.; Dunn, L.A.; Meyer, D.; Palombo, E.A. Investigating the effects of time and temperature on the growth of *Escherichia coli* O157: H7 and *Listeria monocytogenes* in raw cow's milk based on simulated consumer food handling practices. *Int. J. Environ. Res. Public Health* **2019**, *16*, 2691. [CrossRef]
29. Kauppi, K.; Tatini, S.; Harrell, F.; Feng, P. Influence of substrate and low temperature on growth and survival of verotoxigenic *Escherichia coli*. *Food Microbiol.* **1996**, *13*, 397–405. [CrossRef]
30. International Commission on Microbiological Specifications for Foods. *Microorganisms in Foods 5: Characteristics of Microbial Pathogens*, 1st ed.; Kluwer Academic/Plenum Publishers: London, UK, 1996.
31. Wang, G.; Zhao, T.; Doyle, M.P. Survival and growth of *Escherichia coli* O157: H7 in unpasteurized and pasteurized milk. *J. Food Prot.* **1997**, *60*, 610–613. [CrossRef]
32. Dumitrașcu, L.; Nicolau, A.I.; Neagu, C.; Didier, P.; Maître, I.; Nguyen-The, C.; Skuland, S.E.; Møretrø, T.; Langsrud, S.; Truninger, M. Time-temperature profiles and *Listeria monocytogenes* presence in refrigerators from households with vulnerable consumers. *Food Control* **2020**, *111*, 107078. [CrossRef]
33. Roccato, A.; Uyttendaele, M.; Membré, J.-M. Analysis of domestic refrigerator temperatures and home storage time distributions for shelf-life studies and food safety risk assessment. *Food Res. Int.* **2017**, *96*, 171–181. [CrossRef] [PubMed]
34. Joensen, K.G.; Scheutz, F.; Lund, O.; Hasman, H.; Kaas, R.S.; Nielsen, E.M.; Aarestrup, F.M. Real-time whole-genome sequencing for routine typing, surveillance, and outbreak detection of verotoxigenic *Escherichia coli*. *J. Clin. Microbiol.* **2014**, *52*, 1501–1510. [CrossRef]
35. Malberg Tetzschner, A.M.; Johnson, J.R.; Johnston, B.D.; Lund, O.; Scheutz, F. In silico genotyping of *Escherichia coli* isolates for extraintestinal virulence genes by use of whole-genome sequencing data. *J. Clin. Microbiol.* **2020**, *58*, e01269-20. [CrossRef] [PubMed]
36. Camacho, C.; Coulouris, G.; Avagyan, V.; Ma, N.; Papadopoulos, J.; Bealer, K.; Madden, T.L. BLAST+: Architecture and applications. *BMC Bioinform.* **2009**, *10*, 421. [CrossRef]
37. Bortolaia, V.; Kaas, R.S.; Ruppe, E.; Roberts, M.C.; Schwarz, S.; Cattoir, V.; Philippon, A.; Allesoe, R.L.; Rebelo, A.R.; Florensa, A.F. ResFinder 4.0 for predictions of phenotypes from genotypes. *J. Antimicrob. Chemother.* **2020**, *75*, 3491–3500. [CrossRef]
38. Zankari, E.; Allesøe, R.; Joensen, K.G.; Cavaco, L.M.; Lund, O.; Aarestrup, F.M. PointFinder: A novel web tool for WGS-based detection of antimicrobial resistance associated with chromosomal point mutations in bacterial pathogens. *J. Antimicrob. Chemother.* **2017**, *72*, 2764–2768. [CrossRef]
39. Clausen, P.T.; Aarestrup, F.M.; Lund, O. Rapid and precise alignment of raw reads against redundant databases with KMA. *BMC Bioinform.* **2018**, *19*, 307. [CrossRef]
40. Zhou, Z.; Alikhan, N.-F.; Mohamed, K.; Fan, Y.; Achtman, M.; Brown, D.; Chattaway, M.; Dallman, T.; Delahay, R.; Kornschober, C. The EnteroBase user's guide, with case studies on *Salmonella* transmissions, *Yersinia pestis* phylogeny, and *Escherichia* core genomic diversity. *Genome Res.* **2020**, *30*, 138–152. [CrossRef]
41. Joensen, K.G.; Tetzschner, A.M.M.; Iguchi, A.; Aarestrup, F.M.; Scheutz, F. Rapid and easy in silico serotyping of *Escherichia coli* isolates by use of whole-genome sequencing data. *J. Clin. Microbiol.* **2015**, *53*, 2410–2426. [CrossRef]
42. Arndt, D.; Grant, J.R.; Marcu, A.; Sajed, T.; Pon, A.; Liang, Y.; Wishart, D.S. PHASTER: A better, faster version of the PHAST phage search tool. *Nucleic Acids Res.* **2016**, *44*, W16–W21. [CrossRef] [PubMed]
43. Creuzburg, K.; Recktenwald, J.R.; Kuhle, V.; Herold, S.; Hensel, M.; Schmidt, H. The Shiga toxin 1-converting bacteriophage BP-4795 encodes an NleA-like type III effector protein. *J. Bacteriol. Parasitol.* **2005**, *187*, 8494–8498. [CrossRef] [PubMed]

44. Savarino, S.J.; McVeigh, A.; Watson, J.; Cravioto, A.; Molina, J.; Echeverria, P.; Bhan, M.K.; Levine, M.M.; Fasano, A. Enteroaggregative *Escherichia coli* heat-stable enterotoxin is not restricted to enteroaggregative *E. coli*. *J. Infect. Dis.* **1996**, *173*, 1019–1022. [CrossRef] [PubMed]
45. Lorenz, S.C.; Son, I.; Maounounen-Laasri, A.; Lin, A.; Fischer, M.; Kase, J.A. Prevalence of hemolysin genes and comparison of *ehxA* subtype patterns in Shiga toxin-producing *Escherichia coli* (STEC) and non-STEC strains from clinical, food, and animal sources. *Appl. Environ. Microbiol.* **2013**, *79*, 6301–6311. [CrossRef] [PubMed]
46. Veilleux, S.; Dubreuil, J.D. Presence of *Escherichia coli* carrying the EAST1 toxin gene in farm animals. *Vet. Res.* **2006**, *37*, 3–13. [CrossRef]
47. Hua, Y.; Zhang, J.; Jernberg, C.; Chromek, M.; Hansson, S.; Frykman, A.; Xiong, Y.; Wan, C.; Matussek, A.; Bai, X. Molecular Characterization of the Enterohemolysin Gene (*ehxA*) in Clinical Shiga Toxin-Producing *Escherichia coli* Isolates. *Toxins* **2021**, *13*, 71. [CrossRef]
48. Zhou, Y.; Liang, Y.; Lynch, K.; Dennis, J.; Wishart, D. PHAST: A fast phage search tool. *Nucleic Acids Res.* **2011**, *39*, W347–W352. [CrossRef]
49. Melton-Celsa, A.R. Shiga toxin (Stx) classification, structure, and function. *Microbiol. Spectr.* **2014**, *2*, 6. [CrossRef]
50. Fuller, C.A.; Pellino, C.A.; Flagler, M.J.; Strasser, J.E.; Weiss, A.A. Shiga toxin subtypes display dramatic differences in potency. *Infect. Immun.* **2011**, *79*, 1329–1337. [CrossRef]
51. L'Abée-Lund, T.M.; Jørgensen, H.J.; O'Sullivan, K.; Bohlin, J.; Ligård, G.; Granum, P.E.; Lindbäck, T. The highly virulent 2006 Norwegian EHEC O103: H25 outbreak strain is related to the 2011 German O104: H4 outbreak strain. *PLoS ONE* **2012**, *7*, e31413. [CrossRef]
52. Bolton, D.; Byrne, C.; Sheridan, J.; McDowell, D.; Blair, I. The survival characteristics of a non-toxigenic strain of *Escherichia coli* O157: H7. *J. Appl. Microbiol.* **1999**, *86*, 407–411. [CrossRef] [PubMed]
53. Wang, G.; Doyle, M.P. Survival of enterohemorrhagic *Escherichia coli* O157: H7 in water. *J. Food Prot.* **1998**, *61*, 662–667. [CrossRef] [PubMed]
54. Krüger, A.; Lucchesi, P.M. Shiga toxins and stx phages: Highly diverse entities. *Microbiology* **2015**, *161*, 451–462. [CrossRef] [PubMed]
55. Brandal, L.T.; Wester, A.L.; Lange, H.; Løbersli, I.; Lindstedt, B.-A.; Vold, L.; Kapperud, G. Shiga toxin-producing *Escherichia coli* infections in Norway, 1992-2012: Characterization of isolates and identification of risk factors for haemolytic uremic syndrome. *BMC Infect. Dis.* **2015**, *15*, 324. [CrossRef]
56. Muniesa, M.; Blanco, J.E.; De Simón, M.; Serra-Moreno, R.; Blanch, A.R.; Jofre, J. Diversity of *stx2* converting bacteriophages induced from Shiga-toxin-producing *Escherichia coli* strains isolated from cattle. *Microbiology* **2004**, *150*, 2959–2971. [CrossRef]
57. Beutin, L.; Steinrück, H.; Krause, G.; Steege, K.; Haby, S.; Hultsch, G.; Appel, B. Comparative evaluation of the Ridascreen® Verotoxin enzyme immunoassay for detection of Shiga-toxin producing strains of *Escherichia coli* (STEC) from food and other sources. *J. Appl. Microbiol.* **2007**, *102*, 630–639. [CrossRef]
58. Itoh, Y.; Nagano, I.; Kunishima, M.; Ezaki, T. Laboratory investigation of enteroaggregative *Escherichia coli* O untypeable: H10 associated with a massive outbreak of gastrointestinal illness. *J. Clin. Microbiol.* **1997**, *35*, 2546–2550. [CrossRef]
59. Sidhu, J.P.; Ahmed, W.; Hodgers, L.; Toze, S. Occurrence of virulence genes associated with diarrheagenic pathotypes in *Escherichia coli* isolates from surface water. *Appl. Environ. Microbiol.* **2013**, *79*, 328–335. [CrossRef]
60. Heuvelink, A.; Bleumink, B.; Van Den Biggelaar, F.; Te Giffel, M.; Beumer, R.; De Boer, E. Occurrence and survival of verocytotoxin-producing *Escherichia coli* O157 in raw cow's milk in The Netherlands. *J. Food Prot.* **1998**, *61*, 1597–1601. [CrossRef]
61. Claeys, W.L.; Cardoen, S.; Daube, G.; De Block, J.; Dewettinck, K.; Dierick, K.; De Zutter, L.; Huyghebaert, A.; Imberechts, H.; Thiange, P. Raw or heated cow milk consumption: Review of risks and benefits. *Food Control* **2013**, *31*, 251–262. [CrossRef]
62. European Food Safety Authority, panel on Biological Hazards. Scientific opinion on the public health risks related to the consumption of raw drinking milk. *EFSA J.* **2015**, *13*, 3940. [CrossRef]
63. Marklinder, I.; Lindblad, M.; Eriksson, L.; Finnson, A.; Lindqvist, R. Home storage temperatures and consumer handling of refrigerated foods in Sweden. *J. Food Prot.* **2004**, *67*, 2570–2577. [CrossRef] [PubMed]
64. Evans, E.W.; Redmond, E.C. Time-temperature profiling of United Kingdom consumers' domestic refrigerators. *J. Food Prot.* **2016**, *79*, 2119–2127. [CrossRef] [PubMed]
65. Kennedy, J.; Jackson, V.; Blair, I.; McDowell, D.; Cowan, C.; Bolton, D. Food safety knowledge of consumers and the microbiological and temperature status of their refrigerators. *J. Food Prot.* **2005**, *68*, 1421–1430. [CrossRef] [PubMed]
66. Koutsoumanis, K.; Pavlis, A.; Nychas, G.-J.E.; Xanthiakos, K. Probabilistic model for *Listeria monocytogenes* growth during distribution, retail storage, and domestic storage of pasteurized milk. *Appl. Environ. Microbiol.* **2010**, *76*, 2181–2191. [CrossRef]
67. Zech, H.; Echtermeyer, C.; Wöhlbrand, L.; Blasius, B.; Rabus, R. Biological versus technical variability in 2-D DIGE experiments with environmental bacteria. *Proteomics* **2011**, *11*, 3380–3389. [CrossRef]

Article

Study of the Microbiome of the Cretan Sour Cream Staka Using Amplicon Sequencing and Shotgun Metagenomics and Isolation of Novel Strains with an Important Antimicrobial Potential

Konstantinos Papadimitriou [1,*,†], Marina Georgalaki [2,†], Rania Anastasiou [2], Athanasia-Maria Alexandropoulou [2], Eugenia Manolopoulou [2], Georgia Zoumpopoulou [2] and Effie Tsakalidou [2]

[1] Laboratory of Food Quality Control and Hygiene, Department of Food Science and Human Nutrition, Agricultural University of Athens, Iera Odos 75, 11855 Athens, Greece
[2] Laboratory of Dairy Research, Department of Food Science and Human Nutrition, Agricultural University of Athens, Iera Odos 75, 11855 Athens, Greece; mgeor@aua.gr (M.G.); ranastasiou@aua.gr (R.A.); siaal22@yahoo.gr (A.-M.A.); mae@aua.gr (E.M.); gz@aua.gr (G.Z.); et@aua.gr (E.T.)
* Correspondence: kpapadimitriou@aua.gr
† These authors contributed equally to this work.

Abstract: Staka is a traditional Greek sour cream made mostly from spontaneously fermented sheep milk or a mixture of sheep and goat milk. At the industrial scale, cream separators and starter cultures may also be used. Staka is sometimes cooked with flour to absorb most of the fat. In this study, we employed culture-based techniques, amplicon sequencing, and shotgun metagenomics to analyze the Staka microbiome for the first time. The samples were dominated by *Lactococcus* or *Leuconostoc* spp. Most other bacteria were lactic acid bacteria (LAB) from the *Streptococcus* and *Enterococcus* genera or Gram-negative bacteria from the *Buttiauxella*, *Pseudomonas*, *Enterobacter*, *Escherichia-Shigella*, and *Hafnia* genera. *Debaryomyces*, *Kluyveromyces*, or *Alternaria* were the most prevalent genera in the samples, followed by other yeasts and molds like *Saccharomyces*, *Penicillium*, *Aspergillus*, *Stemphylium*, *Coniospotium*, or *Cladosporium* spp. Shotgun metagenomics allowed the species-level identification of *Lactococcus lactis*, *Lactococcus raffinolactis*, *Streptococcus thermophilus*, *Streptococcus gallolyticus*, *Escherichia coli*, *Hafnia alvei*, *Streptococcus parauberis*, and *Enterococcus durans*. Binning of assembled shotgun reads followed by recruitment plot analysis of single reads could determine near-complete metagenome assembled genomes (MAGs). Culture-dependent and culture-independent analyses were in overall agreement with some distinct differences. For example, lactococci could not be isolated, presumably because they had entered a viable but not culturable (VBNC) state or because they were dead. Finally, several LAB, *Hafnia paralvei*, and *Pseudomonas* spp. isolates exhibited antimicrobial activities against oral or other pathogenic streptococci, and certain spoilage and pathogenic bacteria establishing their potential role in food bio-protection or new biomedical applications. Our study may pave the way for additional studies concerning artisanal sour creams to better understand the factors affecting their production and the quality.

Keywords: artisanal; microbial ecosystem; starter culture; bacteriocin; foodborne; oral; medical; pathogen; dead cells; VBNC

Citation: Papadimitriou, K.; Georgalaki, M.; Anastasiou, R.; Alexandropoulou, A.-M.; Manolopoulou, E.; Zoumpopoulou, G.; Tsakalidou, E. Study of the Microbiome of the Cretan Sour Cream Staka Using Amplicon Sequencing and Shotgun Metagenomics and Isolation of Novel Strains with an Important Antimicrobial Potential. *Foods* **2024**, *13*, 1129. https://doi.org/10.3390/foods13071129

Academic Editors: Paolo Formaggioni and Piero Franceschi

Received: 22 January 2024
Revised: 24 March 2024
Accepted: 26 March 2024
Published: 8 April 2024

Copyright: © 2024 by the authors. Licensee MDPI, Basel, Switzerland. This article is an open access article distributed under the terms and conditions of the Creative Commons Attribution (CC BY) license (https://creativecommons.org/licenses/by/4.0/).

1. Introduction

Dairy products with high lipid content can have added nutritious value as milk lipids represent a good dietary source of essential fatty acids, such as linoleic and α-linolenic acid, as well as fat-soluble vitamins, such as retinol, α-tocopherol and β-carotene [1]. Bioactive fatty acids, such as butyric, oleic, and conjugated linoleic acid (CLA), may also play key roles in the prevention of certain diseases [2–4]. Interestingly, although dairy food consumption has been positively correlated with cardiovascular risk in the past, recent observations suggest a potential inverse association of fermented dairy food consumption with cardiovascular problems and type 2 diabetes [5]. Additionally, the fatty acid composition of milk affects the physical properties, organoleptic quality, and oxidative stability of dairy products [6,7].

Fermented milk products of high fat content encompass fermented sour cream and acidified sour cream (18–20% fat), as well as fermented butter (80–90% fat). Fermented cream, also known as ripened cream [8,9], is manufactured from standardized, homogenized, and heat-treated sweet cream after fermentation with lactic acid bacteria (LAB). It is a relatively heavy, viscous product with a delicate, lactic acid taste and a balanced, pleasant, buttery-like aroma associated with the desirable flavor-active compounds of diacetyl, acetoin, δ-decalactone, and 2-methyl-3-furanthiol [10,11]. Additionally, acidified sour cream is prepared with safe and suitable acidifiers, such as lactic acid and citric acid, with or without the use of LAB [12,13]. Finally, fermented butter is made after churning fermented/cultured cream [13].

At the industrial level, mixed strains of mesophilic LAB are being used as starter cultures for sour cream production, mainly belonging to *Lactococcus lactis* subsp. *lactis*, *Lactococcus lactis* subsp. *cremoris*, *Lactococcus lactis* subsp. *lactis* biovar *diacetylactis*, *Leuconostoc mesenteroides* subsp. *cremoris*, and *Leuconostoc citrovorum* [10].These species are used as aroma producers, converting citrate into diacetyl, which is one of the major flavor compounds responsible for the typical sour cream flavor [14]. In many countries, though, artisanal sour cream is produced by keeping cream at a suitable temperature to allow autochthonous LAB to perform spontaneous fermentation [15]. A plethora of traditionally fermented cream and butter types are produced worldwide, such as Créme Fraiche in France, Kaymak in Turkey, Crema Espesa in Mexico, Pomazánkové Máslo in Czech Republic, Jiaoke in Mongolia, China, traditional sour cream in Ukraine, Suero Costeño in Colombia, and smetana/śmietana/śmietanka mainly in Russia and Poland [16–19].

Nowadays, amplicon sequencing and shotgun metagenomics are the main tools used to study the microbiome of artisanal dairy products in depth [20,21]. Amplicon sequencing refers to the sequencing of targeted amplified loci like the 16S rDNA of bacteria or the internal transcribed spacer (ITS) of fungi found in the metagenomic DNA of food samples and it usually resolves the microbiome composition at the genus level. In contrast, shotgun metagenomics relies on the sequencing of the entire metagenomic DNA fraction and can provide species- or even strain-level information of the microbiome under investigation along with in silico predictions of its functional properties. These approaches allow the rapid and comprehensive analysis of the microbial communities in dairy products and can reveal important aspects of their technology, quality, and safety. The microbiology of sour cream products has been reviewed in the recent past to an extent [22]. So far, very few such studies exist, involving only amplicon sequencing approaches. Two studies concern the characterization of the microbiome of traditional sour cream/butter products produced in Russia [23,24], and one concerns the analysis of the microbiome of the traditional "Suero Costeño" sour cream produced in Colombia [19].

Staka, also called "Anthogalo" (milk flower), is a traditional Greek sour cream, which has risen at the surface of the spontaneously fermented sheep or mixtures of sheep and goat milk [25]. The cream is naturally fermented in traditional clay pots called "kounenoi" and/or exposed to air and sun depending on the region and season of the year (personal communication). It can also be produced using separators with the addition of starter cultures. Sometimes, Staka is cooked with flour, and in this case, most of the butter is separated and called "Stakovoutyro" (Staka butter), while the cooked product is also called Staka. Because of its rich taste and creamy mouthfeel, it has several culinary uses, and the "stakopilafo" or "gamopilafo", the special rice dish prepared in weddings, is the most known and appreciated one.

Recently, the microbiota of one Cretan Staka sample was investigated using a combination of culture-dependent methods and MALDI-TOF for the identification of the microbial isolates [25]. Following up these preliminary results, the aim of the current study was to analyze the microbial ecosystem of Staka in depth using a combination of culture-based microbiological analysis, amplicon sequencing, and shotgun metagenomics.

2. Materials and Methods

2.1. Samples

Five Staka samples, deriving from two geographical regions of Crete, namely Chania (Western Crete; Staka 1, 2, 3) and Sitia (Eastern Crete; Staka 4, 5), were included in the present study (Table 1). Four samples (Staka 1, 2, 3, 5) were commercial and purchased from mini markets in Athens or Chania; all were produced from pasteurized milk, while Staka 5 was cooked with wholemeal flour. Staka 4 was homemade, produced from raw milk, and cooked with wheat flour. One sample was analyzed for each Staka. For each Staka sample, three technical replicates were performed in the physicochemical and microbiological analyses.

Table 1. Description and physicochemical characteristics of Staka samples. Values presented are the means ± SD (n = 3).

Staka Sample	Production	Geographical Origin [c]	Milk Type	pH	TA [f]	Concentration (% w/w)				
						Moisture	Dry Matter	Ash	Fat	Protein
1	C [a]	Sfakia, Chania	S [d]	5.2 ± 0.2	0.3 ± 0.1	34.52 ± 1.5	65.48 ± 1.5	0.27 ± 0.07	41.87 ± 0.90	2.31 ± 0.40
2	C	Keramia, Chania	S	4.6 ± 0.3	0.4 ± 0.1	29.54 ± 1.9	70.46 ± 0.9	0.21 ± 0.04	46.47 ± 0.74	2.62 ± 0.15
3	C	Varypetro, Chania	S & G [e]	4.7 ± 0.4	0.5 ± 0.1	36.43 ± 0.8	63.57 ± 2.0	0.22 ± 0.03	40.57 ± 0.88	3.00 ± 0.25
4	H-M [b]	Palaiokastro, Sitia	S & G	5.2 ± 0.2	1.2 ± 0.3	27.05 ± 1.1	72.95 ± 1.7	3.42 ± 0.49	32.68 ± 0.66	16.00 ± 0.78
5	C	Chamezi, Sitia	S & G	5.3 ± 0.1	0.2 ± 0.1	48.25 ± 0.7	51.75 ± 1.2	2.15 ± 0.31	30.31 ± 1.70	6.40 ± 0.11

[a] Commercial; [b] homemade; [c] Crete perfecture; [d] sheep; [e] goat; [f] titratable acidity expressed in lactic acid concentration.

2.2. Physicochemical Analyses

The pH of the Staka samples was measured using a pH meter (827 pH lab, Metrohm Herisau, Switzerland). Titratable acidity was measured by titrating 10 g sample with a standard solution of 0.1 N sodium hydroxide and phenolphthalein (Merck Darmstadt, Germany) as pH indicator and expressed as % lactic acid (w/w). The moisture content was measured according to ISO method [26]; ash content was determined according to [27]; and fat content was determined using the Gerber-van Gulik method [28]. The Kjeldahl method was used for the determination of total nitrogen (TN), which was expressed as protein content (% w/w) [28]. Finally, lactose, organic acids, and ethanol were determined by HPLC analysis [29].

2.3. Microbiological Analysis

The following groups of microorganisms were enumerated: (1) total mesophilic bacteria on plate count agar (PCA, Biokar Diagnostics, Beauvais, France) at 30 °C for 3 days; (2) psychrotrophic bacteria on PCA at 7 °C for 10 days; (3) lipolytic bacteria on PCA containing 1% v/v tributyrin (Sigma Aldrich Co., St. Louis, MO, USA) at 30 °C for 3 days; (4) proteolytic bacteria on PCA containing 10% w/v reconstituted skim milk (Oxoid, Hampshire, UK) at 30 °C for 48 h; (5) thermophilic LAB on MRS agar adjusted at pH 5.4 (presumably lactobacilli) (Biokar Diagnostics) at 42 °C for 48 h anaerobically (double agar layer); (6) mesophilic LAB in MRS agar adjusted at pH 5.4 (presumably lactobacilli) at 22 °C for 3 days anaerobically (double agar layer); (7) thermophilic LAB on M17 agar (presumably cocci) (Biokar Diagnostics) at 42 °C for 48 h; (8) mesophilic LAB on M17 agar (presumably cocci) at 22 °C for 48 h; (9) non-starter lactic acid bacteria (NSLAB) on Rogosa agar (Biokar Diagnostics) at 30 °C for 5 days, anaerobically (double agar layer); (10) enterococci on KAA agar (Merck) at 37 °C for 24 h; (11) coliforms on VRBL agar (Biokar Diagnostics) at 37 °C for 24 h; (12) *Pseudomonas* spp. on Pseudomonas agar base (Biokar Diagnostics), at 30 °C for 48 h; (13) yeasts and molds on YGC agar (Merck) at 25 °C for 3–4 days. Based on morphology (shape, color, and size), colonies were collected from MRS (42 and 22 °C), M17 (42 and 22 °C), Rogosa, KAA, PCA, PCA-tributyrin, and PCA-milk purified by repetitive streaking and stored at −80 °C in nutrient broth containing 20% v/v glycerol for further study.

2.4. Amplicon and Shotgun Metagenomics Sequencing and Analysis

Metagenomic DNA was isolated by the protocol described previously [30]. All the procedures for the DNA amplification, the construction of the different libraries, the Illumina sequencing, and preliminary quality control have been described before [30,31]. Sequencing was performed at Molecular Research DNA (MR DNA, Shallowater, TX, USA). Amplicon sequences and/or shotgun metagenomics were analyzed as described before with the CLC genomics workbench 11.0.1 (Qiagen, Hilden, Germany), the BusyBee web tool [32], and the metagenomics rapid annotation was analyzed using subsystems technology (MG-RAST) server version 4.0.3 [33]. During analysis, reads deriving from *Cellulosimicrobium* sp. Were removed from all sequence datasets due to the contamination by the lyticase used for the lysis of yeasts [34,35]. Finally, shotgun sequencing reads were aligned against different reference genomes in the CLC genomics workbench 11.0.1 and were processed into recruitment plots with the Recplot_4 R package (https://github.com/KGerhardt/Recplot_4, accessed on 1 February 2024) using default parameters. Reference genomes were selected after manually determining the best BLASTn hits of at least four random contigs from each of the top four species in abundance according to results of the BusyBee web tool (i.e., *Lc. lactis*, *Lactococcus raffinolactis*, *Streptococcus thermophilus*, and *Streptococcus gallolyticus*).

2.5. Typing and Identification of Isolates with Rep-PCR Fingerprinting and 16S rDNA Sequencing

A previously reported protocol was employed for bacterial DNA extraction using 2 mL of fresh overnight cultures in the exponential phase and fingerprinting of the new isolates was performed by repetitive element palindromic PCR (rep-PCR) [36]. A SimpliAmp™ Thermal Cycler (ThermoFisher Scientific, Sunnyvale, CA, USA) was used for the PCR. Rep-PCR fingerprint clustering was performed by BioNumerics v. 6.0 (Applied Maths, Ghent, Belgium).

Identification at the species level of bacterial isolates was performed by 16S rDNA sequencing [37]. The NucleoSpin® Gel and PCR Clean-up (Macherey-Nagel, Duren, Germany) were used for DNA purification after electrophoresis. BLASTn was used for the identification at the species level.

2.6. Species Discrimination by Biochemical Tests

To discriminate species with ambiguous 16S rDNA sequencing results, several isolates were subjected to biochemical tests. Isolates were first grown as follows: *Enterococcus* spp. in M17 broth at 37 °C, *Lacticaseibacillus* spp. and *Leuconostoc* spp. in MRS broth at 30 °C, and *Hafnia* spp. in nutrient broth at 30 °C. Discrimination of enterococcal species was performed using mannitol salt agar (Lab M, Heywood, Lancashire, UK), taking into consideration the ability of *E. faecalis* to ferment mannitol [38]. Additionally, to verify the *E. faecium*/*E. faecalis* discrimination, all enterococcal isolates were grown in M17 broth containing ampicillin (final concentration 2 μg mL^{-1} from Sigma Aldrich), as *E. faecalis* is susceptible to this antibiotic [38]. Arabinose fermentation (1% w/v from Sigma Aldrich) was used for a preliminary discrimination of *E. faecium*, as *Enterococcus durans* and *Enterococcus hirae* are not able to ferment this monosaccharide [39]. Similarly, *Lacticaseibacillus casei* and *paracasei* were discriminated using mannitol (1% w/v), which is only fermented by *L. paracasei* [40], while *Leuconostoc mesenteroides*/*paramesenteroides* isolates were differentiated taking into consideration their ability to hydrolyze salicin (1% w/v from Sigma Aldrich) and aesculin (1% w/v from Sigma Aldrich), as *L. paramesenteroides* rather fails to hydrolyze both glucosides [41]. Finally, malonic acid (1% w/v from Sigma Aldrich) was used to discriminate *Hafnia alvei* vs. *paralvei* as it is utilized only by *H. alvei* strains [42].

2.7. Antimicrobial Activity of the Isolated Strains

All isolates were tested for their antimicrobial activity against 20 indicator strains, including four LAB strains, 10 pathogenic streptococci, three *Listeria* spp., two *Bacillus* spp. And one *Pseudomonas* sp. The well diffusion assay (WDA) was performed to assess the antimicrobial activity of the cell-free culture supernatants with pH adjusted to 6.5 [43]. Treatment with ammonium sulfate (60% saturation from AppliChem GmbH, Darmstadt, Germany), was applied in selected active supernatants for 10-fold protein concentration and therefore increase of the antimicrobial activity.

3. Results and Discussion

3.1. Physicochemical Analyses

Physicochemical characteristics of Staka samples, namely pH and titratable acidity, as well as moisture, ash, and fat and protein content, are summarized in Table 1.

The pH values, ranging from 4.6 to 5.3, were similar or higher than those reported for other sour cream samples studied, where the pH ranged between 3.8 and 4.8 [44]. This is an important finding as it has been reported that when the final product pH is very low (e.g., around pH 4.0), the cream has an unpleasant sour flavor [10]. However, the high pH when coupled with high moisture content, as in our study for Staka 1 and 5, may pose spoilage and safety issues and thus shorten the product's shelf-life. Titratable acidity was low in four samples (Staka 1, 2, 3, and 5), ranging from 0.2 to 0.5% w/w with the exception of Staka 4. The higher TA of Sample 4 (1.2 ± 0.3% w/w) is in accordance with the high lactic acid concentration determined (16.3 ± 0.5 g/kg); at the same time, the rather high pH of 5.2 ± 0.2 can be attributed to the buffering capacity of the proteins (16.00 ± 0.78% w/w), which was the highest observed among samples, most probably due to the addition of wheat flour during processing (cooking). These results are in agreement with reported values for sour cream samples ranging from 0.5 to 1.7% [44].

Moisture and ash values ranged from 27.1 to 48.3% w/w and from 0.2 to 3.4% w/w, respectively. Ash values varied from 1.9 to 2.7% w/w for samples of Ispir Kaymak, although this product is not fermented [16].

Fat content values, ranging between 30.3 and 46.5% w/w, were consistent with values reported for fat content of cultured creams, such as full-fat commercial sour creams collected from across the US which ranged between 16.8–33.1% w/w [44], meeting the composition requirements of full fat sour creams, which must have at least either 10% w/v [45] or 18% w/v [8,46] milk fat. Concerning the fat content, Staka samples resembled to the Lithuanian créme fraîche and Ispir Kaymak, which had a fat content around 30–45% [10] and 43–63% [16], respectively. Therefore, Staka is a dairy product with a high fat content that may constitute a rich source of beneficial lipids [47]. Additionally, the fat content expressed as fat in dry matter (FDM) was 64.1 and 56.0%, respectively, for Staka 1 and 3, and these values were consistent to those reported on the labels of the products (min 40.0 and 55.0% w/w, respectively). Fat content was not reported for the rest of the samples.

Protein content of Staka 1, 2, 3 ranged from 2.3 to 3.0% w/w, while values for Staka 4 and 5 were significantly higher (16.0%, and 6.4% w/w, respectively); this can be attributed to the processing (cooking) of these samples with flour, since wheat, no matter wholemeal or not, contains ca. 12% w/w proteins. However, the difference between Samples 4 and 5 cannot be explained since we did not get any further information from the producers about the ratio of milk cream and flour. High protein values ranging from 14.3 to 20.3% w/w have been recorded for samples of Ispir Kaymak [16]. Furthermore, Staka 4 had also the lowest fat content. This was due to the removal of the fat content to produce Stakovoutiro (butter) in parallel to Staka. Such a procedure may alter the overall composition of the particular sample and justify its deviation from the rest of the samples in terms of composition.

HPLC results (Table 2) showed that lactic acid was the prominent organic acid in all samples, as also reported for commercial US sour creams [44], followed by acetic acid (Staka 1, 2, 3, 5). Varying concentrations of lactic acid as well as acetic acid depend on the lactose catabolism pathway that the members of the samples' microbiota follow, i.e., homo- or hetero-fermentative or both.

Table 2. Concentration of lactose, organic acids and ethanol detected in Staka samples as determined by HPLC analysis. Values presented are the means ± SD (n = 3).

Staka Sample	Concentration (g/kg)						
	Lactose	Lactic Acid	Acetic Acid	Succinic Acid	Butyric Acid	Propionic Acid	Ethanol
1	24.7 ± 0.4	8.1 ± 0.5	1.4 ± 0.4	0.3 ± 0.1	nd	nd	0.4 ± 0.2
2	17.2 ± 0.3	6.7 ± 0.5	1.0 ± 0.2	0.3 ± 0.2	0.1 ± 0	nd	nd
3	18.5 ± 0.3	10.0 ± 0.6	0.8 ± 0.2	nd	nd	nd	0.3 ± 0.2
4	40.0 ± 1.1	16.3 ± 0.5	nd	nd	1.3 ± 0.2	1.7 ± 0.3	nd
5	13.6 ± 0.6	2.0 ± 0.3	1.6 ± 0.3	0.1 ± 0.1	nd	nd	0.2 ± 0.1

nd: not detected.

3.2. Culture-Based Microbiological Analysis

Table 3 summarizes the results obtained through culture-based microbiological analysis. Interestingly, Staka 4 and 5 were obviously affected by the heating step in the presence of flour that took place during their production. More specifically, Staka 4 presented low populations of lipolytic bacteria, thermophilic LAB and NSLAB, while none of the microbial groups tested could be detected in Staka 5.

Table 3. Microbial counts (log cfu g^{-1}) of the Staka samples examined. Values presented are the means ± SD (n = 3).

Presumptive Microbial Group	Staka Sample				
	1	2	3	4	5
Total mesophilic bacteria (PCA agar, 30 °C)	8.09 ± 0.09	7.10 ± 0.18	7.88 ± 0.26	0.00	0.00
Thermophilic lactobacilli (MRS agar pH 5.4, 42 °C)	4.51 ± 0.35	6.13 ± 0.17	7.45 ± 0.21	0.00	0.00
Mesophilic lactobacilli (MRS agar pH 5.4, 22 °C)	7.99 ± 0.11	6.88 ± 0.39	7.33 ± 0.22	0.00	0.00
Thermophilic cocci (M17 agar, 42 °C)	4.86 ± 0.19	6.18 ± 0.28	7.67 ± 0.42	3.80 ± 0.22	0.00
Mesophilic cocci (M17 agar, 22 °C)	8.14 ± 0.17	6.44 ± 0.81	7.60 ± 0.19	0.00	0.00
NSLAB (Rogosa agar, 30 °C)	7.84 ± 0.41	6.92 ± 0.16	7.52 ± 0.31	2.80 ± 0.23	0.00
Enterococci (KAA agar, 37 °C)	4.70 ± 0.19	6.03 ± 0.22	7.29 ± 0.26	0.00	0.00
Lipolytic bacteria (PCA-tributyrin, 30 °C)	8.05 ± 0.08	6.97 ± 0.10	7.91 ± 0.04	2.60 ± 0.08	0.00
Proteolytic bacteria (PCA-milk, 30 °C)	7.78 ± 0.35	6.70 ± 0.17	7.95 ± 0.59	0.00	0.00
Psychrotrophic bacteria (PCA, 7 °C)	7.35 ± 0.46	4.13 ± 0.09	5.92 ± 0.16	0.00	0.00
Pseudomonas spp. (Pseudomonas agar base, 30 °C)	7.72 ± 0.25	0.00	5.85 ± 0.30	0.00	0.00
Coliforms (VRBL agar, 37 °C)	7.54 ± 0.34	0.00	5.60 ± 0.26	0.00	0.00
Yeasts and molds (YGC agar, 25 °C)	4.11 ± 0.26	3.41 ± 0.40	5.50 ± 0.35	0.00	0.00

Regarding the rest samples (Staka 1, 2 and 3), which did not receive such treatment, in the case of total mesophilic bacteria as well as the six major groups of presumptive LAB, counts ranging from 6.03 to 8.09 log cfu g^{-1} were observed, and only in Staka 1 lower counts were determined for thermophilic lactobacilli, thermophilic cocci and enterococci (4.51–4.86 log cfu g^{-1}). These differences could be attributed to the microbiota of the milk used, the production environment and utensils employed as well as the technology applied. LAB are known as the main actors in dairy fermented foods; comparable LAB counts have been reported for traditional sour cream from Russia [48], Ukraine [18] and Colombia [19].

Analogous counts (6.70–8.05 log cfu g^{-1}) were enumerated for lipolytic and proteolytic bacteria. Such bacteria can have either beneficial or detrimental impact on fermented foods, depending on the species of microorganisms and the flavor compounds they produce [49]. For instance, LAB, despite being only weakly lipolytic and proteolytic, play an important role in dairy fermented foods as they shape not only their sensory traits but also their nutritional attributes [50]. Regarding psychrotrophic bacteria, *Pseudomonas* spp. and coliforms, higher counts were detected in Staka 1, followed by Staka 3, while in Staka 2, these populations were absent. *Pseudomonas* spp. have been associated with very high levels of lipase and protease activities that can cause off-flavors [51]. Furthermore, the presence of coliforms, which are indicators of poor hygiene conditions during food processing, has also been reported in artisanal Colombian Suero Costeño sour cream [19].

Yeast populations varied, with the highest value corresponding to Staka 3 (5.50 log cfu g^{-1}), followed by Staka 1 and 2 (4.11 and 3.41 log cfu g^{-1}, respectively). The presence of yeasts has also been reported in Suero Costeño sour cream with values even higher than 6.0 log cfu mL^{-1} [19]. Sour cream is one of the cultured milk products which is a favorable medium for the propagation of yeasts, since it exhibits relatively low pH values ranging between 4.0 and 6.0 [52].

3.3. Amplicon Sequencing and Shotgun Metagenomics

Figure 1 presents the findings of the 16S rDNA amplicon sequencing analysis of the Staka samples. In all samples, Firmicutes was the major phylum (>88% abundance), followed by Proteobacteria, except for Staka 1, in which the abundance of the two phyla was reversed with Proteobacteria reaching 72%. At the family level, significant differences in bacterial composition across the samples could be observed. *Streptococcaceae* was the dominant family in Staka 2, 3, and 4 with ca. 85% abundances in the first two and a 52% abundance in the third. *Enterobacteriaceae* exhibited the highest abundance of 50% in Staka 1, while *Leuconostocaceae* dominated Staka 5 with a 94% abundance. *Leuconostocaceae* was the second-highest population in Staka 1 (28% abundance) followed by *Pseudomonadaceae* (22% abundance). *Leuconostocaceae* was also present in Staka 2, 3, and 4 but in much lower abundances (<9%). Low levels of *Pseudomonadaceae* also appeared in the rest of the samples with <5% abundance. Furthermore, *Bacillaceae* was present with 35% abundance as the second-largest population in Staka 4, and *Enterococcaceae* appeared at low abundances (<6% abundance) in Staka 2, 3, and 4. The analysis at the genus level demonstrated that Staka 1 was composed of *Leuconostoc* (28% abundance), *Buttiauxella* (25% abundance), *Pseudomonas* (22% abundance), and *Enterobacter* (21% abundance). Staka 2, 3, and 4 were characterized by the high abundance of *Lactococcus* (up to 84%). The second-largest population in Staka 2 was *Streptococcus* (24% abundance), while in Staka 4, it was *Anoxybacillus* (35% abundance). Staka 5 was practically composed of *Leuconostoc* (94% abundance) and *Pseudomonas* (5% abundance). In different samples, several genera were present in variable and relatively low abundances, including *Leuconostoc* (Staka 2, 3, and 4, <9% abundance), *Pseudomonas* (Staka 2, 3, and 4, <2% abundance), *Enterococcus* (Staka 2, 3, and 4, <6% abundance), *Enterobacteriaceae* ambiguous taxa (Staka 1 and 2, <4% abundance), *Escherichia-Shigella* (Staka 1, 2, and 3, <4% abundance), and *Hafnia* (Staka 1 and 2, <3% abundance).

Alpha-diversity of total number of genus level operational taxonomic units (OTUs) indicated that all samples were sequenced to a sufficient depth (Figure 1D). Staka 2 had the most complex microbiome composition, while Staka 5 had the simplest one. The complexity of the microbiomes of Staka 1, 3, and 4 seemed to be relatively comparable. In addition, beta-diversity could segregate Staka 2, 3, and 4 from Staka 1 and 5 in a principal coordinate analysis (PCoA) using Bray–Curtis distances (Figure 1E). This could be attributed to the dominance of the genus *Lactococcus* in Staka 2, 3, and 4 versus the dominance of *Leuconostoc* accompanied by the absence of *Lactococcus* in Staka 1 and 5.

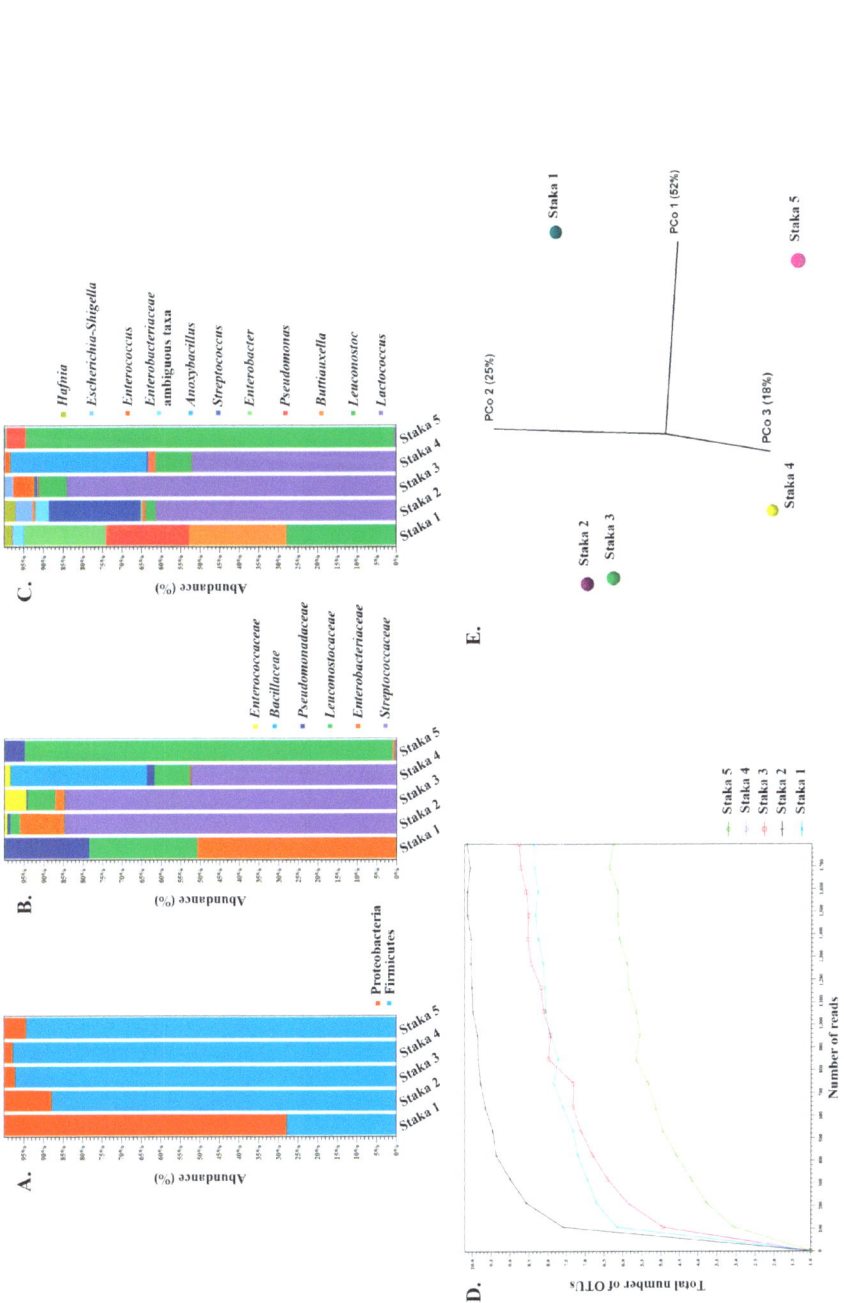

Figure 1. Taxonomic profile of Staka sour cream samples based on 16S rDNA amplicon data at the phylum (**A**), family (**B**), and genus (**C**) levels. Panel (**C**) presents genera with abundance ≥ 1%. Alpha-diversity analysis of 16S rDNA measured using the total number of OTUs at genus level of Staka samples (**D**). Beta-diversity shown through a principal coordinate analysis (PCoA) employing the Bray–Curtis distances for the same samples (**E**).

The findings of the analysis of the 16S rDNA amplicons support the presence of LAB, *Enterococcus* sp., coliforms (e.g., *Buttiauxella* sp., *Enterobacter* sp., *Escherichia* or *Shigella* sp. and *Hafnia* sp.), and psychrotrophs (e.g., *Pseudomonas* sp.) as also determined by the initial culture-based microbiological analysis presented above. Several of these groups/genera may contain strains that are lipolytic and/or proteolytic. It should be highlighted that both Staka 4 and 5 provided 16S rDNA fingerprints, even though the ecosystem of both samples was seriously affected by the heat treatment step employed during their productions. Several bacterial OTUs could be assigned to Staka 4 and 5, even though the first presented very low microbial populations and the second seemed to be sterile. Seemingly, DNA of dead cells or unculturable cells was retained, and thus, 16S rDNA amplicon sequencing could provide retrospective information of the original microbiome composition. In addition, the distribution of genus level OTUs with >1% abundance did not always correlate with that of the culture-based microbiological analysis. For example, enterococci were present in the first three samples (Table 3), but they seemed to be practically absent from the OTUs in Staka 1. This may be an effect of the different populations of enterococci among the samples. As already mentioned, in Staka 1 enterococci had their lowest population ca. 2–3 log lower than their population in Staka 2 and 3 (Table 3), and thus, they may not be presented in the genus level OTUs with >1% abundance. Our findings are partially in agreement with those reported for Suero Costeño [19]. While several bacterial genera are in common between Staka and the Colombian sour cream (e.g., *Lactococcus*, *Leuconostoc*, *Enterobacter*, *Escherichia-Shigella*, etc.), the most abundant OTUs reported for samples of the latter were *Lactobacillus* or *Streptococcus*. This is a major difference of our samples, which were dominated by *Lactococcus* or *Leuconostoc*. Interestingly, in the Staka samples analyzed, lactobacilli were only present with abundances <1%. A prevalence of *Lactococcus* or *Streptococcus* accompanied with a low abundance of lactobacilli has also been reported for Russian sour cream products [23,53]. Of note, *Acetobacter* sp. were reported for both Suero Costeño and some of the Russian sour cream, but they were absent from the Staka samples.

All Staka samples were also analyzed using ITS amplicon sequencing (Figure 2). Ascomycota was the prevailing phylum in all samples with an abundance >81%, reaching up to >99% in Staka 3, 4, and 5 (Figure 2A). With the exception of Staka 4, an additional population unidentified at kingdom level (k_unidentified) was also present in all samples, with abundances reaching 19% in Staka 1 and 4% in Staka 2. Manual analysis with BLASTn indicated that the relevant ITS sequence belongs to the ascomycete fungi *Iodophanus* sp., which could not be identified as such by the Unite database used in this study. At the family level, *Saccharomycetaceae* was the predominant family in Staka 1, 3, and 4 with abundances of 29%, 98%, and 83%, respectively. The *Saccharomycetaceae* family was the second largest in Staka 2 with an abundance of 8% and was also present in Staka 5 with a 3% abundance. In contrast, the most abundant family for Staka 2 was *Debaryomycetaceae* (81% abundance) and *Pleosporaceae* (86% abundance) for Staka 5. The *Debaryomycetaceae* family was the second largest in Staka 1 (26% abundance) and was also found in low abundances in Staka 3, 4, and 5, reaching up to 1% in Staka 3. The *Pleosporaceae* family formed the second largest population in Staka 4 (13% abundance) and was also found in the rest of the samples, reaching an abundance of 9% in Staka 1. Other fungi present in the Staka samples were members of the *Aspergillaceae* family found in Staka 1, 2, and 3 with abundances equal to 14%, 7%, and 1%, respectively. Additionally, the *Cladosporiaceae* family formed the second-largest population in Staka 5 with an 11% abundance, and it was also identified in Staka 1, 2, and 4, reaching up to 3% abundance. The *Herpotrichiellaceae* family was present only in Staka 1 with an abundance of 3%. Except for Staka 4, in all samples the unidentified kingdom mentioned above corresponded to the *Pezizaceae* family (determined manually), reaching 19% abundance in Staka 1. We then performed analysis to establish the prevailing genera with >1% abundance in all samples (Figure 2C). Some of the families described above had only one representative genus, i.e., the *Debaryomyces* for the *Debaryomycetaceae* family, *Coniosporium* for the *Herpotrichiellaceae*, and *Cladosporium* for the *Cladosporiaceae* family. The *Saccharomycetaceae* family was present with genera *Kluyveromyces* and *Saccharomyces*, the *Pleosporaceae* family consisted of *Alternaria* and *Stemphylium* genera, and the *Aspergillaceae* family included the *Penicillium and Aspergillus* genera. As mentioned above, the unidentified genus could be matched to *Iodophanus* sp. of the *Pezizaceae* family.

Alpha-diversity of the total number of genus level OTUs showed that a sufficient sequencing depth of all samples was reached. Beta-diversity based on the genus level using Bray–Curtis distances showed some similarity between pairs of Staka 1 and 2 and Staka 3 and 4, while Staka 5 was found separate from the rest. This could be attributed to the presence of *Debaryomyces* in Staka 1 and 2, *Kluyveromyces* in Staka 3 and 4, and *Alternaria* in Staka 5.

Unfortunately, information about the presence of yeasts or molds in sour cream seems to be rare. *Candida* sp., *Rhodotorula* sp., and *Cryptococcus* sp. have been detected in pasteurized cream (non-sour) [54]. Some studies have clearly indicated that yeast and molds are related to the spoilage of sour cream [55,56]. These microorganisms may originate from raw milk, but they may be also characteristic of the dairy facility. Almost all of the genera identified in our samples are more or less common in fermented dairy products, e.g., *Debaryomyces, Kluyveromyces, Saccharomyces, Alternaria, Cladosporium, Penicillium,* and *Aspergillus* [57–60]. Three genera, i.e., *Iodophanus, Cladosporium,* and *Stemphylium,* seem to be rather rarely related to the dairy environment [61–63], and thus their presence needs further investigation. It should be noted that the uncontrolled presence of genera like *Penicillium* and *Aspergillus* may be problematic, given their ability to produce mycotoxins that can remain unaffected during normal production and storage conditions [59].

Staka 2 was chosen for further investigation of the Staka microbiome with shotgun metagenomics, given the high complexity it showed in the alpha-diversity of both 16S rDNA and ITS OTUs. The analysis revealed the dominance of members of the *Streptococcaceae* family, i.e., *Lc. lactis* with an abundance of 41%, *Lc. raffinolactis* with an abundance of 11.7%, and each of *S. thermophilus* and *S. gallolyticus* with abundances of ca. 6% (Figure 3A). The unidentified species of the *Enterobacteriaceae* family and *Escherichia coli* were present with abundances of 4.4% and 4.2%, respectively. *H. alvei* showed an abundance of 3.2%, followed by *Streptococcus parauberis* with an abundance equal to 2.8%. Unidentified *Streptococcus* spp. and *E. durans* were present with populations of 2.2% and 1.5%, respectively. Other species present in Staka 2 occupied 16.4% abundance in total. Alpha-diversity suggested a sufficient sequencing depth, while the taxonomic findings of the shotgun analysis of Staka 2 are in overall agreement with the 16S rDNA amplicon sequencing results (Figure 3B). The absence of fungi from the shotgun analysis may be due to their low population in the sample as mentioned above.

To identify putative metagenome-assembled genomes (MAGs), contigs from the assembled reads of Staka 2 were further analyzed with the BusyBee server, and three bins were determined (Figure 3C). The quality of the bins was not sufficient, but the taxonomic analysis could assign the majority of contigs in Bin 1 and Bin 3 to *Lc. lactis* and *Lc. raffinolactis*, respectively. Several contigs in Bin 2 could be assigned to *S. gallolyticus*, while contigs assigned to *S. thermophilus* were relatively grouped together but did not form a separate bin. To further aid the identification of chromosomal sequences in the dataset, at least for the four most abundant species in the sample analyzed, and we employed recruitment plots of the sequencing reads against reference genomes. As can be seen in Figure 4, almost full draft genomes could be formed during the recruitment of the reads against each of the reference genomes with relatively short gaps. In fact, an important number of reads could be aligned with an identity > 95% despite the fact that the overall coverage was low, ranging from 8.2× for *Lc. lactis* (the most abundant species) to 2× for *S. thermophilus* (the least abundant species). Thus, recruitment plot analysis provided more comprehensive results than the binning of the contigs.

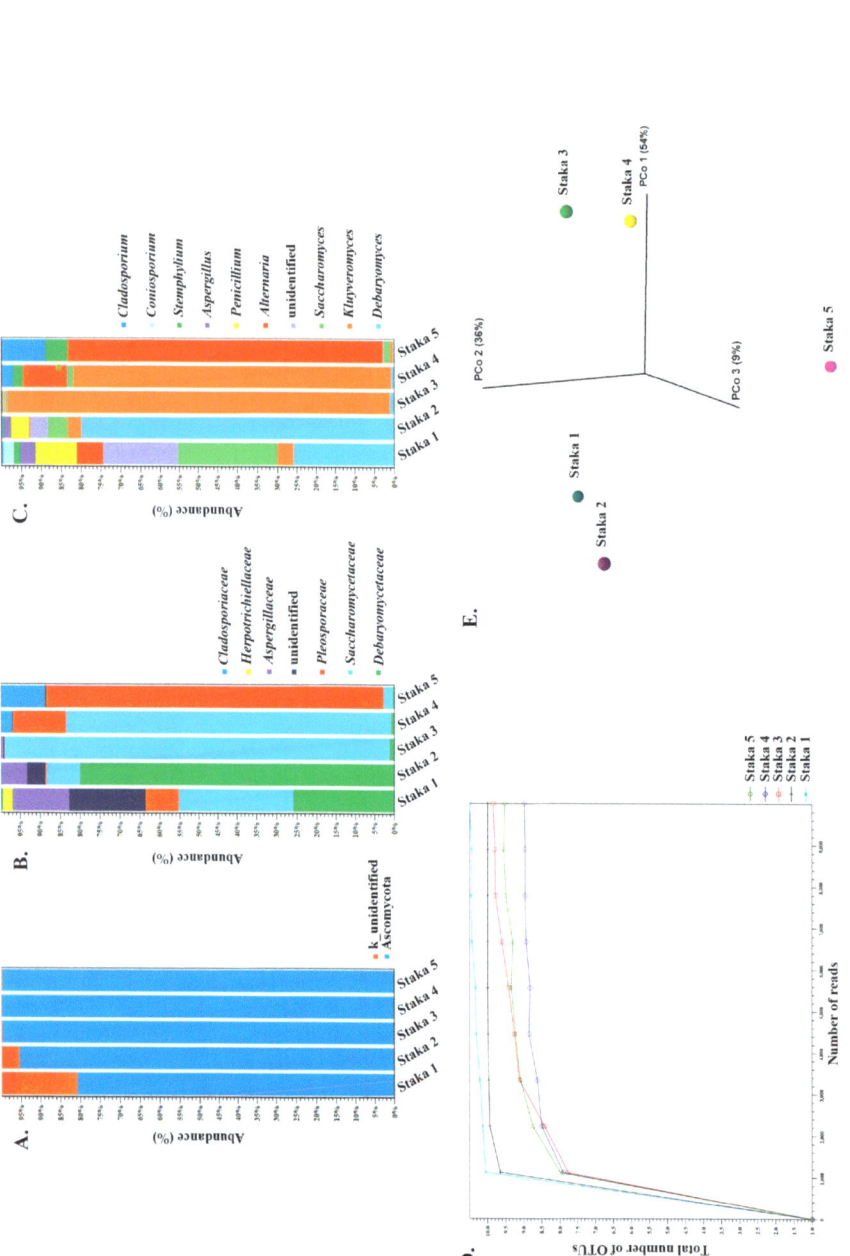

Figure 2. Taxonomic profile of Staka sour cream samples based on ITS amplicon data at the phylum (**A**), family (**B**), and genus (**C**) levels. Panel (**C**) presents genera with abundance ≥ 1%. Alpha-diversity analysis of ITS reads measured using the total number of OTUs at genus level of Staka sour cream samples (**D**). Beta-diversity shown through a principal coordinate analysis (PCoA) employing the Bray–Curtis distances for the same samples (**E**).

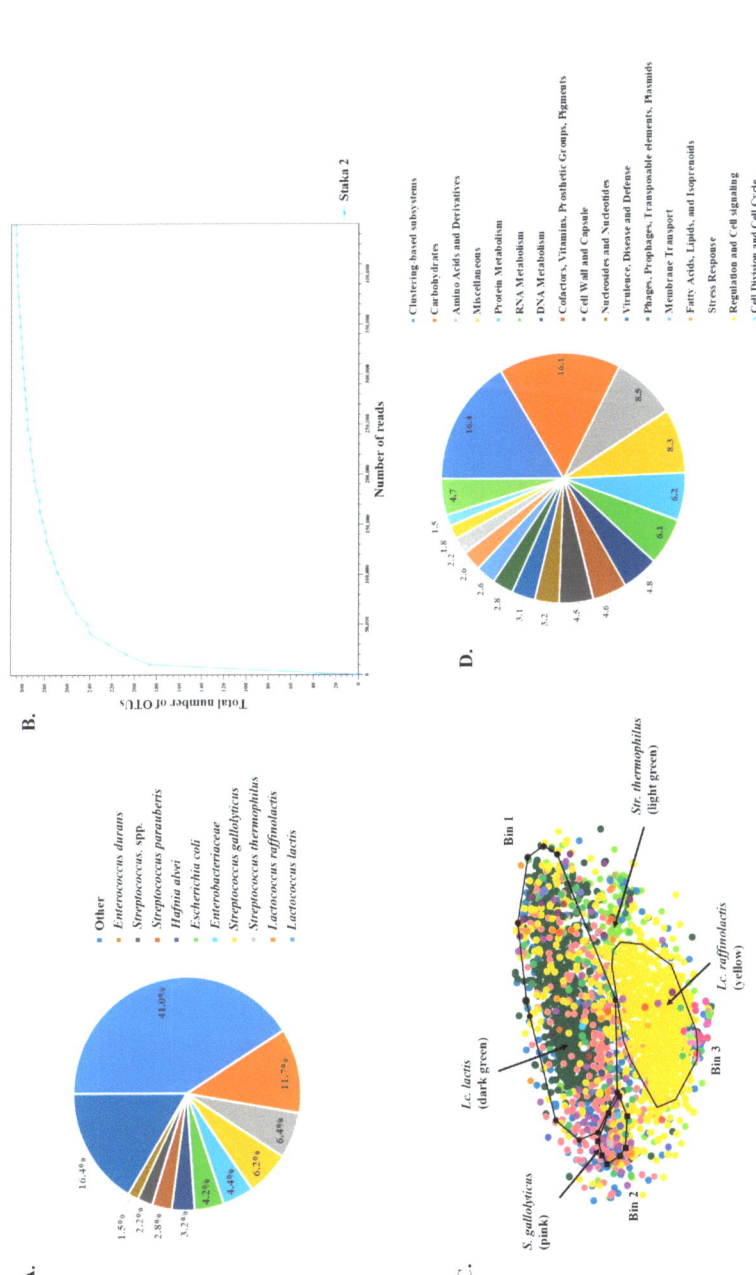

Figure 3. Taxonomic profile of Staka 2 sour cream sample, based on the mapping of shotgun metagenomics reads at the species level (**A**). Alpha-diversity of shotgun reads measured using the total number of OTUs of Staka 2 (**B**). Bins of metagenomics scaffolds of Staka 2 (**C**). Dots of same colors represent scaffolds that originate from the same species, as indicated in the figure. Functional analysis of annotated scaffolds of Staka 2 using the MG-RAST server (**D**).

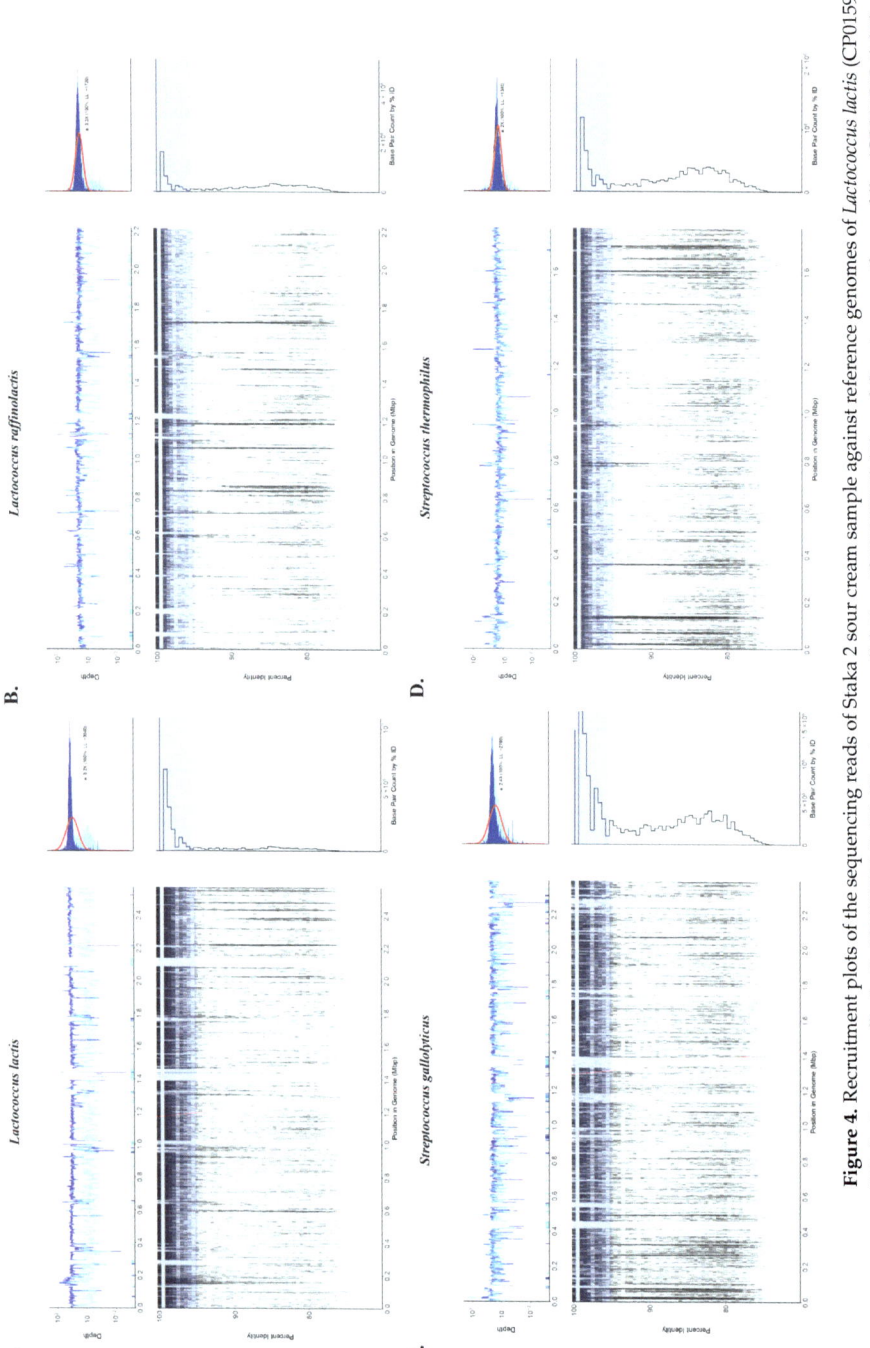

Figure 4. Recruitment plots of the sequencing reads of Staka 2 sour cream sample against reference genomes of *Lactococcus lactis* (CP015902.1) (**A**), *Lactococcus raffinolactis* (CP023392.1) (**B**), *Streptococcus gallolyticus* (CP113954.2) (**C**), and *Streptococcus thermophilus* (CP031545.1) (**D**).

As mentioned above, *Lc. lactis* can be used as a starter for the acidification of the cream [10]. *Lc. lactis* was also present in the full-length 16S rRNA amplicon sequencing of sour cream collected in northeast Asia [53]. *Lc. raffinolactis* has also been reported as a member of the microbiome of sour cream [53]. *S. thermophilus* has been detected in the previous study of Staka [25], but its presence has been verified in other sour creams as well [48]. *Enterobacteriaceae* are frequent member of the microbiome of artisanal dairy products, and some genera of the family have been identified in sour cream [19,53]. To the best of our knowledge, *S. gallolyticus*, *H. alvei*, and *S. parauberis* are reported for the first time in sour cream. Even though the presence of enterococci in sour cream has been reported, *E. durans* has been identified again only in Staka in the past [25].

In order to shed light on the functional potential of the Staka microbiome, functional analysis of the assembled shotgun metagenomes of Staka 2 was performed using the MG-RAST server. Scaffolds were annotated and annotations were assigned to functions. As seen from the Figure 3D, the main part (16.4%) of the Staka 2 functional subsystems belonged to clustering-based subsystems, but their role in the metabolic pathways is yet unknown. Among other subsystems assigned were carbohydrates which occupied the second place with 16.1%, followed by such subsystems as amino-acids and derivatives (8.5%), miscellaneous (8.3%), protein metabolism (6.2%), RNA metabolism (6.1%), and DNA metabolism (4.8%). Additional functional categories could also be identified with decreasing percentages.

3.4. Isolates Typing and Identification

A total of 141 bacterial isolates were collected from the five Staka samples and identified. Based on rep-PCR analysis of bacterial strains, they were clustered in 49 groups (Supplementary Table S1). Representative isolates of all groups were selected and subjected to 16S rDNA sequencing. According to the results obtained, 78 isolates were identified as *Enterococcus* spp., 21 as *Leuconostoc mesenteroides/pseudomesenteroides*, 13 as *Hafnia alvei/paralvei*, 5 as *Pseudomonas* spp., 6 as *Bacillus subtilis*, 6 as *Serratia liquefaciens*, 5 as *Lacticaseibacillus casei/paracasei* (basonym *Lactobacillus casei/paracasei*), 3 as *Companilactobacillus versmoldensis* (basonym *Lactobacillus versmoldensis*), 2 as *Latilactobacillus curvatus* (basonym *Lactobacillus curvatus*), 1 as *Loigolactobacillus coryniformis* (basonym *Lactobacillus coryniformis*), and 1 as *S. thermophilus*.

It should be stressed, however, that 16S rDNA gene sequencing cannot discriminate *L. casei* from *L. paracasei* [64]. The same is valid for closely related enterococcal species, i.e., *E. faecalis*, *E. faecium*, *E. durans* and *E. hirae* [65], *L. mesenteroides* and *L. pseudomesenteroides* [66], and *H. alvei* and *H. paralvei* [42], as well as *Pseudomonas* species [67]. To overcome this obstacle and identify the enterococcal isolates at the species level, mannitol and arabinose fermentation, as well as the ampicillin resistance of the isolates, were considered. According to the results obtained, 24 isolates were identified as *E. faecalis* as they fermented mannitol and were ampicillin susceptible, while 54 were identified as *E. faecium* as they did not ferment mannitol, while they fermented arabinose. Respectively, all *L. casei/paracasei* isolates were identified as *L. paracasei* as they fermented mannitol, all *L. mesenteroides/pseudomesenteroides* isolates were identified as *L. pseudomesenteroides* as they could not hydrolyze salicin and aesculin, and finally, all *Hafnia alvei/paralvei* isolates were identified as *Hafnia paralvei* as they did not use malonic acid. Finally, the presence of Gram-negative and Gram-positive spoilage or opportunistic pathogenic bacteria, i.e., *Pseudomonas* spp. (five isolates in Staka 1), *Serratia liquefaciens* (six isolates in Staka 1), and *B. subtilis* (six isolates in Staka 4) were detected.

Interestingly, no *Lactococcus* spp. could be isolated from the samples during culture-based analysis, in contrast to *Leuconostoc* spp. and *Enterococcus* spp., which were readily isolated. Several isolates, also belonging to the genus *Enterococcus*, were detected in the Staka sample examined by Lappa et al. (2021). In addition, Lappa et al. (2021) reported the isolation of *L. paracasei*, *Lactiplantibacillus plantarum* (basonym *Lactobacillus plantarum*), and *Lactiplantibacillus paraplantarum* (basonym *Lactobacillus paraplantarum*). Clearly, findings about the culturable fraction of the microbial ecosystem of the aforementioned study and the present study cannot be directly correlated with those of the amplicon sequencing and shotgun metagenomics analysis. There are several different putative reasons which may underpin such differences. First of all, colonies in both studies were not picked randomly but on the basis of differences in morphology, and thus, the frequency of taxa defined after cultured-based analysis may not be correlated to that defined after culture independent metagenomics analysis. Second, the conditions prevailing during cell culture may have favored the growth and the subsequent selection of a fraction of the species present in the microbiome of the Staka samples. Finally, the inability to isolate certain species exhibiting high abundances in the Staka samples in silico may indicate that at least some of them could have entered a viable but not culturable state (VBNC) during production and/or storage. Such a scenario is plausible, and there are important data to support it as in the case of *Lc. lactis* [68–71]. In detail, three strains of *Lc. lactis* used to produce a model cheese could be identified as viable by RT-qPCR but could not be recovered by traditional plating on M17 medium through the ripening period [70]. *Lc. lactis* can enter a VBNC state under carbohydrate starvation [72], which may prevail during ripening and/or storage of dairy products. In another study concerning the dynamics of LAB during long-term ripening of cheddar investigated by culture-based analysis, qPCR and 16S rDNA amplicon sequencing, the accumulation of a stable population of *Lactococcus* spp.-permeable cells was demonstrated after treatment of cells with propidium monoazide (PMA) [69]. PMA intercalates in the DNA of cells with membrane damage and inhibits its amplification by PCR. This population was not able to form colonies. Moreover, the presence of *Lc. lactis* cells in the VBNC state was verified in some traditional Lebanese products [68]. While *Lc. lactis* was identified in the metagenomics of some samples, it was not possible to isolate it under typical M17 isolation conditions. Most importantly, adding goat milk in the medium allowed the recovery of *Lc. lactis* colonies, verifying the original VBNC state of these cells. Furthermore, different *Enterococcus* spp. could be identified in the shotgun data of Staka 2 and in the 16S rDNA of the other samples. The same also applies for *Hafnia* spp. These may explain the isolation of *E. faecium* and *E. faecalis*, as well as *H. paralvei*, rather than *E. durans* and *H. alvei*, which were predicted through metagenomics analysis. Similar and even more major deviations between culture-based analysis and metagenomics seem not to be uncommon (Papadimitriou et al. unpublished results; [71]).

3.5. Antimicrobial Activity

All bacterial isolates were examined for antimicrobial activity against 20 indicator strains. Cell-free culture supernatants (CFCSs) of milk as well as the MRS broth (bacilli) or M17 broth (cocci) of overnight cultures were tested. The results for all isolates are summarized in Supplementary Table S2. Positive results concerning the supernatants of selected isolates exhibiting antimicrobial activity, which were further 10-fold concentrated, are presented in Table 4.

Regarding antimicrobial activity against Gram-negative indicators, very low or borderline inhibition was detected against *Pseudomonas aeruginosa* FMCC B-26 by a few strains, namely *L. paracasei* ACA-DC 1119, *L. pseudomesenteroides* ACA-DC 1145, and *Enterococcus faecium* ACA-DC 1117, 1200, 1201, and 1216. *P. aeruginosa* strains are considered responsible for an increased number of nosocomial infections incidents caused by Gram-negative multi-drug resistant (MDR) bacteria [73]; therefore, these, even with low antimicrobial activities, deserve further study so as to clarify if they correspond to bacteriocins. Bacteriocin L-1077 and Enterocin E-760 produced by *Ligilactobacillus salivarius* (basonym *Lactobacillus salivarius*) and *E. faecalis*, respectively, have been previously reported to be active against *P. aeruginosa* [74,75].

Table 4. Antimicrobial activity of 10-fold concentrated supernatants of selected LAB Staka strains.

Producer Strain ACA-DC Number	Target Strain (Growth Medium of the Producer Strain—mm Inhibition Zone)								
	S. mutans LMG 14558[T]	S. oralis LMG 14532[T]	S. pneumoniae LMG 14545[T]	S. agalactiae LMG 14694[T]	S. salivarius LMG 11489[T]	S. sanguinis DSM 20068	S. sobrinus LMG 14641[T]	S. gordonii LMG 14518[T]	S. anginosus LMG 14502[T]
L. curvatus 1135	MRS-10	mye-7t	MRS-10 mye -9	nd	nd	MRS-15	nd	MRS-15	MRS-13
L. coryniformis 1251	nd	nd	MRS-9 mye-7	nd	nd	nd	nd	nd	nd
L. paracasei 1260	nd	nd	MRS-11	nd	nd	mye-9t	nd	nd	MRS-10 mye-bl
C. versmoldensis 1262	MRS-10	nd	MRS-13	nd	nd	nd	nd	nd	MRS-10
L. pseudomesenteroides 1130	mye-10	mye15	MRS-20 mye-15	MRS-8t mye-6t	mye-7t	mye-20	mye-11	mye-20	MRS-14

bl: border line inhibition, t: turbid inhibition zone, mye: milk supplemented with yeast extract, nd: not detected.

Regarding Gram-positive target strains, almost all isolates inhibited the growth of at least one LAB target strain. S. thermophilus ACA-DC 4 was found to be the most sensitive one, followed by *Latilactobacillus sakei* (basonym *Lactobacillus sakei*) ACA-DC 2313. *L. curvatus* ACA-DC 1135, along with three *L. pseudomesenteroides* isolates, namely ACA-DC 1128, 1131, and 1249, were found active against the pathogenic *Streptococcus pneumoniae* LMG 14545[T]. *L. paracasei* ACA-DC 1260 and *C. versmoldensis* ACA-DC 1262 were active against *Streptococcus oralis* LMG 14532[T]. Two strains, namely *L. coryniformis* ACA-DC 1251 and *L. pseudomesenteroides* ACA-DC 1130, were active against both the above pathogenic streptococci. The above-mentioned species have been previously reported to produce bacteriocins, e.g., curvaticin L442 [76], leucocyclicin Q [77], and reuterin from *L. coryniformis* [78]. The antimicrobial activity of the isolated NSLAB strains against pathogenic streptococci reinforces the suitability for their use as adjunct starter cultures and/or nutraceuticals. To our knowledge, there is no report on bacteriocin characterization from *C. versmoldensis*; however, it has been included in a comparative genomics study where bacteriocin production was predicted with bioinformatics tools [79]. Among the isolated enterococci, 13 were found to have antilisterial activity, 10 were active against at least one pathogenic streptococcal isolate (mostly *S. oralis* LMG 14532[T]), and among them, 3 inhibited one pathogenic *Streptococcus* sp. strain (*S. oralis* LMG 14532[T] or *Streptococcus sanguinis* DSM 20068) along with one or three *Listeria* spp. strains.

Finally, various non-LAB isolates, namely *H. paralvei*, *B. subtilis* and *Pseudomonas* spp., exhibited antimicrobial activity against pathogenic streptococcal and *Listeria* species. Interestingly, three *Pseudomonas* spp. isolates inhibited from three to ten pathogenic streptococci, while two inhibited *B. cereus* LMG 6923[T] and *B. subtilis* FMCC B-109 strains and/or the *L. welshimeri* 15008. This antimicrobial activity can be attributed either to antibiotic production, as antibiotics like phenazine-1-carboxylic acid and mupirocin are produced by *Pseudomonas* sp. [80], or to the so-called pyocins, the bacteriocins produced by *P. aeuroginosa* [81]. Pyocins have been previously reported to possess activity against Gram-positive bacterial species, e.g., pyocin SA189 against *S. aureus*, *Streptococcus pyogenes*, and *L. monocytogenes* [82], and pyocin RPU15 against *S. aureus*, *L. monocytogenes*, and *Bacillus cereus* [83]. These results need to be further elucidated, since antibiotics are not impactful anymore against the MDR pathogens' threat, and new approaches are gaining ground to produce food antibacterial compounds.

4. Conclusions

Staka is a traditional Greek sour cream produced in Crete. Based on the physicochemical analysis, it may have a slightly higher pH compared to other sour creams. Its titratable acidity aligns with similar dairy products, indicating a mild sourness. The relatively increased moisture content along with high values of pH could negatively affect its safety and shelf-life. The high fat content, like that of other creamy dairy products, contributes to its mouthful and its potential as a source of beneficial lipids. The protein content varies, especially if Staka is processed with flour, affecting texture and nutritional value. In all samples analyzed, lactic acid was the predominant organic acid detected, while other acids were present in minor concentrations. The culture-based microbiological analysis of Staka revealed how processing variations could influence microbial populations. Heating (with or without flour) may significantly reduce detectable microbes, even leading to no detectable counts. Amplicon sequencing, both 16S rDNA and ITS analyses, provided a comprehensive view of the Staka microbiome concerning bacterial and fungal communities. As mentioned above, the production process affected microbial diversity, as certain heat-treated Staka samples showed reduced or absence of culturable microbial populations. Despite this, amplicon sequencing captured DNA from non-viable or unculturable cells, offering insight into the original microbiome. Shotgun metagenomics analysis of Staka provided a detailed picture of its microbiome, revealing a diverse community predominantly composed of *Streptococcaceae* members like *Lc. lactis* and *Lc. raffinolactis*. The construction of some near-complete MAGs, with both the binning of assembled contigs and reference-based recruitment plots analysis of single reads, further refined the understanding of the microbiome, assigning a majority of contigs and reads to specific bacterial species, enhancing the specificity of microbial identification. Subsequently, isolated bacterial strains were screened for their antimicrobial potential. Most isolates inhibited at least one LAB strain, with various isolates being also active against oral or other pathogenic *Streptococcus* species. Inhibition against Gram-negative *P. aeruginosa* was generally low, but some isolates showed potential for further study due to their antimicrobial activity against this indicator. Additionally, non-LAB isolates like *H. paralvei*, *B. subtilis*, and *Pseudomonas* spp. showed broad antimicrobial activity, possibly due to their ability to produce antibiotics or bacteriocins. These findings highlight that at least some of the isolates could be valuable as starter cultures or as nutraceuticals. Overall, the findings of our study contribute to a deeper knowledge of the microbial ecosystem in Staka, revealing potential influences on its quality, safety, and sensory attributes. Further studies employing multi-omics approaches may shed light on the relation between the microbiome and metabolome of Staka cream which may affect its physicochemical and organoleptic characteristics. In such studies, the cooking step with flour, wholemeal or not, should also be considered since it may have an important effect on the quality of the final product.

Supplementary Materials: The following supporting information can be downloaded at: https://www.mdpi.com/article/10.3390/foods13071129/s1, Supplementary Table S1: Bacteria fingerprinting and identification; Supplementary Table S2: Antimicrobial activity of the Staka bacterial isolates against 20 indicator strains.

Author Contributions: Conceptualization, K.P., M.G. and E.T.; formal analysis, K.P.; investigation, K.P., M.G., R.A., A.-M.A., E.M. and G.Z.; methodology, K.P., M.G. and E.T.; writing—original draft, K.P. and M.G.; writing—review and editing, K.P., M.G., R.A., E.M., G.Z. and E.T. All authors have read and agreed to the published version of the manuscript.

Funding: This research received no external funding.

Institutional Review Board Statement: Not applicable.

Informed Consent Statement: Not applicable.

Data Availability Statement: The data presented in this study are openly available in SRA under BioProject IDs: PRJNA1063285, PRJNA1064923, PRJNA1067114.

Conflicts of Interest: The authors declare no conflicts of interest.

References

1. Butler, G.; Nielsen, J.H.; Slots, T.; Seal, C.; Eyre, M.D.; Sanderson, R.; Leifert, C. Fatty acid and fat-soluble antioxidant concentrations in milk from high- and low-input conventional and organic systems: Seasonal variation. *J. Sci. Food Agric.* **2008**, *88*, 1431–1441. [CrossRef]
2. Banasiewicz, T.; Domagalska, D.; Borycka-Kiciak, K.; Rydzewska, G. Determination of butyric acid dosage based on clinical and experimental studies—A literature review. *Gastroenterol. Rev.* **2020**, *15*, 119–125. [CrossRef] [PubMed]
3. den Hartigh, L.J. Conjugated linoleic acid effects on cancer, obesity, and atherosclerosis: A review of pre-clinical and human trials with current perspectives. *Nutrients* **2019**, *11*, 370. [CrossRef] [PubMed]
4. Santa-María, C.; López-Enríquez, S.; Montserrat-de la Paz, S.; Geniz, I.; Reyes-Quiroz, M.E.; Moreno, M.; Palomares, F.; Sobrino, F.; Alba, G. Update on anti-inflammatory molecular mechanisms induced by oleic acid. *Nutrients* **2023**, *15*, 224. [CrossRef] [PubMed]
5. Nestel, P. Chapter 16—Fermented dairy foods and cardiovascular risk. In *Dairy in Human Health and Disease Across the Lifespan*; Watson, R.R., Collier, R.J., Preedy, V.R., Eds.; Academic Press: Cambridge, MA, USA, 2017; pp. 225–229. [CrossRef]
6. Ledoux, M.; Chardigny, J.-M.; Darbois, M.; Soustre, Y.; Sébédio, J.-L.; Laloux, L. Fatty acid composition of French butters, with special emphasis on conjugated linoleic acid (CLA) isomers. *J. Food Compos. Anal.* **2005**, *18*, 409–425. [CrossRef]
7. Leroy, F.; De Vuyst, L. Lactic acid bacteria as functional starter cultures for the food fermentation industry. *Trends Food Sci. Technol.* **2004**, *15*, 67–78. [CrossRef]
8. 21CFR131.160. Milk and Cream. Available online: https://www.accessdata.fda.gov/scripts/cdrh/cfdocs/cfcfr/CFRSearch.cfm?fr=131.160 (accessed on 22 December 2023).
9. *ISO 27205-IDF 149:2010*; Fermented Milk Products—Bacterial Starter Cultures—Standard of Identity. ISO: London, UK, 2010.
10. Meunier-Goddik, L. *Sour Cream and Crème Fraîche*; CRC Press: Boca Raton, FL, USA, 2004; pp. 235–246. [CrossRef]
11. Niamsiri, N.; Batt, C.A. Dairy Products. In *Encyclopedia of Microbiology*, 3rd ed.; Schaechter, M., Ed.; Academic Press: Oxford, UK, 2009; pp. 34–44.
12. 21CFR131.162. Available online: https://www.accessdata.fda.gov/scripts/cdrh/cfdocs/cfcfr/cfrsearch.cfm?fr=131.162 (accessed on 22 December 2023).
13. Aryana, K.J.; Olson, D.W. A 100-Year Review: Yogurt and other cultured dairy products. *J. Dairy Sci.* **2017**, *100*, 9987–10013. [CrossRef] [PubMed]
14. Monnet, V.C.S.; Cogan, T.M.; Gripon, K.C. Metabolism of starter cultures. In *Dairy Starter Cultures*; Cogan, T.M., Accolas, J.P., Eds.; VCH Publishers: New York, NY, USA, 1995; pp. 47–100.
15. Page, R.A.; Lavalie, V.G. Sour Cream Dairy Product. US Patent No. 2,719,793, 4 October 1955.
16. Cakmakci, S.; Hayaloglu, A.A. Evaluation of the chemical, microbiological and volatile aroma characteristics of Ispir Kaymak, a traditional Turkish dairy product. *Int. J. Dairy Technol.* **2011**, *64*, 444–450. [CrossRef]
17. Dairy, A.M.-G.t.P. A Mini-Guide to Polish Dairy. Available online: https://culture.pl/en/article/a-mini-guide-to-polish-dairy (accessed on 10 November 2023).
18. Garmasheva, I. Isolation and characterization of lactic acid bacteria from Ukrainian traditional dairy products. *AIMS Microbiol.* **2016**, *2*, 372–387. [CrossRef]
19. Motato, K.E.; Milani, C.; Ventura, M.; Valencia, F.E.; Ruas-Madiedo, P.; Delgado, S. Bacterial diversity of the Colombian fermented milk "Suero Costeno" assessed by culturing and high-throughput sequencing and DGGE analysis of 16S rRNA gene amplicons. *Food Microbiol.* **2017**, *68*, 129–136. [CrossRef] [PubMed]
20. Ferrocino, I.; Rantsiou, K.; Cocolin, L. Investigating dairy microbiome: An opportunity to ensure quality, safety and typicity. *Curr. Opin. Biotechnol.* **2022**, *73*, 164–170. [CrossRef] [PubMed]
21. Srinivas, M.; O'Sullivan, O.; Cotter, P.D.; Sinderen, D.V.; Kenny, J.G. The application of metagenomics to study microbial communities and develop desirable traits in fermented foods. *Foods* **2022**, *11*, 3297. [CrossRef] [PubMed]
22. Budhkar, Y.A.; Bankar, S.B.; Singhal, R.S. Milk and milk products | Microbiology of cream and butter. In *Encyclopedia of Food Microbiology*; Academic Press: Cambridge, MA, USA, 2014; pp. 728–737. [CrossRef]
23. Kochetkova, T.V.; Grabarnik, I.P.; Klyukina, A.A.; Zayulina, K.S.; Elizarov, I.M.; Shestakova, O.O.; Gavirova, L.A.; Malysheva, A.D.; Shcherbakova, P.A.; Barkhutova, D.D.; et al. Microbial communities of artisanal fermented milk products from Russia. *Microorganisms* **2022**, *10*, 2140. [CrossRef] [PubMed]
24. Syromyatnikov, M.Y.; Kokina, A.V.; Solodskikh, S.A.; Panevina, A.V.; Popov, E.S.; Popov, V.N. High-Throughput 16S rRNA gene sequencing of butter microbiota reveals a variety of opportunistic pathogens. *Foods* **2020**, *9*, 608. [CrossRef] [PubMed]
25. Lappa, I.K.; Gantzias, C.; Manolopoulou, E.; De Brandt, E.; Aerts, M.; Vandamme, P.; Tsakalidou, E.; Georgalaki, M. MALDI-TOF MS insight into the biodiversity of Staka, the artisanal Cretan soured cream. *Int. Dairy J.* **2021**, *116*, 104969. [CrossRef]
26. *ISO 2920-IDF 58:2004*; Whey Cheese. ISO: London, UK, 2004.
27. AOAC. *Official Methods of Analysis: Official Method for Ash*; Method No. 936.03; AOAC: Rockville, MD, USA, 2000.
28. Ardö, Y.; Polychroniadou, A. *Laboratory Manual for Chemical Analysis of Cheese. Improvement of the Quality of the Production of Raw Milk Cheeses*; Publications Office: Luxembourg, 1999.
29. Angelopoulou, A.; Alexandraki, V.; Georgalaki, M.; Anastasiou, R.; Manolopoulou, E.; Tsakalidou, E.; Papadimitriou, K. Production of probiotic Feta cheese using *Propionibacterium freudenreichii* subsp. *shermanii* as adjunct. *Int. Dairy J.* **2017**, *66*, 135–139. [CrossRef]

30. Papadimitriou, K.; Anastasiou, R.; Georgalaki, M.; Bounenni, R.; Paximadaki, A.; Charmpi, C.; Alexandraki, V.; Kazou, M.; Tsakalidou, E. Comparison of the microbiome of artisanal homemade and industrial Feta cheese through amplicon sequencing and shotgun metagenomics. *Microorganisms* **2022**, *10*, 1073. [CrossRef] [PubMed]
31. Papademas, P.; Aspri, M.; Mariou, M.; Dowd, S.E.; Kazou, M.; Tsakalidou, E. Conventional and omics approaches shed light on Halitzia cheese, a long-forgotten white-brined cheese from Cyprus. *Int. Dairy J.* **2019**, *98*, 72–83. [CrossRef]
32. Schmartz, G.P.; Hirsch, P.; Amand, J.; Dastbaz, J.; Fehlmann, T.; Kern, F.; Müller, R.; Keller, A. BusyBee Web: Towards comprehensive and differential composition-based metagenomic binning. *Nucleic Acids Res.* **2022**, *50*, W132–W137. [CrossRef] [PubMed]
33. Meyer, F.; Bagchi, S.; Chaterji, S.; Gerlach, W.; Grama, A.; Harrison, T.; Paczian, T.; Trimble, W.L.; Wilke, A. MG-RAST version 4—Lessons learned from a decade of low-budget ultra-high-throughput metagenome analysis. *Brief. Bioinform.* **2017**, *20*, 1151–1159. [CrossRef] [PubMed]
34. De Roos, J.; Vandamme, P.; De Vuyst, L. Wort substrate consumption and metabolite production during Lambic beer fermentation and maturation explain the successive growth of specific bacterial and yeast species. *Front. Microbiol.* **2018**, *9*, 2763. [CrossRef] [PubMed]
35. Verce, M.; De Vuyst, L.; Weckx, S. Shotgun metagenomics of a water kefir fermentation ecosystem reveals a novel *Oenococcus* species. *Front. Microbiol.* **2019**, *10*, 479. [CrossRef] [PubMed]
36. Georgalaki, M.; Zoumpopoulou, G.; Mavrogonatou, E.; Van Driessche, G.; Alexandraki, V.; Anastasiou, R.; Papadelli, M.; Kazou, M.; Manolopoulou, E.; Kletsas, D.; et al. Evaluation of the antihypertensive angiotensin-converting enzyme inhibitory (ACE-I) activity and other probiotic properties of lactic acid bacteria isolated from traditional Greek dairy products. *Int. Dairy J.* **2017**, *75*, 10–21. [CrossRef]
37. Ntougias, S.; Zervakis, G.I.; Ehaliotis, C.; Kavroulakis, N.; Papadopoulou, K.K. Ecophysiology and molecular phylogeny of bacteria isolated from alkaline two-phase olive mill wastes. *Res. Microbiol.* **2006**, *157*, 376–385. [CrossRef] [PubMed]
38. Quiloan, M.L.G.; Vu, J.; Carvalho, J. *Enterococcus faecalis* can be distinguished from *Enterococcus faecium* via differential susceptibility to antibiotics and growth and fermentation characteristics on mannitol salt agar. *Front. Biol.* **2012**, *7*, 167–177. [CrossRef]
39. Manero, A.B.R.A. Identification of *Enterococcus* spp. with a Biochemical Key. *Appl. Environ. Microbiol.* **1999**, *65*, 4425–4430. [CrossRef] [PubMed]
40. Charteris, W.P.; Kelly, P.M.; Morelli, L.; Collins, J.K. Quality control *Lactobacillus* strains for use with the API 50CH and API ZYM systems at 37 °C. *J. Basic Microbiol.* **2001**, *41*, 241–251. [CrossRef]
41. Garvie, E.I. The growth factor and amino acid requirements of species of the genus *Leuconostoc*, including *Leuconostoc paramesenteroides* (sp.nov.) and *Leuconostoc Oenos*. *J. Gen. Microbiol.* **1967**, *48*, 439–447. [CrossRef] [PubMed]
42. Huys, G.; Cnockaert, M.; Abbott, S.L.; Janda, J.M.; Vandamme, P. *Hafnia paralvei* sp. nov., formerly known as *Hafnia alvei* hybridization group 2. *Int. J. Syst. Evol. Microbiol.* **2010**, *60*, 1725–1728. [CrossRef]
43. Zoumpopoulou, G.; Tzouvanou, A.; Mavrogonatou, E.; Alexandraki, V.; Georgalaki, M.; Anastasiou, R.; Papadelli, M.; Manolopoulou, E.; Kazou, M.; Kletsas, D.; et al. Probiotic features of lactic acid bacteria isolated from a diverse pool of traditional Greek dairy products regarding specific strain-host interactions. *Probiotics Antimicrob. Proteins* **2018**, *10*, 313–322. [CrossRef] [PubMed]
44. Shepard, L.; Miracle, R.E.; Leksrisompong, P.; Drake, M.A. Relating sensory and chemical properties of sour cream to consumer acceptance. *J. Dairy Sci.* **2013**, *96*, 5435–5454. [CrossRef] [PubMed]
45. *CXS 288-1976*; Standard for Cream and Prepared Creams. WHO: Geneva, Switzerland, 1976.
46. USDA-AMS. *Specifications for Sour Cream and Acidified Sour Cream*; Agricultural Marketing Service: Washington, DC, USA, 2005.
47. Ribar, S.; Karmelić, I.; Mesarić, M. Sphingoid bases in dairy products. *Food Res. Int.* **2007**, *40*, 848–854. [CrossRef]
48. Yu, J.; Wang, H.M.; Zha, M.S.; Qing, Y.T.; Bai, N.; Ren, Y.; Xi, X.X.; Liu, W.J.; Menghe, B.L.; Zhang, H.P. Molecular identification and quantification of lactic acid bacteria in traditional fermented dairy foods of Russia. *J. Dairy Sci.* **2015**, *98*, 5143–5154. [CrossRef] [PubMed]
49. Shelley, A.; Deeth, H.; MacRae, I. Review of methods of enumeration, detection and isolation of lipolytic microorganisms with special reference to dairy applications. *J. Microbiol. Methods* **1987**, *6*, 123–137. [CrossRef]
50. Smacchi, E.; Gobbetti, M. Bioactive peptides in dairy products: Synthesis and interaction with proteolytic enzymes. *Food Microbiol.* **2000**, *17*, 129–141. [CrossRef]
51. Fox, P.F. Proteolysis during cheese manufacture and ripening. *J. Dairy Sci.* **1989**, *72*, 1379–1400. [CrossRef]
52. Seiler, H. Yeasts in milk and dairy products. *Encycl. Dairy Sci.* **2002**, *4*, 2761–2769.
53. Yu, Z.; Peng, C.; Kwok, L.; Zhang, H. The Bacterial diversity of spontaneously fermented dairy products collected in Northeast Asia. *Foods* **2021**, *10*, 2321. [CrossRef] [PubMed]
54. Fleet, G.H.; Mian, M.A. The occurrence and growth of yeasts in dairy products. *Int. J. Food Microbiol.* **1987**, *4*, 145–155. [CrossRef]
55. Kosikowski, F.V.; Brown, D.P. Influence of carbon dioxide and nitrogen on microbial populations and shelf life of cottage cheese and sour cream. *J. Dairy Sci.* **1973**, *56*, 12–18. [CrossRef]
56. Lu, M.; Wang, N.S. Chapter 7—Spoilage of milk and dairy products. In *The Microbiological Quality of Food*; Bevilacqua, A., Corbo, M.R., Sinigaglia, M., Eds.; Woodhead Publishing: Sawston, UK, 2017; pp. 151–178. [CrossRef]

57. Geronikou, A.; Srimahaeak, T.; Rantsiou, K.; Triantafillidis, G.; Larsen, N.; Jespersen, L. Occurrence of yeasts in white-brined cheeses: Methodologies for identification, spoilage potential and good manufacturing practices. *Front. Microbiol.* **2020**, *11*, 582778. [CrossRef] [PubMed]
58. Jakobsen, M.; Narvhus, J. Yeasts and their possible beneficial and negative effects on the quality of dairy products. *Int. Dairy J.* **1996**, *6*, 755–768. [CrossRef]
59. Kure, C.F.; Skaar, I. The fungal problem in cheese industry. *Curr. Opin. Food Sci.* **2019**, *29*, 14–19. [CrossRef]
60. Zheng, X.; Shi, X.; Wang, B. A Review on the general cheese processing technology, flavor biochemical pathways and the influence of yeasts in cheese. *Front. Microbiol.* **2021**, *12*, 703284. [CrossRef] [PubMed]
61. Islam, S.M.R.; Tanzina, A.Y.; Foysal, M.J.; Hoque, M.N.; Rumi, M.H.; Siddiki, A.M.A.M.Z.; Tay, A.C.-Y.; Hossain, M.J.; Bakar, M.A.; Mostafa, M.; et al. Insights into the nutritional properties and microbiome diversity in sweet and sour yogurt manufactured in Bangladesh. *Sci. Rep.* **2021**, *11*, 22667. [CrossRef] [PubMed]
62. Kamilari, E.; Tsaltas, D.; Stanton, C.; Ross, R.P. Metataxonomic mapping of the microbial diversity of Irish and eastern mediterranean cheeses. *Foods* **2022**, *11*, 2483. [CrossRef] [PubMed]
63. Lavoie, K.; Touchette, M.; St-Gelais, D.; Labrie, S. Characterization of the fungal microflora in raw milk and specialty cheeses of the province of Quebec. *Dairy Sci. Technol.* **2012**, *92*, 455–468. [CrossRef] [PubMed]
64. Naser, S.M.; Dawyndt, P.; Hoste, B.; Gevers, D.; Vandemeulebroecke, K.; Cleenwerck, I.; Vancanneyt, M.; Swings, J. Identification of lactobacilli by *pheS* and *rpoA* gene sequence analyses. *Int. J. Syst. Evol. Microbiol.* **2007**, *57*, 2777–2789. [CrossRef] [PubMed]
65. Naser, S.; Thompson, F.L.; Hoste, B.; Gevers, D.; Vandemeulebroecke, K.; Cleenwerck, I.; Thompson, C.C.; Vancanneyt, M.; Swings, J. Phylogeny and identification of enterococci by *atpA* gene sequence analysis. *J. Clin. Microbiol.* **2005**, *43*, 2224–2230. [CrossRef] [PubMed]
66. Chun, B.H.; Kim, K.H.; Jeon, H.H.; Lee, S.H.; Jeon, C.O. Pan-genomic and transcriptomic analyses of *Leuconostoc mesenteroides* provide insights into its genomic and metabolic features and roles in kimchi fermentation. *Sci. Rep.* **2017**, *7*, 11504. [CrossRef] [PubMed]
67. Ercolini, D.; Russo, F.; Blaiotta, G.; Pepe, O.; Mauriello, G.; Villani, F. Simultaneous detection of *Pseudomonas fragi*, *P. lundensis*, and *P. putida* from meat by use of a multiplex PCR assay targeting the *carA* gene. *Appl. Environ. Microbiol.* **2007**, *73*, 2354–2359. [CrossRef] [PubMed]
68. Ammoun, I.; Kothe, C.I.; Mohellibi, N.; Beal, C.; Yaacoub, R.; Renault, P. Lebanese fermented goat milk products: From tradition to meta-omics. *Food Res. Int.* **2023**, *168*, 112762. [CrossRef] [PubMed]
69. Barzideh, Z.; Siddiqi, M.; Mohamed, H.M.; LaPointe, G. Dynamics of starter and non-starter lactic acid bacteria populations in long-ripened cheddar cheese using propidium monoazide (PMA) treatment. *Microorganisms* **2022**, *10*, 1669. [CrossRef] [PubMed]
70. Ruggirello, M.; Giordano, M.; Bertolino, M.; Ferrocino, I.; Cocolin, L.; Dolci, P. Study of *Lactococcus lactis* during advanced ripening stages of model cheeses characterized by GC-MS. *Food Microbiol.* **2018**, *74*, 132–142. [CrossRef] [PubMed]
71. Wang, J.; Hao, S.; Ren, Q. Analysis of bacterial diversity in fermented grains of Baijiu based on culturomics and amplicon sequencing. *Fermentation* **2023**, *9*, 260. [CrossRef]
72. Ganesan, B.; Stuart, M.R.; Weimer, B.C. Carbohydrate starvation causes a metabolically active but nonculturable state in *Lactococcus lactis*. *Appl. Environ. Microbiol.* **2007**, *73*, 2498–2512. [CrossRef] [PubMed]
73. Ghodhbane, H.; Elaidi, S.; Sabatier, J.M.; Achour, S.; Benhmida, J.; Regaya, I. Bacteriocins active against multi-resistant gram negative bacteria implicated in nosocomial infections. *Infect. Disord. Drug Targets* **2015**, *15*, 2–12. [CrossRef] [PubMed]
74. Line, J.E.; Svetoch, E.A.; Eruslanov, B.V.; Perelygin, V.V.; Mitsevich, E.V.; Mitsevich, I.P.; Levchuk, V.P.; Svetoch, O.E.; Seal, B.S.; Siragusa, G.R.; et al. Isolation and purification of enterocin E-760 with Broad Antimicrob. Act. Against gram-positive and gram-negative bacteria. *Antimicrob. Agents Chemother.* **2008**, *52*, 1094–1100. [CrossRef] [PubMed]
75. Svetoch, E.A.; Eruslanov, B.V.; Levchuk, V.P.; Perelygin, V.V.; Mitsevich, E.V.; Mitsevich, I.P.; Stepanshin, J.; Dyatlov, I.; Seal, B.S.; Stern, N.J. Isolation of *Lactobacillus salivarius* 1077 (NRRL B-50053) and characterization of its bacteriocin, including the antimicrobial activity spectrum. *Appl. Environ. Microbiol.* **2011**, *77*, 2749–2754. [CrossRef] [PubMed]
76. Xiraphi, N.; Georgalaki, M.; Driessche, G.V.; Devreese, B.; Beeumen, J.V.; Tsakalidou, E.; Metaxopoulos, J.; Drosinos, E.H. Purification and characterization of curvaticin L442, a bacteriocin produced by *Lactobacillus curvatus* L442. *Antonie Van Leeuwenhoek* **2006**, *89*, 19–26. [CrossRef] [PubMed]
77. Masuda, Y.; Ono, H.; Kitagawa, H.; Ito, H.; Mu, F.; Sawa, N.; Zendo, T.; Sonomoto, K. Identification and characterization of leucocyclicin Q, a novel cyclic bacteriocin produced by *Leuconostoc mesenteroides* TK41401. *Appl. Environ. Microbiol.* **2011**, *77*, 8164–8170. [CrossRef]
78. Martin, R.; Olivares, M.; Marin, M.L.; Xaus, J.; Fernandez, L.; Rodriguez, J.M. Characterization of a reuterin-producing *Lactobacillus coryniformis* strain isolated from a goat's milk cheese. *Int. J. Food Microbiol.* **2005**, *104*, 267–277. [CrossRef] [PubMed]
79. Sun, Z.; Harris, H.M.; McCann, A.; Guo, C.; Argimon, S.; Zhang, W.; Yang, X.; Jeffery, I.B.; Cooney, J.C.; Kagawa, T.F.; et al. Expanding the biotechnology potential of lactobacilli through comparative genomics of 213 strains and associated genera. *Nat. Commun.* **2015**, *6*, 8322. [CrossRef] [PubMed]
80. Matthijs, S.; Vander Wauven, C.; Cornu, B.; Ye, L.; Cornelis, P.; Thomas, C.M.; Ongena, M. Antimicrobial properties of *Pseudomonas* strains producing the antibiotic mupirocin. *Res. Microbiol.* **2014**, *165*, 695–704. [CrossRef] [PubMed]
81. Ghequire, M.G.; Kemland, L.; Anoz-Carbonell, E.; Buchanan, S.K.; De Mot, R. A natural chimeric *Pseudomonas* bacteriocin with novel pore-forming activity parasitizes the ferrichrome transporter. *mBio* **2017**, *8*, e01961-16. [CrossRef] [PubMed]

82. Naz, S.A.; Jabeen, N.; Sohail, M.; Rasool, S.A. Biophysicochemical characterization of Pyocin SA189 produced by *Pseudomonas aeruginosa* SA189. *Braz. J. Microbiol.* **2015**, *46*, 1147–1154. [CrossRef] [PubMed]
83. Mohamed, A.A.; Elshawadfy, A.M.; Amin, G.; Askora, A. Characterization of R-pyocin activity against Gram-positive pathogens for the first time with special focus on *Staphylococcus aureus*. *J. Appl. Microbiol.* **2021**, *131*, 2780–2792. [CrossRef] [PubMed]

Disclaimer/Publisher's Note: The statements, opinions and data contained in all publications are solely those of the individual author(s) and contributor(s) and not of MDPI and/or the editor(s). MDPI and/or the editor(s) disclaim responsibility for any injury to people or property resulting from any ideas, methods, instructions or products referred to in the content.

Article

Investigation of the Microbiome of Industrial PDO Sfela Cheese and Its Artisanal Variants Using 16S rDNA Amplicon Sequencing and Shotgun Metagenomics

Natalia Tsouggou [1], Aleksandra Slavko [1], Olympia Tsipidou [2], Anastasios Georgoulis [3], Svetoslav G. Dimov [4], Jia Yin [5,6], Constantinos E. Vorgias [3], John Kapolos [1], Marina Papadelli [1] and Konstantinos Papadimitriou [2,*]

1. Department of Food Science and Technology, University of the Peloponnese, 24100 Kalamata, Greece; n.tsouggou@go.uop.gr (N.T.); a.slavko@go.uop.gr (A.S.); i.kapolos@uop.gr (J.K.); m.papadelli@uop.gr (M.P.)
2. Laboratory of Food Quality Control and Hygiene, Department of Food Science and Human Nutrition, 18855 Athens, Greece; olibiatsip@gmail.com
3. Department of Biochemistry and Molecular Biology, National and Kapodistrian University of Athens, Panepistimioupolis-Zographou, 15784 Athens, Greece; tgeorgoulis@med.uoa.gr (A.G.); cvorgias@biol.uoa.gr (C.E.V.)
4. Faculty of Biology, Sofia University "St. Kliment Ohridski", 8, Dragan Tzankov Blvd., 1164 Sofia, Bulgaria; svetoslav@biofac.uni-sofia.bg
5. Hunan Provincial Key Laboratory of Animal Intestinal Function and Regulation, College of Life Sciences, Hunan Normal University, Changsha 410081, China; jiayin@hunnu.edu.cn
6. Hunan International Joint Laboratory of Animal Intestinal Ecology and Health, College of Life Science, Hunan Normal University, Changsha 410081, China
* Correspondence: kpapadimitriou@aua.gr

Abstract: Sfela is a white brined Greek cheese of protected designation of origin (PDO) produced in the Peloponnese region from ovine, caprine milk, or a mixture of the two. Despite the PDO status of Sfela, very few studies have addressed its properties, including its microbiology. For this reason, we decided to investigate the microbiome of two PDO industrial Sfela cheese samples along with two non-PDO variants, namely Sfela touloumotiri and Xerosfeli. Matrix-assisted laser desorption/ionization–time of flight mass spectrometry (MALDI-TOF MS), 16S rDNA amplicon sequencing and shotgun metagenomics analysis were used to identify the microbiome of these traditional cheeses. Cultured-based analysis showed that the most frequent species that could be isolated from Sfela cheese were *Enterococcus faecium*, *Lactiplantibacillus plantarum*, *Levilactobacillus brevis*, *Pediococcus pentosaceus* and *Streptococcus thermophilus*. Shotgun analysis suggested that in industrial Sfela 1, *Str. thermophilus* dominated, while industrial Sfela 2 contained high levels of *Lactococcus lactis*. The two artisanal samples, Sfela touloumotiri and Xerosfeli, were dominated by *Tetragenococcus halophilus* and *Str. thermophilus*, respectively. *Debaryomyces hansenii* was the only yeast species with abundance > 1% present exclusively in the Sfela touloumotiri sample. Identifying additional yeast species in the shotgun data was challenging, possibly due to their low abundance. Sfela cheese appears to contain a rather complex microbial ecosystem and thus needs to be further studied and understood. This might be crucial for improving and standardizing both its production and safety measures.

Keywords: Sfela cheese; 16S rDNA; shotgun metagenomics; binning; lactic acid bacteria; microbiota

1. Introduction

White brined cheeses constitute an essential category of cheese produced worldwide and they are an important part of the diet of Mediterranean and Balkan regions. The white brined cheeses, also known as "pickled cheeses", owe their name to the distinctive white color and the high salt brines used during their manufacturing [1,2]. These cheeses lack rinds and vary in texture and moisture levels, ranging from soft to semi-hard, with a pleasantly acidic and salty taste that becomes piquant as they ripen. Their unique characteristics are mainly attributed to the lactic acid fermentation process taking place in the initial steps of production, as well as during ripening in brine, which may last for several months. These traits have been developed over centuries of household or artisanal practices.

Sfela cheese has been recognized as PDO (protected designation of origin) according to the Official Government Gazette of the Hellenic Parliament (25/18.01.94). It is a Greek white brined cheese produced in the Messinia and Laconia regional units, located in the southern Peloponnese. According to the PDO regulation, milk used for Sfela production can be ovine, caprine, or a mixture of both. Traditional rennet or enzymes can be added, while in the case of pasteurized milk, cultures of lactic acid bacteria (LAB) can be used. The coagulation of the milk is conducted at 30–32 °C, and the resulting curd is cut into small pieces and cooked at a temperature between 38°C and 40 °C with continuous stirring. The curd is transferred to cheesecloth to drain the whey and then placed on a cheese rack where it is slightly pressed. Subsequently, the curd is sliced into strips called "sfelides", dry salting is applied, and it is moved into barrels or metal containers filled with a 20 Bé brine. Finally, the cheese is left at room temperature for one month before being transferred to cold rooms (4–6 °C), where it ripens for at least three months. When kept sealed under brine, the cheese can be stored for up to two years [3]. The desired characteristics of Sfela are a maximum moisture content of 45% and a minimum fat-in-dry matter of 40%. Based on studies, the chemical composition of the cheese is estimated to be approximately 28% fat, 21% protein, 6% salt and 12% salt in moisture [4]. Sfela is known for its distinctively spicy and salty taste. Interestingly, it has been suggested that when it comes to marketing, this cheese has limited export potential [5].

Today, manufacturing of white brined cheeses may include the addition of starter cultures. Several starters have been evaluated, including *Streptococcus thermophilus*, *Lactobacillus delbrueckii* ssp. *bulgaricus* and *Lactococcus lactis* [6,7]. Mixtures of mesophilic and thermophilic LAB strains are also frequently employed as they have the ability to cause fast acidification of the milk, which is necessary for cheese manufacturing. The selection of the starter cultures is an essential step in the production of white brined cheeses using pasteurized milk to ensure safety while maintaining consistent quality and sensory properties. Besides that, adjunct cultures have been suggested as potential additives to further enhance the flavor, functional attributes, or safety properties of white brined cheeses. Overall, adjunct cultures may offer the potential to improve cheese quality and even create a probiotic dynamic [8–11]. Additional members of the microbiome of white brined cheeses may include various species and may be shifted throughout the ripening process. For instance, *Enterococcus faecalis*, *Enterococcus durans*, *Pediococcus pentosaceus* and *Pediococcus acidilactici* were initially present in high numbers in fresh Feta cheese but declined during ripening and were outgrown by lactobacilli [12]. Counts for lactobacilli may increase during the first stage of ripening. Lactobacilli are favored by low pH and high salt content, with *Lactiplantibacillus plantarum* found to be the predominant species in Feta cheese [13]. These bacteria are non-starter LAB (NSLAB) and can be detected in raw milk, enduring the initial heat treatment, or can be transferred to the cheese and brine from the surrounding environment [14]. Species of *Leuconostoc* (*Leuconostoc lactis* and *Leuconostoc mesenteroides* subsp. *dextranicum*) have also been found in Beyaz Peynir cheese. Coliforms can also be present in brined cheeses, but their numbers have been found to decrease during the reduction of pH [15]. Micrococci can contribute to the development of flavor and aroma. Still, their numbers are usually low in white brined cheeses due to the inhibitory effect of the low pH of the cheese and brine [16]. The microbiome of white brined cheeses may additionally contain various yeasts. Species including *Saccharomyces cerevisiae*, *Debaryomyces hansenii*, *Candida famata*, *Pichia membranifaciens*, *Torulaspora delbrueckii*, *Kluyveromyces marxianus*, *Candida sake*, and *Kluyveromyces lactis* have been previously found in Feta cheese [17–19]. In Domiati cheese, some of the yeast species that have been previously identified are *Issatchenkia orientalis*, *Candida albicans*, *Clavispora lusitaniae* (*Candida lusitaniae*), *Kodamaea ohmeri* (*Pichia ohmeri*), *K. marxianus*, and *Candida catenulate* [20]. It is worth noting that, some yeast species, such as *C. albicans* and other common yeast pathogens, which are found in fresh cheese and brines, are probably unable to survive the maturation process [21]. An overgrowth of yeasts may cause defects to the white brined cheeses like early blowing, characterized by the presence of gas holes [22].

Furthermore, next generation sequencing (NGS) approaches are gaining momentum as they can provide a more precise identification of rare or difficult-to-culture microbial populations. Metagenomics, a culture-independent approach, is widely used for studying microbial ecosystems in various foods, including traditional dairy and fermented products [23]. Currently, two main tools are employed: amplicon sequencing for identifying the microbial populations mostly at the genus level and shotgun metagenomics analysis for insights into the composition and functional properties of species (or even subspecies) in a niche. Both techniques have been previously applied in studies conducted for the characterization of various white brined cheese microbiomes. More specifically, amplicon-based sequencing for the identification of both bacterial and yeasts genera has been employed in the case of various Turkish white cheeses including Tulum and Ezine [24–26], Greek artisanal cheeses like Kalathaki Limnou, Gidotyri and Batzos, [27–29], Brazilian Minas cheese [30] and the Cypriot cheeses Halloumi and Halitzia [31,32]. Feta, which is the most popular Greek cheese, has been selected as the basis for multiple studies employing amplicon sequencing and/or shotgun metagenomics to explore its microbiome under different technological conditions [33–35].

While several studies have been conducted on the microbiology of Greek white brined cheeses, to our knowledge, there is a notable gap in the study of Sfela cheese, despite its PDO status and its significance for the local producers in the Peloponnese. Therefore, in this study, we combined a culture-based approach with 16S rDNA amplicon-based sequencing and shotgun metagenomics to initiate a first investigation of the microbiome of Sfela cheese, which has not been reported in the literature before. Throughout this study, two different variants of Sfela were also investigated, namely Sfela touloumotiri and Xerosfeli, which were included to provide insight into these traditional cheeses' microbial ecosystems.

2. Materials and Methods

2.1. Cheese Sampling

In this study, a total of four cheese samples were selected for analysis, including two PDO industrial Sfela samples and two non-PDO artisanal variants (Xerosfeli and Sfela touloumotiri). All samples analyzed were from different producers. For industrial Sfela 1 and the two artisanal variants, a mixture of ovine and caprine milk was used, while industrial Sfela 2 was produced from ovine milk solely. In all cases, milk was pasteurized with the exception of Sfela touloumotiri that was thermised. Both the industrial PDO samples and the two Sfela variants were produced in locations that fulfilled the geographical requirements specified for PDO Sfela cheese. Furthermore, the industrial Sfela cheese samples had been ripened for at least three months, as stipulated by PDO regulations. The non-PDO Xerosfeli was produced in a similar manner to PDO Sfela, but with a shorter ripening duration that may be distributed without being immersed in brine ("xero" means dry in Greek). The Sfela touloumotiri cheese was ripened in a touloum (animal skin bag) for an unknown period of time. All the samples were carefully transferred to the laboratory and stored at a temperature of 4 °C until analysis.

2.2. Growth Conditions and Strain Isolation

To initiate the isolations of the different Sfela strains, ten grams (10 g) of the cheese samples were homogenized in 90 mL of Buffered Peptone Water (BPW, Condalab, Madrid, Spain). Serial dilutions were prepared, and plate pouring was performed using De Man, Rogosa and Sharpe agar (MRS, Condalab, Madrid, Spain) and M17 agar (Himedia Laboratories, Mumbai, India). MRS and M17 were supplemented with 0.1% cycloheximide (AppliChem, Darmstadt, Germany), and they were incubated at 28 °C and 42 °C under anaerobic conditions using AnaeroGen™ Sachet (Thermo Fisher Scientific Oxoid, Basingstoke, UK). Colonies were selected from each medium, based on their distinct shape and color characteristics and then were further purified. The purified colonies were then stored at $-80\ °C$ with 20% (v/v) glycerol for future use.

2.3. Identification by MALDI-TOF MS

Isolated strains were subjected to MALDI-TOF MS analysis for identification. Stored strains were subcultured once using the respective medium each time. After incubation at the appropriate temperature for 24 h, the cultures were streaked on MRS agar (Condalab, Madrid, Spain) or M17 agar (Himedia Laboratories, Mumbai, India), depending on the specific medium in which each strain was initially isolated during the sampling process and incubated again for 24 h. Using a sterile toothpick, single colonies were carefully spotted on a stainless-steel target plate, and 1 µL matrix of saturated solution of α-HCCA (α-cyano-4-hydroxycinnamic acid in 50% acetonitrile and 2.5% trifluoroacetic acid) (Bruker Daltonics, Bremen, Germany) was added on top of the spots. The samples were left to dry at room temperature and then analyzed with the MALDI-TOF MS AutoflexIII (Bruker Daltonics, Bremen, Germany) using the default parameter settings within the MALDI Biotyper software ver. 3.1 (Bruker Daltonics, Bremen, Germany). To ensure reproducibility, each isolate was spotted two times on the plate. The identification of the isolates relied on log (score) values ranging from 0 to 3, which were acquired from a search within the reference database. Values of ≥ 2.0 signified identification at the species level, values of ≥ 1.7 represented identification at the genus level, and values < 1.7 were considered as unidentifiable.

2.4. DNA Extraction from Sfela Cheese Samples

DNA extraction from Sfela cheese samples was carried out using the DNeasy PowerFood Microbial Kit (Qiagen, Hilden, Germany) according to the manufacturer's protocol with slight modifications. To summarize the procedure, 0.5 g of Sfela cheese was collected in ten sterile Eppendorf tubes (1.5 mL), and 750 µL of trisodium citrate [2% (m/v), pH 7.4] was added to each tube. The suspensions were then centrifuged at $12{,}000\times g$ for 10 min at $4\ °C$, and the resulting pellet was washed several times with the citrate solution. After each centrifugation step, any remaining fat was removed, and all the pellets from each sample were combined into one tube. Next, the final pellets of each Sfela sample were treated with 500 µL of lysozyme (25 mg mL^{-1} in Tris-EDTA buffer at pH 8.0), 20 µL of mutanolysin (5 U mL^{-1}), 15µL of RNase (10 mg mL^{-1}) and 5 µL of lyticase (1 U µL^{-1}) (all reagents from Sigma-Aldrich ChemieGmbH, Munich, Germany). The samples were then incubated at $37\ °C$ for 1.5 h while vortexing every 15 min. After centrifugation at $10{,}500\times g$ for two minutes, DNA extraction from the pellet was carried out following the manufacturer's instructions provided with the kit. The resulting eluted DNA was stored at $-20\ °C$. Subsequently, DNA concentration and the ratios of 260/280 nm and 260/230 nm were determined using NanoDrop™ One/OneC Microvolume UV-Vis Spectrophotometer (Thermo Fisher Scientific, Cleveland, OH, USA). Furthermore, DNA integrity was assessed through agarose gel electrophoresis.

2.5. Sequencing and Bioinformatics Analysis

Library preparation, Illumina sequencing and quality control for 16S rDNA and shotgun metagenomics were conducted by Beijing Genomics Institute (BGI) and Novogene (Beijing, China). Data was imported into the CLC genomics workbench 23.0.5 (Qiagen, Hilden, Germany), and at first, chimera removal and operational taxonomic unit (OTU) clustering were performed with the default parameters, as previously described [35]. The 16S rDNA OTU tables were constructed using the Silva database version 138.1, with 99% identity. A closed OTU picking was selected to avoid the high number of unidentified taxa. Rarefaction of reads was performed using default settings. Alpha and beta diversities were calculated using the total number of OTUs and principal coordinate analysis (PCoA) based on Bray–Curtis distances, respectively. For shotgun metagenomics, the initial taxonomic profiling of samples was conducted by aligning the shotgun reads against a microbial genome database after creating a taxonomic profiling index in the CLC genomics workbench with default parameters. Subsequently, the shotgun reads were assembled using the de novo assembly function with a minimum length of 1Kbp for all the cheese samples with the default settings. Bin identification and taxonomic assignment was performed by the BusyBee web server [36,37]. For this analysis, the default parameters were chosen as well. *Cellulosimicrobium cellulans* reads were excluded from all the datasets during the analysis as their presence is due to the lyticase used for yeast lysis during DNA extraction [38,39]. The reference genomes in RefSeq of the most abundant species were employed to construct recruitment plots using the RecruitPlotEasy software with the default settings [40]. The initial step of mapping the sequence reads against the reference genomes was performed with the CLC genomics workbench. The specific reference genomes chosen for the analysis were those of *Str. thermophilus* CIRM 65, *L. lactis* LAC460, and *Tetragenococcus halophilus* MJ4. Finally, functional analysis of the assembled metagenomic reads was conducted using the CLC genomics workbench, for annotation and functional characterization of genetic sequences using the Gene Ontology (GO) molecular function terms. GO molecular function terms were visualized through heatmap construction using Euclidean distance for the first 25 molecular functions [41].

3. Results and Discussion

3.1. Identification of Isolates via MALDI-TOF MS Analysis

A total of 140 isolates from all the samples were selected based on their morphologies and growth at different temperatures. These isolates were subsequently identified using MALDI-TOF MS analysis, as shown in Table 1. The results revealed that the isolated species originated from various genera within LAB. The identification was considered successful, as all the species matches accounted for a score near or higher than 2.0. Among the total microbes analyzed from the four Sfela samples, the most frequent species were *E. faecium* (44.3%), *Lcb. plantarum* (23.6%), *Lvl. brevis* (10%), *P. pentosaceus* (8.6%), *Str. thermophilus* (5%), *Leu. mesenteroides* (3.6%) and *Lct. paracasei* (1.4%). Other bacteria were also identified, including *Enterococcus gallinarum*, *Latilactobacillus curvatus*, *Lacticaseibacillus rhamnosus*, *Streptococcus gallolyticus* and *Staphylococcus epidermidis*, but with lower abundances (<1%). Our results are in general accordance with previous studies conducted to identify LAB in white brined cheeses using MALDI-TOF MS analysis. More specifically, *E. faecium*, *Leu. mesenteroides* and *P. pentosaceus* have been isolated from white brined Serbian artisanal cheeses [42]. Furthermore, in the same study there were additional bacterial species that were absent from Sfela cheese, like *Macrococcus caseolyticus* and *E. faecalis* among others. In another study on the Spanish cheeses type "Torta", researchers also identified some common bacteria shared with Sfela cheese, which included *Ltl. curvatus*, *Lct. paracasei*, *Lcb. plantarum*, *Lct. rhamnosus* and *Leu. mesenteroides* [43]. *Lcb. plantarum* is a rather frequent bacterial species of white brined cheeses that has been also identified with MALDI-TOF MS in Touloumotyri [44]. Finally, the findings most closely resembling ours were those of Feta cheese which was examined for its microbial shift during ripening [45]. During the initial ripening, some of the dominant lactobacilli species were *Lcb. plantarum*, *Lvl. brevis*, *Lct. rhamnosus*, *Lct. paracasei* and *Ltl. curvatus*, while the most frequently encountered *Enterococcus* species were *E. faecalis*, *E. faecium* and *E. durans*. *Str. thermophilus* and *P. pentosaceus* were also present in significant quantities during the first three months of ripening, while *Staph. epidermidis* was only detected after six months [45]. It is evident that common LAB species may be present in different white brined cheeses, but variations can also occur.

Table 1. Species identification via MALDI-TOF MS.

Identified Species via MALDI-TOF MS	Total Number
Enterococcus faecium	62/140 (44.3%)
Lactiplantibacillus plantarum	33/140 (23.6%)
Levilactobacillus brevis	14/140 (10%)
Pediococcus pentosaceus	12/140 (8.6%)
Streptococcus thermophilus	7/140 (5%)
Leuconostoc mesenteroides	5/140 (3.6%)
Lacticaseibacillus paracasei	2/140 (1.4%)
Enterococcus gallinarum	1/140 (0.7%)
Latilactobacillus curvatus	1/140 (0.7%)
Lacticaseibacillus rhamnosus	1/140 (0.7%)
Streptococcus gallolyticus	1/140 (0.7%)
Staphylococcus epidermidis	1/140 (0.7%)

3.2. 16S rDNA Amplicon Sequencing

The next step of our analysis was the 16S rDNA amplicon sequencing of all four cheese samples. The clustering analysis of 16S rDNA reads, as shown in Figure 1A, reveals that the Firmicutes phylum was predominant across all samples, indicating that the Sfela cheese environment provides favorable conditions for the proliferation of Gram-positive bacteria. Some examples of Firmicutes commonly found in white brined cheeses include members of the *Lactobacillaceae* family and *Streptococcus* sp. The LAB are known for their role in the fermentation of dairy products and their contribution to the development of specific flavors, textures, and aromas [46]. Previous research conducted on Feta cheese has also reported the prevalence of the Firmicutes phylum in the bacterial microbiota [33]. In addition to Feta, Firmicutes was described to clearly dominate in various Greek PDO cheeses, including Anevato, Batzos, Galotiri, Kalathaki Limnou and Kopanisti [29].

The bacterial composition was more diverse at the family level, as shown in Figure 1B. More specifically, *Streptococcaceae* emerged as the dominant family in industrial Sfela 2 and Sfela touloumotiri, accounting for 79% and 43.7%, respectively. In industrial Sfela 1 and Xerosfeli, it also appeared with high percentages, particularly 39% and 48%, but in both samples, it was the second most prevalent family. The family that predominated in industrial Sfela 1 (61%) and Xerosfeli (49.9%) was *Lactobacillaceae*, which was the second highest population in industrial Sfela 2 (21%) and in Sfela touloumotiri (30%). Sfela touloumotiri contained two additional families, namely *Enterococcaceae* and *Staphylococcaceae*, with abundances of 23.3% and 3%, respectively. *Lachnospiraceae* is a family that only appeared in Xerosfeli but with a low abundance of 2%.

The differences between the samples became even more noticeable when compared for the most dominant genera (Figure 1C). In the industrial Sfela 1 and the Xerosfeli samples, the primary member of the *Streptococcaceae* family was the genus *Streptococcus* with 43% and 53% abundances, respectively. In contrast, the most abundant genus of the same family in the other two samples was *Lactococcus* representing 78% of the industrial Sfela 2 and 45% of Sfela touloumotiri. In these two samples, *Streptococcus* was present in low percentages (<1% in industrial Sfela 2 and 3.9% in Sfela touloumotiri), as was also observed in the case of *Lactococcus* detected in industrial Sfela 1 and Xerosfeli (<1%). *Lactiplantibacillus* was present in industrial Sfela 1, industrial Sfela 2 and Xerosfeli with 18.9%, 10.9% and 2% abundance, respectively. *Lacticaseibacillus* was identified only in industrial Sfela 1, occupying a notable percentage of 28.2% of the total sample. The genus *Companilactobacillus* appeared only in Sfela touloumotiri with a relatively high abundance (27.9%), while in the rest of the samples, its presence was not significant (<1%). On the contrary, *Latilactobacillus* appeared in industrial Sfela 1, industrial Sfela 2 and Xerosfeli with varying abundances of 2–9%, while in Sfela touloumotiri with <1%. *Lactobacillus* is another genus of the same family, which was detected in industrial Sfela 1, accounting for 2.4% and in Xerosfeli 36.9%. Finally, *Weissella* was identified in industrial Sfela 2, Sfela touloumotiri and Xerosfeli with 3.1%, 2.1% and 3.5%, respectively.

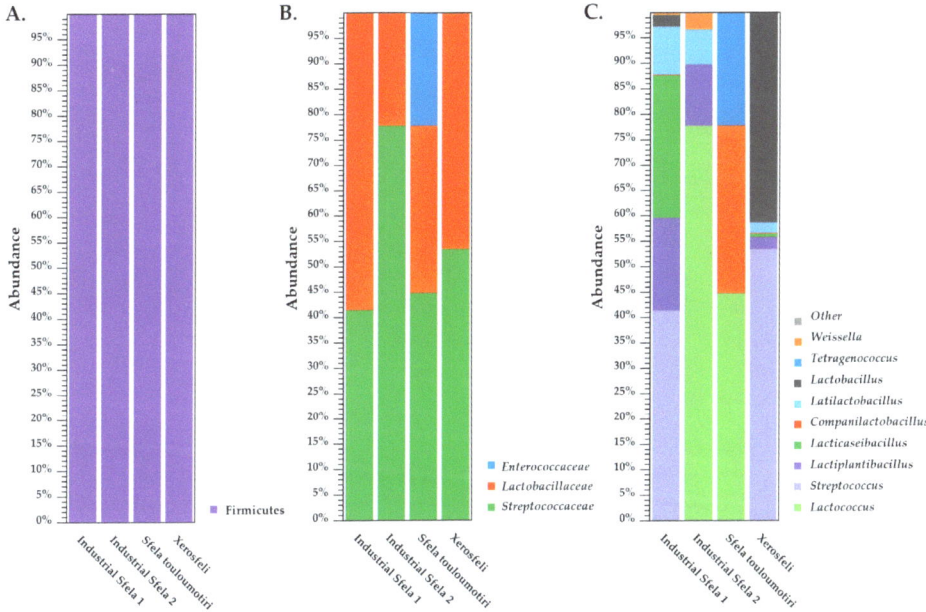

Figure 1. Illustration of the taxonomic distribution of industrial Sfela cheese and its artisanal variants, Sfela touloumotiri and Xerosfeli, using 16S rDNA amplicon sequence data. The samples were analyzed at the phylum (**A**), family (**B**), and genus (**C**) levels.

Regarding the most prevalent bacterial genera reported in Sfela samples, the results are in accordance with studies employing amplicon sequencing on various types of white brined cheeses. More specifically, *Lactococcus*, *Streptococcus* and *Lactobacillus* were consistently identified as the significant genera in Tulum, Minas Gerais, Feta, Halloumi and Halitzia cheeses [25,30–32,34,47]. Additionally to these three frequent cheese genera, *Lactiplantibacillus* and *Lacticaseibacillus* were also identified with 16S rDNA amplicon sequencing in Feta [35] and *Weissella* in Minas Gerais and Tulum cheeses [25,30]. Nevertheless, the distribution of these bacteria varied between the cheeses, and there were also differences concerning the presence of less abundant genera, including *Bifidobacterium*, *Pediococcus* and *Pseudomonas* which uniquely appeared in specific studies [25,30,34,35]. As already mentioned, white brined cheeses usually contain NSLAB that originate from raw milk or contamination occurring after heat-treatment. The mesophilic lactobacilli are the predominant type of NSLAB found in white brined cheeses [12,48], which could explain the high frequency of these genera in Sfela samples. Interestingly, the genus *Tetragenococcus* of the *Enterococcaceae* family was exclusively found in Sfela touloumotiri, showing a significant abundance of 23.3%. This genus consists of halophilic LAB, thriving in high-salt environments, which makes them well-suited for white brined cheese environments, and they have also been identified in cheese brines [49,50]. Some species show proteolytic and lipolytic activities, contributing to the development of sensory characteristics in ripened cheese products [51]. In a previous study, *T. halophilus* has been suggested as a potential starter culture for fermented foods with high salinity [52].

The analysis at the phylum level did not reveal any taxonomic differences, but the family and genus levels provided a deeper insight into the bacterial distribution among the samples. The alpha diversity analysis (Figure 2A) confirmed that the sequence depth was adequate for each sample. Furthermore, beta diversity revealed that some samples could be clustered together when assessing the variation at the genus level. More specifically, industrial Sfela 1 and Xerosfeli were classified together and separately from industrial Sfela 2 and Sfela touloumotiri according to PCo1 (Figure 2B).

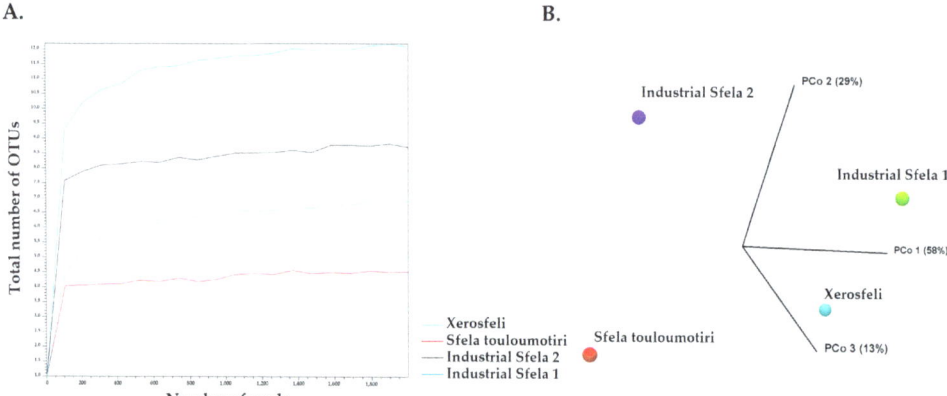

Figure 2. Alpha diversity analysis of 16S rDNA reads for a maximum depth of 18,000 read counts based on the total number of OTUs at genus level (**A**) and beta diversity presented in a Principal Coordinate Analysis (PCoA) employing the Bray—Curtis distances for the samples (**B**).

3.3. Identification of Sfela Microbiome at Species Level Using Shotgun Metagenomics Analysis

Species-level identification of the Sfela cheese microbiome was achieved by shotgun metagenomics. We compared the reads from these samples against a representative database of bacterial and fungal genomes (Figure 3). The species-level analysis revealed that industrial Sfela 1 was mainly characterized by the presence of *Str. thermophilus* (40.74% abundance), which appeared in even higher abundance in Xerosfeli at 50.5% in particular.

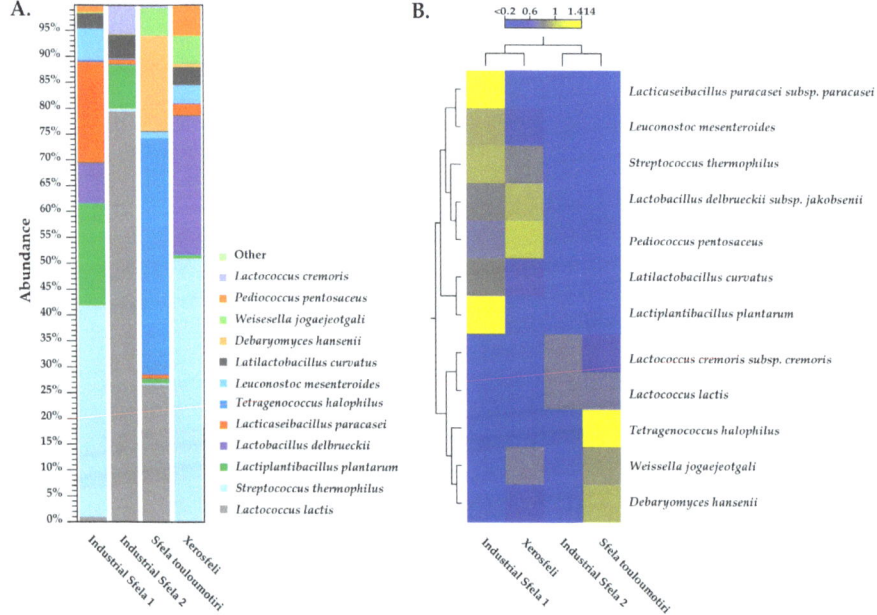

Figure 3. Illustration of the taxonomic distribution of industrial Sfela cheese and its artisanal variants, Sfela touloumotiri and Xerosfeli, using shotgun metagenomics sequence data. The samples were analyzed at the species level (**A**). Heat-map analysis of the species identified in industrial Sfela 1, industrial Sfela 2, Sfela touloumotiri and Xerosfeli (**B**).

L. lactis predominated in industrial Sfela 2, reaching up to 79.2%, and it also appeared in Sfela touloumotiri with 27.1% abundance. *T. halophilus*, the most frequent species in Sfela touloumotiri, occupied 45.9% of the total sample, while it could not be detected in the rest of the samples (<1% abundance). *Lcb. plantarum* was another species that exhibited significant variations among the samples, and in particular, it reached 19.7% in industrial Sfela 1 and 8.3% in industrial Sfela 2, while it appeared in less than 1% in the two artisanal cheeses. Industrial Sfela 1 and Xerosfeli shared two additional species, namely *Lb. delbrueckii* reaching 8.1% and 26.6% and *Lct. paracasei* with 19.7% and 2.4% abundance in industrial Sfela 1 and Xerosfeli, respectively. *Ltl. curvatus*, was found in notable amounts in industrial Sfela 1 (2.9%), industrial Sfela 2 (4.5%) and Xerosfeli (3.3%). *Leu. mesenteroides* was also detected in substantial abundancies of 6.4%, 1.2% and 3.3% in industrial Sfela 1, Sfela touloumotiri and Xerosfeli, respectively. Several other bacterial species were detected in varying abundances, including *P. pentosaceus*, *Lactococcus cremoris* and *Weissella jogaejeotgali*. Of note, some *Bifidobacterium* spp. were detected but their abundance was low (<1%). *D. hansenii* was the only yeast species found in abundance exceeding 1%. Notably, it appeared in a high percentage of 13.5% in Sfela touloumotiri. It was also present in the remaining samples, occupying < 1% of the microbial population. Additional yeast species found in the samples, but with low abundances (<1%), included *Neurospora crassa*, *Tor. delbrueckii* and *Debaryomyces nepalensis*.

As previously described, the diverse array of bacterial species found across samples verifies the rich and dynamic microbial profile of white brined cheeses like Sfela. *Str. thermophilus* has been previously identified in various cheese types, particularly in those produced through fermentation processes involving high temperatures. Its prevalence and significance are noteworthy in Feta cheese [35,45] and other Mediterranean cheeses, like Italian Grana Padano [53] and Parmigiano-Reggiano [54]. Its high abundance can be attributed to the potential addition of starter cultures. The dominance of *L. lactis* in the industrial Sfela 2 is in agreement with its widespread occurrence in various white brined cheeses, including Feta cheese [35,55]. 16S rDNA amplicon sequencing and shotgun metagenomics consistently identify *L. lactis*, *Str. thermophilus*, and *Lb. delbrueckii* as the three most prevailing species in white brined cheeses [24,26,27]. *Leu. mesenteroides* has been previously identified in the traditional Montenegrin brine cheese [56] and in Feta cheese through shotgun metagenomics analysis [35]. *W. jogaejeotgali*, initially found in traditionally fermented Korean jogae jeotgal [57], exhibits resistance to osmotic stress and tolerance to acidic conditions. Recent discoveries have placed *W. jogaejeotgali* as part of the cheese microbiome [26,58], though other species of this genus have been also isolated from dairy environments in previous studies [51,59]. Furthermore, our findings align with previous studies on Feta, Domiati, and Halloumi cheeses, indicating a consistent presence of non-starter lactobacilli in mature samples [12,60,61].

The results from the shotgun analysis are generally in accordance with the 16S rDNA amplicon sequencing analysis, which provided insights into the bacterial composition of the samples up to the genus level. Moreover, the shotgun analysis at the species level is partially consistent with the culture-based analysis and the identification of the isolates with MALDI-TOF MS after culture-based isolation. In particular, both approaches revealed the presence of *Str. thermophilus*, *Lcb. plantarum*, *Lct. paracasei* and *Leu. mesenteroides*. However, it is worth noting that the data regarding the species with abundances exceeding 1% did not always match the culture-based analysis results. For instance, enterococci were found in high populations among the identified isolates, in contrast to the metagenomics results in which they were <1% abundant. Shotgun analysis revealed the presence of highly abundant species that could not be isolated and characterized by the cultured-based analysis. This discrepancy may be attributed to the selection of colonies based on morphological variations rather than a randomized selection. Consequently, the identified taxa through culture-based analysis may appear with different frequencies than those found through metagenomics analysis. Another aspect to consider is that both amplicon and shotgun sequencing techniques can detect bacteria even if they are no longer viable or able to be cultivated, as previously suggested [62]. Furthermore, during our study, cell culture conditions might have provided an environment favoring the growth and subsequent selection of certain species from the microbiome present in the Sfela samples. In addition, the inability to isolate some taxa with high abundances, as indicated by the shotgun metagenomics and the 16S amplicon sequencing, may suggest that some of them could have been in a state where they are "viable but not culturable" (VBNC). For instance, in one study, strains of *L. lactis* used in cheese making were found to be viable with RT-qPCR, but they could not be grown on M17 throughout the ripening period [63]. Another study supporting the VBNC condition of *L. lactis* in Lebanese fermented milk products demonstrated that despite the inability to isolate the species on M17, the addition of goat milk in the culture medium facilitated its recovery [64]. This phenomenon could potentially be attributed to the depletion of carbohydrates, possibly occurring during ripening or storage, leading to *L. lactis* entering a VBNC state [65].

Comparing the most dominant species within the four samples revealed that the bacterial diversity in industrial Sfela 1 and Xerosfeli exhibited similarity and industrial Sfela 2 was grouped together with Sfela touloumotiri, as illustrated in Figure 3B. Among the bacterial species, the presence of *Str. thermophilus*, *Lb. delbrueckii* and *P. pentosaceus* were responsible for grouping industrial Sfela 1 and Xerosfeli. The grouping of industrial Sfela 2 and Sfela touloumotiri relied on the presence of two distinct lactococci, i.e., *L. lactis* and *L. cremoris*. In addition, *T. halophilus* was the characteristic species in Sfela touloumotiri. Furthermore, *Lcb. plantarum*, *Lactiplantibacillus paraplantarum*, and *Lct. paracasei* were prevalent in industrial Sfela 1.

Alpha diversity was also calculated for the shotgun metagenomic reads. As indicated in Figure 4A, the depth of the sequencing was again adequate for each sample. As shown in Figure 4B, the beta diversity analysis of the shotgun metagenomics sequencing segregated the samples in a manner similar to the heatmap analysis of the most dominant species. More precisely, industrial Sfela 1 and Xerosfeli were grouped together and distinct from industrial Sfela 2 and Sfela touloumotiri according to PCo1.

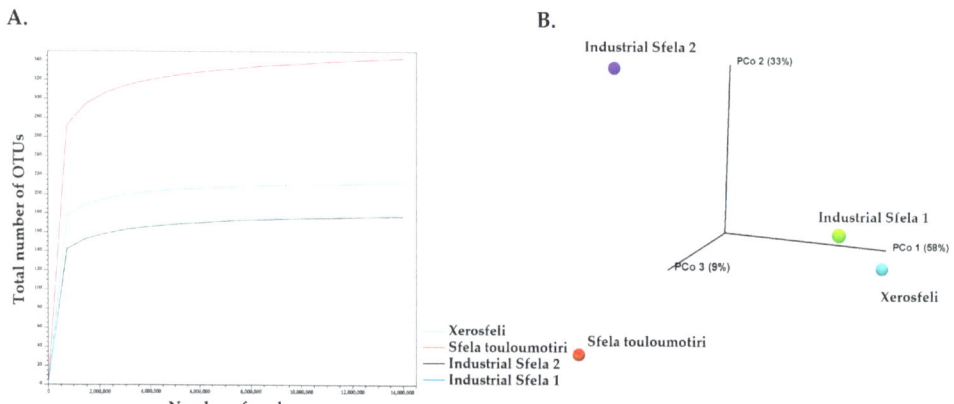

Figure 4. Alpha diversity analysis of shotgun reads for a maximum depth of 14,000,000 read counts based on the total number of OTUs (**A**) and beta diversity presented in PCoA employing the Bray—Curtis distances for the samples (**B**).

3.4. Binning of Metagenome-Assembled Contigs of Sfela Cheese Samples—Construction of Recruitment Plots

To identify metagenome-assembled genomes (MAGs), we performed binning using the assembled contigs of each sample (Figure 5). As shown in Table 2, we were able to get several bins with high completeness. Still, in most cases, these were accompanied by high levels of contamination and/or strain heterogeneity. However, there were some exceptions of high-quality bins, like Bin 12 in industrial Sfela 1, as well as Bin 4, Bin 6 and Bin 8 in Xerosfeli. These bins were identified as *Lb. delbrueckii* (Bin 12 and Bin 4), *M. caseolyticus* (Bin 6) and *Lct. paracasei* (Bin 8).

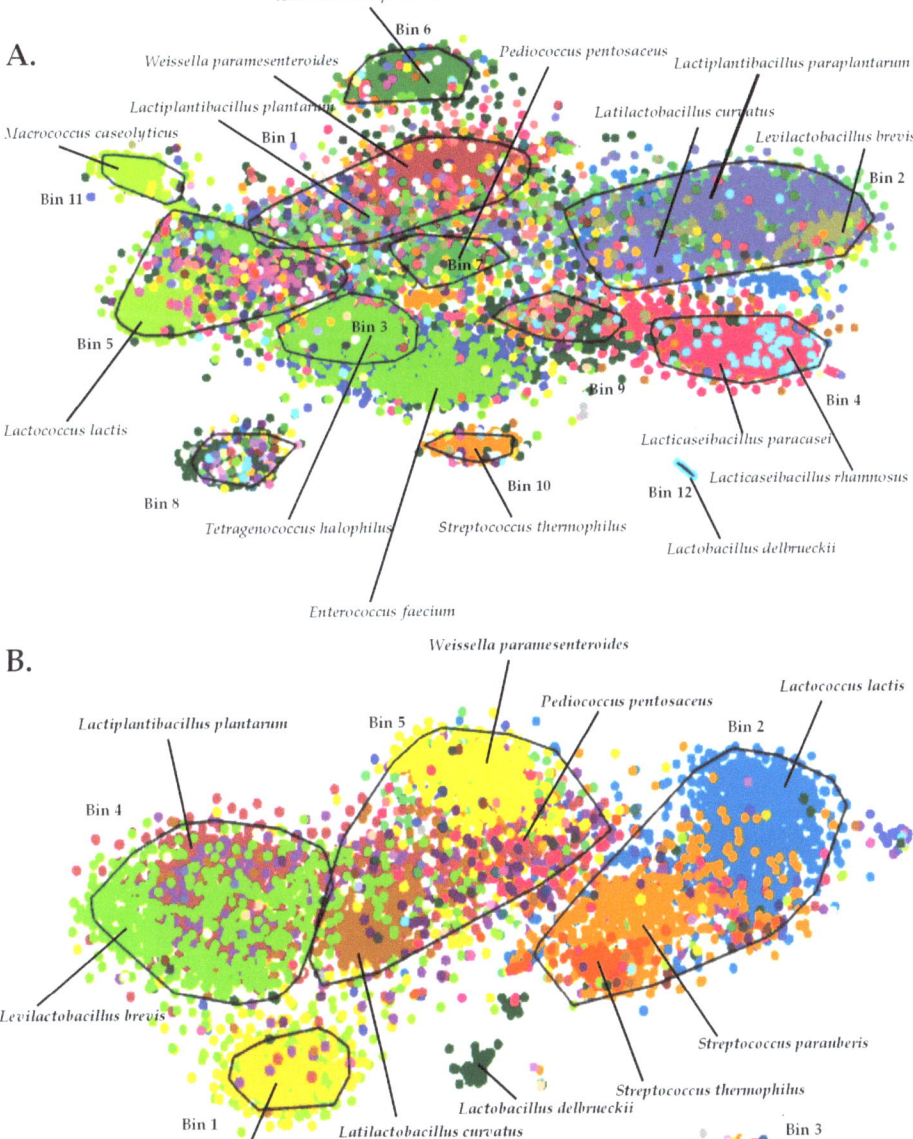

Figure 5. *Cont.*

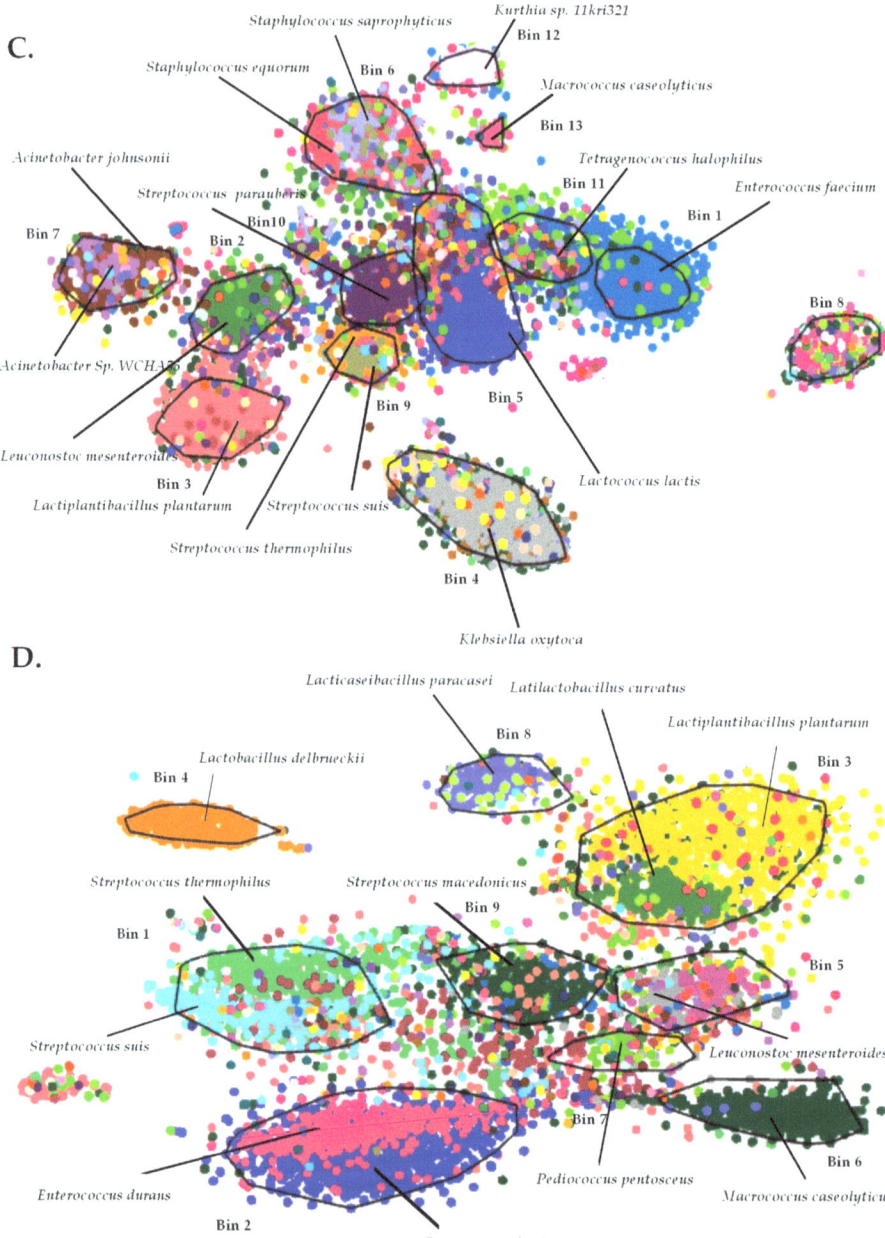

Figure 5. Bins of metagenomics scaffolds of industrial Sfela 1 (**A**), industrial Sfela 2 (**B**), Sfela touloumotiri (**C**) and Xerosfeli (**D**). The figure indicates the identified species' contigs distributed in different bins represented by dots of the same color.

Table 2. Quality of bins obtained from the assembled metagenomes of Sfela cheese samples.

Sample	Bin	Completeness *	Contamination *	Strain Heterogeneity *
Industrial Sfela 1	1	93.69	127.03	7.69
	2	95.50	225.23	9.83
	3	80.18	36.94	8.77
	4	82.88	9.91	60.00
	5	87.39	95.50	2.67
	6	81.08	17.12	13.04
	7	72.97	9.91	7.69
	8	29.73	0.90	0.00
	9	8.11	0.00	0.00
	10	45.95	0.00	0.00
	11	91.89	63.06	7.80
	12	91.89	0.00	0.00
Industrial Sfela 2	1	20.72	1.80	50.00
	2	79.28	122.52	14.80
	3	11.71	3.60	100.00
	4	93.69	80.18	12.73
	5	94.59	161.26	10.29
Sfela touloumotiri	1	48.65	11.71	40.00
	2	94.59	207.21	6.35
	3	57.66	14.41	18.75
	4	82.88	8.11	33.33
	5	67.57	25.23	28.57
	6	91.89	81.08	62.04
	7	86.49	12.61	50.00
	8	93.69	74.77	6.52
	9	72.07	18.02	40.00
	10	54.95	5.41	44.44
	11	91.89	46.85	8.33
	12	95.50	36.94	12.24
	13	72.97	0.90	0.00
Xerosfeli	1	54.95	29.73	26.32
	2	94.59	87.39	63.79
	3	95.50	85.59	10.17
	4	90.09	0.00	0.00
	5	95.50	138.74	3.70
	6	93.69	9.01	25.00
	7	68.47	2.70	0.00
	8	95.50	3.60	0.00
	9	61.26	18.02	39.13

* determined through analysis with the BusyBee web tool.

The taxonomic evaluation of all the bins and contigs agreed with our species identification findings by the shotgun metagenomics analysis of single reads presented above. However, BusyBee analysis demonstrated the presence of bins of contigs corresponding to species appearing with less than 1% abundance in the shotgun analysis. This was the case for the species Lcb. paraplantarum, E. durans, Streptococcus parauberis, M. caseolyticus and other less frequent species found dispersed among the bins of the different samples. Staphylococcus equorum was one of the aforementioned species, which is a non-pathogenic staphylococcal member and was found to contribute to the intense flavor profiles found in cheeses [66] and has been suggested as a starter culture for the manufacturing of semi-hard cheeses [67]. The binning process also revealed the presence of some additional bacteria, including Acinetobacter johnsonii and Acinetobacter sp. (TTH0-4, WCHA55), which have been related to cheese spoilage [68] and can act as opportunistic pathogens [69]. Despite the fact that Str. parauberis has been previously reported in cheese environments [33,70] and sheep milk [71], its presence may be unwanted as it has been connected with cases of bovine mastitis [72]. *Streptococcus suis* is another bacterium that has been identified in cheese samples [73], but its occurrence may pose safety concerns as a zoonotic pathogen [74]. Detecting these bacteria in Sfela samples can raise concerns regarding the hygienic conditions during the cheese production and it underlines the importance of an improved surveillance and monitoring of the process.

To surpass the problem of low-quality bins within our dataset and more specifically, to determine the MAGs of the most prevalent species in each sample, we constructed recruitment plots of sequencing reads against the reference genomes of *Str. thermophilus* (for industrial Sfela 1 and Xerosfeli), *L. lactis* (for industrial Sfela 2) and *T. halophilus* (for Sfela touloumotiri). As shown in Figure 6, this approach enabled the assembly of nearly complete draft genomes for each of the reference genomes. Interestingly, a significant number of reads exhibited alignments with an identity exceeding 95%, while also displaying a high average depth of sequencing coverage.

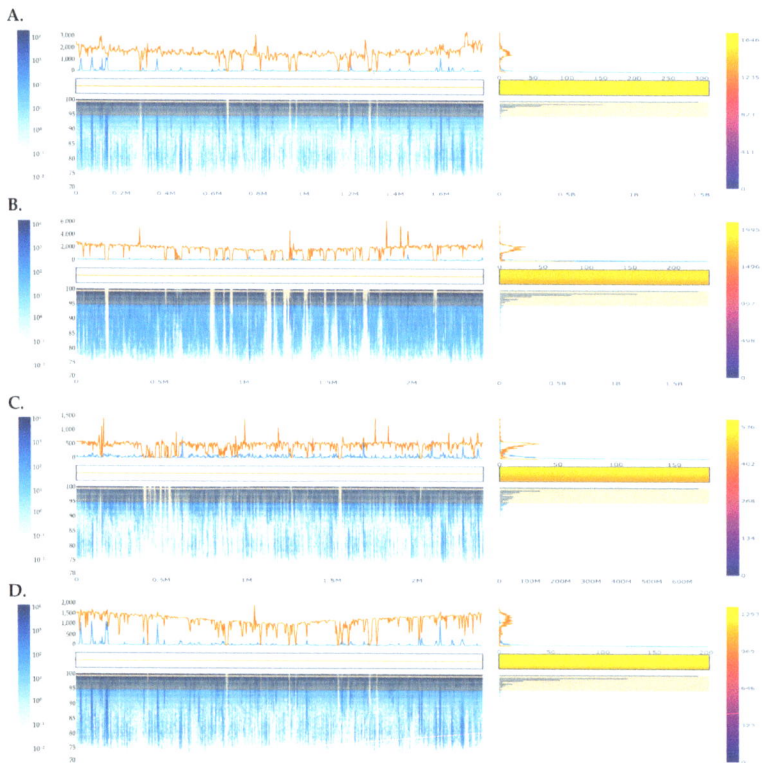

Figure 6. Recruitment plots of metagenomic reads from industrial Sfela 1 against the sequence of the reference genome of *Str. thermophilus* (CIRM 65) (**A**), industrial Sfela 2 against the genomic sequence of the reference strain of *L. lactis* (LAC460) (**B**), Sfela touloumotiri against the genomic sequence of the reference strain of *T. halophilus* (MJ4) (**C**), and Xerosfeli against the genomic sequence of the reference strain of *Str. thermophilus* (CIRM 65) (**D**). Bottom left panel: reads that have been aligned to the genome. The *X*-axis represents the position in the genome, while the *Y*-axis represents the percent of identity. The color of the data points corresponds to the number of reads that match a particular position on the genome with a specific level of identity indicated from the left bar. Upper left panel: sequencing depth across different regions of the genome. Bottom right panel: number of nucleotide bases at different levels of nucleotide identity. The *Y*-axis represents the nucleotide identity, and the *X*-axis represents the number of bases. Upper right panel: histogram of the sequencing depth. Orange and blue lines represent matches with nucleotide identity above and below the 95% cutoff, respectively.

3.5. Functional Analysis of Metagenome Contigs

Functional analysis of the assembled shotgun metagenomes was performed in the CLC genomics workbench. Heatmap analysis demonstrated that different functions had different distributions in each sample (Figure 7). The grouping between industrial Sfela 2 and Sfela touloumotiri was again evident. In contrast, industrial Sfela 1 and Xerosfeli did not form a distinct group. It should be mentioned that our analysis concerns the presence/absence of functions rather than their actual activity during the production of these cheese samples, given that they derive from a metagenomics rather than a metatranscriptomics dataset. Further research is required to explore the roles of the identified activities and to develop a more comprehensive image of the molecular functions contributing to the production of the different Sfela cheese types.

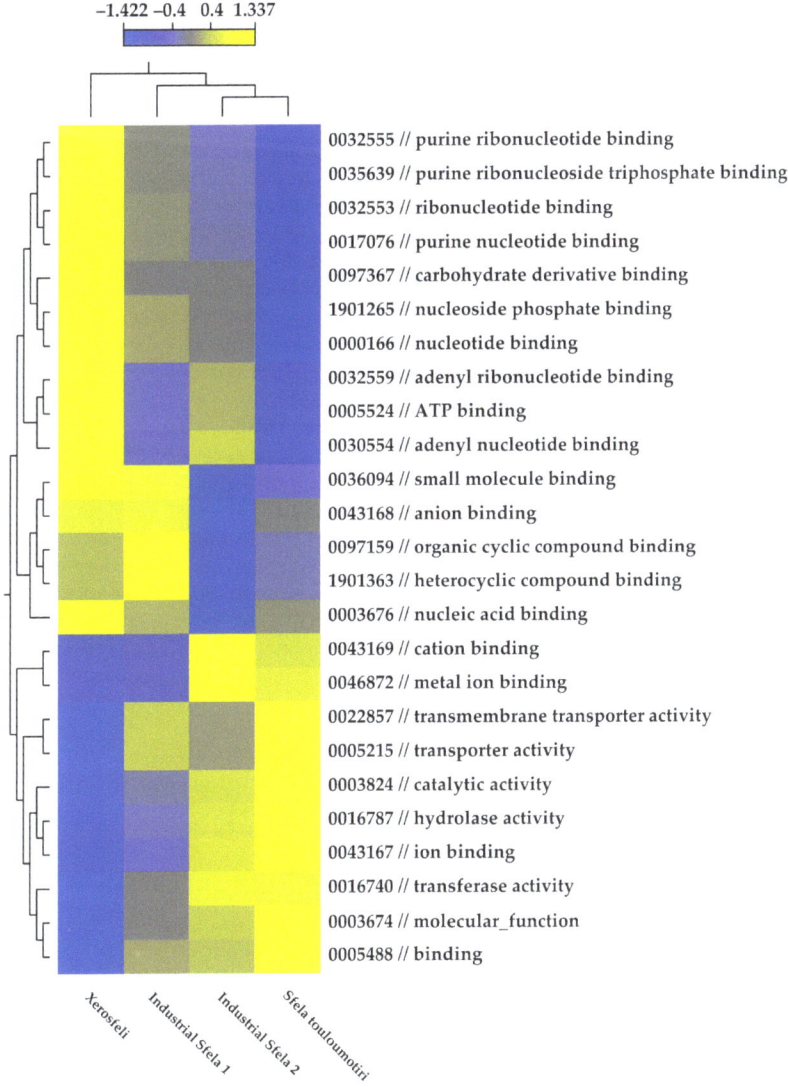

Figure 7. Heat map representing the results of functional analysis conducted on Sfela cheese samples according to GO molecular function. The analysis was conducted using the Euclidean clustering distance to group functional properties and the different cheese samples.

4. Conclusions and Future Perspectives

Sfela cheese has a PDO status, signifying that its distinct organoleptic characteristics are firmly connected to the Messinia and Laconia regions in the Peloponnese. In our study, we utilized a combination of 16S rDNA amplicon sequencing and shotgun metagenomics to evaluate four different Sfela samples: two produced on an industrial scale under PDO regulation and two non-PDO variants of Sfela cheese. Notably, industrial Sfela 1 and Xerosfeli were dominated by *Str. thermophilus* and *Lb. delbrueckii* subsp. *bulgaricus*, whereas industrial Sfela 2 and Sfela touloumotiri were characterized both by high populations of *L. lactis*, while *T. halophilus* prevailed in Sfela touloumotiri. As already mentioned, bacteria employed as starter cultures may persist in the first stages of cheese fermentation, while NSLAB may prevail at the later ripening stages due to the low pH and the elevated salt content. The dominance of *Str. thermophilus* and *Lb. delbrueckii* subsp. *bulgaricus* in industrial Sfela 1 and Xerosfeli may be attributed to production under strict hygiene conditions, enabling these bacteria to thrive during ripening against other microbes [35]. Metagenomics analysis also revealed that NSLAB were detected in all samples, but they were not among the most abundant species. The presence of yeast was rather scarce, as indicated by shotgun metagenomics. The binning of scaffolds did provide several sequence bins of varying completeness and quality. Notably, this analysis unveiled the presence of *Acinetobacter* spp., *Str. parauberis* and *Str. suis* in Sfela. This suggests the need for stricter surveillance during production, ensuring adherence to safety protocols, and compliance with PDO regulations to ensure food safety and consumers' health. Standardizing the production of traditional cheese can also potentially enhance their demand [75,76]. Omics approaches are critical as they can provide a very detailed picture of the microorganisms participating throughout the whole procedure of cheese production. Culture-dependent techniques also provide essential information about the product's dominant microbiome, and thus the combination of both methods can shed light on the dynamic changes that the microbial composition undergoes along with the functional pathways they carry. The information presented here provides a first framework concerning the intricate microbial communities that may exist in Sfela cheese. Further exploration may be needed to unveil a more comprehensive map of the Sfela cheese microbiome and its impact on sensory attributes. The construction of a pool of naturally occurring strains with desirable properties may lead to the isolation of novel strains for use as valuable starters or adjuncts specific to the production of Sfela cheese. Such research will not only expand our understanding of the Sfela cheese microbiome but also holds potential for enhancing its quality and safety in the future. Our study may also encourage the application of shotgun metagenomics to study the microbiomes of other types of white brined cheese produced in different parts of the world.

Author Contributions: Conceptualization, J.K., M.P. and K.P.; formal analysis, N.T., A.S., O.T. and K.P.; funding acquisition, J.K., M.P. and K.P.; investigation, N.T., A.S., O.T., A.G. and K.P.; methodology, J.K., M.P. and K.P.; supervision, K.P.; writing—original draft, N.T., A.S. and K.P.; writing—review and editing, N.T., A.S., O.T., A.G., S.G.D., J.Y., C.E.V., J.K., M.P. and K.P. All authors have read and agreed to the published version of the manuscript.

Funding: This research has been co-financed by the European Union and Greek national funds through the Operational Program Competitiveness, Entrepreneurship and Innovation under the call SUPPORT FOR REGIONAL EXCELLENCE (MIS 5047289).

Institutional Review Board Statement: Not applicable.

Informed Consent Statement: Not applicable.

Data Availability Statement: Publicly available datasets were analyzed in this study. This data can be found here: [SRA under BioProject IDs: PRJNA1072025, PRJNA1072084].

Conflicts of Interest: The authors declare no conflicts of interest.

References

1. McSweeney, P.L.H.; Ottogalli, G.; Fox, P.F. Diversity and classification of cheese varieties: An overview. In *Cheese*, 4th ed.; McSweeney, P.L.H., Fox, P.F., Cotter, P.D., Everett, D.W., Eds.; Academic Press: San Diego, CA, USA, 2017; pp. 781–808. [CrossRef]
2. Hayaloglu, A.A. Cheese varieties ripened under brine. In *Cheese*, 4th ed.; McSweeney, P.L.H., Fox, P.F., Cotter, P.D., Everett, D.W., Eds.; Academic Press: San Diego, CA, USA, 2017; pp. 997–1040. [CrossRef]
3. Moatsou, G.; Govaris, A. White brined cheeses: A diachronic exploitation of small ruminants milk in Greece. *Small Rumin. Res.* **2011**, *101*, 113–121. [CrossRef]
4. Nega, A.; Moatsou, G. Proteolysis and related enzymatic activities in ten Greek cheese varieties. *Dairy Sci. Technol.* **2011**, *92*, 57–73. [CrossRef]
5. Vakrou, A.; Fotopoulos, C.; Mattas, K. Location effects in the production and marketing of traditional Greek cheeses. In Proceedings of the European Association of Agricultural Economists: 52nd Seminar, Parma, Italy, 19–21 June 1997. [CrossRef]
6. Pappas, C.P.; Kondyli, E.; Voutsinas, L.P.; Mallatou, H. Effects of starter level, draining time and aging on the physicochemical, organoleptic and rheological properties of feta cheese. *IJDT* **1996**, *49*, 73–78. [CrossRef]
7. Litopoulou-Tzanetaki, E.; Tzanetakis, N.; Vafopoulou-Mastrojiannaki, A. Effect of the type of lactic starter on microbiological-chemical and sensory characteristics of Feta cheese. *Food Microbiol.* **1993**, *10*, 31–41. [CrossRef]
8. Angelopoulou, A.; Alexandraki, V.; Georgalaki, M.; Anastasiou, R.; Manolopoulou, E.; Tsakalidou, E.; Papadimitriou, K. Production of probiotic Feta cheese using *Propionibacterium freudenreichii* subsp. *shermanii* as adjunct. *Int. Dairy J.* **2017**, *66*, 135–139. [CrossRef]
9. Dimitrellou, D.; Kandylis, P.; Sidira, M.; Koutinas, A.A.; Kourkoutas, Y. Free and immobilized *Lactobacillus casei* ATCC 393 on whey protein as starter cultures for probiotic Feta-type cheese production. *J. Dairy Sci.* **2014**, *97*, 4675–4685. [CrossRef] [PubMed]
10. Mantzourani, I.; Terpou, A.; Alexopoulos, A.; Chondrou, P.; Galanis, A.; Bekatorou, A.; Bezirtzoglou, E.; Koutinas, A.A.; Plessas, S. Application of a novel potential probiotic *Lactobacillus paracasei* strain isolated from kefir grains in the production of Feta-type cheese. *Microorganisms* **2018**, *6*, 121. [CrossRef] [PubMed]
11. Papadopoulou, O.S.; Argyri, A.A.; Varzakis, E.E.; Tassou, C.C.; Chorianopoulos, N.G. Greek functional Feta cheese: Enhancing quality and safety using a *Lactobacillus plantarum* strain with probiotic potential. *Food Microbiol.* **2018**, *74*, 21–33. [CrossRef] [PubMed]
12. Tzanetakis, N.; Litopoulou-Tzanetaki, E. Changes in numbers and kinds of lactic acid bacteria in Feta and Teleme, two greek cheeses from ewes' milk. *J. Dairy Sci.* **1992**, *75*, 1389–1393. [CrossRef]
13. Bintsis, T.; Litopoulou-Tzanetaki, E.; Davies, R.O.N.; Robinson, R. Microbiology of brines used to mature Feta cheese. *Int. J. Dairy Technol.* **2000**, *53*, 106–112. [CrossRef]
14. Anastasiou, R.; Kazou, M.; Georgalaki, M.; Aktypis, A.; Zoumpopoulou, G.; Tsakalidou, E. Omics approaches to assess flavor development in cheese. *Foods* **2022**, *11*, 188. [CrossRef] [PubMed]
15. Hayaloglu, A.A.; Guven, M.; Fox, P.F. Microbiological, biochemical and technological properties of Turkish white cheese 'Beyaz Peynir'. *Int. Dairy J.* **2002**, *12*, 635–648. [CrossRef]
16. Bhowmik, T.; Marth, E.H. Rote of *Micrococcus* and *Pediococcus* Species in cheese ripening: A review. *J. Dairy Sci.* **1990**, *73*, 859–866. [CrossRef]
17. Fadda, M.E.; Cosentino, S.; Deplano, M.; Palmas, F. Yeast populations in Sardinian feta cheese. *Int. J. Food Microbiol.* **2001**, *69*, 153–156. [CrossRef]
18. Psomas, E.; Andrighetto, C.; Litopoulou-Tzanetaki, E.; Lombardi, A.; Tzanetakis, N. Some probiotic properties of yeast isolates from infant faeces and Feta cheese. *Int. J. Food Microbiol.* **2001**, *69*, 125–133. [CrossRef] [PubMed]
19. Rantsiou, K.; Urso, R.; Dolci, P.; Comi, G.; Cocolin, L. Microflora of Feta cheese from four Greek manufacturers. *Int. J. Food Microbiol.* **2008**, *126*, 36–42. [CrossRef]
20. El-Sharoud, W.M.; Belloch, C.; Peris, D.; Querol, A. Molecular identification of yeasts associated with traditional Egyptian dairy products. *J. Food Sci.* **2009**, *74*, 341–346. [CrossRef]
21. Jacques, N.; Casaregola, S. Safety assessment of dairy microorganisms: The hemiascomycetous yeasts. *Int. J. Food Microbiol.* **2008**, *126*, 321–326. [CrossRef]
22. Seiler, H.; Busse, M. The yeasts of cheese brines. *Int. J. Food Microbiol.* **1990**, *11*, 289–303. [CrossRef]
23. Ferrocino, I.; Rantsiou, K.; Cocolin, L. Investigating dairy microbiome: An opportunity to ensure quality, safety and typicity. *Curr. Opin. Biotechnol.* **2022**, *73*, 164–170. [CrossRef]
24. Kahraman-Ilıkkan, Ö.; Bağdat, E. Metataxonomic sequencing to assess microbial safety of Turkish white cheeses. *Braz. J. Microbiol.* **2022**, *53*, 969–976. [CrossRef] [PubMed]
25. Gezginc, Y.; Karabekmez-Erdem, T.; Tatar, H.D.; Dağgeçen, E.C.; Ayman, S.; Akyol, İ. Metagenomics and volatile profile of Turkish artisanal Tulum cheese microbiota. *Food Biosci.* **2022**, *45*, 101497. [CrossRef]
26. Ozturk, H.I.; Demirci, T.; Akin, N.; Ogul, A. Elucidation of the initial bacterial community of Ezine PDO cheese using next-generation sequencing. *Arch. Microbiol.* **2022**, *204*, 656. [CrossRef] [PubMed]
27. Nelli, A.; Venardou, B.; Skoufos, I.; Voidarou, C.C.; Lagkouvardos, I.; Tzora, A. An Insight into goat cheese: The tales of artisanal and industrial Gidotyri microbiota. *Microorganisms* **2023**, *11*, 123. [CrossRef] [PubMed]
28. Kamilari, E.; Tsaltas, D.; Stanton, C.; Ross, R.P. Metataxonomic mapping of the microbial diversity of Irish and Eastern Mediterranean cheeses. *Foods* **2022**, *11*, 2483. [CrossRef]

29. Michailidou, S.; Pavlou, E.; Pasentsis, K.; Rhoades, J.; Likotrafiti, E.; Argiriou, A. Microbial profiles of Greek PDO cheeses assessed with amplicon metabarcoding. *Food Microbiol.* **2021**, *99*, 103836. [CrossRef]
30. Sant'Anna, F.M.; Wetzels, S.U.; Cicco, S.H.S.; Figueiredo, R.C.; Sales, G.A.; Figueiredo, N.C.; Nunes, C.A.; Schmitz-Esser, S.; Mann, E.; Wagner, M.; et al. Microbial shifts in Minas artisanal cheeses from the Serra do Salitre region of Minas Gerais, Brazil throughout ripening time. *Food Microbiol.* **2019**, *82*, 349–362. [CrossRef]
31. Kamilari, E.; Anagnostopoulos, D.A.; Papademas, P.; Kamilaris, A.; Tsaltas, D. Characterizing Halloumi cheese's bacterial communities through metagenomic analysis. *LWT* **2020**, *126*, 109298. [CrossRef]
32. Papademas, P.; Aspri, M.; Mariou, M.; Dowd, S.E.; Kazou, M.; Tsakalidou, E. Conventional and omics approaches shed light on Halitzia cheese, a long-forgotten white-brined cheese from Cyprus. *Int. Dairy J.* **2019**, *98*, 72–83. [CrossRef]
33. Papadakis, P.; Konteles, S.; Batrinou, A.; Ouzounis, S.; Tsironi, T.; Halvatsiotis, P.; Tsakali, E.; Van Impe, J.F.M.; Vougiouklaki, D.; Strati, I.F.; et al. Characterization of bacterial microbiota of P.D.O. Feta cheese by 16S metagenomic analysis. *Microorganisms* **2021**, *9*, 2377. [CrossRef]
34. Spyrelli, E.; Stamatiou, A.; Tassou, C.; Nychas, G.-J.; Doulgeraki, A. Microbiological and metagenomic analysis to assess the effect of container material on the microbiota of Feta cheese during ripening. *Fermentation* **2020**, *6*, 12. [CrossRef]
35. Papadimitriou, K.; Anastasiou, R.; Georgalaki, M.; Bounenni, R.; Paximadaki, A.; Charmpi, C.; Alexandraki, V.; Kazou, M.; Tsakalidou, E. Comparison of the microbiome of artisanal homemade and industrial Feta cheese through amplicon sequencing and shotgun metagenomics. *Microorganisms* **2022**, *10*, 1073. [CrossRef] [PubMed]
36. Schmartz, G.P.; Hirsch, P.; Amand, J.; Dastbaz, J.; Fehlmann, T.; Kern, F.; Muller, R.; Keller, A. BusyBee Web: Towards comprehensive and differential composition-based metagenomic binning. *Nucleic Acids Res.* **2022**, *50*, W132–W137. [CrossRef] [PubMed]
37. Laczny, C.C.; Kiefer, C.; Galata, V.; Fehlmann, T.; Backes, C.; Keller, A. BusyBee Web: Metagenomic data analysis by bootstrapped supervised binning and annotation. *Nucleic Acids Res.* **2017**, *45*, 171–179. [CrossRef] [PubMed]
38. De Roos, J.; Vandamme, P.; De Vuyst, L. Wort substrate consumption and metabolite production during lambic beer fermentation and maturation explain the successive growth of specific bacterial and yeast species. *Front. Microbiol.* **2018**, *9*, 2763. [CrossRef] [PubMed]
39. Verce, M.; De Vuyst, L.; Weckx, S. Shotgun metagenomics of a water kefir fermentation ecosystem reveals a novel oenococcus species. *Front. Microbiol.* **2019**, *10*, 479. [CrossRef] [PubMed]
40. Gerhardt, K.; Ruiz-Perez, C.A.; Rodriguez, R.L.; Conrad, R.E.; Konstantinidis, K.T. RecruitPlotEasy: An advanced read recruitment plot tool for assessing metagenomic population abundance and genetic diversity. *Front. Bioinform.* **2021**, *1*, 826701. [CrossRef] [PubMed]
41. Metsalu, T.; Vilo, J. ClustVis: A web tool for visualizing clustering of multivariate data using principal component analysis and heatmap. *Nucleic. Acids Res.* **2015**, *43*, 566–570. [CrossRef] [PubMed]
42. Ledina, T.; Golob, M.; Djordjević, J.; Magas, V.; Colovic, S.; Bulajic, S. MALDI-TOF mass spectrometry for the identification of Serbian artisanal cheeses microbiota. *JCF* **2018**, *13*, 309–314. [CrossRef]
43. Sanchez-Juanes, F.; Teixeira-Martin, V.; Gonzalez-Buitrago, J.M.; Velazquez, E.; Flores-Felix, J.D. Identification of species and subspecies of lactic acid bacteria present in Spanish cheeses type "Torta" by MALDI-TOF MS and pheS gene analyses. *Microorganisms* **2020**, *8*, 301. [CrossRef]
44. Gantzias, C.; Lappa, I.K.; Aerts, M.; Georgalaki, M.; Manolopoulou, E.; Papadimitriou, K.; De Brandt, E.; Tsakalidou, E.; Vandamme, P. MALDI-TOF MS profiling of non-starter lactic acid bacteria from artisanal cheeses of the Greek island of Naxos. *Int. J. Food Microbiol.* **2020**, *323*, 108586. [CrossRef] [PubMed]
45. Tzora, A.; Nelli, A.; Voidarou, C.; Fthenakis, G.; Rozos, G.; Theodorides, G.; Bonos, E.; Skoufos, I. Microbiota "fingerprint" of Greek Feta cheese through ripening. *Appl. Sci.* **2021**, *11*, 5631. [CrossRef]
46. Coelho, M.C.; Malcata, F.X.; Silva, C.C.G. Lactic acid bacteria in raw-milk cheeses: From starter cultures to probiotic functions. *Foods* **2022**, *11*, 2276. [CrossRef] [PubMed]
47. Laranjo, M.; Potes, M.E. Chapter 6—Traditional Mediterranean cheeses: Lactic acid bacteria populations and functional traits. In *Lactic Acid Bacteria in Food Biotechnology*; Ray, R.C., Paramithiotis, S., de Carvalho Azevedo, V.A., Montet, D., Eds.; Elsevier: Amsterdam, The Netherlands, 2022; pp. 97–124. [CrossRef]
48. Özer, B. CHEESE | Microflora of white-brined cheeses. In *Encyclopedia of Food Microbiology*, 2nd ed.; Batt, C.A., Tortorello, M.L., Eds.; Academic Press: Oxford, MS, USA, 2014; pp. 402–408. [CrossRef]
49. Marino, M.; Innocente, N.; Maifreni, M.; Mounier, J.; Cobo-Diaz, J.F.; Coton, E.; Carraro, L.; Cardazzo, B. Diversity within Italian cheesemaking brine-associated bacterial communities evidenced by massive parallel 16S rRNA gene tag sequencing. *Front. Microbiol.* **2017**, *8*, 2119. [CrossRef] [PubMed]
50. Vermote, L.; Verce, M.; De Vuyst, L.; Weckx, S. Amplicon and shotgun metagenomic sequencing indicates that microbial ecosystems present in cheese brines reflect environmental inoculation during the cheese production process. *Int. Dairy J.* **2018**, *87*, 44–53. [CrossRef]
51. Morales, F.; Morales, J.I.; Hernandez, C.H.; Hernandez-Sanchez, H. Isolation and partial characterization of halotolerant lactic acid bacteria from two Mexican cheeses. *Appl. Biochem. Biotechnol.* **2011**, *164*, 889–905. [CrossRef] [PubMed]
52. Kim, D.H.; Kim, S.A.; Jo, Y.M.; Seo, H.; Kim, G.Y.; Cheon, S.W.; Yang, S.H.; Jeon, C.O.; Han, N.S. Probiotic potential of *Tetragenococcus halophilus* EFEL7002 isolated from Korean soy Meju. *BMC Microbiol.* **2022**, *22*, 149. [CrossRef] [PubMed]

53. Fornasari, M.E.; Rossetti, L.; Carminati, D.; Giraffa, G. Cultivability of *Streptococcus thermophilus* in Grana Padano cheese whey starters. *FEMS Microbiol. Lett.* **2006**, *257*, 139–144. [CrossRef]
54. Sola, L.; Quadu, E.; Bortolazzo, E.; Bertoldi, L.; Randazzo, C.L.; Pizzamiglio, V.; Solieri, L. Insights on the bacterial composition of Parmigiano Reggiano natural whey starter by a culture-dependent and 16S rRNA metabarcoding portrait. *Sci. Rep.* **2022**, *12*, 17322. [CrossRef]
55. Bozoudi, D.; Torriani, S.; Zdragas, A.; Litopoulou-Tzanetaki, E. Assessment of microbial diversity of the dominant microbiota in fresh and mature PDO Feta cheese made at three mountainous areas of Greece. *LWT-Food Sci. Technol.* **2016**, *72*, 525–533. [CrossRef]
56. Ruppitsch, W.; Nisic, A.; Hyden, P.; Cabal, A.; Sucher, J.; Stoger, A.; Allerberger, F.; Martinovic, A. Genetic diversity of *Leuconostoc mesenteroides* isolates from traditional Montenegrin brine cheese. *Microorganisms* **2021**, *9*, 1612. [CrossRef] [PubMed]
57. Lee, S.H.; Ku, H.J.; Ahn, M.J.; Hong, J.S.; Lee, S.H.; Shin, H.; Lee, K.C.; Lee, J.S.; Ryu, S.; Jeon, C.O.; et al. *Weissella jogaejeotgali* sp. nov., isolated from jogae jeotgal, a traditional Korean fermented seafood. *Int. J. Syst. Evol. Microbiol.* **2015**, *65*, 4674–4681. [CrossRef] [PubMed]
58. Rustemoglu, M.; Erkan, M.E.; Cengiz, G.; Hajyzadeh, M. Bacterial metagenome profiling of hand-made herby cheese samples utilizing high-throughput sequencing to detect geographical indication and marketing potential. *Heliyon* **2023**, *9*, 13334. [CrossRef] [PubMed]
59. Morea, M.; Baruzzi, F.; Cappa, F.; Cocconcelli, P.S. Molecular characterization of the *Lactobacillus* community in traditional processing of Mozzarella cheese. *Int. J. Food Microbiol.* **1998**, *43*, 53–60. [CrossRef] [PubMed]
60. El-Salam, M.H.A.; Alichanidis, E.; Zerfiridis, G.K. Domiati and Feta type cheeses. In *Cheese: Chemistry, Physics and Microbiology: Volume 2 Major Cheese Groups*; Fox, P.F., Ed.; Springer: Boston, MA, USA, 1993; pp. 301–335. [CrossRef]
61. Papademas, P.; Robinson, R.K. A comparison of the chemical, microbiological and sensory characteristics of bovine and ovine Halloumi cheese. *Int. Dairy J.* **2000**, *10*, 761–768. [CrossRef]
62. Yap, M.; O'Sullivan, O.; O'Toole, P.W.; Cotter, P.D. Development of sequencing-based methodologies to distinguish viable from non-viable cells in a bovine milk matrix: A pilot study. *Front. Microbiol.* **2022**, *13*, 1036643. [CrossRef]
63. Ruggirello, M.; Giordano, M.; Bertolino, M.; Ferrocino, I.; Cocolin, L.; Dolci, P. Study of *Lactococcus lactis* during advanced ripening stages of model cheeses characterized by GC-MS. *Food Microbiol.* **2018**, *74*, 132–142. [CrossRef] [PubMed]
64. Ammoun, I.; Kothe, C.I.; Mohellibi, N.; Beal, C.; Yaacoub, R.; Renault, P. Lebanese fermented goat milk products: From tradition to meta-omics. *Food Res. Int.* **2023**, *168*, 112762. [CrossRef]
65. Ganesan, B.; Stuart, M.R.; Weimer, B.C. Carbohydrate starvation causes a metabolically active but nonculturable state in *Lactococcus lactis*. *Appl. Environ. Microbiol.* **2007**, *73*, 2498–2512. [CrossRef]
66. Irlinger, F.; Loux, V.; Bento, P.; Gibrat, J.F.; Straub, C.; Bonnarme, P.; Landaud, S.; Monnet, C. Genome sequence of *Staphylococcus equorum* subsp. *equorum* Mu2, isolated from a French smear-ripened cheese. *J. Bacteriol.* **2012**, *194*, 5141–5142. [CrossRef]
67. Place, R.B.; Hiestand, D.; Gallmann, H.R.; Teuber, M. *Staphylococcus equorum* subsp. *linens*, subsp. nov., a starter culture component for surface ripened semi-hard cheeses. *Syst. Appl. Microbiol.* **2003**, *26*, 30–37. [CrossRef]
68. Yang, X. Moraxellaceae. In *Encyclopedia of Food Microbiology*, 2nd ed.; Batt, C.A., Tortorello, M.L., Eds.; Elsevier Ltd.: Amsterdam, The Netherlands, 2014; pp. 826–833. [CrossRef]
69. Seifert, H.; Strate, A.; Schulze, A.; Pulverer, G. Vascular catheter-related bloodstream infection due to *Acinetobacter johnsonii* (formerly *Acinetobacter calcoaceticus* var. *lwoffi*): Report of 13 cases. *Clin. Infect. Dis.* **1993**, *17*, 632–636. [CrossRef] [PubMed]
70. Bereswill, S.; Fuka, M.M.; Wallisch, S.; Engel, M.; Welzl, G.; Havranek, J.; Schloter, M. Dynamics of bacterial communities during the ripening process of different croatian cheese types derived from raw ewe's milk cheeses. *PLoS ONE* **2013**, *8*, e80734. [CrossRef]
71. de Garnica, M.L.; Sáez-Nieto, J.A.; González, R.; Santos, J.A.; Gonzalo, C. Diversity of gram-positive catalase-negative cocci in sheep bulk tank milk by comparative 16S rDNA sequence analysis. *Int. Dairy J.* **2014**, *34*, 142–145. [CrossRef]
72. Pitkala, A.; Koort, J.; Bjorkroth, J. Identification and antimicrobial resistance of *Streptococcus uberis* and *Streptococcus parauberis* isolated from bovine milk samples. *J. Dairy Sci.* **2008**, *91*, 4075–4081. [CrossRef] [PubMed]
73. Fortina, M.G.; Ricci, G.; Acquati, A.; Zeppa, G.; Gandini, A.; Manachini, P.L. Genetic characterization of some lactic acid bacteria occurring in an artisanal protected denomination origin (PDO) Italian cheese, the Toma piemontese. *Food Microbiol.* **2003**, *20*, 397–404. [CrossRef]
74. Lun, Z.R.; Wang, Q.P.; Chen, X.G.; Li, A.X.; Zhu, X.Q. *Streptococcus suis*: An emerging zoonotic pathogen. *Lancet Infect. Dis.* **2007**, *7*, 201–209. [CrossRef]
75. Danezis, G.P.; Tsiplakou, E.; Pappa, E.C.; Pappas, A.C.; Mavrommatis, A.; Sotirakoglou, K.; Georgiou, C.A.; Zervas, G. Fatty acid profile and physicochemical properties of Greek protected designation of origin cheeses, implications for authentication. *Eur. Food Res. Technol.* **2020**, *246*, 1741–1753. [CrossRef]
76. Pappa, E.C.; Kondyli, E. Descriptive characteristics and cheesemaking technology of Greek cheeses not listed in the EU geographical indications registers. *Dairy* **2023**, *4*, 43–67. [CrossRef]

Disclaimer/Publisher's Note: The statements, opinions and data contained in all publications are solely those of the individual author(s) and contributor(s) and not of MDPI and/or the editor(s). MDPI and/or the editor(s) disclaim responsibility for any injury to people or property resulting from any ideas, methods, instructions or products referred to in the content.

Article

Associations between the Bacterial Composition of Farm Bulk Milk and the Microbiota in the Resulting Swedish Long-Ripened Cheese

Li Sun [1,*], Annika Höjer [2], Monika Johansson [1], Karin Hallin Saedén [2], Gun Bernes [3], Mårten Hetta [3], Anders H. Gustafsson [4], Johan Dicksved [5] and Åse Lundh [1]

[1] Department of Molecular Sciences, Swedish University of Agricultural Sciences, SE-750 07 Uppsala, Sweden; monika.johansson@slu.se (M.J.); ase.lundh@slu.se (Å.L.)
[2] Norrmejerier Ek. Förening, Mejerivägen 2, SE-906 22 Umeå, Sweden; annika.hojer@norrmejerier.se (A.H.); karin.hallin-saeden@norrmejerier.se (K.H.S.)
[3] Department of Animal Nutrition and Management, Swedish University of Agricultural Sciences, SE-901 83 Umeå, Sweden; gun.bernes@slu.se (G.B.); marten.hetta@slu.se (M.H.)
[4] Växa Sverige, Ulls väg 29 A, SE-751 05 Uppsala, Sweden; anders.h.gustafsson@vxa.se
[5] Department of Animal Nutrition and Management, Swedish University of Agricultural Sciences, SE-750 07 Uppsala, Sweden; johan.dicksved@slu.se
* Correspondence: li.sun@slu.se

Citation: Sun, L.; Höjer, A.; Johansson, M.; Saedén, K.H.; Bernes, G.; Hetta, M.; Gustafsson, A.H.; Dicksved, J.; Lundh, Å. Associations between the Bacterial Composition of Farm Bulk Milk and the Microbiota in the Resulting Swedish Long-Ripened Cheese. *Foods* **2023**, *12*, 3796. https://doi.org/10.3390/foods12203796

Academic Editors: Paolo Formaggioni and Piero Franceschi

Received: 11 September 2023
Revised: 10 October 2023
Accepted: 12 October 2023
Published: 16 October 2023

Copyright: © 2023 by the authors. Licensee MDPI, Basel, Switzerland. This article is an open access article distributed under the terms and conditions of the Creative Commons Attribution (CC BY) license (https:// creativecommons.org/licenses/by/ 4.0/).

Abstract: The maturation of a traditional Swedish long-ripened cheese has shown increasing variation in recent years and the ripening time is now generally longer than in the past. While the cheese is reliant on non-starter lactic acid bacteria for the development of its characteristic flavour, we hypothesised that the observed changes could be due to variations in the microbiota composition and number of bacteria in the raw milk used for production of the cheese. To evaluate associations between microbiota in the raw milk and the resulting cheese, three clusters of commercial farms were created to increase variation in the microbiota of dairy silo milk used for cheese production. Cheese production was performed in three periods over one year. Within each period, milk from the three farm clusters was collected separately and transported to the cheese production facility. Following pasteurisation, the milk was processed into the granular-eyed cheese and matured at a dedicated cheese-ripening facility. For each cheese batch, farm bulk and dairy silo milk samples, a starter culture, early process samples and cheese samples from different stages of maturation (7–20 months) were collected and their microbiota characterised using 16S rRNA amplicon sequencing. The microbiota in the farm bulk milk differed significantly between periods and clusters. Differences in microbiota in dairy silo milk were observed between periods, but not between farm clusters, while the cheese microbiota differed between periods and clusters. The top 13 amplicon sequence variants were dominant in early process samples and the resulting cheese, making up at least 93.3% of the relative abundance (RA). *Lactococcus* was the dominant genus in the early process samples and, together with *Leuconostoc*, also dominated in the cheese samples. Contradicting expectations, the RA of the aroma-producing genus *Lactobacillus* was low in cheese during ripening and there was an unexpected dominance of starter lactic acid bacteria even at the later stages of cheese ripening. To identify factors behind the recent variations in ripening time of this cheese, future studies should address the effects of process variables and the dairy environment.

Keywords: farm and dairy silo milk microbiota; starter and non-starter lactic acid bacteria; *Lactobacillus*; *Lactococcus*; cheese ripening

1. Introduction

The production of long-ripened extra-hard cheese is a complex process in which microbial and biochemical processes contribute to the characteristic flavour and texture of the cheese [1]. The metabolic activity in the cheese core during ripening derives from native enzymes in the milk, added rennet, added lactic acid bacteria (LAB) starter culture and adventitious non-starter lactic acid bacteria (NSLAB). The major LAB metabolic pathways involved in flavour formation in cheese during ripening can be categorised into primary (lipolysis, proteolysis and metabolism of residual lactose, lactate and citrate) and secondary (metabolism of fatty acids and amino acids) pathways [1]. Proteolysis is of the greatest importance for the final texture and flavour of the cheese, and peptides and amino acids resulting from activities of the starter culture provide the main nutritional compounds for NSLAB. The amino acids also act as precursors in a series of catabolic reactions, resulting in volatile aroma compounds, e.g., alcohols, aldehydes, ketones, esters and phenolic and sulphur compounds [2].

The NSLAB in cheese is believed to originate from the milk, either by surviving pasteurisation or by contaminating the pasteurised milk later in the manufacturing process via the dairy environment, e.g., a facility-specific "house" microbiota [3]. NSLAB is a heterogeneous group of mesophilic bacteria consisting of, e.g., *Lactobacillus*, *Pediococcus* and *Enterococcus*. Species of facultative heterofermentative lactobacilli, e.g., *Lactobacillus casei*, *Lactobacillus paracasei*, *Lactobacillus plantarum*, *Lactobacillus curvatus* and *Lactobacillus rhamnosus*, predominate in Cheddar-type cheeses [4]. *L. rhamnosus*, *L. casei* and *L. paracasei*, commonly referred to as the *L. casei* group, are highly abundant during the ripening of extra-hard, cooked cheeses such as Grana Padano [5]. These bacteria are present in low numbers and do not grow well in raw milk, but dominate the viable bacterial population in the maturing cheese [6]. However, dependence on adventitious NSLAB introduces variability into the ripening process, resulting in differences between cheeses produced at the same cheese-making plant on different days, and even in different batches on the same day [7].

In recent years, there has been increasing variation in the maturation of a Swedish traditional long-ripened cheese and the ripening time is now generally longer than in the past. Stricter hygiene on farms, including pre-milking, udder-cleaning, and teat-disinfection routines, have likely contributed to reducing the total number of bacteria in bulk milk and therefore also the number of NSLAB in the milk, with consequences for the cheese maturation process. In a recent study, we explored farm-related factors contributing to the variation in the microbial community in bulk milk from different dairy farms [8]. The results revealed an effect of routines associated with teat preparation and cleaning of the milking equipment on bulk milk microbiota, with milk from farms using an automated milking system (AMS) having different microbial composition than milk from tie-stall farms. We also observed a difference in the microbial composition between milk from AMS farms using different brands of milking robots, which was likely explained by differences in the performance of the robots. In contrast, Doyle et al. [9] concluded that teat preparation has a limited impact on raw milk microbiota and that the herd habitat is the major driver of milk microbiota composition. Gagnon et al. [10] investigated silage as a contamination source of facultative heterofermentative LAB in milk and concluded that silage is probably a minor contributor. We found similar results in a recent study (unpublished data). In practice, NSLAB are difficult to control and their pathways to the milk are diverse. For this reason, knowledge of the diversity and abundance of NSLAB species and factors affecting their presence in the raw milk is essential for the successful production of long-ripened cheese.

The aim of this study was to analyse the microbiota in bulk milk and determine its contribution to the microbiota in dairy silo milk and in the resulting long-ripened cheese.

2. Materials and Methods

2.1. Participating Farms

The participating farms were located in a region between 64°1′ to 64°9′ N and 20°5′ to 21°4′ E in the county of Västerbotten, Sweden. All selected farms routinely delivered their milk to the participating cheese-making facility. To introduce greater variation in the milk used for cheese-making, three different clusters of farms (A, B and C) producing milk differing in various aspects were created. In total, 18 farms were selected based on data on detailed milk composition and farm characteristics obtained in a previous study [11]. Cluster A consisted of farms delivering milk of average quality in terms of composition, total bacteria count and microbiota. Cluster B consisted of mostly tie-stall farms delivering milk characterised by higher fat and protein content and fewer clostridia. Milk from farms in cluster C typically had higher total numbers of bacteria and larger proportions of lactic acid bacteria (Table 1).

Table 1. Milk selection criteria and basic characteristics associated with farms in clusters A–C. Farms belonging to cluster B had significantly smaller herd size than farms in clusters A and C and fed a significantly higher proportion of silage as round bales than farms in cluster A.

Farm Cluster and Selection Criteria	Number of Farms	Average Number of Cows per Herd (Min–Max) [1]	Dominant Cow Breed [2] on the Farm (Number of Farms)	Milking System (Number of Farms)	Proportion of Silage Fed as Round Bales [1]
A: Average milk quality	4	110 [a] (68–180)	SH (3) Mixed (1)	AMS (3) Milking parlour (1)	29% [b]
B: Higher fat and protein content, fewer clostridia in milk	9	49 [b] (17–90)	SH (1) SRB (2) SKB (1) SJB (1) Mixed (4)	Tie-stall (6) AMS (2) Milking parlour (1)	73% [a]
C: Higher total bacteria number, higher abundance of lactic acid bacteria in milk	5	84 [a] (41–127)	SH (4) SKB (1)	AMS (3) Milking parlour (2)	60% [ab]

[1] Values within columns with different superscript letters differ significantly ($p < 0.05$). [2] Dominant cow breed, i.e., >70% of cows in the herd: SH = Swedish Holstein, SRB = Swedish Red, SKB = Swedish Mountain breed or SJB = Swedish Jersey. AMS = automatic milking system.

2.2. Cheese Production Practice

Cheese production was performed in three periods: November 2017 (period 1), February–March 2018 (period 2) and September 2018 (period 3). In each period, bulk milk from the farms belonging to each of the three clusters was collected separately and transported to a dedicated dairy silo at the cheese-making facility. This was repeated on two separate days during one week per period. The total volume of dairy silo milk originating from each farm cluster was approximately 15,000 L, which was the volume needed for full-scale production of one batch of cheese. After pasteurisation (72 °C, 15 s), the milk was processed into a Swedish traditional, granular-eyed long-ripened cheese, using a mesophilic starter and a coagulant consisting of 75/25 bovine chymosin and pepsin (180 IMCU, Sacco System Nordic [Kemikalia], Skurup, Sweden). Cheese making comprised long cooking periods (several hours) at temperatures above 40 °C. The cheeses were produced in 18 kg cylinders (16 cm high), brine-salted to a salt content of around 1.2%, waxed and ripened at specific temperatures between 10° and 13 °C at a dedicated cheese-ripening facility. Thus, in each of the three periods, milk from each farm cluster was used for one batch of full-scale cheese production on two separate days, resulting in a total of 18 cheese batches (3 periods × 3 clusters × 2 production days × 1 batch per day).

2.3. Sampling of Farm Bulk Milk, Dairy Silo Milk and the Resulting Cheese

On each day of cheese production, individual farm bulk milk (250 mL) was sampled by the tanker driver during milk collection on the farms. These milk samples were kept in a cool box during transportation to the dairy facility. The raw dairy silo milk (250 mL) used for cheese production was sampled before processing. During cheese production, samples of freshly propagated bulk starter (50 mL), milk gel (50 mL) and cheese grains before pressing (40–50 g) were collected in the early cheese-making process (early process samples, EPSs). Fresh cheese at 24 h, cheese matured for 7 months and cheese matured for 12–20 months (with 2-month intervals) were also sampled. On each day of production, approximately 80 cheese wheels were produced from the milk collected from a farm cluster and of these, at least four cheese wheels produced in the middle of the batch were reserved for this study. On each cheese sampling occasion, one drill core of cheese (25–30 g) was extracted for microbiota analysis. An individual cheese was never sampled on more than four occasions.

The farm bulk and dairy silo milk samples were stored at 4 °C at the dairy facility until transportation to SLU, Uppsala, Sweden. Samples were transported at ambient temperature using cooling pads, and upon arrival, all milk samples were aliquoted into 2-mL tubes and stored at −80 °C until DNA extraction. The maximum time from sampling of bulk and silo milk to storage at −80 °C was 24 h. The EPSs and cheese samples were stored at −60 °C at the dairy facility until transportation to SLU, Uppsala. Upon arrival, all samples were stored at −80 °C until DNA extraction. In addition, fresh cheese samples aged for 12 to 20 months (with 2 months interval) were transported refrigerated to SLU, Uppsala, and stored at 4 °C until analysis of total bacteria count. The maximum time from sampling of drill cores of the cheese to analysis of the fresh samples was 48 h. Due to practical circumstances, the EPSs and fresh cheese samples at 24 h were missing in period 2 for cluster B on cheese production day 2.

2.4. Total Bacteria Count in Farm Bulk Milk, Dairy Silo Milk and Cheese

Total bacteria count in farm bulk milk was analysed at Eurofins Food & Feed Testing Sweden AB (Jönköping, Sweden) using Bactoscan FC (Foss, Hilleroed, Denmark), while bacteria in the dairy silo milk samples were determined by culturing on plate count agar (PCA; Casein-peptone Dextrose Yeast Agar, Merck KGaA, Darmstadt, Germany) followed by incubation at 30 °C for 3 days at the cheese-making facility. Total bacteria count in fresh cheese samples was analysed at SLU, Uppsala. In brief, 25 g of cheese was homogenised in 100 mL of phosphate-buffered saline (PBS, pH 7.4) for 2 min at normal speed in a stomacher bag (400 Classic, Seward, West Sussex, UK) in a stomacher blender (Stomacher 400, Seward, AK, USA) at room temperature. Decimal dilutions in PBS were prepared, 0.1 mL aliquots of the appropriate dilutions were inoculated in duplicate on plate count agar and incubated at 30 °C for 48 h to obtain viable counts as colony forming unit (cfu) per gram of cheese.

2.5. DNA Extraction from Bacteria in Milk and Cheese

Liquid samples (milk, starter culture and milk gel) were thawed at 37 °C for 15 min in a water bath. Solid samples (cheese granules, cheese of different ages) were thawed at room temperature for 15 min. Milk gel and solid cheese samples (25 g) were homogenised as previously described for analysis of total bacteria count. For cheese grains before pressing and cheese aged for 24 h, the homogenisation time was extended to 6 min. DNA extraction was conducted using a PowerFood DNA isolation kit (Qiagen AB, Sollentuna, Sweden) according to a customised protocol. In brief, a 1.8 mL portion of whole milk or a homogenised aliquot was centrifuged at $13,000\times g$ for 15 min at 4 °C, and then incubated on ice for 5 min. The resulting cell pellets with carefully collected fat layer were resuspended in 450 µL of MBL buffer (provided with kit). The resuspended mixture was transferred to MicroBead tubes (provided with kit). Cell lysis was conducted by incubating the tubes at 65 °C for 10 min, after which they were processed in a Fastprep 24 instrument (MP Biomedicals, Santa Ana, CA, USA) at 5.0 speed for 60 s, repeated two times with a 5-min pause. The tubes were then centrifuged at $13,000\times g$ for 15 min at 4 °C, followed by incubation on ice for 5 min. The supernatant excluding the fat layer was transferred to new 2 mL collection tubes and the remaining steps were carried out according to the manufacturer's protocol. The resulting DNA was eluted with 50 µL of buffer EB (provided with kit) and stored at −20 °C until use.

2.6. Illumina Amplicon Library Construction, Sequencing and Bioinformatic Analysis

The DNA extracted from the milk, EPSs and cheese samples was used to construct a 16S rRNA gene library with primers 515F and 805R [12]. Illumina adaptors and barcodes were used for amplification, following a two-step PCR approach previously described [13]. The 16S rRNA gene library was sequenced using the Illumina Miseq platform at SciLifeLab (Stockholm, Sweden). The raw sequencing data have been deposited in the Sequence Read Archive at the National Center for Biotechnology Information database (http://www.ncbi.nlm.nih.gov/sra, (accessed on 10 September 2023)), under accession number PRJNA1010645. Bioinformatic data processing was performed using Quantitative Insights into Microbial Ecology 2 (Core 2020.11) [14]. The raw demultiplexed reads were trimmed using Cutadapt to remove primer sequences [15]. Sequencing base with quality score below 30 was trimmed off from the 3′ end. A read was discarded if it contained N base or did not contain primer sequences. The trimmed reads were further processed using DADA2 to de-noise, de-replicate reads, merge pair end reads and remove chimeras [16], using a truncation length of 210 and 160 bp for forward and reverse reads, respectively. A phylogenetic tree was built using FastTree and MAFFT alignment [17,18]. The SSU Ref NR 99 138 dataset was first trimmed to the corresponding primer region and used for training the classify-sklearn taxonomy classifier [19–21]. Amplicon sequence variants (ASVs) were assigned taxonomy using the resulting classifier. The weighted UniFrac distance matrix and alpha rarefaction were generated using the QIIME2 diversity plugin [14]. To identify the dominant ASVs present in farm milk, silo milk, EPS and cheese, the top 30 ASVs with relative abundance (RA) higher than 0.04% in the total sequencing pool were selected for analysis.

2.7. Statistical Analysis

Microbiota analyses were performed with the q2-diversity plugin. The rarefied ASV table was used to calculate the number of observed ASVs. Kruskal–Wallis rank test [22] and Benjamini and Hochberg (B-H) correction [23] were used to identify statistical differences in RA of the ASVs between groups, i.e., periods and clusters. Principal coordinate analysis (PCoA) was used to visualise differences in microbial composition based on the generalised UniFrac distances. Permutational multivariate analysis of variance (PERMANOVA) testing of generalised UniFrac distance matrix with (B-H) correction [24] was conducted to evaluate differences between groups.

3. Results

The alpha diversity of the microbiota in farm bulk and dairy silo milk was higher ($p < 0.001$) than that of the microbiota in EPSs and cheese in terms of the number of observed ASVs (Figure 1a) and the Shannon index ($p < 0.001$). The beta diversity of the microbiota present in farm bulk and dairy silo milk was also different from that in EPSs and cheese ($p < 0.001$), as revealed using the generalised UniFrac distance (Figure 1b).

3.1. Exploring the Microbiota Present in Farm Bulk and Dairy Silo Milk

The microbiota in farm bulk milk showed significant differences between periods and clusters (Figure 2a,b, both $p < 0.001$). A pairwise comparison confirmed that each period was different from the others ($p < 0.001$). In the pairwise comparison of the microbiota in bulk milk from the different clusters, a difference was observed between clusters A and B ($p < 0.05$), and between clusters B and C ($p < 0.01$). However, there was no difference between clusters A and C ($p > 0.05$) and no clear separation in the PCoA plot (Figure 2b). On comparing the microbiota in the dairy silo milk samples, differences between periods were still evident (Figure 3, $p < 0.05$). However, in the pairwise comparison of the microbiota in dairy silo milk during different periods, the only significant difference observed was between periods 1 and 2 ($p < 0.05$). There was no significant difference in microbiota in dairy silo milk between the farm clusters.

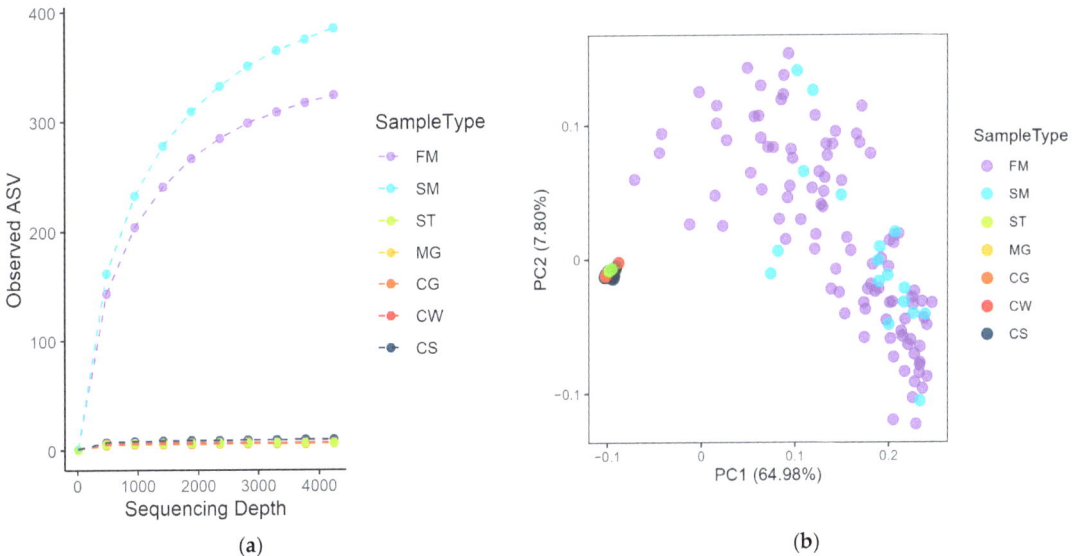

Figure 1. (a) Rarefaction curves of observed amplicon sequence variants (ASVs) and (b) principal coordinate analysis plot of generalised UniFrac distance associated with farm bulk milk (FM), dairy silo milk (SM), starter culture (ST), milk gel (MG), cheese grains (CG), fresh cheese aged for 24 h (CW) and cheese samples ripened for 7–20 months (CS).

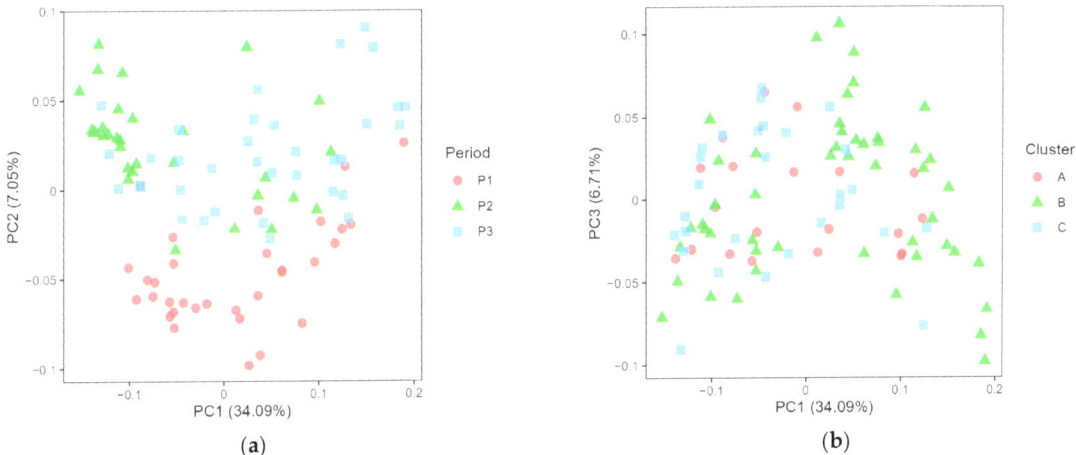

Figure 2. Principal coordinate analysis plot of generalised UniFrac distance comparing the microbiota of farm bulk milk according to (a) the different cheese production periods (P1–P3) and (b) the different farm clusters (A–C).

The microbiota in the dairy silo milk resulted from a combination of volume, total bacteria count and microbiota composition of the bulk milk delivered from the individual farms in the same cluster. This was illustrated in Figure 4 for farm no. 20 (F20) in cluster C, with an elevated total bacteria count and a high relative abundance (RA) of *Streptococcus* bd2e in the bulk milk used on the second cheese-making day (D2) in period 1. Since the volume of milk delivered from this farm made up 40% of the total volume of the dairy silo, *Streptococcus* bd2e also had a high RA in the dairy silo milk (SM).

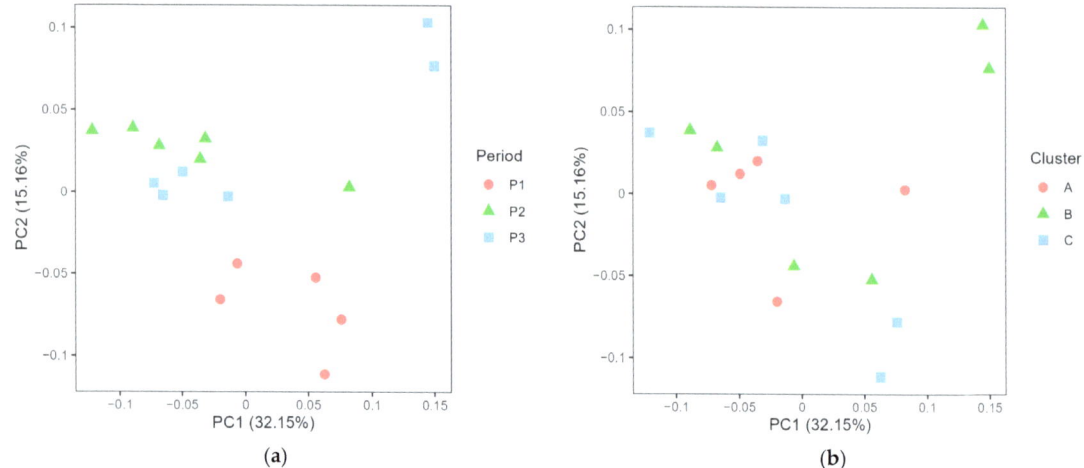

Figure 3. Principal coordinate analysis plot of generalised UniFrac distance comparing the microbiota of dairy silo milk according to (**a**) the different cheese production periods (P1–P3) and (**b**) the different farm clusters (A–C). One of the dairy silo milk samples for period 1, cluster A, is missing.

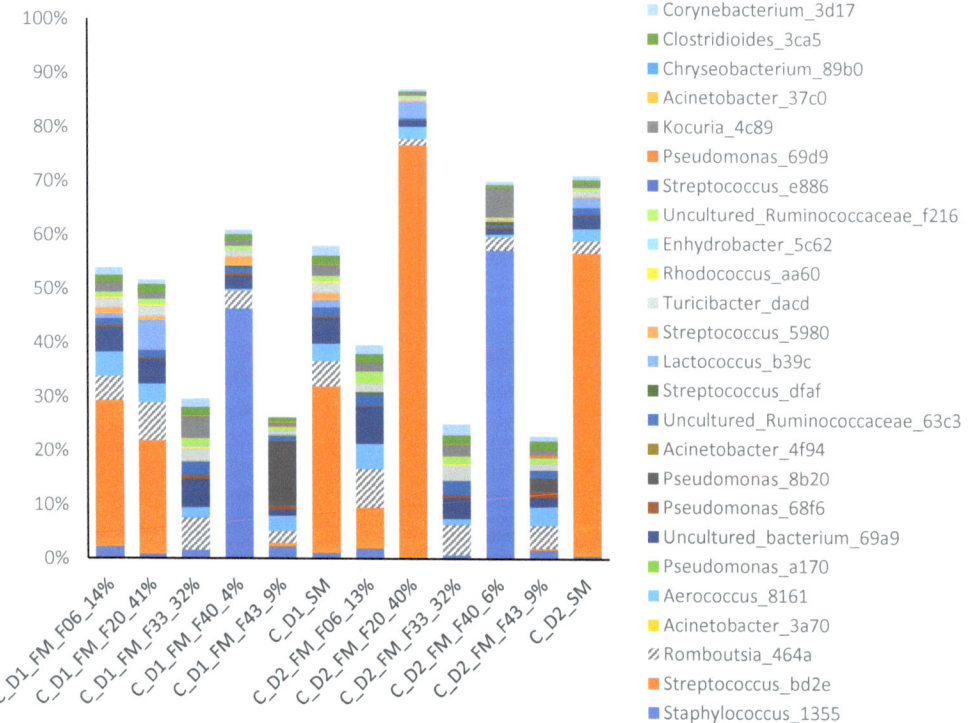

Figure 4. Relative abundance (RA, %) of the top 20 ASVs in bulk milk (FM) samples from the five farms (F06, F20, F33, F40 and F43) in cluster C, and the corresponding dairy silo milk (SM) samples on cheese-making days 1 and 2 (D1, D2) in period 1. The percentage given in the sample labels is the relative proportion of bulk milk from each individual farm out of the total milk volume in the dairy silo.

3.2. Exploring the Microbiota in Cheese

The PCoA of generalised UniFrac distance showed significant differences in cheese microbiota between the periods (Figure 5a, $p < 0.001$), and the pairwise comparison confirmed that all periods differed from each other ($p < 0.001$). There was also a significant difference in cheese microbiota between the farm clusters (Figure 5b, $p < 0.001$), with pairwise comparisons confirming that all clusters differed from each other ($p < 0.01$).

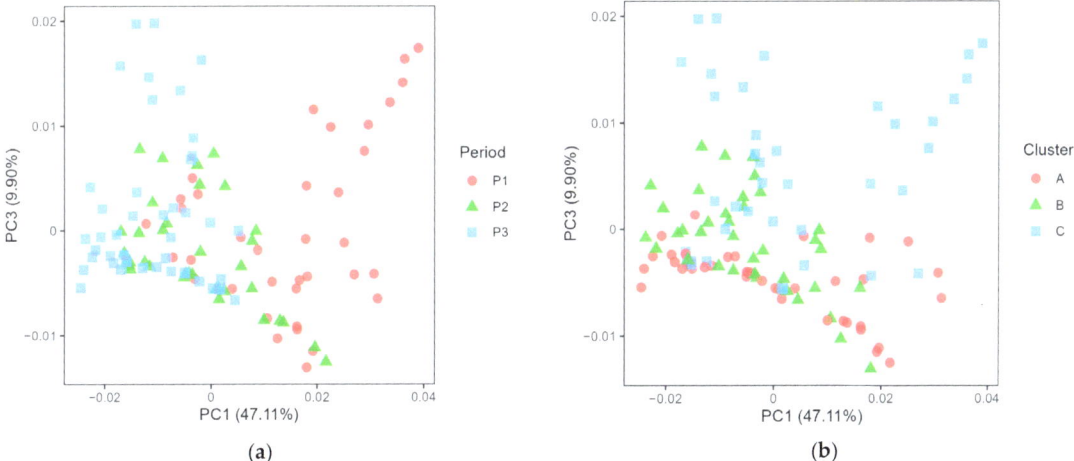

Figure 5. Principal coordinate analysis plot of generalised UniFrac distance comparing the microbiota in different cheeses depending on (**a**) cheese production period (P1–P3) and (**b**) farm cluster (A–C).

The total bacteria count in cheese samples aged for 12 to 20 months varied from log 2.82 to 6.74 cfu/g cheese (Figure 6). There was fluctuation in total bacteria counts in cheese during this stage of maturation, but there was a clear trend suggesting that cluster A had a lower total bacteria count in all three periods.

To investigate the development of the microbiota in the early stages of the cheese-making process and during ripening, samples of the starter culture, milk gel, cheese grains and cheese aged for 24 h and 7–20 months were analysed (Figure 7). The top 13 ASVs were dominant in both early process and cheese samples, making up at least 93.3% of the RA. *Lactococcus* was the dominant genus in EPSs and in cheese aged for 24 h. In cheese samples aged for 7 to 20 months, *Lactococcus* and *Leuconostoc* were the two dominant genera. In some batches of cheese, *Leuconostoc* already started to increase in RA in the young cheese (24 h). In contrast, the genus *Lactobacillus* was not among the most abundant genera in most samples. However, this genus was detected in all cheeses, on at least one sampling occasion for each batch of cheese and at a higher RA in some batches of cheese, especially in period 1. *Streptococcus* was more common in cheese from cluster C and the RA of this genus was higher in cheese from periods 1 and 3. *Acinetobacter* was occasionally detected, with a higher RA in one batch of cheese produced in period 1 (cluster C). Several ASVs within the genera *Lactococcus*, *Leuconostoc* and *Lactobacillus* were found. Within the genus *Lactobacillus*, ASV c982 was the most commonly observed ASV, while ASV 2653 and 3486 were present at a high RA in a few cheese batches, in most cases associated with cluster B. *Lactococcus* b39c was dominant compared with the other four ASVs in the same genus, i.e., d90e, 072f, bb15 and cc1f. It is worth noting that *Lactococcus* bb15 and cc1f were more commonly found in periods 1 and 2. For the genus *Leuconostoc*, in most batches ASV 7f62 had a higher RA than ASV 24cd, although the latter was present at a high RA in some of the batches, e.g., in period 2, cluster B.

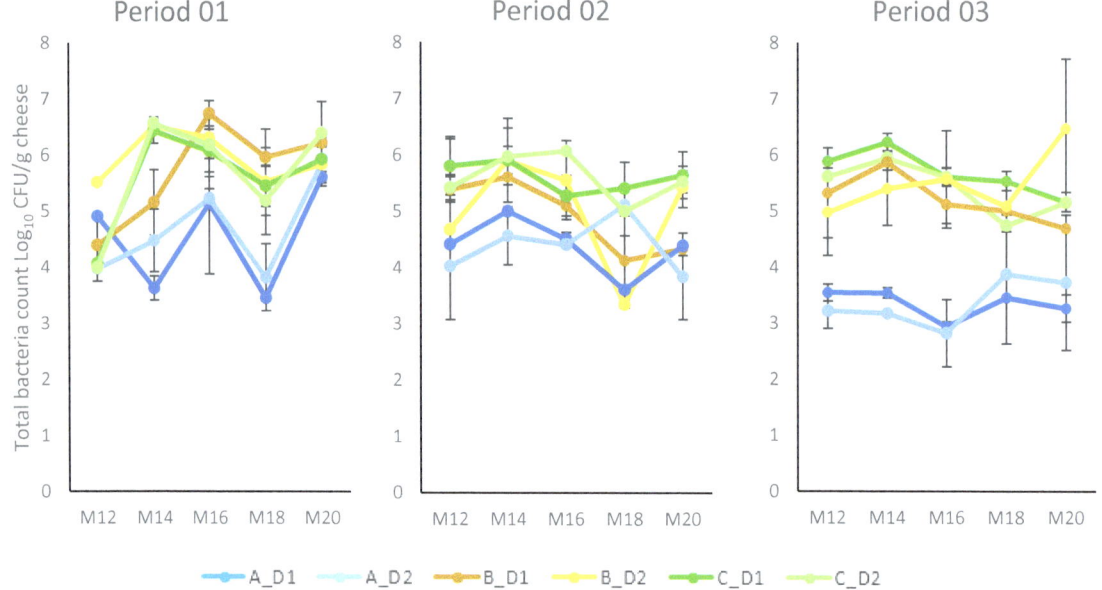

Figure 6. Total bacteria count in cheese aged for 12 to 20 months (M12–M20). The cheeses were produced using dairy silo milk from three different farm clusters (A–C) in a full-scale production setting. Cheese production was performed in three periods (P1–P3) over one year, and within each period milk from the three farm clusters was collected, and cheese production was performed, on two separate days (D1, D2).

3.3. Common ASVs Found in Different Sample Types

Nine of the top 13 ASVs found in early process and cheese samples (Figure 7) were also identified in the dairy silo milk samples (Figure 8). Most of the ASVs that were common in cheese had rather low RA in the milk. *Streptococcus* bd2e was the exception, and this ASV had a higher RA in dairy silo milk samples from cluster C than in samples from the other two clusters in all three periods. As previously mentioned, cheese samples associated with cluster C had a higher RA of this ASV. The four ASVs observed in cheese but not in dairy silo milk were *Acinetobacter* 6933, *Lactobacillus* 2653, *Lactococcus* bb15 and *Lactococcus* 072f.

To avoid the risk of overlooking ASVs common to both silo milk and cheese due to sequencing limitations, e.g., poor sequencing depth, screening for the ASVs common to cheese and the associated farm bulk milk, dairy silo milk and EPSs was conducted (Table 2). In this screening, the *Lactobacillus* 0b36, *Lactobacillus* 57c1 and *Tetragenococcus* 49a8 present in the cheese were also found in the farm bulk milk, but not in the dairy silo milk or EPS, and only at a very low frequency. Screening for the ASVs present in both EPSs and cheese showed that the previously mentioned *Lactococcus* bb15 and *Lactococcus* 072f were present at high frequencies in EPSs, including in the starter culture. *Acinetobacter* 6933 and *Lactobacillus* bf19 were identified in EPSs, but not in farm bulk or dairy silo milk, and were limited to one batch of production (period 1, cluster C) where *Lactobacillus* was identified in the starter culture. However, *Lactobacillus* 2653, *Lactococcus* 6fa5 and *Lactococcus* ffd9 were only identified in the cheese.

Figure 7. Relative abundance (RA, %) of the top 13 ASVs present in early process and cheese samples. Cheese production was performed in three periods (P1–P3) over one year. Within each period, milk from three farm clusters (A–C) was collected, and cheese was produced on two separate days within one week (D1, D2). Samples were taken at different time points in the cheese production process, including starter culture (ST), milk gel (MG), cheese grains (CG), fresh cheese at 24 h (CW) and cheese after ripening for 7 up to 20 months (M07–M20).

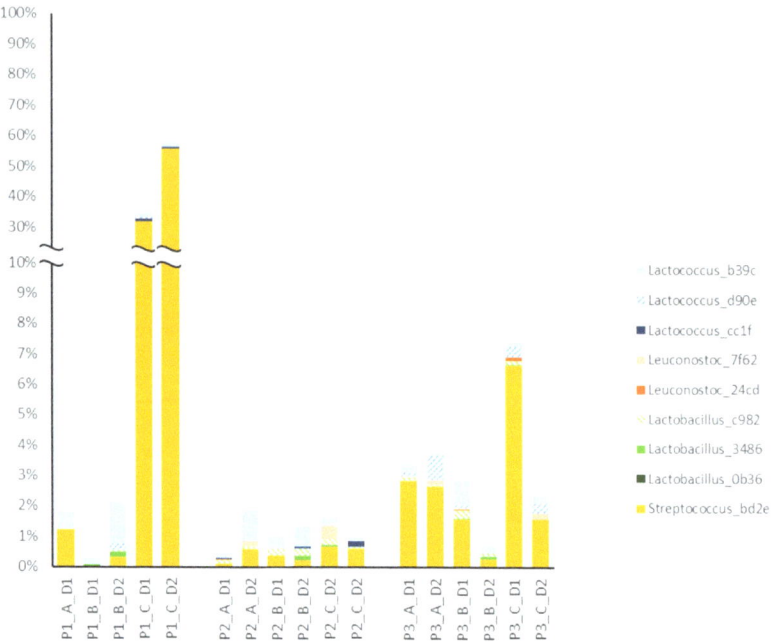

Figure 8. Relative abundance (RA, %) in dairy silo milk samples of nine ASVs that were among the top 13 ASVs found in cheese and also identified in the silo milk. Cheeses were produced using dairy silo milk from three different clusters of farms (A–C) in a full-scale production setting in three periods (P1–P3) distributed over one year. In each period, cheese was produced on two separate days (D1, D2). One of the silo milk samples for period 1, cluster A, day 2 (D2) is missing.

Table 2. Top 30 amplicon sequence variants (ASVs) and their relative abundance (RA) in 18 batches of cheese aged for 7 to 20 months, and number of batches in which ASV found in cheese were also detected in the corresponding farm bulk milk (FM), dairy silo milk (SM) and early process samples (EPS).

ASV	Batches with ASV Found in Cheese and in Samples from Early Stages in the Process			Average RA (%) in Cheese Sampled after Ageing for 7–20 Months
	FM	SM [1]	EPS [2]	
Lactococcus b39c	17	17	17	42.1
Leuconostoc 7f62	10	6	17	41.7
Leuconostoc 24cd	11	2	15	6.1
Lactococcus d90e	16	8	17	3.2
Streptococcus bd2e	18	16	9	1.8
Lactobacillus c982	17	12	11	1.3
Lactobacillus 2653	0	0	0	0.75
Lactobacillus 3486	6	4	2	0.61
Lactococcus 072f	0	0	14	0.52
Lactococcus bb15	0	0	11	0.51
Lactococcus cc1f	8	5	11	0.45
Lactobacillus 0b36	1	0	0	0.34
Acinetobacter 6933	0	0	1	0.17
Streptococcus 0c46	2	1	1	0.15
Lactobacillus bf19	0	0	1	0.13
Lactobacillus 57c1	1	0	0	0.064
Lactococcus 6fa5	0	0	0	0.041

Table 2. Cont.

ASV	Batches with ASV Found in Cheese and in Samples from Early Stages in the Process			Average RA (%) in Cheese Sampled after Ageing for 7–20 Months
	FM	SM [1]	EPS [2]	
Romboutsia 464a	15	14	0	0.013
Tetragenococcus 49a8	1	0	0	0.011
Turicibacter dacd	12	11	0	0.011
Paeniclostridium 69a9	14	13	0	0.010
Streptococcus 84e3	1	1	0	0.010
Lactobacillus 61ee	2	2	1	0.008
Staphylococcus 1355	10	9	0	0.005
Clostridium ddf8	0	0	1	0.005
Trueperella b116	4	3	0	0.005
Enterococcus 028e	5	3	0	0.004
Lactococcus ffd9	0	0	0	0.003
Unclassified Enterobacteriaceae 0315	2	1	2	0.002
Unclassified Enterobacteriaceae 0942	0	0	1	0.0004

[1] One silo milk sample missing. [2] EPS missing for one batch. EPS included freshly propagated starter culture, milk gel and cheese grains before pressing.

4. Discussion

In this study, 16S rRNA gene amplicon sequencing was used to investigate the microbiota of raw bulk milk collected at farms, the resulting silo milk at the cheese-making facility, samples collected early in the cheese-making process and samples taken repeatedly during cheese ripening in order to investigate changes in the composition of the microbiota throughout the cheese manufacturing process. A rarefaction curve and Shannon index values revealed that the microbiota in farm bulk and dairy silo milk samples had a much higher alpha diversity than that in EPSs and cheese samples (Figure 1). The low alpha diversity in EPSs and cheese samples in our study is a consequence of the dairy processes, i.e., pasteurisation of the raw milk and the fermentation taking place after addition of the starter culture. Species that cannot compete with the active lactic acid bacteria, which ferment lactose into lactic acid, thereby lowering pH, will be eliminated and disappear. This was also illustrated by Choi et al. [25], who investigated the overall microbial community shift during Cheddar cheese production from the raw milk to the aged cheese. SLAB inoculation decreased the microbial richness by inhibiting the growth of other bacteria present in the milk. In contrast, species richness increased with ripening time for the non-pasteurised Parmigiano Reggiano [26].

Comparing the diversity associated with farm bulk and dairy silo milk, the higher number of ASVs in dairy silo milk was probably explained by the fact that the dairy silo comprised bulk milk from several farms, with each farm contributing a different microbiota. A higher alpha diversity of the microbiota in farm bulk milk than in samples representing the subsequent stages in cheese production was previously reported by McHugh, et al. [27], who explored changes in the dairy microbiome from the farm through transportation and processing to skimmed milk powder. Although the microbiota composition of the bulk milk was not the only criterion used to create the three farm clusters in the present study, the microbiota in the bulk milk from cluster B was different from that in the bulk milk from clusters A and C (Figure 2). The period had a more pronounced effect on the bulk milk microbiota, with each period being different from the other two in this respect. Similarly, Skeie et al. [28] reported shifts in the microbiota composition of raw farm bulk milk during a three-month monitoring period and suggested random variation due to variable sources of contamination at the farm level. A follow-up study on the same farms two years later concluded that temporal within-farm changes in bulk milk microbiota are mostly driven by mastitis-related genera [29]. For some farms, there were major shifts over time in milk microbiota which were not correlated with changes in management, indicating that other factors, e.g., the weather during the harvesting season, contributed to the observed differences [29]. Investigating the variation in the microbiota in bulk milk collected monthly on 42 dairy farms over one year, we found that the type of dairy farm (milking system) had a strong effect on the bulk milk microbiota, an effect likely associated with hygiene routines during the pre-milking and cleaning of the milking equipment [8]. The 18 farms used in the present study were selected from among those original 42 farms and an effect of milking system on bulk milk microbiota was also clear in the present study.

In contrast to farm bulk milk, there were no significant differences in the microbiota in dairy silo milk from the three farm clusters, while the difference between periods persisted at the dairy silo level (Figure 3). Under certain conditions, the microbiota composition of the dairy silo milk was strongly affected by the farm bulk milk microbiota. This is likely explained by differences in the volumes of milk delivered to the dairy by the individual farms. It was particularly obvious in period 1 (Figure 4), where one farm belonging to cluster C contributed a large proportion of the milk in the dairy silo, and this had a clear impact on the dairy silo milk microbiota. While *Streptococcus salivarius* ASV bd2e was dominant in the bulk milk from this farm, it also became dominant in the dairy silo milk, and it was one major ASV in the resulting cheese. This highlights the importance of good herd management, especially on larger farms, to maintain high hygiene standards and keep the total bacteria count in the bulk milk at low levels. The genera *Romboutsia*, commonly associated with the intestinal tract, and *Staphylococcus*, a major mastitis pathogen associated with the mammary gland tissue, had a relatively high RA in the bulk and silo milk samples, but had low RA (<0.1%) in the cheese samples, which was explained by the pasteurisation of the raw milk before cheese making.

Although there was no clear effect of farm cluster on the microbiota of the dairy silo milk used for cheese production, the cheese microbiota showed significant differences among both periods and clusters (Figure 4). In general, *Lactobacillus*, a NSLAB species known from previous studies to be important for flavour development in this traditional Swedish cheese, was present at a higher RA in cheeses produced in period 1 compared with periods 2 and 3 (Figure 7). *Acinetobacter* was also more frequently observed in cheese produced in period 1 associated with cluster C and day 1. Considering that the *Acinetobacter* ASV 6933 was not detected in our farm or silo milk samples, contamination may have occurred in the dairy environment. Within the genus *Lactococcus*, a single ASV was completely dominant in period 3, while several minor ASVs of this genus were observed in periods 1 and 2. Screening of farm bulk milk, dairy silo milk and EPSs to identify ASVs common to both milk and cheese revealed that the dominant ASVs found in cheese (average RA > 1%) were also present in milk and EPS samples (Table 2, Figures 7 and 8). This confirms that the LAB of importance for the cheese manufacturing process are common in both farm and dairy environments. However, the dominant LAB species found in cheese were usually present at low RAs in farm bulk and dairy silo milk, while NSLAB, specifically *Lactobacillus*, showed very low RA in both milk and EPSs. It is, therefore, difficult to draw conclusions regarding the origin of the *Lactobacillus* ASV found in the cheese; it could stem from the raw milk, the dairy environment or both. More sensitive methods, e.g., qPCR using primers targeting specific NSLAB, are needed for the identification of the origin of these bacteria.

The starter LAB were present in the EPS and cheese samples at a relatively high RA. Some variation in the starter LAB was observed, e.g., the three less dominant *Lactococcus* ASVs (d90e, 072f and bb15) were present at much lower RAs in the cheese from period 3 than in the cheese from periods 1 and 2 (Figure 7). Different steps in the cheese-making process may have a major effect on the cheese microbiota, including changes in the abundance of NSLAB. As an example, a previous study by Porcellato and Skeie [30] reported that a small change (2 °C) in scalding temperature can significantly influence the cheese microbiota. Stress and bacteriophage attack may also give rise to variation in the LAB starter between different production periods, illustrating the importance of using starter cultures with high strain diversity, e.g., undefined, mixed-strain starter cultures, which are generally considered more resilient [31]. *Lactococcus* d90e was identified in the screening of ASVs passing from milk to cheese and from EPSs to cheese, while 072f and bb15 were only observed in EPSs and cheese. The identification of ASVs which were only associated with EPSs suggests a potential enrichment of specific LAB in the cheese production process, e.g., contamination by facility-specific "house" microflora from processing equipment and the dairy environment [32]. According to Bokulich and Mills [3], this "house" microbiota plays an important role in shaping site-specific product characteristics. In addition, three other ASVs (*Lactobacillus* 2653, *Lactococcus* 6fa5 and *Lactococcus* ffd9) were only detected in cheese, and were present neither in milk nor in EPS samples.

Comparing cheese aged for 12 months or longer, cheese from cluster A milk had a consistently lower total bacterial count in all three periods (Figure 5). The numbers of viable bacteria in the cheese were generally lower than expected, in most samples ranging between 10^5 and 10^6 cfu/g cheese and in cluster A samples exhibited even lower counts, i.e., 10^3–10^5 cfu/g. In a previous study on the same type of cheese [33], the number of LAB cultured on MRS agar was in the order of 10^6–10^7 cfu/g cheese from 12 weeks until 56 weeks of ripening. The use of plate count agar instead of MRS for bacterial counting in our study likely explains the difference in observed CFU levels. The sequencing results revealed that cheese from cluster A had a lower RA of *Lactobacillus* in all periods, but especially in periods 2 and 3 (Figure 7). This suggests that the low number of viable bacteria in cheese from cluster A was likely associated with a low number of NSLAB.

A decline in starter LAB and proliferation of NSLAB usually take place during the early stage of ripening. In a culture-based study on Cheddar cheese, Dasen et al. [34] found that the *Lactobacillus* count on *Lactobacillus*-selective medium reached a peak after only 3 months, while counts on M17 medium declined from the beginning of ripening, reaching the lowest value at around 9 months. In a study by Rehn et al. [33], starter LAB counts were higher than NSLAB counts up to 3 weeks of ripening and decreased between 3 and 12 weeks, while after 26 weeks of ripening NSLAB dominated in the cheese. The high RA of starter LAB (i.e., *Lactococcus* and *Leuconostoc*) in cheese from 7 until 20 months of ripening was therefore unexpected (Figure 7). Although several studies have confirmed the presence of viable *Lactococcus* at later stages of cheese ripening using RNA-based analysis [35,36], their contribution at this stage needs further investigation. By using retentostat cultures, van Mastrigt and co-workers mimicked the cheese ripening conditions and quantified the aroma profile of *Lactococcus lactis* [37,38]. One possible explanation for the high abundance of starter LAB in our study could be that the DNA from dead starter bacteria was amplified and sequenced. In a study examining the effects of the cheese-making process on the cheese microbiota in a Dutch-type cheese, treating bacteria cells with PMA (propidium monoazide) to selectively detect viable bacteria, Porcellato and Skeie [30] found that the RA of *Lactobacillus* spp. increased and the RA of *Lactococcus lactis* decreased in cheese ripened for 1 and 3 months. Barzideh et al. [39] found that after PMA treatment the RA of *Lactococcus* in a long-ripened Cheddar cheese was below 20% after 7 months, whereas non-PMA treated *Lactococcus* had RA of 30% at the same timepoint. In our study, *Leuconostoc* reached high RA at month 7 of ripening and in two batches had already increased to 25% and 39% RA, respectively, at 24 h (Figure 7, period 3, clusters A and B). Although *Leuconostoc* was included in the DL-type starter culture used in this study, its growth dynamic during cheese ripening was similar to that of a typical NSLAB, in agreement with observations in the case of Cheddar cheese [39].

The RA of *Streptococcus* (ASV bd2e) increased during ripening, and it was present at a higher RA in the cheese from cluster C (Figure 7). It was also present at a high RA in the bulk milk from one of the larger individual farms in cluster C; milk from this farm contributed with 40% of the total volume of dairy silo milk in period 1 (Figure 4). Some strains of *Streptococcus*, e.g., *S. thermophilus*, are used in thermophilic starter cultures for the production of different types of cheese [40,41]. Since a mesophilic starter culture was used to produce the cheese, the *Streptococcus* most likely originated from raw bulk milk, surviving pasteurisation, and suggesting the passage of the *Streptococcus* ASV from farm bulk to dairy silo milk, EPSs, and the resulting cheese (Table 2). According to Johnson et al. [42], the *Streptococcus* group is commonly found in aged Cheddar cheese owing to the long production schedule in modern cheese manufacturing, which creates conditions that support the growth of microorganisms in the processing environment. This may introduce differences between cheese batches produced early and late during the same day, due to bacterial growth on food contact surfaces and changes in the starter culture [42]. Due to the short read length of the sequencing platform used, the type of *Streptococcus* could not be identified in detail, so we cannot be sure that passage of the ASV truly reflected the transfer of this group of bacteria from bulk milk to cheese. It is also worth noting that due to the short read length, observing the same ASV in different sample types did not necessarily confirm the presence of the same bacteria strain.

In contrast to our hypothesis that the increasing variation and longer ripening time of the investigated cheese would reflect the variation in the raw milk microbiota, the results showed little association between the milk microbiota and cheese ripening. Future studies should therefore address facility-specific aspects, e.g., by characterising the in-house microbiota, isolating and characterising the NSLAB from successful cheeses, evaluating the possibility to use isolated NSLAB as an adjunct culture or assessing the effect of slight differences in heat treatment of the milk used for cheese making.

5. Conclusions

Through careful selection of participating farms, we managed to create three clusters with some variation in bulk milk microbiota between clusters, but this variation was less apparent in the resulting dairy silo milk. Variation over time, i.e., between periods, was more profound, both in farm bulk and dairy silo milk. The cheese microbiota varied between periods and clusters, but the variation showed little association with the microbiota in the dairy silo milk used for cheese-making. The microbiota in the ripened cheese was generally dominated by starter LAB, while the RA of NSLAB was generally low. The dominant bacteria ASVs present in the ripened cheese were also found in farm bulk and dairy silo milk and in EPS, while some minor bacteria ASVs were only identified in EPSs and cheese, or only in cheese, suggesting that the dairy environment likely enriched such bacteria. Based on these results, the focus in future studies should be shifted from the milk microbiota to the effects of process variables and the dairy environment on cheese maturation.

Author Contributions: Conceptualisation, A.H., K.H.S., M.H., A.H.G., J.D. and Å.L.; methodology, L.S., A.H., G.B., J.D. and Å.L.; software, L.S. and J.D.; formal analysis, L.S., A.H. and G.B.; resources, A.H., K.H.S., M.H., A.H.G., J.D. and Å.L.; data curation, L.S., A.H. and G.B.; writing—original draft preparation, L.S.; writing—review and editing, A.H., M.J., K.H.S., G.B., M.H., A.H.G., J.D. and Å.L.; visualisation, L.S.; supervision, A.H., J.D. and Å.L.; project administration, Å.L. All authors have read and agreed to the published version of the manuscript.

Funding: This study was conducted with financial support from the Swedish Farmers' Foundation for Agricultural Research, Stockholm, Sweden (project O-16-20-764) and the Kamprad Family Foundation, Växjö, Sweden (project 20160098).

Data Availability Statement: Data are contained within the article.

Acknowledgments: The authors acknowledge all dairy farmers who participated in the study.

Conflicts of Interest: Author A.H. Gustafsson was employed by the economic association Vaxa Sverige. Vaxa Sverige is a farmer-owned organization doing; development and research, consulting and supporting cattle farmers in breeding, feeding, animal health, milk recording and management. Author A.H. was employed by the company Norrmejerier Ek. Forening. Author K.H.S. was employed by the company Norrmejerier Ek. Forening. The remaining authors declare that the research was conducted in the absence of any commercial or financial relationships that could be construed as a potential conflict of interest.

References

1. McSweeney, P.L.H. Biochemistry of cheese ripening. *Int. J. Dairy Technol.* **2004**, *57*, 127–144. [CrossRef]
2. Ardö, Y. Flavour formation by amino acid catabolism. *Biotechnol. Adv.* **2006**, *24*, 238–242. [CrossRef]

3. Bokulich, N.A.; Mills, D.A. Facility-specific "house" microbiome drives microbial landscapes of artisan cheesemaking plants. *Appl. Environ. Microbiol.* **2013**, *79*, 5214–5223. [CrossRef]
4. Broadbent, J.R.; Houck, K.; Johnson, M.; Oberg, C. Influence of adjunct use and cheese microenvironment on nonstarter bacteria in reduced-fat Cheddar-type cheese. *J. Dairy Sci.* **2003**, *86*, 2773–2782. [CrossRef]
5. Levante, A.; De Filippis, F.; La Storia, A.; Gatti, M.; Neviani, E.; Ercolini, D.; Lazzi, C. Metabolic gene-targeted monitoring of non-starter lactic acid bacteria during cheese ripening. *Int. J. Food Microbiol.* **2017**, *257*, 276–284. [CrossRef]
6. Beresford, T.P.; Fitzsimons, N.A.; Brennan, N.L.; Cogan, T.M. Recent advances in cheese microbiology. *Int. Dairy J.* **2001**, *11*, 259–274. [CrossRef]
7. Williams, A.G.; Choi, S.-C.; Banks, J.M. Variability of the species and strain phenotype composition of the non-starter lactic acid bacterial population of cheddar cheese manufactured in a commercial creamery. *Food Res. Int.* **2002**, *35*, 483–493. [CrossRef]
8. Sun, L.; Lundh, Å.; Höjer, A.; Bernes, G.; Nilsson, D.; Johansson, M.; Hetta, M.; Gustafsson, A.H.; Saedén, K.H.; Dicksved, J. Milking system and premilking routines have a strong effect on the microbial community in bulk tank milk. *J. Dairy Sci.* **2022**, *105*, 123–139. [CrossRef] [PubMed]
9. Doyle, C.J.; Gleeson, D.; O'Toole, P.W.; Cotter, P.D. Impacts of seasonal housing and teat preparation on raw milk microbiota: A high-throughput sequencing study. *Appl. Environ. Microbiol.* **2017**, *83*, e02694-16. [CrossRef] [PubMed]
10. Gagnon, M.; Ouamba, A.J.; LaPointe, G.; Chouinard, P.Y.; Roy, D. Prevalence and abundance of lactic acid bacteria in raw milk associated with forage types in dairy cow feeding. *J. Dairy Sci.* **2020**, *103*, 5931–5946. [CrossRef]
11. Priyashantha, H.; Lundh, Å.; Höjer, A.; Bernes, G.; Nilsson, D.; Hetta, M.; Saedén, K.H.; Gustafsson, A.H.; Johansson, M. Composition and properties of bovine milk: A study from dairy farms in northern Sweden; Part I. *Eff. Dairy Farming System. J. Dairy Sci.* **2021**, *104*, 8582–8594.
12. Hugerth, L.W.; Wefer, H.A.; Lundin, S.; Jakobsson, H.E.; Lindberg, M.; Rodin, S.; Engstrand, L.; Andersson, A.F. DegePrime, a Program for Degenerate Primer Design for Broad-Taxonomic-Range PCR in Microbial Ecology Studies. *Appl. Environ. Microbiol.* **2014**, *80*, 5116–5123. [CrossRef]
13. Sun, L.; Dicksved, J.; Priyashantha, H.; Lundh, Å.; Johansson, M. Distribution of bacteria between different milk fractions, investigated using culture-dependent methods and molecular-based and fluorescent microscopy approaches. *J. Appl. Microbiol.* **2019**, *127*, 1028–1037. [CrossRef] [PubMed]
14. Bolyen, E.; Rideout, J.R.; Dillon, M.R.; Bokulich, N.A.; Abnet, C.C.; Al-Ghalith, G.A.; Alexander, H.; Alm, E.J.; Arumugam, M.; Asnicar, F.; et al. Reproducible, interactive, scalable and extensible microbiome data science using QIIME 2. *Nat. Biotechnol.* **2019**, *37*, 852–857. [CrossRef] [PubMed]
15. Martin, M. Cutadapt removes adapter sequences from high-throughput sequencing reads. *EMBnet J.* **2011**, *17*, 10–12. [CrossRef]
16. Callahan, B.J.; McMurdie, P.J.; Rosen, M.J.; Han, A.W.; Johnson, A.J.A.; Holmes, S.P. DADA2: High-resolution sample inference from Illumina amplicon data. *Nat. Methods* **2016**, *13*, 581–583. [CrossRef]
17. Katoh, K.; Misawa, K.; Kuma, K.-i.; Miyata, T. MAFFT: A novel method for rapid multiple sequence alignment based on fast Fourier transform. *Nucleic Acids Res.* **2002**, *30*, 3059–3066. [CrossRef]
18. Price, M.N.; Dehal, P.S.; Arkin, A.P. FastTree 2—Approximately Maximum-Likelihood Trees for Large Alignments. *PLoS ONE* **2010**, *5*, e9490. [CrossRef]
19. Pedregosa, F.; Varoquaux, G.; Gramfort, A.; Michel, V.; Thirion, B.; Grisel, O.; Blondel, M.; Prettenhofer, P.; Weiss, R.; Dubourg, V. Scikit-learn: Machine learning in Python. *J. Mach. Learn. Res.* **2011**, *12*, 2825–2830.
20. Quast, C.; Pruesse, E.; Yilmaz, P.; Gerken, J.; Schweer, T.; Yarza, P.; Peplies, J.; Glöckner, F.O. The SILVA ribosomal RNA gene database project: Improved data processing and web-based tools. *Nucleic Acids Res.* **2013**, *41*, D590–D596. [CrossRef]
21. Bokulich, N.A.; Kaehler, B.D.; Rideout, J.R.; Dillon, M.; Bolyen, E.; Knight, R.; Huttley, G.A.; Gregory Caporaso, J. Optimizing taxonomic classification of marker-gene amplicon sequences with QIIME 2's q2-feature-classifier plugin. *Microbiome* **2018**, *6*, 90. [CrossRef]
22. Kruskal, W.H.; Wallis, W.A. Use of ranks in one-criterion variance analysis. *J. Am. Stat. Assoc.* **1952**, *47*, 583–621. [CrossRef]
23. Benjamini, Y.; Hochberg, Y. Controlling the False Discovery Rate: A Practical and Powerful Approach to Multiple Testing. *J. R. Stat. Soc. Ser. B Stat. Methodol.* **1995**, *57*, 289–300. [CrossRef]
24. Anderson, M.J. A new method for non-parametric multivariate analysis of variance. *Austral Ecol.* **2001**, *26*, 32–46.
25. Choi, J.; In Lee, S.; Rackerby, B.; Frojen, R.; Goddik, L.; Ha, S.-D.; Park, S.H. Assessment of overall microbial community shift during Cheddar cheese production from raw milk to aging. *Appl. Microbiol. Biotechnol.* **2020**, *104*, 6249–6260. [CrossRef] [PubMed]
26. Bottari, B.; Levante, A.; Bancalari, E.; Sforza, S.; Bottesini, C.; Prandi, B.; De Filippis, F.; Ercolini, D.; Nocetti, M.; Gatti, M. The Interrelationship between Microbiota and Peptides during Ripening as a Driver for Parmigiano Reggiano Cheese Quality. *Front. Microbiol.* **2020**, *11*, 581658. [CrossRef]
27. McHugh, A.J.; Feehily, C.; Fenelon, M.A.; Gleeson, D.; Hill, C.; Cotter, P.D. Tracking the Dairy Microbiota from Farm Bulk Tank to Skimmed Milk Powder. *mSystems* **2020**, *5*, e00226-20. [CrossRef] [PubMed]
28. Skeie, S.B.; Håland, M.; Thorsen, I.M.; Narvhus, J.; Porcellato, D. Bulk tank raw milk microbiota differs within and between farms: A moving goalpost challenging quality control. *J. Dairy Sci.* **2019**, *102*, 1959–1971. [CrossRef]
29. Porcellato, D.; Smistad, M.; Bombelli, A.; Abdelghani, A.; Jørgensen, H.J.; Skeie, S.B. Longitudinal study of the bulk tank milk microbiota reveals major temporal shifts in composition. *Front. Microbiol.* **2021**, *12*, 616429. [CrossRef]

30. Porcellato, D.; Skeie, S.B. Bacterial dynamics and functional analysis of microbial metagenomes during ripening of Dutch-type cheese. *Int. Dairy J.* **2016**, *61*, 182–188. [CrossRef]
31. de Vos, W.M. Systems solutions by lactic acid bacteria: From paradigms to practice. *Microb. Cell Factories* **2011**, *10*, S2. [CrossRef] [PubMed]
32. Somers, E.; Johnson, M.; Wong, A. Biofilm formation and contamination of cheese by nonstarter lactic acid bacteria in the dairy environment. *J. Dairy Sci.* **2001**, *84*, 1926–1936. [CrossRef]
33. Rehn, U.; Petersen, M.A.; Saedén, K.H.; Ardö, Y. Ripening of extra-hard cheese made with mesophilic DL-starter. *Int. Dairy J.* **2010**, *20*, 844–851. [CrossRef]
34. Dasen, A.; Berthier, F.; Grappin, R.; Williams, A.G.; Banks, J. Genotypic and phenotypic characterization of the dynamics of the lactic acid bacterial population of adjunct-containing Cheddar cheese manufactured from raw and microfiltered pasteurised milk. *J. Appl. Microbiol.* **2003**, *94*, 595–607. [CrossRef] [PubMed]
35. Ruggirello, M.; Cocolin, L.; Dolci, P. Fate of *Lactococcus lactis* starter cultures during late ripening in cheese models. *Food Microbiol.* **2016**, *59*, 112–118. [CrossRef]
36. Desfossés-Foucault, É.; LaPointe, G.; Roy, D. Dynamics and rRNA transcriptional activity of lactococci and lactobacilli during Cheddar cheese ripening. *Int. J. Food Microbiol.* **2013**, *166*, 117–124. [CrossRef]
37. van Mastrigt, O.; Gallegos Tejeda, D.; Kristensen, M.N.; Abee, T.; Smid, E.J. Aroma formation during cheese ripening is best resembled by *Lactococcus lactis* retentostat cultures. *Microb. Cell Factories* **2018**, *17*, 104. [CrossRef]
38. van Mastrigt, O.; Abee, T.; Lillevang, S.K.; Smid, E.J. Quantitative physiology and aroma formation of a dairy *Lactococcus lactis* at near-zero growth rates. *Food Microbiol.* **2018**, *73*, 216–226. [CrossRef]
39. Barzideh, Z.; Siddiqi, M.; Mohamed, H.M.; LaPointe, G. Dynamics of Starter and Non-Starter Lactic Acid Bacteria Populations in Long-Ripened Cheddar Cheese Using Propidium Monoazide (PMA) Treatment. *Microorganisms* **2022**, *10*, 1669. [CrossRef]
40. Hols, P.; Hancy, F.; Fontaine, L.; Grossiord, B.; Prozzi, D.; Leblond-Bourget, N.; Decaris, B.; Bolotin, A.; Delorme, C.; Dusko Ehrlich, S. New insights in the molecular biology and physiology of *Streptococcus thermophilus* revealed by comparative genomics. *FEMS Microbiol. Rev.* **2005**, *29*, 435–463.
41. Iyer, R.; Tomar, S.; Maheswari, T.U.; Singh, R. *Streptococcus thermophilus* strains: Multifunctional lactic acid bacteria. *Int. Dairy J.* **2010**, *20*, 133–141. [CrossRef]
42. Johnson, J.; Selover, B.; Curtin, C.; Waite-Cusic, J. Nonstarter Bacterial Communities in Aged Cheddar Cheese: Patterns on Two Timescales. *Appl. Environ. Microbiol.* **2022**, *88*, e01939-21. [CrossRef] [PubMed]

Disclaimer/Publisher's Note: The statements, opinions and data contained in all publications are solely those of the individual author(s) and contributor(s) and not of MDPI and/or the editor(s). MDPI and/or the editor(s) disclaim responsibility for any injury to people or property resulting from any ideas, methods, instructions or products referred to in the content.

Review

Mild Heat Treatment and Biopreservatives for Artisanal Raw Milk Cheeses: Reducing Microbial Spoilage and Extending Shelf-Life through Thermisation, Plant Extracts and Lactic Acid Bacteria

Beatriz Nunes Silva [1,2,3,*], José António Teixeira [3,4], Vasco Cadavez [1,2] and Ursula Gonzales-Barron [1,2]

1. Centro de Investigação de Montanha (CIMO), Instituto Politécnico de Bragança, Campus de Santa Apolónia, 5300-253 Bragança, Portugal; vcadavez@ipb.pt (V.C.); ubarron@ipb.pt (U.G.-B.)
2. Laboratório Associado para a Sustentabilidade e Tecnologia em Regiões de Montanha (SusTEC), Instituto Politécnico de Bragança, Campus de Santa Apolónia, 5300-253 Bragança, Portugal
3. CEB—Centre of Biological Engineering, University of Minho, 4710-057 Braga, Portugal; jateixeira@deb.uminho.pt
4. LABBELS—Associate Laboratory, 4710-057 Braga, Portugal
* Correspondence: beatrizsilva@ceb.uminho.pt; Tel.: +351-912-073-401

Abstract: The microbial quality of raw milk artisanal cheeses is not always guaranteed due to the possible presence of pathogens in raw milk that can survive during manufacture and maturation. In this work, an overview of the existing information concerning lactic acid bacteria and plant extracts as antimicrobial agents is provided, as well as thermisation as a strategy to avoid pasteurisation and its negative impact on the sensory characteristics of artisanal cheeses. The mechanisms of antimicrobial action, advantages, limitations and, when applicable, relevant commercial applications are discussed. Plant extracts and lactic acid bacteria appear to be effective approaches to reduce microbial contamination in artisanal raw milk cheeses as a result of their constituents (for example, phenolic compounds in plant extracts), production of antimicrobial substances (such as organic acids and bacteriocins, in the case of lactic acid bacteria), or other mechanisms and their combinations. Thermisation was also confirmed as an effective heat inactivation strategy, causing the impairment of cellular structures and functions. This review also provides insight into the potential constraints of each of the approaches, hence pointing towards the direction of future research.

Keywords: natural preservatives; inactivation; antimicrobial; dairy; food safety; pathogens

1. Introduction

Cheese is a highly nutritious food, with hundreds of varieties that have different colours, odours, flavours and textures, depending on the type of milk used, production and maturation processes, and age, for example [1].

Artisanal raw milk cheeses are particularly appreciated for their unique sensory characteristics (namely, texture, aroma and flavour) when compared to other types of cheeses, and there has been a growing demand for specialty and artisanal cheeses due to the number of consumers who currently prefer minimally processed foods that provide a feeling of authenticity [2].

The unique sensory characteristics of artisanal raw milk cheeses result, among other factors, from the use of unpasteurised milk. In fact, despite having numerous advantages, such as reducing the bacterial load and extending the shelf-life of milk, pasteurisation causes, among other heat-induced changes, denaturation of whey proteins and complex interactions among denatured whey proteins, casein micelles, minerals and fat globules [3]. These modify the biochemistry and microbiology of milk acidification and cheese ripening,

---

Citation: Silva, B.N.; Teixeira, J.A.; Cadavez, V.; Gonzales-Barron, U. Mild Heat Treatment and Biopreservatives for Artisanal Raw Milk Cheeses: Reducing Microbial Spoilage and Extending Shelf-Life through Thermisation, Plant Extracts and Lactic Acid Bacteria. *Foods* **2023**, *12*, 3206. https://doi.org/10.3390/foods12173206

Academic Editors: Piero Franceschi and Paolo Formaggioni

Received: 25 July 2023
Revised: 22 August 2023
Accepted: 24 August 2023
Published: 25 August 2023

Copyright: © 2023 by the authors. Licensee MDPI, Basel, Switzerland. This article is an open access article distributed under the terms and conditions of the Creative Commons Attribution (CC BY) license (https://creativecommons.org/licenses/by/4.0/).

and, consequently, the characteristic flavour, aroma and texture of raw milk cheeses cannot be achieved using pasteurised milk [4].

Nonetheless, consumption of raw milk cheeses may pose health safety issues due to the possible presence of pathogenic bacteria in raw milk that can remain viable during manufacture and through ripening [5–7]. The consumption of this type of dairy product has caused a few outbreaks [8–12], thus highlighting the need for preservation strategies to improve the microbial safety of raw milk cheeses.

Chemical preservatives would not be suitable for artisanal cheeses, as they would disregard the appeal of a traditional product derived from cultural heritage and produced using only natural, healthy ingredients. Furthermore, they would be an outdated preservation strategy, as the mishandling and extensive consumption of some chemical additives have been shown to induce gut microbiota dysbiosis, which is a contributing factor to various diseases, including neurodegenerative ones [13–15]. Finally, current consumer expectations are increasingly towards "clean-label", chemical preservative-free food products, and consequently, the food industry and scientific community are compelled to investigate novel food preservation methods [16].

Between other techniques, advanced non-thermal technologies (high pressure, cold plasma, pulsed light, and ultrasound) and packaging systems (bioactive films, coating, and modified atmospheric packaging) are among the innovative cheese preservation approaches developed to inactivate microorganisms in milk and extend the shelf life of raw milk cheeses [17]. However, these are not easily implementable for artisanal producers, mainly because of the need for specific and costly equipment, as well as the need for training to operate such technologies.

On the other hand, the incorporation of natural antimicrobial agents in artisanal cheese production is more feasible since starter cultures (lactic acid bacteria, LAB), plant extracts, essential oils and propolis [17,18] can be easily purchased and added directly to the milk, cheese curd, or final product. Another alternative would be to implement a mild thermal process such as thermisation, which uses sub-pasteurisation temperatures to reduce bacterial load while avoiding large heat-induced changes in milk that would affect the final typical organoleptic characteristics of raw milk cheeses [19,20]. This technology would also be easy for artisanal producers to implement since it does not require specialised equipment. Figure 1 displays a schematic diagram of the general cheesemaking process, and the steps at which thermisation, addition of plant extracts, and addition of a starter culture may be implemented are highlighted.

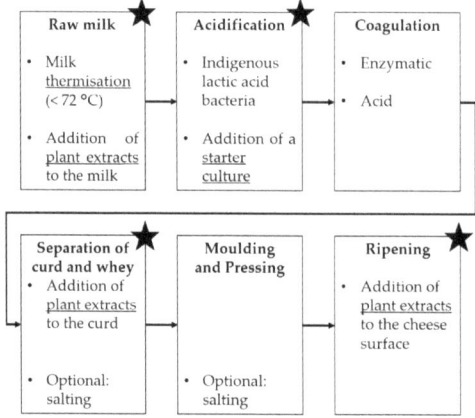

Figure 1. General cheesemaking process. Star symbols indicate at which steps thermisation, addition of plant extracts and/or addition of a starter culture may be implemented.

Considering the above-mentioned possibilities, this review presents an overview of the existing information on LAB and plant extracts as biopreservative strategies, as well as thermisation as a mild heat treatment, to be used in raw milk cheeses.

The main microorganisms involved in cheese spoilage are described, and for each biopreservation strategy, the various targets, mechanisms of antimicrobial action, limitations and, when applicable, relevant commercial applications are discussed.

2. Spoilage Microorganisms in Raw Milk and Raw Milk Cheeses

The most prevalent spoilage fungi genera identified in raw milk and cheeses are *Candida, Cryptococcus, Debaryomyces, Geotrichum, Kluyveromyces, Trichosporon, Pichia*, and *Rhodotorula* spp. (yeasts), and *Penicillium, Aspergillus, Cladosporium, Mucor, Fusarium* and *Alternaria* spp. (moulds) [21–24]. *Candida rugosa, Geotrichum candidum, Torulaspora delbrueckii, Kluyveromyces marxianus* and *Yarrowia lipolytica* are among the common yeast species found in raw milk, while *Penicillium commune* is one of the most frequently occurring mould species [22,24,25].

In the case of excessive yeast growth, cheese softening and unpleasant flavours and odours may occur, as well as the formation of brown spots on the cheese surface (caused by *Y. lipolytica*, for example) and cheese blowing, caused by early gas formation (as a result of high counts of *T. delbrueckii*, for instance) [22]. Yeasts are also able to promote an increase in the pH of the cheese surface, enabling the growth of pathogenic bacteria [22]. Moulds can also produce off-flavours and undesired pigments, as well as synthesise toxic metabolites, such as mycotoxins [23].

Psychrotrophic bacteria dominate the microflora of raw milk, particularly species of the genera *Pseudomonas, Acinetobacter, Aeromonas, Serratia, Bacillus, Lactococcus, Microbacterium*, and *Staphylococcus* [21,26]. Other bacteria associated with cheese spoilage are *Enterobacteriaceae* and clostridial species (*E. cloacae, E. agglomerans, E. zakazakii, C. tyrobutyricum, C. butyricum, C. sporogenes*, and *C. beijerinckii* have been isolated from milk) [27,28]. Clostridial species may produce excessive amounts of gas and butyric acid during growth, resulting in cheese blowing [27]. Likewise, some *Enterobacteriaceae* species can also cause early blowing in cheese, as well as negatively affect the organoleptic features of cheeses [29]. Moreover, the *Enterobacteriaceae* family includes a wide range of pathogenic bacteria.

The main pathogens of concern that have been detected in raw milk cheeses include enterotoxin-producing *Staphylococcus aureus*, Shiga toxin-producing *Escherichia coli* (STEC; *E. coli* O157:H7, for example), *L. monocytogenes, Salmonella* spp., *Brucella* spp. and *Campylobacter* spp. [2,6,10,21,30]. These pathogens may be shed directly into milk via the udder by a diseased or infected animal or may enter milk from the external surfaces of animals, the environment, the milking environment, equipment or from personnel (operators' hands, for example) [6,30]. *L. monocytogenes* and STEC have been identified as especially high-risk pathogens owing to the severity of illness and potential lethality associated with each [19].

To reduce spoilage in dairy products, adequate cleaning and sanitation of the processing environment is imperative, but on-farm interventions (to reduce the concentration of spores and pathogens in bulk tank raw milk) and processing technologies (such as bactofugation and microfiltration) may also be used [31]. However, in the context of artisanal cheese production, the referred processing technologies are generally not used.

3. Biopreservation Strategies

3.1. Plant Extracts

The use of plants and herbs as colouring and flavouring agents in cheese manufacture is not new, with some traditional herb-flavoured cheeses having centuries of history [32]. However, plants may be used for more than their organoleptic and decorative properties, owing to their phytochemical constituents that have been shown to have antimicrobial activity [18,33]. The addition of plants and herbs to cheese can be carried out by incorporating them into milk (before cheese making), into cheese curd, or by rolling the cheese into crushed herbs, for example [18].

Plant extracts can be obtained from a multitude of plants using various solvents and extraction methodologies. However, if intended for human consumption, they must be obtained using non-toxic solvents authorised for the industrial production of foodstuffs and food ingredients [34], such as water, ethanol, or their combination.

Conventional extraction procedures include maceration, percolation, infusion, decoction, reflux extraction, Soxhlet extraction and hydro-distillation (which can be subcategorised into steam-, water-distillation, or a combination of both) [35–37]. While these may still be widely used, nowadays, it is crucial to consider the ecological impact of extraction methods and those that are more sustainable and "green", reducing the amount of solvents used and waste generated, and optimising the recovery of bioactive compounds with high added value, should be preferred [38]. To this, techniques such as subcritical water extraction, supercritical fluid extraction, enzyme-, microwave- and ultrasound-assisted extractions, pulsed electric field extraction and accelerated solvent extraction can be used, among many other modern procedures [35–37,39]. Moreover, as the extraction method, temperature, solvent and pressure, for example, influence the chemical profile of the extracts produced, the most appropriate extraction parameters should be selected, considering the desired compounds and bioactivity [38,40]. In addition, the plant genotype, geographical location, and environmental and agronomic conditions, among other factors, also contribute to variations in the chemical composition of plant extracts [41].

Based on their structure, plant-derived chemicals may be classified as alkaloids, organosulfur compounds, phenolic compounds, coumarins and terpenes [42]. Generally, phenolic compounds are found in higher concentrations in plants [43] and are assumed to be the main antimicrobial agents [43–45], although the remaining compounds have also shown this capacity [42]. With respect to the chemical structure of the bioactive compounds, it has been demonstrated that functional groups such as hydroxyl groups and the number of double bonds can influence antimicrobial strength [37].

The exact targets of plant antimicrobials are often difficult to define, considering the many interacting reactions taking place simultaneously [33] and the various compounds found in plant extracts, each exerting its own effect [42]. Nonetheless, several mechanisms have been suggested to explain the antimicrobial mode of action of plant extracts. These include inhibition of efflux pumps (implicated in the export of harmful substances from within the cell into the external environment) [42] and permeabilisation or disruption of the cell membrane, which allows, respectively, the passage of compounds or the release of intracellular contents (especially potassium, calcium, and sodium ions [34]), adding to the loss of cellular integrity [33,37,43,45,46]. Disruption of the cell membrane may be prompted, for example, by the interaction of phenolics with membrane proteins, inducing alterations in their structure and function, namely in terms of electron transport, nutrient uptake, synthesis of proteins and nucleic acids, and enzyme activity [37]. Additionally, plant extracts may also inhibit DNA and protein synthesis [42], inactivate cellular enzymes (including ATPase) [45,46], and dissipate cellular energy in the form of ATP [33].

Different mechanisms of action have been reported for distinct groups of compounds. In fact, while membrane disruption is associated with the action of terpenoids and phenolics, the antimicrobial properties of phenols and flavonoids seem related to their chelating properties, complexing metal ions that are essential for bacterial growth, whereas coumarin and alkaloids seem to produce effects on genetic material [33,41]. In turn, the antimicrobial activity of some organosulfur compounds, such as onion and garlic isothiocyanates, is due to the inactivation of extracellular enzymes through oxidative dissociation of -S-S- bonds [37]. The mechanism of action may also be dependent on the concentration of the compounds, as it has been shown that at a low concentration, phenols inhibit microbial enzyme activity, whereas at high concentrations, they induce protein denaturation [44].

Irrespective of the mode of action, it is recurrent that Gram-positive bacteria are more susceptible to plant extracts and phenolic compounds than Gram-negative, whose greater resistance is due to the existence of lipopolysaccharides in their outer membranes [37,43,47].

Considering that cheese is a fermented product that contains natural and, sometimes, artificially added microbial populations of LAB, which are a group of Gram-positive organisms, it is reasonable to question if using plants and plant extracts as preservatives may influence bacterial metabolism and/or inhibit this beneficial set of bacteria, potentially compromising the fermentation process. Some studies have reported on this drawback [48–51], including that of Shori et al., who observed a reduction in peptides content and free amino acids of cheeses in the presence of three different types of plant extracts (*I. verum*, *C. longa*, and *P. guajava*), caused by the impairment of LAB growth and, consequently, LAB proteolytic activity [52].

Nonetheless, the ability of herbal extracts to impact LAB is determined by a number of variables, including the genus, species and strain of the LAB, as well as the plant species and the extraction method used, for example [53]. Various studies have shown that when selected plant extracts are employed in appropriate amounts, they may be able to promote the growth of desired microorganisms, or at least not affect them negatively while avoiding the development of harmful bacteria [53]. For example, Mohamed et al. [54] reported the inhibitory effect of ethanolic and aqueous extracts of *Moringa oleifera* leaves against numerous pathogens in vitro, stressing that these did not inhibit LAB growth. In addition, Ziarno et al. [53] investigated the effect of seven plant extracts (valerian, sage, chamomile, cistus, linden blossom, ribwort plantain and marshmallow) with known antimicrobial activity against pathogens on the activity and growth of LAB and observed that the addition of such extracts up to 3% in milk did not hinder the growth of LAB in fermented milk drinks such as yoghurts. Likewise, Chouchouli et al. [55] supplemented yoghurts but with grape seed extracts and did not observe any effect on pH or the viability of *Lactobacilli*.

Considering the distinct results described in the literature, it is important to establish if a particular plant extract can be successfully used in cheeses by evaluating its impact on the growth and the technological properties of LAB populations, whether they are endogenous raw milk flora or intentionally added starter cultures.

Other issues that should be considered when adding plant extracts to cheeses are, for instance, the loss of bioactive compounds during cheesemaking and storage and the interactions between plant extracts and the cheese matrix, which can have an impact on the texture and organoleptic characteristics of the novel cheese [17]. Even though the sensory attributes of the novel cheese may be different from the corresponding "traditional" cheese right after manufacture, it seems that they do not change drastically during ripening and storage, behaving similarly to cheeses without plant extracts [56,57].

The food matrix is an important factor as interactions with food ingredients occur, resulting in reduced biological activities of the natural compounds when comparing the results of in vitro and in situ (cheese) studies. More specifically, it is generally accepted that high concentrations of lipids or proteins limit the antimicrobial efficacy of plant extracts [58–61]. Studies regarding the effects of carbohydrates on the antimicrobial activity of plant extracts are scarce [62], as most of the literature focuses on the interaction between carbohydrates and plant essential oils. In this case, different authors report contrasting results: Gutierrez et al. [63] observed a reduction in oregano and thyme essential oils efficacy when testing 5% and 10% starch concentrations, whereas Shelef et al. [64] reported that carbohydrates in foods do not protect bacteria from the antimicrobial action of essential oils, at least not as much as fat and protein. The complexity of the food structure also plays an important role in the biological activity of plant extracts in food, as well as the changing variables during cheese production (namely water activity, pH, microflora composition, temperature and nutrient composition) [43].

Natural compounds can be lost during cheese making or storage as a result of their sensibility to environmental factors (including light, temperature, oxygen and pH [43,65], which can cause the epimerisation of bioactive components [65]), solubility in whey [66] or solubility of hydrophobic active molecules in lipidic phases [18]. Aqueous phases are generally the preferred ones for cell growth [67], not lipidic phases, although some bacteria have been reported to prefer the fat–water interface in emulsion systems [68–70].

Although not as intense as essential oils, plant extracts may still negatively affect the sensory characteristics of the food product, especially if the concentrations needed to inactivate pathogens and ensure food safety are higher than those that lead to acceptable sensory properties of the treated products [63]. Tayel et al. [71], for example, evaluated the impact of flavouring with plant extracts on the sensory attributes of processed cheese, and the trained panellists scored the taste and odour of cheeses with extracts of cloves, cress and sage lower than those of the control cheeses. On the other hand, the same study reported enhancements in terms of odour, taste, colour, and overall quality when cinnamon, lemon grass, and oregano extracts were added to the cheese [71].

Other studies have also reported improved sensory quality of cheeses containing plant extracts [72–75], thus showing that the sensory issue does not always arise and that it is dependent on the type of extract used and the dose applied, as well as other factors. For example, Lee et al. [72] observed that the addition of *Inula britannica* flower extract (0.25% to 1%) did not significantly affect the odour and taste of Cheddar-type cheeses; and Abd El-Aziz et al. [73] also evaluated the flavour of ginger extract-fortified (0.15 and 0.3%) soft cheeses and found no differences compared to the controls. In turn, Mahajan et al. [74] reported significantly higher scores for flavour and texture when applying pine needle extract (2.5% and 5.0%) to cheeses.

To avoid interactions with food components, degradation and loss of bioactive compounds, as well as the unpleasant taste of polyphenols, bio-based functional packaging materials incorporating natural active compounds and ingredients may be used (for example, coatings and edible films using nano- and microencapsulation techniques) [41,43,65].

Other concerns that must be considered include: (i) the effects of plant extracts and their natural compounds on human health, as typical toxicological information such as "acceptable daily intake" or "no observed adverse effect level" are usually not available [33]; and (ii) the economic costs, legislation, and practical effectiveness [43] of using plant extracts as preservatives in the food industry.

The potential toxicity of plant extracts is generally difficult to define when considering the problems in terms of their standardisation due to the great variability in their composition between batches [33]. In terms of economic costs and legislation, it is crucial that the price of natural preservatives is competitive in comparison to that of synthetic compounds providing comparable antimicrobial effect and that plant additions in and on foods comply with the existing legislation [76,77], which nonetheless is still limited and must be improved (for example, natural additives are legislated in the same manner as synthetic ones, making it sometimes difficult to understand how production is carried and what is their source [78]).

Overall, it is clear that plant extracts can be useful as antimicrobial agents in foods, including raw milk cheeses, although further scientific and legal grounds are needed to motivate and simplify the use of such additives.

3.2. Lactic Acid Bacteria (LAB)

Traditional raw milk cheeses exhibit a complex microbiota, including LAB naturally occurring in milk and purposefully introduced LAB [79]. They comprise a large and heterogeneous group, and bacterial communities differ vastly among raw milk cheeses, but usually, the main genera identified in raw milk artisanal cheeses include *Lactococcus*, *Lactobacillus*, *Enterococcus*, *Streptococcus* and *Leuconostoc* [79,80].

LAB can be relevant for their role as starter cultures, which promote the rapid acidification of milk (crucial for adequate fermentation and production of high-quality cheeses) through the production of organic acids (primarily lactic and acetic acids) [79]. Starter cultures and adjunct cultures (also called non-starter LAB) can also contribute to the maturation of cheese and the development of desirable texture, flavour, aroma and nutritional value as a result of their metabolic features [79,81]. Various selected LAB strains or mixtures of strains are commercially available as starter cultures for cheese production, and the most frequently used species are *Lc. lactis* (particularly subspecies *lactis* and *cremoris*), *S. salivarius* subsp. *thermophilus*, *L. helveticus*, and *L. delbrueckii* [79,82].

Furthermore, LAB may also have probiotic potential, meaning that they can offer health-promoting benefits to consumers. These include immune system modulation [83], improvement of mental health via the gut–brain axis [83], degradation of nutrient-damaging compounds, such as biogenic amines [84] and cholesterol [85], and increase the quantity of beneficial compounds, such as antihypertensive peptides [86], short-chain fatty acids [87], γ-aminobutyric acid and conjugated linoleic acid [88].

Besides their role in successful fermentations, contribution to textural and sensorial characteristics, and health-promoting properties, some LAB species and strains can also act as antimicrobial agents during and after fermentation throughout the maturation/storage step. This can be due to competition for the adherence site [89], competition for nutrients (i.e., Jameson effect [90]), ability to acidify the environment, and ability to produce antimicrobial metabolites during fermentation, which remain in the final product (except for volatile compounds) [79,91]. In fact, some studies have screened the antimicrobial properties of these microorganisms as a strategy to improve the safety of cheeses and successfully used cocktails of LAB strains to hinder the growth of pathogenic bacteria [91–93].

The antimicrobial metabolites produced by LAB that reduce the risk of pathogen growth and survival include organic acids, hydrogen peroxide, diacetyl, fatty acids, reuterin and bacteriocins [79].

Acidification of the environment by organic acids creates adverse conditions for the growth of spoilage and pathogenic microorganisms [94]. *S. aureus*, for example, is strongly inhibited by lactic and acetic acids, as most Gram-negative and neutrophilic bacteria [95]. Undissociated organic acids can diffuse across the cell membrane of pathogens when $pH_{environment}$ < pKa and dissociate within the cell (due to the higher cellular pH), which lowers the cytoplasmic pH [94]. This affects various metabolic processes, promotes the accumulation of toxic anions, dysregulates cell homeostasis, and neutralises the electrochemical proton gradient, disrupting the substrate transport systems and the cell membrane, which potentially leads to the death of the organism [37,79,94]. The concentrations and types of organic acids produced during fermentation are specie- and strain-dependent and also vary with matrix composition and growing conditions [96].

Hydrogen peroxide can be produced by LAB in the presence of oxygen through the action of flavoprotein oxidases or NADH peroxidases [96]. Since LAB cannot degrade this compound, it accumulates in the medium, exerting its bactericidal effect through the destruction of basic molecular structures of cell proteins, denaturation of metabolic enzymes (by oxidation of sulfhydryl groups), and peroxidation of membrane lipids, which increases cell membrane permeability [94,95]. Hydrogen peroxide may also serve as a precursor to the DNA-damaging superoxide ($O_2^{\bullet-}$) and hydroxyl ($^\bullet$OH) free radicals [94]. In milk, hydrogen peroxide activates the lactoperoxidase system, which has proven bacteriostatic and/or bactericidal activity against various Gram-positive and Gram-negative bacteria [96,97].

Diacetyl is an aromatic compound produced by some LAB strains in the presence of organic acids such as citrate, which is converted via pyruvate into diacetyl (citrate fermentation) [94,96]. Figure 2 displays a schematic representation of the diacetyl production via citrate fermentation.

Lactobacilli and *Enterococci* are the genera associated with high diacetyl production, whereas *Leuconostoc* strains produce none or low amounts of diacetyl from citrate [99]. Jay [100] showed that diacetyl was much more effective against Gram-negative bacteria, yeasts, and moulds than against Gram-positive bacteria, while LAB and clostridia were virtually unaffected. The same study also showed that the inhibitory activity of diacetyl against Gram-negative bacteria was related to its interference with arginine utilisation in the periplasmic space, and that pH has an inverse synergistic effect on diacetyl's bioactivity (lower pH, higher bioactivity) [100], statements corroborated by the research of Tan et al. [95].

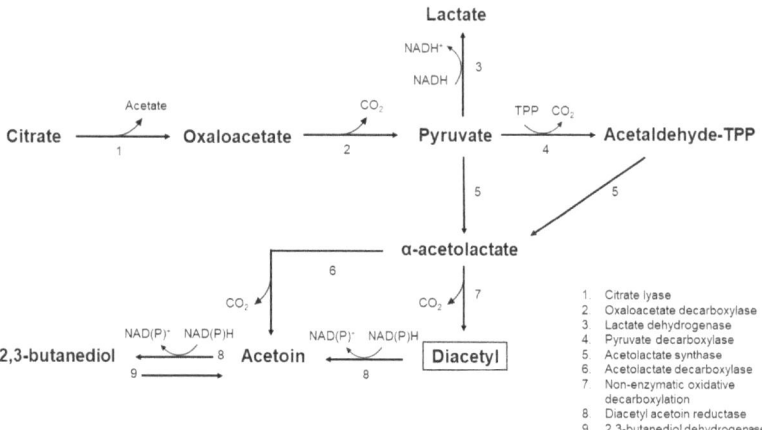

Figure 2. Diacetyl synthesis via citrate metabolism in lactic acid bacteria. TPP: thiamine pyrophosphate. Based on García-Quintans et al. [98].

LAB can produce various fatty acids that improve the sensory attributes of fermented products while potentially exerting antibacterial and antifungal activity [79]. The antibacterial mechanisms of action of these compounds include DNA/RNA replication inhibition, cell wall biosynthesis inhibition in Gram-positive bacteria, inhibition of protein synthesis, cytoplasmic membrane disruption and inhibition of metabolic pathways [101]. The literature available reports that both unsaturated and saturated fatty acids have antibacterial properties towards Gram-positive and Gram-negative bacteria [101], but that fatty acids with medium and long carbon chains, such as lauric (12C) and capric (10C) acids, provide higher inhibitory effects than short chain fatty acids (<8C) [101,102]. For a scheme of the possible cell targets and mechanisms of antimicrobial action of free fatty acids, please refer to Desbois and Smith [102].

Lactobacillus reuterin strains can anaerobically convert glycerol into 3-hydroxypropionaldehyde (3-HPA), which in aqueous solutions exists in equilibrium as a dynamic system of 3-HPA, 3-HPA hydrate, 3-HPA dimer and acrolein [96,103,104], as depicted in Figure 3. This multi-compound system is commonly known as reuterin [96,103,104].

Effective against Gram-positive and Gram-negative bacteria, yeasts, moulds and protozoa [44], this broad-spectrum antimicrobial aldehyde can also be produced by other LAB, including *L. brevis*, *L. buchneri*, *L. collinoids*, and *L. coryniformis* [105]. The antimicrobial activity of reuterin has been linked to the ability of 3-HPA to cause depletion of free thiol groups in glutathione, proteins and enzymes, resulting in an imbalance of the cellular redox status and leading to bacterial cell death [106]. However, the work of Engels et al. [103] suggested, for the first time, that acrolein, and not 3-HPA, is the active compound responsible for the antimicrobial activity attributed to reuterin, and the proposed mechanism of action is schematically represented in their work [103]. The high potential of reuterin as a food biopreservative is supported by its hydrosolubility, stability over a wide range of pH and temperatures, and resistance to degradation by proteolytic and lipolytic enzymes [37,104]. Moreover, reuterin has a wider range of antimicrobial activity than bacteriocins and other non-bacteriocin antimicrobial compounds [104]. However, due to legislative and regulatory requirements, reuterin is not yet commercially available [107].

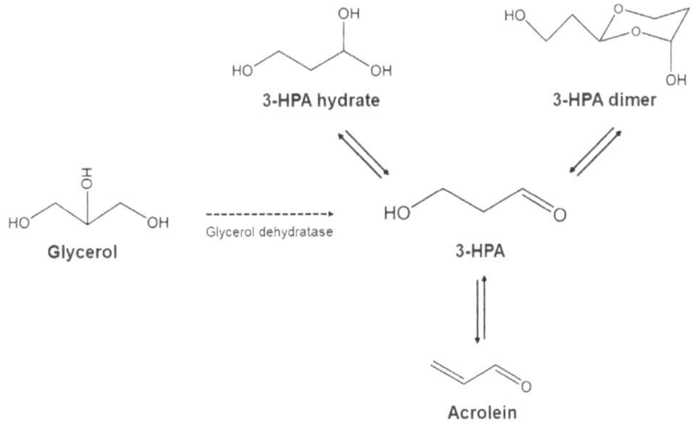

Figure 3. Formation of the reuterin system (comprised of 3-HPA, 3-HPA hydrate, 3-HPA dimer and acrolein) as a result of the dehydration of glycerol by the enzyme glycerol dehydratase. ⟶ Enzymatic reaction; ⇌ Equilibrium reactions.

To that, bacteriocins are extracellularly released bioactive peptides or peptide complexes synthetised in ribosomes [49]. They have a narrow-to-broad antimicrobial effect against bacteria in the same species or across genera, respectively [108], and the producer cell exhibits specific immunity to the action of its own bacteriocin [94]. The majority of bacteriocins produced by LAB are active only against LAB and other Gram-positive bacteria [109,110], but some studies reported on their effectiveness also against Gram-negative bacteria [110,111]. Antifungal bacteriocins have also been reported, with *Lactobacillus* species being the most predominant isolates associated with such compounds [112]. Bacteriocin-producing LAB include *Lactococcus*, *Lactobacillus*, *Pediococcus*, *Streptococcus* and *Enterococcus* strains [108]. The mechanism of action of bacteriocins depends on their primary structure [111]. In bacteria, while some bacteriocins can promote the formation of pores in the phospholipidic bilayer of the cytoplasmic membrane, causing the dissipation of the proton motive force and loss of cell contents, others can inhibit cell wall synthesis or enter the cytoplasm, and affect gene expression and protein synthesis [111]. The antifungal mode of action of protein compounds by LAB, however, remains somewhat unclear, requiring further studies [112,113]. Bacteriocins maintain activity at high temperatures and over a large pH range, and as they are rapidly hydrolysed in the human gastrointestinal tract by digestive proteases, they pose no negative impacts on the gut microbiota [79]. Currently, and although other LAB bacteriocins have shown potential to be used as biopreservatives, only nisin A, produced by *Lc. lactis*, and pediocin PA-1, produced by *P. acidilactici*, have been approved as food preservatives for industrial application and are commercially available [79,81].

Considering the vast diversity of LAB species and antimicrobial metabolites available, there are numerous possibilities for improving food safety and preventing microbial food spoilage. Nonetheless, it is important to consider any potential limiting factors that might reduce the antimicrobial activity of LAB or its compounds. In this sense, the food matrix and its inherent microflora [114], environmental conditions (such as temperature and pH), aerobic conditions, LAB growth phase and load [89], and pathogen content, for example, are among the factors that should not be disregarded when aiming to use such biopreservatives in foods.

4. Thermisation

Thermisation is the standard description for a range of sub-pasteurisation heat treatments of milk, generally from 57 to 68 °C with a holding time between 5 s and 30 min, that is able to reduce bacterial contamination by 3 to 4 log [19,21,115–118].

Unlike pasteurisation, thermisation causes minimum collateral heat damage to milk constituents, has a mild effect on the raw milk flora and the functionality of milk caseins and salts, and has a reduced impact on the sensory profile of the final cheeses [20,115–117,119]. For example, since the heat load is lower compared to that used in pasteurisation, enzymes involved in cheese flavour development, such as lipoprotein lipase, are less inactivated [117]. For this reason, this process may be suitable for the production of artisanal cheeses as it reduces microbial counts and simultaneously enables the profile of the heat-treated milk to be closer to that of raw milk, thus allowing the desirable sensory properties of typical raw milk cheeses to develop [20].

The mechanisms for heat inactivation of mesophilic microorganisms have been extensively studied, and while the ultimate cause leading to cell inactivation by heat remains uncertain, it is clear that heat can affect a wide range of cellular structures and functions, generally known as cellular targets [120,121]. Focusing on non-sporulating bacteria, the cellular targets most affected by heat treatments are the outer and inner membrane, the peptidoglycan cell wall, the nucleoid, the cell's RNA, the ribosomes, and the proteins [120].

Damage to the outer cell layers of bacteria (cell wall for Gram-positive bacteria; outer membrane for Gram-negative bacteria) has been reported by several researchers: in Gram-negative bacteria, damage to the outer membrane after mild thermal treatment can be verified by loss of outer membrane lipopolysaccharides [122] and morphological and structural changes [123] in membrane integrity and permeability, which leads to the release of periplasmic proteins and sensitivity to hydrophobic antibiotics, for example [124,125]. The cell wall of Gram-positive bacteria is also susceptible to heat, but these organisms are generally more heat-resistant due to the high content and extent of cross-linked peptidoglycan of the cell wall [120].

Damage to the cytoplasmic or inner membrane (of Gram-positive and Gram-negative bacteria, respectively) by heat injury can be detected through the loss of intracytoplasmic material leaked from the heated cells, including RNA, DNA, proteins, enzymes, amino acids, and potassium ions, for example [126–128]. Furthermore, the formation of membrane vesicles and loss of membrane material and integrity after heat treatments have also been reported [120,129,130].

Although DNA has high thermostability [131], less intense heat treatments can still modify the nucleoid structure and damage the DNA molecule during and after the treatment [120,130]. Heat-induced DNA damage is manifested by single- or double-strand breaks, as well as increased mutation frequency in surviving populations after heat exposure [132,133]. Moreover, single-strand denaturation induces the action of deoxyribonucleases, which further degrades DNA via the hydrolysis of its phosphodiester backbone [134].

RNA and ribosomes, on the other hand, are more heat-sensitive than DNA [135]. In that sense, mild temperatures have been reported to cause degradation of ribosomes and ribosomal RNA (rRNA), with associated leakage of substances from the metabolic pool (free amino acids and proteins, for example) [136–138] that precedes loss of cell viability. Denaturation of 70S ribosomes and 30S and 50S ribosomal subunits can be a consequence of membrane heat damage and subsequent depletion of magnesium ions from within the cell, as they are essential for the maintenance of the coupled ribosome subunits [136].

Proteins, whether structural or functional (enzymes, for example), may undergo denaturation when bacterial cells are thermally stressed [121]. Protein pumps and channels are also heat-sensitive [120], and, as a response to misfolding and denaturation, protein aggregation may also occur [139]. Rosenberg et al. found a correlation between the thermodynamic parameters of protein denaturation and the death rates of several bacteria [140]. Nevertheless, irreversible denaturation of some proteins might not be lethal to the cell if they can be resynthesised after the heat treatment. On the other hand, it is hypothesised that irreversible denaturation of all copies of RNA polymerase, for example, would represent a lethal event, as this enzyme could not be resynthesised by a cell lacking a single copy [141]. Research has shown that proteins irreversibly denatured by heat are governed by chemical modifications, including deamination of Asn/Gln residues, hydrolysis of peptide bonds at Asp-X residues (X being a small hydrophobic residue), and disulphide bond scrambling [141].

To summarise, the most relevant cellular events that can occur after heat exposure include permeabilisation of membranes, DNA and RNA alterations, loss of ribosome or protein conformation and loss of intracellular components [120]. These events are represented in Figure 4.

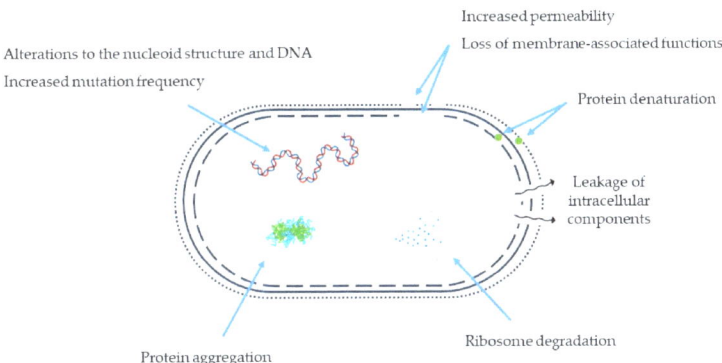

Figure 4. Main cellular events that occur in a vegetative bacterial cell (Gram-positive or Gram-negative) after heat exposure. Based on Cebrián et al. [120].

As microbial inactivation by heat is a multi-target phenomenon, these events may be interconnected and are likely to occur simultaneously [120]. In any case, the lethality of heat treatment is contingent on the alteration of at least one critical component (one whose destruction triggers cell death) beyond a critical threshold, which can be a result of the direct effect of heat on the critical cellular target itself or a consequence of a parallel alteration of another cellular target [120]. It is also crucial to consider that the resistance of each cell target depends on the environmental conditions and the type of microorganism (pH and water activity of the medium during the heat treatment, for example, Gram-positive vs. Gram-negative bacteria, as mentioned before in this section) [120]. Additionally, exposure to sublethal thermal stresses can mediate adaptive responses in bacteria, including the induction of heat shock proteins, which are determinant for protein folding, repair and degradation, and the prevention of aggregation, thus promoting increased heat resistance and, consequently, bacterial survival [142,143].

Thermisation has been noted for both psychrotroph and pathogen control [30,144,145]. Nevertheless, and as previously described, different microorganisms may respond differently to heat treatments, depending on a variety of factors [19]. In this sense, a few authors have reported the survival of some yeasts [146], that some pathogens may remain viable at the lower end of the thermisation temperature range, where the lethal effect is more reduced [30,117], and that thermisation may not be enough to significantly reduce the population of vegetative cells of the more heat-resistant bacterial species (*Enterococcus*, for example) [117,145]. Besides the possibility of some bacteria remaining viable in thermised milk, other shortcomings associated with this thermal treatment are the possible germination of spores present in milk during subsequent cold storage (for example, thermisation at 65 °C for 10 s may be sufficient to stimulate the germination of *B. cereus* spores [115]) and the possible selection for heat-resistant microorganisms such as *M. tuberculosis* and *C. burnetii*, by enabling their survival while reducing competitive flora [2,116]. Thermisation may also have a negative impact on LAB strains and the biodiversity of raw milk bacteria. To this, Sameli et al. [119] observed that thermisation at 60 °C for 30 s reduced the total number of *Leuconostoc*, *Lactococcus* and mesophilic *Lactobacillus*, while producing an enterococcal selecting effect. To avoid such negative effects, it is important that thermisation parameters are carefully selected, aiming to target pathogens while preserving LAB as much as possible.

Moreover, the addition of a starter culture post heat treatment to counteract the reduction in LAB numbers may also be recommended.

5. Conclusions

Artisanal raw milk cheeses may pose health risks to consumers, considering that the manufacturing processes are not standardised, and good manufacturing practices are not always followed, which can lead to undesirable microbiological quality of the cheeses. To avoid pasteurisation and the use of chemical preservatives, which are unfit for this niche product, this work collected and discussed the main antimicrobial action mechanisms, bacterial targets, advantages, limitations and, whenever possible, relevant commercial applications of two biopreservatives, plant extracts and lactic acid bacteria, as well as a mild heat treatment of milk, thermisation, with the goal of promoting their use in cheese production. The literature currently available is supportive of the use of these strategies for the improvement of the microbiological quality of artisanal raw milk cheeses, although some considerations, such as their impact on the sensory characteristics of the product and on the natural microflora, must be carefully assessed, as referred in this review.

Author Contributions: Conceptualization: B.N.S., V.C. and U.G.-B.; writing—original draft: B.N.S.; writing—review and editing. U.G.-B.; visualization, B.N.S.; supervision, J.A.T. and U.G.-B.; funding acquisition: V.C. and U.G.-B. All authors have read and agreed to the published version of the manuscript.

Funding: This work was supported by the Foundation for Science and Technology (FCT, Portugal) for financial support through national funds FCT/MCTES (PIDDAC) to CIMO (UIDB/00690/2020 and UIDP/00690/2020) and SusTEC (LA/P/0007/2020). The authors are also grateful to the EU PRIMA program and FCT for funding the ArtiSaneFood project (PRIMA/0001/2018). This study was supported by FCT under the scope of the strategic funding of the UIDB/04469/2020 unit and BioTecNorte operation (NORTE-01-0145-FEDER-000004) funded by the European Regional Development Fund under the scope of Norte2020—Programa Operacional Regional do Norte. B.N. Silva acknowledges the financial support provided by FCT through the Ph.D. grant SFRH/BD/137801/2018.

Data Availability Statement: No new data were created or analysed in this study. Data sharing is not applicable to this article.

Conflicts of Interest: The authors declare no conflict of interest.

References

1. Reis, P.J.M.; Malcata, F.X. Current state of Portuguese dairy products from ovine and caprine milks. *Small Rumin. Res.* **2011**, *10*, 122–133. [CrossRef]
2. Arias-Roth, E.; Bachmann, H.-P.; Fröhlich-Wyder, M.-T.; Schmidt, R.S.; Wechsler, D.; Beuvier, E.; Buchin, S.; Delbès, C. Raw Milk Cheeses. In *Encyclopedia of Dairy Sciences*, 3rd ed.; McSweeney, P.L.H., McNamara, J.P., Eds.; Elsevier Ltd.: London, UK, 2022; pp. 299–308. [CrossRef]
3. Singh, H.; Waungana, A. Influence of heat treatment of milk on cheesemaking properties. *Int. Dairy J.* **2001**, *11*, 543–551. [CrossRef]
4. Grappin, R.; Beuvier, E. Possible implications of milk pasteurization on the manufacture and sensory quality of ripened cheese. *Int. Dairy J.* **1997**, *7*, 751–761. [CrossRef]
5. Gonzales-Barron, U.; Gonçalves-Tenório, A.; Rodrigues, V.; Cadavez, V. Foodborne pathogens in raw milk and cheese of sheep and goat origin: A meta-analysis approach. *Curr. Opin. Food Sci.* **2017**, *18*, 7–13. [CrossRef]
6. Costanzo, N.; Ceniti, C.; Santoro, A.; Clausi, M.T.; Casalinuovo, F. Foodborne Pathogen Assessment in Raw Milk Cheeses. *Int. J. Food Sci.* **2020**, *2020*, 3616713. [CrossRef] [PubMed]
7. Salazar, J.K.; Gonsalves, L.J.; Natarajan, V.; Shazer, A.; Reineke, K.; Mhetras, T.; Sule, C.; Carstens, C.K.; Schill, K.M.; Tortorello, M.L. Population Dynamics of *Listeria monocytogenes*, *Escherichia coli* O157:H7, and Native Microflora during Manufacture and Aging of Gouda Cheese Made with Unpasteurized Milk. *J. Food Prot.* **2020**, *83*, 266–276. [CrossRef]
8. Johler, S.; Giannini, P.; Jermini, M.; Hummerjohann, J.; Baumgartner, A.; Stephan, R. Further Evidence for Staphylococcal Food Poisoning Outbreaks Caused by egc-Encoded Enterotoxins. *Toxins* **2015**, *7*, 997–1004. [CrossRef] [PubMed]
9. Boyd, E.; Trmcic, A.; Taylor, M.; Shyng, S.; Hasselback, P.; Man, S.; Tchao, C.; Stone, J.; Janz, L.; Hoang, L.; et al. *Escherichia coli* O121 outbreak associated with raw milk Gouda-like cheese in British Columbia, Canada, 2018. *Can. Commun. Dis. Rep.* **2021**, *47*, 11. [CrossRef]

10. Anses. AVIS de l'Anses Relatif Aux Modalités de Maîtrise du Risque Lié à la pRESENCE de Dangers Microbiologiques dans les Fromages et Autres Produits Laitiers Fabriqués à Partir de Lait Cru. 2022. Available online: https://www.anses.fr/fr/system/files/BIORISK2019SA0033.pdf (accessed on 28 December 2022).
11. Food Safety News. Raw Milk Cheese Recalled Because of Risk of Listeria monocytogenes. Available online: https://www.foodsafetynews.com/2019/07/raw-milk-cheese-recalled-because-of-risk-of-listeria-monocytogenes/ (accessed on 28 December 2022).
12. Food Safety News. More Ill in Listeria, E. coli Outbreaks Linked to Raw Milk Cheese. Available online: https://www.foodsafetynews.com/2019/05/more-ill-in-listeria-e-coli-outbreaks-linked-to-raw-milk-cheese/ (accessed on 28 December 2022).
13. Abiega-Franyutti, P.; Freyre-Fonseca, V. Chronic consumption of food-additives lead to changes via microbiota gut-brain axis. *Toxicology* **2021**, *464*, 153001. [CrossRef] [PubMed]
14. Luo, M.; Zhou, D.D.; Shang, A.; Gan, R.Y.; Li, H.B. Influences of food contaminants and additives on gut microbiota as well as protective effects of dietary bioactive compounds. *Trends Food Sci. Technol.* **2021**, *113*, 180–192. [CrossRef]
15. Peterson, C.T. Dysfunction of the Microbiota-Gut-Brain Axis in Neurodegenerative Disease: The Promise of Therapeutic Modulation with Prebiotics, Medicinal Herbs, Probiotics, and Synbiotics. *J. Evid.-Based Integr. Med.* **2020**, *25*. [CrossRef]
16. Bulajic, S. Biopreservation of traditional raw milk cheeses with an emphasis on Serbian artisanal cheeses and their historical production. *Meat Technol.* **2017**, *58*, 52–61.
17. Ali, A.M.M.; Sant'Ana, A.S.; Bavisetty, S.C.B. Sustainable preservation of cheese: Advanced technologies, physicochemical properties and sensory attributes. *Trends Food Sci. Technol.* **2022**, *129*, 306–326. [CrossRef]
18. Dupas, C.; Métoyer, B.; El Hatmi, H.; Adt, I.; Mahgoub, S.A.; Dumas, E. Plants: A natural solution to enhance raw milk cheese preservation? *Food Res. Int.* **2020**, *130*, 108883. [CrossRef] [PubMed]
19. Lindsay, D.; Robertson, R.; Fraser, R.; Engstrom, S.; Jordan, K. Heat induced inactivation of microorganisms in milk and dairy products. *Int. Dairy J.* **2021**, *121*, 105096. [CrossRef]
20. Giaccone, D.; Revello-Chion, A.; Galassi, L.; Bianchi, P.; Battelli, G.; Coppa, M.; Tabacco, E.; Borreani, G. Effect of milk thermisation and farming system on cheese sensory profile and fatty acid composition. *Int. Dairy J.* **2016**, *59*, 10–19. [CrossRef]
21. Dash, K.K.; Fayaz, U.; Dar, A.H.; Shams, R.; Manzoor, S.; Sundarsingh, A.; Deka, P.; Khan, S.A. A comprehensive review on heat treatments and related impact on the quality and microbial safety of milk and milk-based products. *Food Chem. Adv.* **2022**, *1*, 100041. [CrossRef]
22. Bintsis, T. Yeasts in different types of cheese. *AIMS Microbiol.* **2021**, *7*, 447. [CrossRef]
23. Sørhaug, T. Yeasts and Molds: Spoilage Molds in Dairy Products. In *Encyclopedia of Dairy Sciences*, 2nd ed.; Fuquay, J.W., Ed.; Elsevier Ltd.: London, UK, 2011; pp. 780–784. [CrossRef]
24. Kure, C.F.; Skaar, I. The fungal problem in cheese industry. *Curr. Opin. Food Sci.* **2019**, *29*, 14–19. [CrossRef]
25. Lavoie, K.; Touchette, M.; St-Gelais, D.; Labrie, S. Characterization of the fungal microflora in raw milk and specialty cheeses of the province of Quebec. *Dairy Sci. Technol.* **2012**, *92*, 455. [CrossRef]
26. Yuan, L.; Sadiq, F.A.; Burmølle, M.; Wang, N.; He, G. Insights into psychrotrophic bacteria in raw milk: A review. *J. Food Prot.* **2019**, *82*, 1148–1159. [CrossRef]
27. Podrzaj, L.; Burtscher, J.; Küller, F.; Domig, K.J. Strain-Dependent Cheese Spoilage Potential of *Clostridium tyrobutyricum*. *Microorganisms* **2020**, *8*, 1836. [CrossRef] [PubMed]
28. Robinson, R.K.; Tamime, A.Y.; Wszolek, M. Microbiology of fermented milks. In *Dairy Microbiology Handbook, the Microbiology of Milk and Milk Products*; Robinson, R.K., Ed.; John Wiley and Sons, Inc.: New York, NY, USA, 2002; pp. 367–430.
29. Tabla, R.; Gómez, A.; Simancas, A.; Rebollo, J.E.; Molina, F.; Roa, I. *Enterobacteriaceae* species during manufacturing and ripening of semi-hard and soft raw ewe's milk cheese: Gas production capacity. *Small Rumin. Res.* **2016**, *145*, 123–129. [CrossRef]
30. Food Standards Australia New Zealand. Microbiological Risk Assessment of Raw Milk Cheeses—Risk Assessment Microbiology Section. 2009. Available online: https://www.foodstandards.gov.au/code/proposals/documents/P1007%20PPPS%20for%20raw%20milk%201AR%20SD3%20Cheese%20Risk%20Assessment.pdf (accessed on 2 March 2023).
31. Martin, N.H.; Torres-Frenzel, P.; Wiedmann, M. *Invited review*: Controlling dairy product spoilage to reduce food loss and waste. *J. Dairy Sci.* **2021**, *104*, 1251–1261. [CrossRef] [PubMed]
32. Hayaloglu, A.A.; Farkye, N.Y. Cheese | Cheese with Added Herbs, Spices and Condiments. In *Encyclopedia of Dairy Sciences*, 2nd ed.; Fuquay, J.W., Ed.; Elsevier Ltd.: London, UK, 2011; pp. 783–789. [CrossRef]
33. Negi, P.S. Plant extracts for the control of bacterial growth: Efficacy, stability and safety issues for food application. *Int. J. Food Microbiol.* **2012**, *156*, 7–17. [CrossRef]
34. European Union. Directive 2009/32/EC of the European Parliament and of the Council of 23 April 2009 on the approximation of the laws of the Member States on extraction solvents used in the production of foodstuffs and food ingredients. *Off. J. Eur. Union* **2009**, *5*, 3–9.
35. Azwanida, N.N. A Review on the Extraction Methods Use in Medicinal Plants, Principle, Strength and Limitation. *Med. Aromat. Plants* **2015**, *4*, 196. [CrossRef]
36. Zhang, Q.W.; Lin, L.G.; Ye, W.C. Techniques for extraction and isolation of natural products: A comprehensive review. *Chin. Med.* **2018**, *13*, 1–26. [CrossRef]
37. Pisoschi, A.M.; Pop, A.; Georgescu, C.; Turcuş, V.; Olah, N.K.; Mathe, E. An overview of natural antimicrobials role in food. *Eur. J. Med. Chem.* **2018**, *143*, 922–935. [CrossRef]

38. Ferreira-Santos, P.; Genisheva, Z.; Botelho, C.; Santos, J.; Ramos, C.; Teixeira, J.A.; Rocha, C.M.R. Unravelling the biological potential of pinus pinaster bark extracts. *Antioxidants* **2020**, *9*, 334. [CrossRef]
39. Cheng, Y.; Xue, F.; Yu, S.; Du, S.; Yang, Y. Subcritical Water Extraction of Natural Products. *Molecules* **2021**, *26*, 4004. [CrossRef] [PubMed]
40. Sasidharan, S.; Chen, Y.; Saravanan, D.; Sundram, K.M.; Latha, L.Y. Extraction, Isolation and Characterization of Bioactive Compounds From Plants' Extracts. *Afr. J. Tradit. Complement. Altern. Med.* **2010**, *8*, 1–10. [CrossRef] [PubMed]
41. Martillanes, S.; Rocha-Pimienta, J.; Cabrera-Bañegil, M.; Martín-Vertedor, D.; Delgado-Adámez, J. Application of Phenolic Compounds for Food Preservation: Food Additive and Active Packaging. In *Phenolic Compounds—Biological Activity*; Soto-Hernández, M., Palma-Tenango, M., García-Mateos, R., Eds.; InTechOpen: Vienna, Austria, 2017; pp. 39–58. [CrossRef]
42. Khameneh, B.; Iranshahy, M.; Soheili, V.; Bazzaz, B.S.F. Review on plant antimicrobials: A mechanistic viewpoint. *Antimicrob. Resist. Infect. Control.* **2019**, *8*, 1–28. [CrossRef]
43. Ritota, M.; Manzi, P. Natural Preservatives from Plant in Cheese Making. *Animals* **2020**, *10*, 749. [CrossRef] [PubMed]
44. Hayek, S.A.; Gyawali, R.; Ibrahim, S.A. Antimicrobial Natural Products. In *Microbial Pathogens and Strategies for Combating Them: Science, Technology And Education*; Méndez-Vilas, A., Ed.; Formatex Research Center: Badajoz, Spain, 2013; pp. 910–921.
45. Gouvea, F.S.; Rosenthal, A.; Ferreira, E.H.R. Plant extract and essential oils added as antimicrobials to cheeses: A review. *Cienc. Rural* **2017**, *47*, 1–9. [CrossRef]
46. Cetin-Karaca, H. Evaluation of Natural Antimicrobial Phenolic Compounds against Foodborne Pathogens. Master's Thesis, University of Kentucky, Lexington, KY, USA, 2011. Available online: https://uknowledge.uky.edu/gradschool_theses/652 (accessed on 2 March 2023).
47. Giner, M.J.; Vegara, S.; Funes, L.; Marti, N.; Saura, D.; Micol, V.; Valero, M. Antimicrobial activity of food-compatible plant extracts and chitosan against naturally occurring micro-organisms in tomato juice. *J. Sci. Food Agric.* **2012**, *92*, 1917–1923. [CrossRef]
48. Hołderna-Kędzia, E.; Kędzia, B. Działanie preparatów pochodzenia roślinnego na drobnoustroje probiotyczne. *Post. Fitoter.* **2012**, *2*, 72–77.
49. Ziarno, M.; Kozłowska, M.; Ratusz, K.; Hasalliu, R. Effect of the Addition of Selected Herbal Extracts on the Quality Characteristics of Flavored Cream and Butter. *Foods* **2023**, *12*, 471. [CrossRef]
50. Zaika, L.L.; Kissinger, J.C.; Wasserman, A.E. Inhibition of Lactic Acid Bacteria by Herbs. *J. Food Sci.* **1983**, *48*, 1455–1459. [CrossRef]
51. Behrad, S.; Yusof, M.Y.; Goh, K.L.; Baba, A.S. Manipulation of Probiotics Fermentation of Yogurt by Cinnamon and Licorice: Effects on Yogurt Formation and Inhibition of *Helicobacter pylori* Growth in vitro. *Int. J. Nutr. Food Eng.* **2009**, *3*, 563–567. [CrossRef]
52. Shori, A.B.; Yong, Y.S.; Baba, A.S. Effects of medicinal plants extract enriched cheese with fish collagen on proteolysis and in vitro angiotensin-I converting enzyme inhibitory activity. *LWT—Food Sci. Technol.* **2022**, *159*, 113218. [CrossRef]
53. Ziarno, M.; Kozłowska, M.; Scibisz, I.; Kowalczyk, M.; Pawelec, S.; Stochmal, A.; Szleszyński, B. The Effect of Selected Herbal Extracts on Lactic Acid Bacteria Activity. *Appl. Sci.* **2021**, *11*, 3898. [CrossRef]
54. Mohamed, F.A.E.F.; Salama, H.H.; El-Sayed, S.M.; El-Sayed, H.S.; Zahran, H.A.H. Utilization of Natural Antimicrobial and Antioxidant of Moringa oleifera Leaves Extract in Manufacture of Cream Cheese. *J. Biol. Sci.* **2018**, *18*, 92–106. [CrossRef]
55. Chouchouli, V.; Kalogeropoulos, N.; Konteles, S.J.; Karvela, E.; Makris, D.P.; Karathanos, V.T. Fortification of yoghurts with grape (*Vitis vinifera*) seed extracts. *LWT—Food Sci. Technol.* **2013**, *53*, 522–529. [CrossRef]
56. Aktypis, A.; Christodoulou, E.D.; Manolopoulou, E.; Georgala, A.; Daferera, D.; Polysiou, M. Fresh ovine cheese supplemented with saffron (*Crocus sativus* L.): Impact on microbiological, physicochemical, antioxidant, color and sensory characteristics during storage. *Small Rumin. Res.* **2018**, *167*, 32–38. [CrossRef]
57. Hassanien, M.F.R.; Mahgoub, S.A.; El-Zahar, K.M. Soft cheese supplemented with black cumin oil: Impact on food borne pathogens and quality during storage. *Saudi J. Biol. Sci.* **2014**, *21*, 280–288. [CrossRef]
58. Gavriil, A.; Zilelidou, E.; Papadopoulos, A.-E.; Siderakou, D.; Kasiotis, K.M.; Haroutounian, S.A.; Gardeli, C.; Giannenas, I.; Skandamis, P.N. Evaluation of antimicrobial activities of plant aqueous extracts against *Salmonella* Typhimurium and their application to improve safety of pork meat. *Sci. Rep.* **2021**, *11*, 21971. [CrossRef]
59. Tassou, C.C.; Nychas, G.J.E. Inhibition of *Staphylococcus aureus* by olive phenolics in broth and in a model food system. *J. Food Prot.* **1994**, *57*, 120–124. [CrossRef]
60. Boziaris, I.S.; Proestos, C.; Kapsokefalou, M.; Komaitis, M. Antimicrobial Effect of *Filipendula ulmaria* Plant Extract Against Selected Foodborne Pathogenic and Spoilage Bacteria in Laboratory Media, Fish Flesh and Fish Roe Product. *Food Technol. Biotechnol.* **2011**, *49*, 263–270.
61. Del Campo, J.; Amiot, M.-J.; Nguyen-The, C. Antimicrobial effect of rosemary extracts. *J. Food Prot.* **2000**, *63*, 1359–1368. [CrossRef]
62. Oulahal, N.; Degraeve, P. Phenolic-Rich Plant Extracts With Antimicrobial Activity: An Alternative to Food Preservatives and Biocides? *Front. Microbiol.* **2022**, *1*, 753518. [CrossRef]
63. Gutierrez, J.; Barry-Ryan, C.; Bourke, P. The antimicrobial efficacy of plant essential oil combinations and interactions with food ingredients. *Int. J. Food Microbiol.* **2008**, *124*, 91–97. [CrossRef] [PubMed]
64. Shelef, L.A.; Jyothi, E.K.; Bulgarelii, M.A. Growth of Enteropathogenic and Spoilage Bacteria in Sage-Containing Broth and Foods. *J. Food Sci.* **1984**, *49*, 737–740. [CrossRef]
65. Tosif, M.M.; Najda, A.; Bains, A.; Krishna, T.C.; Chawla, P.; Dyduch-Siemińska, M.; Klepacka, J.; Kaushik, R. A Comprehensive Review on the Interaction of Milk Protein Concentrates with Plant-Based Polyphenolics. *Int. J. Mol. Sci.* **2021**, *22*, 13548. [CrossRef] [PubMed]

66. Moro, A.; Librán, C.M.; Berruga, M.I.; Carmona, M.; Zalacain, A. Dairy matrix effect on the transference of rosemary (*Rosmarinus officinalis*) essential oil compounds during cheese making. *J. Sci. Food Agric.* **2015**, *95*, 1507–1513. [CrossRef]
67. Brocklehurst, T.F.; Wilson, P.D.G. The role of lipids in controlling microbial growth. *Grasas Aceites* **2000**, *51*, 66–73. [CrossRef]
68. Verheyen, D.; Xu, X.M.; Govaert, M.; Baka, M.; Skåra, T.; Van Impe, J.F. Food microstructure and fat content affect growth morphology, growth kinetics, and preferred phase for cell growth of Listeria monocytogenes in fish-based model systems. *Appl. Environ. Microbiol.* **2019**, *85*, e00707-19. [CrossRef]
69. Peng, K.; Liu, W.; Xiong, Y.; Lu, L.; Liu, J.; Huang, X. Emulsion microstructural evolution with the action of environmentally friendly demulsifying bacteria. *Colloids Surf.* **2018**, *553*, 528–538. [CrossRef]
70. Lopez, C.; Maillard, M.B.; Briard-Bion, V.; Camier, B.; Hannon, J.A. Lipolysis during ripening of emmental cheese considering organization of fat and preferential localization of bacteria. *J. Agric. Food Chem.* **2006**, *54*, 5855–5867. [CrossRef]
71. Tayel, A.A.; Hussein, H.; Sorour, N.M.; El-Tras, W.F. Foodborne Pathogens Prevention and Sensory Attributes Enhancement in Processed Cheese via Flavoring with Plant Extracts. *J. Food Sci.* **2015**, *80*, M2886–M2891. [CrossRef]
72. Lee, N.K.; Jeewanthi, R.K.C.; Park, E.H.; Paik, H.D. Short communication: Physicochemical and antioxidant properties of Cheddar-type cheese fortified with *Inula britannica* extract. *J. Dairy Sci.* **2016**, *99*, 83–88. [CrossRef]
73. Abd El-Aziz, M.; Mohamed, S.H.S.; Seleet, F.L. Production and Evaluation of Soft Cheese Fortified with Ginger Extract as a Functional Dairy Food. *Polish J. Food Nutr. Sci.* **2012**, *62*, 77–83. [CrossRef]
74. Mahajan, D.; Bhat, Z.F.; Kumar, S. Pine needles (*Cedrus deodara* (Roxb.) Loud.) extract as a novel preservative in cheese. *Food Pack. Shelf Life* **2016**, *7*, 20–25. [CrossRef]
75. Evstigneeva, T.; Skvortsova, N.; Yakovleva, R. The application of green tea Extract as a source of antioxidants in the processing of dairy products. *Agron.Res.* **2016**, *14*, 1284–1298.
76. European Union. Regulation No 1334/2008 of the European Parliament and of the Council of 16 December 2008 on flavourings and certain food ingredients with flavouring properties for use in and on foods and amending Council Regulation (EEC) No 1601/91, Regulations (EC) No 2232/96 and (EC) No 110/2008 and Directive 2000/13/EC. *Off. J. Eur. Union* **2008**, *354*, 34–49.
77. Code of Federal Regulations. eCFR: 21 CFR 172.510—Natural Flavoring Substances and Natural Substances Used in Conjunction with Flavors. Available online: https://www.ecfr.gov/current/title-21/chapter-I/subchapter-B/part-172/subpart-F/section-17 2.510 (accessed on 2 March 2023).
78. Carocho, M.; Morales, P.; Ferreira, I.C.F.R. Natural food additives: Quo vadis? *Trends Food Sci.Technol.* **2015**, *45*, 284–295. [CrossRef]
79. Coelho, M.C.; Malcata, F.X.; Silva, C.C.G. Lactic Acid Bacteria in Raw-Milk Cheeses: From Starter Cultures to Probiotic Functions. *Foods* **2022**, *11*, 2276. [CrossRef]
80. Bettera, L.; Levante, A.; Bancalari, E.; Bottari, B.; Gatti, M. Lactic acid bacteria in cow raw milk for cheese production: Which and how many? *Front. Microbiol.* **2023**, *13*, 5413. [CrossRef]
81. Silva, C.C.G.; Silva, S.P.M.; Ribeiro, S.C. Application of bacteriocins and protective cultures in dairy food preservation. *Front. Microbiol.* **2018**, *9*, 594. [CrossRef]
82. Beresford, T.P.; Fitzsimons, N.A.; Brennan, N.L.; Cogan, T.M. Recent advances in cheese microbiology. *Int. Dairy J.* **2001**, *11*, 259–274. [CrossRef]
83. Tsai, Y.T.; Cheng, P.C.; Pan, T.M. The immunomodulatory effects of lactic acid bacteria for improving immune functions and benefits. *Appl. Microbiol. Biotechnol.* **2012**, *96*, 853–862. [CrossRef]
84. Capozzi, V.; Russo, P.; Ladero, V.; Fernández, M.; Fiocco, D.; Alvarez, M.A.; Grieco, F.; Spano, G. Biogenic amines degradation by *Lactobacillus plantarum*: Toward a potential application in wine. *Front. Microbiol.* **2012**, *3*, 122. [CrossRef]
85. Ma, C.; Zhang, S.; Lu, J.; Zhang, C.; Pang, X.; Lv, J. Screening for Cholesterol-Lowering Probiotics from Lactic Acid Bacteria Isolated from Corn Silage Based on Three Hypothesized Pathways. *Int. J. Mol. Sci.* **2019**, *20*. [CrossRef]
86. Damodharan, K.; Palaniyandi, S.A.; Yang, S.H.; Suh, J.W. Functional probiotic characterization and in vivo cholesterol-lowering activity of *Lactobacillus helveticus* isolated from fermented cow milk. *J. Microbiol. Biotechnol.* **2016**, *26*, 1675–1686. [CrossRef] [PubMed]
87. Hati, S.; Patel, M.; Mishra, B.K.; Das, S. Short-chain fatty acid and vitamin production potentials of *Lactobacillus* isolated from fermented foods of Khasi Tribes, Meghalaya, India. *Ann. Microbiol.* **2019**, *69*, 1191–1199. [CrossRef]
88. Renes, E.; Linares, D.M.; González, L.; Fresno, J.M.; Tornadijo, M.E.; Stanton, C. Production of conjugated linoleic acid and gamma-aminobutyric acid by autochthonous lactic acid bacteria and detection of the genes involved. *J. Funct. Foods* **2017**, *34*, 340–346. [CrossRef]
89. Łepecka, A.; Szymański, P.; Rutkowska, S.; Iwanowska, K.; Kołożyn-Krajewska, D. The Influence of Environmental Conditions on the Antagonistic Activity of Lactic Acid Bacteria Isolated from Fermented Meat Products. *Foods* **2021**, *10*, 2267. [CrossRef]
90. Mellefont, L.A.; McMeekin, T.A.; Ross, T. Effect of relative inoculum concentration on *Listeria monocytogenes* growth in co-culture. *Int. J. Food Microbiol.* **2008**, *121*, 157–168. [CrossRef] [PubMed]
91. Gonzales-Barron, U.; Campagnollo, F.B.; Schaffner, D.W.; Sant'Ana, A.S.; Cadavez, V.A.P. Behavior of *Listeria monocytogenes* in the presence or not of intentionally-added lactic acid bacteria during ripening of artisanal Minas semi-hard cheese. *Food Microbiol.* **2020**, *91*, 103545. [CrossRef]

92. Campagnollo, F.B.; Margalho, L.P.; Kamimura, B.A.; Feliciano, M.D.; Freire, L.; Lopes, L.S.; Alvarenga, V.O.; Cadavez, V.A.P.; Gonzales-Barron, U.; Schaffner, D.W.; et al. Selection of indigenous lactic acid bacteria presenting anti-listerial activity, and their role in reducing the maturation period and assuring the safety of traditional Brazilian cheeses. *Food Microbiol.* **2018**, *73*, 288–297. [CrossRef]
93. Callon, C.; Picque, D.; Corrieu, G.; Montel, M.C. Ripening conditions: A tool for the control of *Listeria monocytogenes* in uncooked pressed type cheese. *Food Control* **2011**, *22*, 1911–1919. [CrossRef]
94. Ammor, S.; Tauveron, G.; Dufour, E.; Chevallier, I. Antibacterial activity of lactic acid bacteria against spoilage and pathogenic bacteria isolated from the same meat small-scale facility. 1-Screening and characterization of the antibacterial compounds. *Food Control* **2006**, *17*, 454–461. [CrossRef]
95. Tan, P.L.; Peh, K.K.; Gan, C.Y.; Liong, M.T. Bioactive dairy ingredients for food and non-food applications. *Acta Aliment.* **2014**, *43*, 113–123. [CrossRef]
96. Lindgren, S.E.; Dobrogosz, W.J. Antagonistic activities of lactic acid bacteria in food and feed fermentations. *FEMS Microbiol. Rev.* **1990**, *7*, 149–163. [CrossRef] [PubMed]
97. Elliot, R.M.; McLay, J.C.; Kennedy, M.J.; Simmonds, R.S. Inhibition of foodborne bacteria by the lactoperoxidase system in a beef cube system. *Int. J. Food Microbiol.* **2004**, *91*, 73–81. [CrossRef] [PubMed]
98. García-Quintans, N.; Blancato, V.; Repizo, G.; Magni, C.; López, P. Citrate metabolism and aroma compound production in lactic acid bacteria. In *Molecular Aspects of Lactic Acid Bacteria for Traditional and New Applications*; Mayo, B., López, P., Pérez-Martínez, G., Eds.; Research Signpost: Kerala, India, 2008.
99. Domingos-Lopes, M.F.P.; Stanton, C.; Ross, P.R.; Dapkevicius, M.L.E.; Silva, C.C.G. Genetic diversity, safety and technological characterization of lactic acid bacteria isolated from artisanal Pico cheese. *Food Microbiol.* **2017**, *63*, 178–190. [CrossRef] [PubMed]
100. Jay, J.M. Antimicrobial properties of diacetyl. *Appl. Environ. Microbiol.* **1982**, *44*, 525. [CrossRef]
101. Casillas-Vargas, G.; Ocasio-Malavé, C.; Medina, S.; Morales-Guzmán, C.; García, R.; Valle, D.; Carballeira, N.M.; Sanabria-Ríos, D.J. Antibacterial fatty acids: An update of possible mechanisms of action and implications in the development of the next-generation of antibacterial agents. *Prog. Lipid Res.* **2021**, *82*, 101093. [CrossRef] [PubMed]
102. Desbois, A.P.; Smith, V.J. Antibacterial free fatty acids: Activities, mechanisms of action and biotechnological potential. *Appl. Microbiol. Biotechnol.* **2010**, *85*, 1629–1642. [CrossRef]
103. Engels, C.; Schwab, C.; Zhang, J.; Stevens, M.J.A.; Bieri, C.; Ebert, M.-O.; Mcneill, K.; Sturla, S.J.; Lacroix, C. Acrolein contributes strongly to antimicrobial and heterocyclic amine transformation activities of reuterin. *Sci. Rep.* **2016**, *6*, 36246. [CrossRef]
104. Kumar, N.; Kumar, V.; Mohsin Waheed, S.; Pradhan, D. Efficacy of Reuterin and Bacteriocins Nisin and Pediocin in the Preservation of Raw Milk from Dairy Farms. *Food Technol. Biotechnol.* **2020**, *58*, 359–369. [CrossRef] [PubMed]
105. Dalié, D.K.D.; Deschamps, A.M.; Richard-Forget, F. *Lactic acid bacteria*—Potential for control of mould growth and mycotoxins: A review. *Food Control* **2010**, *21*, 370–380. [CrossRef]
106. Schaefer, L.; Auchtung, T.A.; Hermans, K.E.; Whitehead, D.; Borhan, B.; Britton, R.A. The antimicrobial compound reuterin (3-hydroxypropionaldehyde) induces oxidative stress via interaction with thiol groups. *Microbiology* **2010**, *156*, 1589. [CrossRef]
107. Soltani, S.; Couture, F.; Boutin, Y.; Said, L.B.; Cashman-Kadri, S.; Subirade, M.; Biron, E.; Fliss, I. In vitro investigation of gastrointestinal stability and toxicity of 3-hyrdox-ypropionaldehyde (reuterin) produced by *Lactobacillus reuteri*. *Toxicol. Rep.* **2021**, *8*, 740–746. [CrossRef] [PubMed]
108. Kaškonienė, V.; Stankevičius, M.; Bimbiraitė-Survilienė, K.; Naujokaitytė, G.; Šernienė, L.; Mulkytė, K.; Malakauskas, M.; Maruška, A. Current state of purification, isolation and analysis of bacteriocins produced by lactic acid bacteria. *Appl. Microbiol. Biotechnol.* **2017**, *101*, 1323–1335. [CrossRef] [PubMed]
109. Cotter, P.D.; Ross, R.P.; Hill, C. Bacteriocins—A viable alternative to antibiotics? *Nat. Rev. Microbiol.* **2013**, *11*, 95–105. [CrossRef]
110. Line, J.E.; Svetoch, E.A.; Eruslanov, B.V.; Perelygin, V.V.; Mitsevich, E.V.; Mitsevich, I.P.; Levchuk, V.P.; Svetoch, O.E.; Seal, B.S.; Siragusa, G.R.; et al. Isolation and purification of enterocin E-760 with broad antimicrobial activity against gram-positive and gram-negative bacteria. *Antimicrob. Agents Chemother.* **2008**, *52*, 1094–1100. [CrossRef]
111. Hernández-González, J.C.; Martínez-Tapia, A.; Lazcano-Hernández, G.; García-Pérez, B.E.; Castrejón-Jiménez, N.S. Bacteriocins from Lactic Acid Bacteria. A Powerful Alternative as Antimicrobials, Probiotics, and Immunomodulators in Veterinary Medicine. *Animals* **2021**, *11*, 979. [CrossRef]
112. Crowley, S.; Mahony, J.; Van Sinderen, D. Current perspectives on antifungal lactic acid bacteria as natural bio-preservatives. *Trends Food Sci. Technol.* **2013**, *33*, 93–109. [CrossRef]
113. Abouloifa, H.; Hasnaoui, I.; Rokni, Y.; Bellaouchi, R.; Ghabbour, N.; Karboune, S.; Brasca, M.; Abousalham, A.; Jaouadi, B.; Saalaoui, E.; et al. Antifungal activity of lactic acid bacteria and their application in food biopreservation. *Adv. Appl. Microbiol.* **2022**, *120*, 33–77. [CrossRef]
114. Ibrahim, S.A.; Ayivi, R.D.; Zimmerman, T.; Siddiqui, S.A.; Altemimi, A.B.; Fidan, H.; Esatbeyoglu, T.; Bakhshayesh, R.V. Lactic Acid Bacteria as Antimicrobial Agents: Food Safety and Microbial Food Spoilage Prevention. *Foods* **2021**, *10*, 3131. [CrossRef] [PubMed]
115. Rukke, E.O.; Sørhaug, T.; Stepaniak, L. Heat Treatment of Milk: Thermization of Milk. In *Encyclopedia of Dairy Sciences*; Fuquay, J.W., Ed.; Elsevier Ltd.: London, UK, 2011; pp. 693–698. [CrossRef]

116. Panthi, R.R.; Jordan, K.N.; Kelly, A.L.; Sheehan, J.J.D. Selection and Treatment of Milk for Cheesemaking. In *Cheese: Chemistry, Physics and Microbiology*, 4th ed.; McSweeney, P.L.H., Fox, P.F., Cotter, P.D., Everett, D.W., Eds.; Elsevier Ltd.: London, UK, 2017; pp. 23–50. [CrossRef]
117. Eugster, E.; Jakob, E. Pre-treatments of Milk and their Effect on the Food Safety of Cheese. *Milk Sci. Int.* **2019**, *72*, 45–52.
118. Codex Committee on Food Hygiene. Code of Hygienic Practice for Milk and Milk Products (CAC/RCP 57-2004). 2009. Available online: https://www.fao.org/fileadmin/user_upload/livestockgov/documents/CXP_057e.pdf (accessed on 1 June 2023).
119. Samelis, J.; Lianou, A.; Kakouri, A.; Delbès, C.; Rogelj, I.; Bogovic-Matijasić, B.; Montel, M.-C. Changes in the Microbial Composition of Raw Milk Induced by Thermization Treatments Applied Prior to Traditional Greek Hard Cheese Processing. *J. Food Prot.* **2009**, *72*, 783–790. [CrossRef] [PubMed]
120. Cebrián, G.; Condón, S.; Mañas, P. Physiology of the Inactivation of Vegetative Bacteria by Thermal Treatments: Mode of Action, Influence of Environmental Factors and Inactivation Kinetics. *Foods* **2017**, *6*, 107. [CrossRef]
121. Russell, A.D. Lethal effects of heat on bacterial physiology and structure. *Sci. Prog.* **2003**, *86*, 115–137. [CrossRef]
122. Tsuchido, T.; Katsui, N.; Takeuchi, A.; Takano, M.; Shibasaki, I. Destruction of the outer membrane permeability barrier of *Escherichia coli* by heat treatment. *Appl. Environ. Microbiol.* **1985**, *50*, 298. [CrossRef]
123. Baumgarten, T.; Sperling, S.; Seifert, J.; von Bergen, M.; Steiniger, F.; Wick, L.Y.; Heipieper, H.J. Membrane vesicle formation as a multiple-stress response mechanism enhances *Pseudomonas putida* DOT-T1E cell surface hydrophobicity and biofilm formation. *Appl. Environ. Microbiol.* **2012**, *78*, 6217–6224. [CrossRef]
124. Katsui, N.; Tsuchido, T.; Hiramatsu, R.; Fujikawa, S.; Takano, M.; Shibasaki, I. Heat-induced blebbing and vesiculation of the outer membrane of *Escherichia coli*. *J. Bacteriol.* **1982**, *151*, 1523. [CrossRef]
125. Mackey, B.M. Changes in antibiotic sensitivity and cell surface hydrophobicity in *Escherichia coli* injured by heating, freezing, drying or gamma radiation. *FEMS Microbiol. Lett.* **1983**, *20*, 395–399. [CrossRef]
126. Beuchat, L.R. Injury and Repair of Gram-Negative Bacteria, with Special Consideration of the Involvement of the Cytoplasmic Membrane. *Adv. Appl. Microbiol.* **1978**, *23*, 219–243. [CrossRef]
127. Witter, L.D. Thermal Injury and Recovery of Selected Microorganisms. *J. Dairy Sci.* **1979**, *64*, 174–177. [CrossRef]
128. Ebrahimi, A.; Csonka, L.N.; Alam, M.A. Analyzing Thermal Stability of Cell Membrane of Salmonella Using Time-Multiplexed Impedance Sensing. *Biophys. J.* **2018**, *114*, 609. [CrossRef] [PubMed]
129. Li, J.; Suo, Y.; Liao, X.; Ahn, J.; Liu, D.; Chen, S.; Ye, X.; Ding, T. Analysis of *Staphylococcus aureus* cell viability, sublethal injury and death induced by synergistic combination of ultrasound and mild heat. *Ultrason. Sonochem.* **2017**, *39*, 101–110. [CrossRef] [PubMed]
130. Mitsuzawa, S.; Deguchi, S.; Horikoshi, K. Cell structure degradation in *Escherichia coli* and *Thermococcus* sp. strain Tc-1-95 associated with thermal death resulting from brief heat treatment. *FEMS Microbiol. Lett.* **2006**, *260*, 100–105. [CrossRef]
131. Karni, M.; Zidon, D.; Polak, P.; Zalevsky, Z.; Shefi, O. Thermal Degradation of DNA. *DNA Cell Biol.* **2013**, *32*, 298–301. [CrossRef] [PubMed]
132. Hanlin, J.H.; Lombardi, S.J.; Slepecky, R.A. Heat and UV light resistance of vegetative cells and spores of *Bacillus subtilis* Rec-mutants. *J. Bacteriol.* **1985**, *163*, 774. [CrossRef]
133. Zamenhof, S. Gene unstabilization induced by heat and by nitrous acid. *J. Bacteriol.* **1961**, *81*, 111. [CrossRef]
134. Varela-Ramirez, A.; Abendroth, J.; Mejia, A.A.; Phan, I.Q.; Lorimer, D.D.; Edwards, T.E.; Aguilera, R.J. Structure of acid deoxyribonuclease. *Nucleic Acids Res.* **2017**, *45*, 6217. [CrossRef]
135. Earnshaw, R.G.; Appleyard, J.; Hurst, R.M. Understanding physical inactivation processes: Combined preservation opportunities using heat, ultrasound and pressure. *Int. J. Food Microbiol.* **1995**, *28*, 197–219. [CrossRef]
136. Tolker-Nielsen, T.; Molin, S. Role of ribosome degradation in the death of heat-stressed *Salmonella typhimurium*. *FEMS Microbiol. Lett.* **1996**, *142*, 155–160. [CrossRef]
137. Allwood, M.C.; Russell, A.D. Thermally Induced Ribonucleic Acid Degradation and Leakage of Substances from the Metabolic Pool in *Staphylococcus aureus*. *J. Bacteriol.* **1968**, *95*, 345. [CrossRef] [PubMed]
138. Allwood, M.C.; Russell, A.D. Mechanism of Thermal Injury in *Staphylococcus aureus*: I. Relationship Between Viability and Leakage. *Appl. Microbiol.* **1967**, *15*, 1266–1269.
139. Weids, A.J.; Ibstedt, S.; Tamás, M.J.; Grant, C.M. Distinct stress conditions result in aggregation of proteins with similar properties. *Sci. Rep.* **2016**, *6*, 24554. [CrossRef] [PubMed]
140. Rosenberg, B.; Kemeny, G.; Switzer, R.C.; Hamilton, T.C. Quantitative evidence for protein denaturation as the cause of thermal death. *Nature* **1971**, *232*, 471–473. [CrossRef]
141. Nguyen, H.T.T.; Corry, J.E.L.; Miles, C.A. Heat Resistance and Mechanism of Heat Inactivation in Thermophilic Campylobacters. *Appl. Environ. Microbiol.* **2006**, *72*, 908–913. [CrossRef]
142. Mackey, B.M.; Derrick, C.M. Elevation of the heat resistance of Salmonella typhimurium by sublethal heat shock. *J. Appl. Bacteriol.* **1986**, *61*, 389–393. [CrossRef] [PubMed]
143. D'Amico, D.J.; Donnelly, C.W. Growth and Survival of Microbial Pathogens in Cheese. In *Cheese: Chemistry, Physics and Microbiology*, 4th ed.; McSweeney, P.L.H., Fox, P.F., Cotter, P.D., Everett, D.W., Eds.; Elsevier Ltd.: London, UK, 2017; pp. 573–594. [CrossRef]
144. Engstrom, S.K.; Mays, M.F.; Glass, K.A. Determination and validation of D-values for *Listeria monocytogenes* and Shiga toxin–producing *Escherichia coli* in cheese milk. *J. Dairy Sci.* **2012**, *104*, 12332–12341. [CrossRef]

145. Peng, S.; Hummerjohann, J.; Stephan, R.; Hammer, P. Short communication: Heat resistance of *Escherichia coli* strains in raw milk at different subpasteurization conditions. *J. Dairy Sci.* **2013**, *96*, 3543–3546. [CrossRef]
146. Van Den Tempel, T.; Jakobsen, M. Yeasts associated with Danablu. *Int. Dairy J.* **1998**, *8*, 25–31. [CrossRef]

Disclaimer/Publisher's Note: The statements, opinions and data contained in all publications are solely those of the individual author(s) and contributor(s) and not of MDPI and/or the editor(s). MDPI and/or the editor(s) disclaim responsibility for any injury to people or property resulting from any ideas, methods, instructions or products referred to in the content.

Article

Microbiological, Physicochemical and Nutritional Properties of Fresh Cow Milk Treated with Industrial High-Pressure Processing (HPP) during Storage

Shu Huey Lim [1], Nyuk Ling Chin [1,*], Alifdalino Sulaiman [1], Cheow Hwang Tay [2] and Tak Hiong Wong [2]

1. Department of Process and Food Engineering, Faculty of Engineering, Universiti Putra Malaysia (UPM), Serdang 43400, Selangor, Malaysia
2. F&N Global Marketing Pte. Ltd., Singapore 119958, Singapore
* Correspondence: chinnl@upm.edu.my; Tel.: +60-3-97696353

Abstract: The safety, shelf life, and quality of fresh cow milk treated using industrial High-Pressure Processing (HPP) treatment at 600 MPa for 10 min was studied to identify the novelty of this non-thermal technology in milk processing. Changes in microbiological and physicochemical properties, including nutritional values of vitamins and amino acid profiles, were measured for a 60-day storage period at 6 °C +/− 1 °C. The HPP treatment produced milk that met all microbial safety requirements and exhibited a shelf life beyond 60 days in a hot and humid region. High physicochemical stability was achieved, with consistent pH and undetectable titratable acidity. The HPP treatment successfully retained all vitamins and minerals, including calcium (99.3%), phosphorus (99.4%), and magnesium (99.1%). However, the 60-day storage caused some degradation of Vitamin A (25%), B3 (91%), B5 (35%), B6 (80%), and C (85%), and minerals, including potassium (5%) and zinc (18%) when compared with fresh milk. This research has shown that the adoption of advanced treatment with HPP is very beneficial to the dairy industry in preserving milk quality in terms of its physicochemical and nutritional properties and extending its storage shelf life beyond 60 days.

Keywords: milk processing; microbiological properties; physicochemical properties; vitamins; minerals; storage

Citation: Lim, S.H.; Chin, N.L.; Sulaiman, A.; Tay, C.H.; Wong, T.H. Microbiological, Physicochemical and Nutritional Properties of Fresh Cow Milk Treated with Industrial High-Pressure Processing (HPP) during Storage. *Foods* **2023**, *12*, 592. https://doi.org/10.3390/foods12030592

Academic Editors: Piero Franceschi and Paolo Formaggioni

Received: 31 December 2022
Revised: 16 January 2023
Accepted: 21 January 2023
Published: 31 January 2023

Copyright: © 2023 by the authors. Licensee MDPI, Basel, Switzerland. This article is an open access article distributed under the terms and conditions of the Creative Commons Attribution (CC BY) license (https://creativecommons.org/licenses/by/4.0/).

1. Introduction

Milk is a balanced and nutritive food that is important for human health. However, the high moisture and nutrition contents of milk cause rapid microorganism proliferation and result in spoilage. The main microorganisms present in fresh milk include *Staphylococcus aureus*, *Salmonella spp.*, *Listeria monocytogenes*, *Escherichia coli* O157:H7, and *Campylobacter* [1], which could pose a severe hazard to humans if they unknowingly consume contaminated milk. Thermal treatments, including low-temperature long-time pasteurisation (LTLT, 63 °C/30 min), high-temperature short-time pasteurisation (HTST, 72–75 °C for 15–20 s), ultra-pasteurisation/extended shelf-life treatment (ESL, 125–128 °C for 2–4 s), and ultra-high temperature treatment (UHT, 135–140 °C for 1–3 s) are common techniques used for extending the shelf life of milk [2]. However, common LTLT or HTST pasteurisation only provide an average of 10 days of shelf life [3], while ultra-pasteurisation or extended shelf-life (ESL) milk products may have a shelf life between 15 and 30 days [3,4]. In addition to limited shelf life, thermal treatments also have detrimental effects on milk quality, such as whey protein denaturation [5], reduction of calcium, vitamins (thiamine, B12, and C), and changes in organoleptic properties [6]. In this regard, non-thermal technologies such as high-pressure processing (HPP), pulse electric field, ultrasonication, and irradiation are innovated with aims to minimise loss of nutrients by thermal treatment while still being effective in destroying pathogenic and spoilage microorganisms in milk [7]. An investigation has demonstrated that HPP offers promising potential in processing high-quality milk [8]. It normally employs pressure of 300–600 MPa at room temperature for 2–30 min to eliminate pathogenic microorganisms and extends the shelf life of milk with minimal alteration of nutritional and sensorial attributes [9]. High isostatic pressure only disrupts the non-covalent bonds (hydrogen, ionic, and hydrophobic bonds), causing alterations in cell morphology and membrane that eventually leads to cell death and achieves microbial inactivation [10]. Hence, the nutritional and sensory properties of the products are not affected. HPP also has other advantages such as low contamination risk and is declared an environmentally friendly processing technology [10]. Although HPP offers many benefits, its use by manufacturers is still limited due to high equipment costs.

Most HPP research has been performed at lab-scale, with García–Risco et al. (1998) [11] reporting the shelf life of HPP milk (45 days) when treated at 400 MPa for 30 min using an HPP apparatus (Model 900, Eurotherm Automation, Lyon, France), while Mussa and Ramaswamy (1997) [12] reported that milk subjected to HPP at 350 MPa for 32 min using an ABB Isostatic Press (ABB Autoclave System, Autoclave Engineers, Erie, PA, USA) had a shelf-life of 12–25 days when stored at 0–10 °C. At the industrial scale, Stratakos et al. (2019) [13] observed a storage life of 28 days at a storage temperature of 4 °C when milk was treated at 600 MPa for 3 min using a 35 L commercial-scale high-pressure press. Studies on the industrial scale of HPP are still limited as scale-up of HPP treatment for milk preservation from lab-scale to pilot or industrial scale is a huge challenge due to high costs and equipment availability. Tan et al. (2020) [14] reported that fresh milk's shelf life was extended to 22 days using an industrial-scale HPP treatment (600 MPa, 5–7 min), but the data were available for only microbiological evaluations. To date, no complete study has been found reporting all microbiological, physicochemical, and nutritional contents of vitamins and minerals of HPP-treated milk at an industrial scale. This research was conducted to evaluate the effectiveness of industrial HPP treatment of locally produced fresh milk in Malaysia in a hot and humid climate with storage issues while overcoming detrimental effects of milk quality using heat treatments.

2. Materials and Methods

2.1. Processing of Milk

Fresh cow milk was collected from the university farm, Ladang 16, Universiti Putra Malaysia. Milk was mixed evenly, cooled to 4 °C, and poured into 350 mL polyethylene terephthalate (PET) bottles. Milk was treated in a 55 L high-pressure processing unit (Hiperbaric 55; Hiperbaric High-Pressure Technologies, Burgos, Spain) at 600 MPa for 10 min at 10 ± 2 °C using water as the transmitting fluid. Fresh milk was used as a control. High-pressure treated milk was labelled as HPP milk, and untreated milk as fresh milk. All milk samples were stored at 6 ± 1 °C for storage analysis. Samples were analysed for microbiological and physicochemical properties and vitamin and mineral contents from day 0. Fresh milk was kept for analysis until day 6, while treated milk until day 60. For fresh milk, microbiology analysis was conducted daily, with a 3-day interval for physicochemical, vitamin, and mineral analysis. The HPP milk samples were analysed at intervals of 5 days for microbiological, 6 days for physicochemical analysis, and 15 days for vitamin and mineral analysis during the 60-day storage.

2.2. Microbiological Analysis

Total plate count (TPC) and total yeast and mold count in milk were determined using procedures described in Chapters 3 and 18 of the FDA Bacteriological Analytical Manual (BAM) [15]. The total coliform count was calculated using the AOAC Official Method 991.14 by incubating for 24 ± 2 h at 35 ± 1 °C; then, the coliforms appeared as red colonies with one or more gas bubbles counted promptly using the standard colony counter [15]. Mesophilic, thermophilic aerobic spore count, and psychrotrophic bacteria count were determined following Chapters 23, 26, and 13, respectively, in the *Compendium of Methods for the Microbiological Examination of Foods* (CMMEF) [16] with different incubating times and temperatures, where mesophilic incubated anaerobically at 35 °C for 3 to 5 days, thermophilic aerobic at 55 ± 2 °C for 72 h while psychrotrophic bacteria at 17 ± 1 °C for 16 h, followed by 3 more days at 7 ± 1 °C. Psychrotrophic spore count was determined. *Bacillus cereus, Staphylococcus aureus,* and *Listeria monocytogenes* were determined following FDA BAM Chapters 14, 12, and 10, respectively [15]. *Escherichia coli (E. coli)* and *Clostridium perfringens* were determined by AOAC Official Method 991.14 [15] and 976.30 [15], respectively. For *E. coli*, the plates were incubated for 48 ± 4 h at 35 ± 1 °C. After the incubation period, *E. coli* colonies appeared as blue colonies associated with gas bubbles and were counted promptly. For *Clostridium perfringens,* tryptose sulfite cycloserine (TSC) agar was used for the incubation of black colony for 20 h at 35 °C. *Salmonella* spp. was determined using the enzyme-linked immunosorbent assay (ELISA) method as described by Veling et al. [17]. Colonies were enumerated, and results were expressed as the logarithm of colony-forming units ($\log_{10}$CFU/mL). Colony-forming unit (CFU) is a unit used in microbiology to measure the number of viable microorganisms (bacteria, fungi, viruses, etc.) in a sample that can reproduce through binary fission under controlled conditions.

2.3. Physicochemical Analysis

The pH of cow milk samples was measured using a pH meter (Mi805, Milwaukee, Hungary). Titratable acidity (TA) was determined according to the Association of Official Analytical Chemists (AOAC) Official Method 947.05 using phenolphthalein indicator to titrate with 0.1M NaOH and expressed as a percentage of lactic acid. Specific gravity was determined following the method for the specific gravity of water in AOAC Official Method 955.37, using a pycnometer. Total protein was determined using formaldehyde titration described by Pyne's method [18], in which oxalate was used as an alkali for the titration, and total protein was calculated using a factor of 1.74. The total solid of cow milk was determined using the oven drying method at 102 ± 2 °C for 2.5 h in accordance with AOAC Official Method 16.032 [19]. Fat content was determined using the Roese–Gottlieb Method in AOAC Official Method 905.02, which involved extraction, distillation, and drying. Solid-non-fat (SNF) was determined by subtracting the total solid from the total fat content.

2.4. Nutritional Analysis

2.4.1. Vitamin Content Analysis

Vitamin A (Retinol) and E (alpha-tocopherol) were determined using high-performance liquid chromatographic methods (HPLC) after alkaline saponification. Vitamin B1, B2, and B6 were analysed following the method described by Agostini-Costa et al. [20]. Vitamin B5 was determined based on Food Chemistry (2000), similar to the method reported by Woollard et al. [21]. Vitamin B7, B9, and B12 were analysed based on AOAC Official Method 960.46 using a microtiter assay [22]. In this test, the test vitamin extract was pipetted to the walls of a microtiter plate coated with microorganisms, followed by incubation at 37 °C for 44–48 h until all vitamins were consumed. After the incubation period, the absorbances were measured using a microtiter plate reader at 610–630 nm. Vitamin C (ascorbic acid) was determined according to AOAC 967.21 using 2,6-dichlorophenolindophenol (DCPIP) titration [23]. Vitamin K was determined based on AOAC Official Method 999.15, where extraction was performed using hexene, followed by HPLC determination [24]. Vitamin D was determined based on AOAC Official Method 995.05 via a single liquid–liquid extraction followed by saponification, solid-phase extraction, and evaporation, as described by Silva and Sanders [25].

2.4.2. Mineral Content Analysis

Milk mineral content, including Calcium (Ca), Potassium (K), Magnesium (Mg), Phosphorus (P), Zinc (Zn), and Selenium (Se), first underwent acid digestion as reported by Kira and Maihara [26], followed by mineral content analysis based on AOAC Official Method 984.27 (final action 1986) using inductively coupled plasma optical emission spectrometry,

ICP-OES (Varian 720-ES, Varian Inc., Walnut Creek, CA, USA) with a solution of yttrium as internal standard.

2.5. Statistical Analysis

All measurements were performed in triplicate with samples prepared from the same batch of milk, except for logistic reasons, Vitamin K was without repetition. Mean values were compared using analysis of variance (ANOVA) using the tool XLSTAT by Addinsoft and Microsoft Excel software. Tukey's test was performed to compare the differences between the mean values at a confidence level of 0.05.

3. Results and Discussion

3.1. Microbiological Analysis

HPP treatment successfully decreased TPC by 75% and other bacteria counts in milk by 100% (Table 1), meeting food safety limits and local industrial standards [27]. Figure 1 shows that storage of HPP-treated milk for 60 days gave high microbiological stability as its microbial activities were under control when compared to the definite unsafe condition of fresh milk with the continuous growth of all microbiological bacteria stored at 6 ± 1 °C for its first 6 days (Appendix A).

Table 1. Effects of HPP on microbiological properties of fresh milk.

Bacteria Count ($\log_{10}$ CFU/mL)	Fresh Milk	HPP	Safety Limit
Total Plate	5.13 ± 0.06 [b]	1.26 ± 0.04 [a]	<5
Total Coliform	2.56 ± 0.02	<1	<1.7
Escherichia coli	1.53 ± 0.12	<1	<2
Yeast and Molds	1.61 ± 0.15	<1	<1 *
Psychrotrophic Bacteria	6.81 ± 0.05	<1	<5
Psychrotrophic Spore	0.99 ± 0.15	<1	<1 *
Mesophilic Aerobic Spore	0.72 ± 0.06	<1	<5
Thermophilic Aerobic Spore	<1	<1	<1 *
Staphylococcus aureus	2.96 ± 0.03	<1	<1.3
Bacillus cereus	<1	<1	<5
Clostridium perfringens	<1	<1	<4
Bacteria Existence (Present/Absent)	**Fresh Milk**	**HPP**	**Safety Limit**
Listeria monocytogenes per 25 gm	Absent	Absent	Absent
Salmonella spp. per 25 gm	Absent	Absent	Absent

[ab] Mean ± standard deviation in same row with different superscripts letters are significantly different at $p < 0.05$. Safety Limit values were compiled from Laws of Malaysia P.U.(A) 437 Of 1985 Food Act 1983 Food Regulations 1985 Arrangement of Regulations, Regasa et al. (2019), Official Journal of the European Communities (1990), Gilbert et al. (2000) and Food Standards Australia New Zealand, (2018) except for * which was obtained from industry (Fraser and Neave).

Total Plate Count (TPC) is a method of estimating the total number of microorganisms in products commonly used by dairy manufacturers for determining the microbiological quality of milk. In the Malaysian Food Act [27], the limit of TPC is $\leq 10^5$ CFU/mL for safe consumption, but the TPC of fresh milk was recorded at 5.13 $\log_{10}$CFU/mL, and it increased significantly ($p < 0.05$) to 6.19 $\log_{10}$CFU/mL after 6 days of storage while HPP treated milk was successfully reduced to 1.26 $\log_{10}$CFU/mL and maintained below 2 $\log_{10}$CFU/mL for 60 days of storage (Figure 1a), meeting the safety limit (Table 1). A lab-scale HPP study by Liepa et al. [28] reported a pronounced reduction in TPC by 99.7% using HPP (STANSTED fluid power LTD, Stansted, Harlow, UK) at 550 MPa for 3 min, while Razali et al. [29] reported HPP (Avure 2L-700 HPP Laboratory Food Processing System, Avure, Kent, Washington, DC, USA) at 400 Mpa for 5 min reduced TPC to an undetectable level in milk. In a similar study using the same commercial unit (Hiperbaric 55, Hiperbaric, Spain), Tan et al. [14] reported a high TPC reduction of 99.98% at 600 Mpa for 5–7 min. Stratakos et al. [13] showed that commercial scale HPP (Quintus 35 L, Avure Technologies, Kent, Washington,

DC, USA) at 600 Mpa for 3 min reduced TPC by 83.33% and prolonged the shelf life of milk to 28 days.

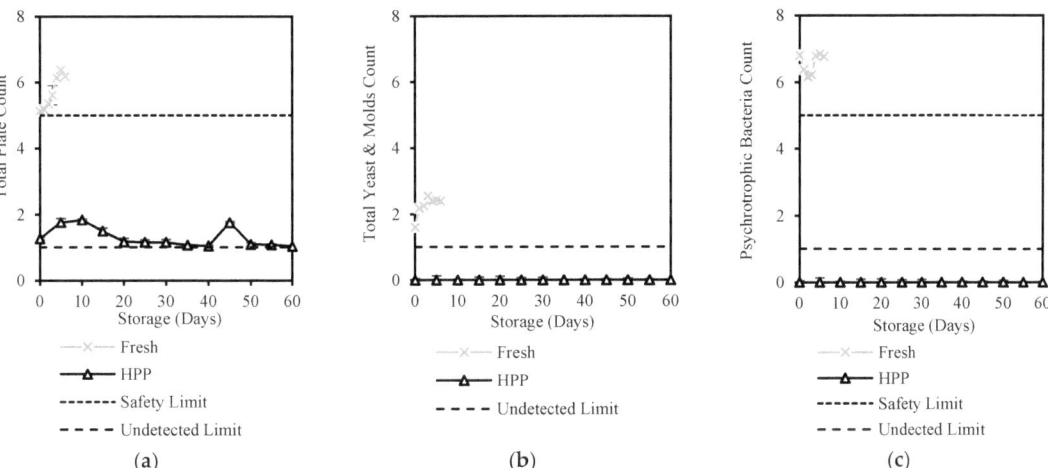

Figure 1. Changes in (**a**) Total Plate Count, (**b**) Total Yeast and Molds Count, and (**c**) Psychrotrophic Bacteria Count of HPP and ESL milk during 60-day storage (log10 CFU/mL).

Yeasts and mould can cause the development of off-flavours in milk due to the generation of toxic metabolites by mycotoxins which restrict the shelf life of milk and pose a potential health risk [30]. Fresh milk used in this study had yeast and molds count above the safety limit of <1 $\log_{10}$CFU/mL [30]. HPP treatment reduced the count to safe and undetectable levels (Figure 1b). A similar result was reported by Tan et al. [14], where most vegetative yeasts and moulds in cow and goat milk were inactivated within 5–7 min at 450–600 Mpa. The effectiveness of HPP in controlling yeast and mould count in this study has shown stability at an undetectable level throughout the 60 days of storage (Figure 1b) which is longer than studies by Tan et al. [14] for 22 days.

Spore-forming bacteria, which include the psychrotrophic, mesophilic, and thermophilic spore formers, are common in dairy products due to their ability to survive under different temperatures [31]. Psychrotrophic microorganisms constitute a major cause of milk spoilage [32] due to their ability to produce heat-resistant enzymes such as proteases, lipases, and phospholipases under refrigeration [33,34]. The psychrotrophic bacteria count of high-quality fresh milk should be <10^5 CFU/mL based on the Malaysian Food Act [27], but this study shows that the psychrotrophic bacteria count in fresh milk was higher at 6.81 ± 0.05 $\log_{10}$ CFU/mL (Table 1). The high psychrotrophic bacteria was reduced to an undetectable level after HPP treatment, as were its spores and both the mesophilic and thermophilic aerobic spores. This result is supported by Tan et al. [14], who reported that psychrotrophic and mesophilic spores were not detected in HPP milk (600 Mpa, 7 min). The psychrotrophic bacteria count of HPP milk was also not affected by long refrigeration storage, which was maintained at undetectable levels throughout the 60 days of storage (Figure 1c).

Other microbiological properties with counts that were all well-kept below the safety limits during the 60 days of storage after HPP treatment, although they had counts initially exceeding safety levels before treatment, are the coliform (1.7 $\log_{10}$CFU/mL) and *Staphylococcus aureus* (1.3 $\log_{10}$CFU/m). This suggests that HPP treatment had high efficiency in ensuring treated milk is safe despite unavoidable conditions in the milking process and farms causing microbial contamination. Several researchers who have reported similar findings on the effectiveness of HPP treatment in reducing these bacterial counts include Stratakos et al. [13] on *E. coli* reduction by 5.6 and 6.8 $\log_{10}$CFU/mL at 600 Mpa for 3 and

5 min, respectively, Liu et al. [35] on *E. coli* reduction by 2.9 $\log_{10}$CFU/mL at 600 Mpa for 5 min, and Tan et al. [14] on total coliform reduction by 1.6 $\log_{10}$CFU/mL at 450–600 Mpa for 5–7 min.

Pathogenic bacteria such as *Bacillus cereus, Clostridium perfringens, Listeria monocytogenes,* and *Salmonella spp.* were all undetected in fresh milk and also throughout the storage for HPP milk. For safety consumption, *Bacillus cereus* and *Clostridium perfringens* in food should be <10^5 CFU/mL [36] and <10^4 CFU/g [37], respectively, while *Listeria monocytogenes* and *Salmonella spp.* should not be detected in a 25g sample [38].

3.2. Physicochemical Analysis

Figure 2 shows slight changes in all physicochemical properties of HPP milk during 60 days of storage except for acidity, which was reduced from initially 0.20% to <0.1% after HPP treatment (Figure 2b). Acidity reduction accompanied by pH value increase after HPP treatment at 600 MPa was also reported in other studies [15,29]. During HPP, the casein micelle disaggregation alters minerals distribution and raises the concentration of ionic calcium in milk, resulting in an increase in the phosphate concentration of the milk serum and pH [28].

HPP milk had high pH stability throughout 60 days of storage, indicating it has low bacteria load to produce lactic acid and generate free fatty acid through fat lipolysis [39]. Tan et al. similarly reported a pH decrease in HPP milk stored for 22 days [14]. It is common to have an acidity increase in spoilt milk, as observed in fresh milk with 6 days of storage (Appendix B), normally associated with an increase in free fatty acids [40], age gelation [41], or browning reactions [42] during storage.

HPP did not cause many changes to the protein and fat contents of milk during its 60 days of storage (Figure 2c,d). This result was consistent with Tan et al. [14], who reported no significant changes ($p > 0.05$) in the total protein and fat content of HPP milk samples during 22 days of storage. High pressure will only influence the bonds which stabilise the spatial structure of proteins and cause reversible or non-reversible denaturation depending on the pressure level [43]. Most studies have proven the effectiveness of HPP in preserving milk fat, as high pressure does not damage milk fat globule membranes; thus, lipolysis is prevented, and fat content is retained [44,45].

HPP treatment was also found to cause no significant changes ($p > 0.05$) to the total solids (Figure 2e) and non-fat milk solids (Figure 2f), similar to those reported by Tan et al. [14]. However, these contents changed significantly ($p < 0.05$) in an inconsistent manner during 60 days of storage. The increase in total solids and non-fat solids of HPP-treated milk could be due to the precipitation fraction of large casein–casein and the formation of casein–fat aggregates during HPP [46] or induced fat crystallisation due to the duration of pressure treatment and storage [47]. Specific gravity was not affected throughout this study; it was retained at 1.02–1.03 g/mL.

3.3. Nutritional Analysis
3.3.1. Vitamin Content

Most of the vitamin contents of milk in the present study, which includes fresh milk, were found in relatively low levels compared to commercially fortified milk, probably due to the nature of milk in which vitamins are not present as a primary source of milk. Some of these vitamins, which include beta-carotene and Vitamin A, B1, B2, B3, B6, C, and E, were too low and were below the limit of reporting (LOR) or below the minimum concentration of a substance in a sample that can be reliably detected by a laboratory, hence levels below this limit can cause variation and affect accuracy. Despite the low vitamin levels, HPP treatment did not cause significant changes ($p > 0.05$) in milk vitamins except for the slight Vitamin C increase (Table 2). This was unexpected and can be explained by the variations due to the detection level below LOR. Sierra and Vidal–Valverde [48] found no significant ($p > 0.05$) losses of Vitamin B1 and B6 in whole milk after HPP (400 MPa, 30 min), while Moltó-Puigmartí et al. [49] also reported no loss in the total Vitamin C level in human milk after HPP (600 MPa, 5 min, 22–27 °C). Retention of vitamins in milk was probably due to the property of vitamins, which consists of small molecules and covalent bonds that were not affected by high pressure [50]. The storage study here shows that vitamins in HPP milk reduced gradually over the 60-day storage (Figure 3). Most vitamins in HPP milk, including Vitamin A (25%), B3 (91%), B5 (35%), B6 (80%), and C (85%), fell below their original level at the end of storage, except for Vitamin B7 (25%), B9 (100%), and B12 (20%) with increment. There is no comparison of other work on vitamin deterioration in HPP-treated milk during storage, but records of vitamin deterioration in fresh milk during 6 days of refrigeration storage were found for Vitamin A, B6, B12, and C (Appendix C). It is similarly reported that although vitamins were well-retained in juices after high-pressure treatment, they were also degraded during refrigeration storage [51,52].

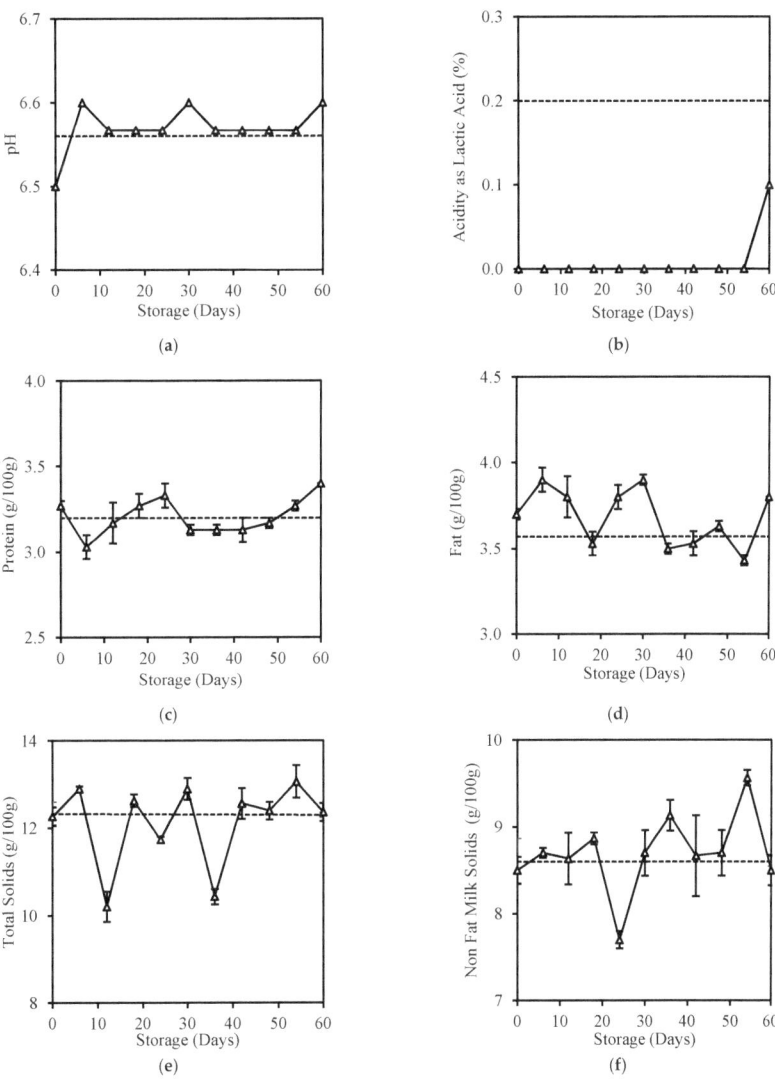

Figure 2. Changes in (**a**) pH, (**b**) Acidity as Lactic Acid, (**c**) Protein, (**d**) Fat, (**e**) Total Solid, and (**f**) Non-Fat Milk Solids of HPP milk during 60-day storage. Dotted lines represent the fresh milk before treatment.

Table 2. Effects of HPP on vitamin content of fresh milk.

Vitamin Content	LOR Unit	Fresh Milk	HPP
Vitamin A (as Retinol) *	0.1 mg/100 mL	0.04 ± 0.00 [a]	0.04 ± 0.00 [a]
Beta-carotene *	1 mg/100 mL	0.02 ± 0.00 [a]	0.02 ± 0.00 [a]
Vitamin B1 *	0.25 mg/100 mL	0.04 ± 0.00 [a]	0.03 ± 0.00 [a]
Vitamin B2 *	0.25 mg/100 mL	0.17 ± 0.00 [a]	0.18 ± 0.00 [a]
Vitamin B3 (as Niacin) *	0.25 mg/100 mL	0.11 ± 0.00 [a]	0.11 ± 0.00 [a]
Vitamin B5	0.25 mg/100 mL	0.40 ± 0.00 [a]	0.40 ± 0.00 [a]
Vitamin B6 *	0.25 mg/100 mL	0.05 ± 0.01 [a]	0.06 ± 0.00 [a]
Vitamin B7 (Biotin)	1 µg/100 mL	4.00 ± 0.00 [a]	4.00 ± 0.00 [a]
Vitamin B9 (Folic Acid)	1 µg/100 mL	6.00 ± 0.00 [a]	6.33 ± 0.33 [a]
Vitamin B12 (Cyanocobalamin)	1 µg/100 mL	1.67 ± 0.58 [a]	2.00 ± 0.00 [a]
Vitamin C *	2 mg/100 mL	1.00 ± 0.00 [a]	1.67 ± 0.33 [ab]
Vitamin D	0.1 µg/100 mL	2.50 ± 0.00 [a]	2.50 ± 0.00 [a]
Vitamin E (Alpha-Tocopherol) *	0.1 mg/100 mL	0.07 ± 0.00 [a]	0.07 ± 0.00 [a]
Vitamin K	0.1 µg/100 g	0.3	0.65

[ab] Mean ± standard deviation in same row with different superscripts letters are significantly different at $p < 0.05$; * Results showing vitamin content lower than LOR. LOR is the limit of reporting.

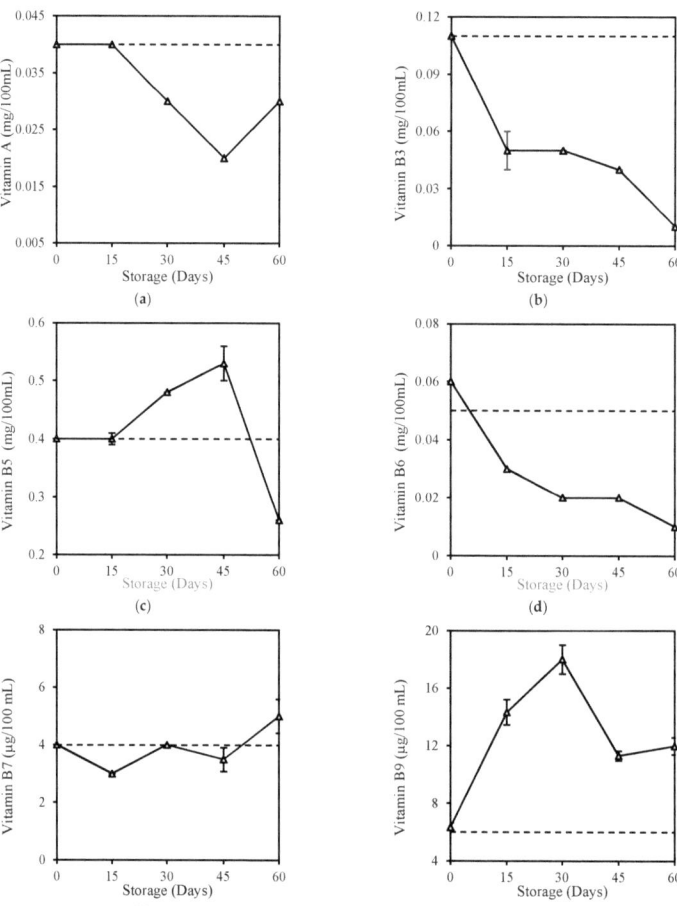

Figure 3. Cont.

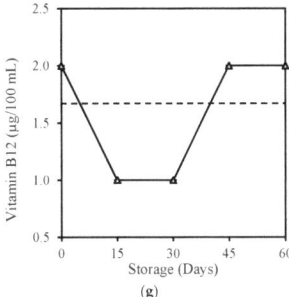
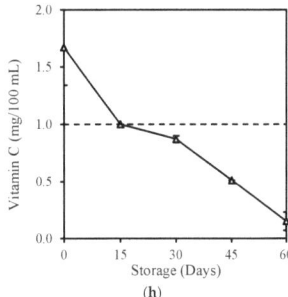

(g) (h)

Figure 3. Changes in (**a**) Vitamin A, (**b**) Vitamin B3, (**c**) Vitamin B5, (**d**) Vitamin B6, (**e**) Vitamin B7, (**f**) Vitamin B9, (**g**) Vitamin B12, and (**h**) Vitamin C of HPP and ESL milk during 60-day storage. Dotted lines represent the fresh milk before treatment.

3.3.2. Mineral Content

The milk samples used in this study had the highest mineral content of potassium, followed by calcium, phosphorus, magnesium, and zinc (Figure 4). There was no significant difference ($p > 0.05$) in the mineral contents of milk after HPP treatment (Day 0). It successfully retained calcium, phosphorus, and magnesium contents by 99.3%, 99.4%, and 99.1%, respectively. However, significant changes ($p < 0.05$) were observed in the mineral contents of HPP milk over the 60 days of storage in an inconsistent manner (Figure 5). Significant loss ($p < 0.05$) was observed in potassium (5.3%) and zinc (18.4%), whereas calcium, phosphorus, and magnesium contents increased by 1.6%, 1.1%, and 13.1%, respectively, after storage when compared with the original fresh milk. Selenium remained undetected at levels <0.2 mg/kg throughout this study. Andrés et al. [53] observed no significant ($p > 0.05$) changes in mineral profiles of potassium, calcium, magnesium, and zinc in milk after HPP (450–650 MPa, 3 min, 20 °C) during the 45 days of storage at 4 °C. The inconsistent changes were more prominent for fresh milk kept refrigerated for 6 days of storage, with a higher degradation of potassium (4.3%), calcium (7.2%), and phosphorus (14.8%), except for selenium which also remained at undetectable level (Appendix D).

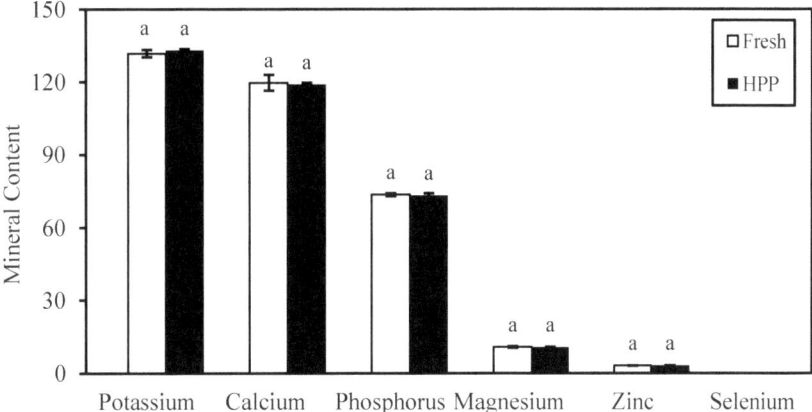

Figure 4. Changes in mineral content of fresh milk after HPP treatment. Potassium, Calcium, Phosphorus, and Magnesium in unit mg/100 mL; Zinc and Selenium in unit mg/kg. Alphabet "a" represent no significant difference between bar at $p > 0.05$.

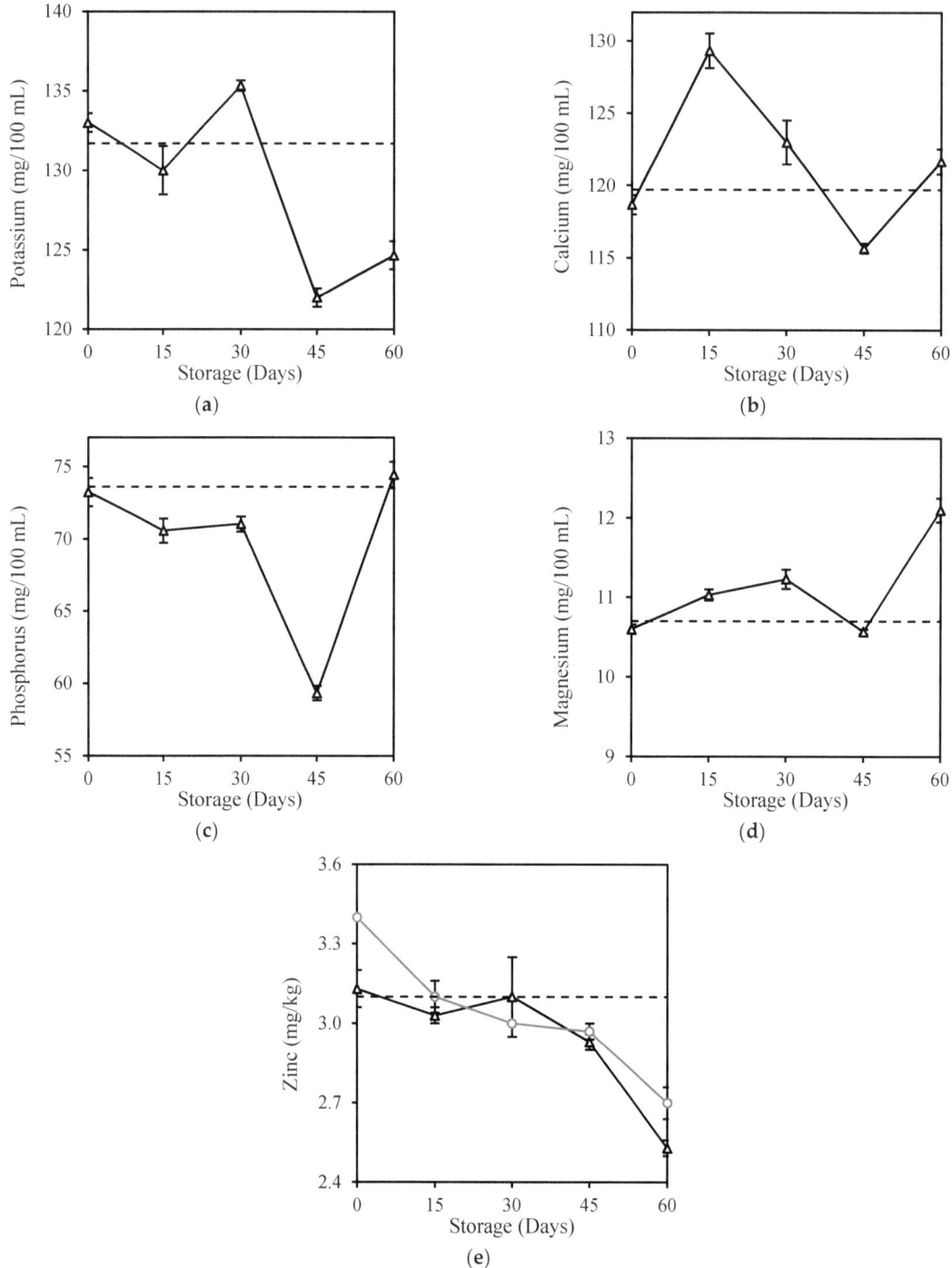

Figure 5. Changes in (**a**) Potassium, (**b**) Calcium, (**c**) Phosphorus, (**d**) Magnesium, and (**e**) Zinc contents of HPP and ESL milk during 60-day storage. Dotted lines represent the fresh milk before treatment.

4. Conclusions

This research has shown the potential of HPP treatment in preserving milk quality for the dairy industry. This novel technology gives promising results in terms of not only product safety and nutritional properties but also extending shelf life significantly when compared to conventionally heat-treated milk. HPP-treated milk had a storage shelf life beyond 60 days, with all microbial testing meeting permitted safety levels. It had high stability for physicochemical properties with consistent pH and acidity during the entire storage. HPP treatment has successfully retained calcium, phosphorus, magnesium, and zinc contents by 99.3, 99.4, 99.1, and 100%, respectively. The HPP treatment itself did not cause much vitamin and mineral deterioration. However, some changes were observed for the vitamin and mineral contents at the end of 60-day storage. Degradation was observed for Vitamin A, B3, B5, B6, and C, and the minerals potassium and zinc, while an increase in Vitamin B7, B9, B12, calcium, phosphorus, and magnesium contents were detected. Future HPP research focusing on milk vitamins and minerals to overcome this study limitation is suggested.

Author Contributions: S.H.L. designed the experiments, collected and analysed the data, and wrote the manuscript. N.L.C. conceptualized the study, supervised the research, and edited and revised the manuscript. A.S. supervised the experiments and provided technical know-how in running the industrial HPP unit. C.H.T. and T.H.W. gave inputs on industrial practice and parameters. All authors have read and agreed to the published version of the manuscript.

Funding: This research was funded by F&N Global Marketing Pte. Ltd., a subsidiary company of Fraser and Neave Limited, via project no. 2020-0093/1133 in Universiti Putra Malaysia.

Institutional Review Board Statement: Not applicable.

Informed Consent Statement: Informed consent was obtained from all subjects involved in the study.

Data Availability Statement: The data are contained within the article.

Acknowledgments: S.H.L acknowledges Universiti Putra Malaysia for Graduate Research Assistantship for her Ph.D. studies.

Conflicts of Interest: All authors declare no conflicts of interest, including C.H.T. and T.H.W., who are attached to the sponsor company. They agreed to the proposed conception of work but did not have roles in the execution and interpretation of the study. They had roles in the funding acquisition, final review, editing of the paper, and final approval of the version to be published.

Appendix A Microbiological Analysis Data

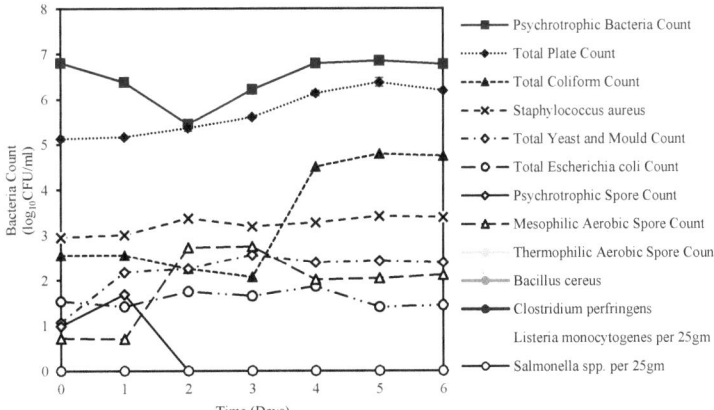

Figure A1. Changes in microbiological properties of fresh milk during 6 days of storage at 6 ± 1 °C.

Appendix B Physicochemical Analysis Data

Table A1. Changes in physicochemical properties of fresh milk during 6 days of storage at 6 ± 1 °C.

Properties	Storage (Days)		
	0	3	6
pH Value (pH)	6.57 ± 0.03 [ab]	6.60 ± 0.00 [a]	6.50 ± 0.00 [b]
Acidity as Lactic Acid (%)	0.20 ± 0.00 [a]	0.20 ± 0.00 [a]	0.37 ± 0.03 [b]
Fat (g/100 g)	3.57 ± 0.12 [a]	3.50 ± 0.06 [a]	3.57 ± 0.12 [a]
Protein (g/100 g)	3.10 ± 0.00 [a]	3.27 ± 0.09 [a]	3.10 ± 0.00 [a]
Total Solids (g/100 mL)	12.07 ± 0.12 [a]	12.57 ± 0.09 [b]	12.37 ± 0.12 [ab]
Non-Fat Milk Solids (g/100 g)	8.40 ± 0.15 [a]	8.77 ± 0.03 [a]	8.60 ± 0.00 [a]
Specific Gravity (g/ mL)	1.02 ± 0.00	1.02 ± 0.00	1.02 ± 0.00

[ab] Mean ± standard deviation in same row with different superscripts letters are significantly different at $p < 0.05$.

Appendix C Vitamin Content Analysis Data

Table A2. Changes in vitamin content of fresh milk during 6 days of storage at 6 ± 1 °C.

Vitamin Content	LOR Unit	Storage (Days)		
		0	3	6
Vitamin A (as Retinol) *	0.1 mg/100 mL	0.04 ± 0.00 [b]	0.02 ± 0.00 [a]	0.02 ± 0.00 [a]
Beta-carotene *	1 mg/100 mL	0.02 ± 0.00 [a]	0.02 ± 0.00 [a]	0.02 ± 0.00 [a]
Vitamin B1 *	0.25 mg/100 mL	0.04 ± 0.00 [a]	0.04 ± 0.00 [a]	0.04 ± 0.00 [a]
Vitamin B2 *	0.25 mg/100 mL	0.17 ± 0.00 [a]	0.18 ± 0.01 [a]	0.17 ± 0.01 [a]
Vitamin B3 (as Niacin) *	0.2 mg/100 mL	0.11 ± 0.00	<0.2	<0.2
Vitamin B5	0.25 mg/100 mL	0.40 ± 0.00 [a]	0.58 ± 0.01 [b]	0.58 ± 0.00 [b]
Vitamin B6 *	0.25 mg/100 mL	0.05 ± 0.01 [b]	0.02 ± 0.00 [a]	0.02 ± 0.00 [a]
Vitamin B7 (Biotin)	1 µg/100 mL	4.00 ± 0.00 [a]	4.33 ± 0.58 [a]	4.33 ± 0.58 [a]
Vitamin B9 (Folic Acid)	1 µg/100 mL	6.00 ± 0.00 [a]	7.50 ± 0.50 [b]	7.50 ± 0.50 [b]
Vitamin B12 (Cyanocobalamin)	1 µg/100 mL	1.67 ± 0.58	<1	<1
Vitamin C *	2 mg/100 mL	1.00 ± 0.00 [c]	0.47 ± 0.06 [b]	0.20 ± 0.10 [a]
Vitamin D	0.1 µg/100 mL	2.50 ± 0.00 [a]	2.52 ± 0.01 [a]	2.51 ± 0.01 [a]
Vitamin E (Alpha-Tocopherol) *	0.1 mg/100 mL	0.07 ± 0.00 [a]	0.08 ± 0.00 [b]	0.08 ± 0.00 [b]
Vitamin K	0.1 µg/100 g	0.30	0.45	0.46

[abc] Mean ± standard deviation in same row with different superscripts letters are significantly different at $p < 0.05$. * Results showing vitamin content lower than limit of reporting (LOR).

Appendix D Mineral Content Analysis Data

Table A3. Changes in mineral content of fresh milk during 6 days of storage at 6 ± 1 °C.

Mineral Content	Storage (Days)		
	0	3	6
Potassium (mg/100 mL)	131.67 ± 0.88 [a]	145.33 ± 2.03 [b]	126.00 ± 0.58 [a]
Calcium (mg/100 mL)	119.67 ± 1.86 [b]	147.67 ± 2.33 [c]	111.00 ± 1.53 [a]
Phosphorus (mg/100 mL)	73.63 ± 0.37 [b]	86.10 ± 1.78 [c]	62.73 ± 1.27 [a]
Magnesium (mg/100 mL)	10.73 ± 0.17 [a]	12.97 ± 0.18 [b]	10.90 ± 0.10 [a]
Zinc (mg/kg)	3.13 ± 0.07 [b]	2.37 ± 0.03 [a]	3.27 ± 0.24 [b]
Selenium (mg/kg)	<0.2	<0.2	<0.2

[abc] Mean ± standard deviation in same row with different superscripts letters are significantly different at $p < 0.05$.

References

1. Claeys, W.L.; Cardoen, S.; Daube, G.; De Block, J.; Dewettinck, K.; Dierick, K.; De Zutter, L.; Huyghebaert, A.; Imberechts, H.; Thiange, P.; et al. Raw or heated cow milk consumption: Review of risks and benefits. *Food Control.* **2013**, *31*, 251–262. [CrossRef]
2. Tamime, A.Y. *Milk Processing and Quality Management*; Wiley-Blackwell: Oxford, UK, 2008.
3. Lorenzen, P.C.; Clawin-Rädecker, I.; Einhoff, K.; Hammer, P.; Hartmann, R.; Hoffmann, W.; Martin, D.; Molkentin, J.; Walte, H.G.; Devrese, M. A survey of the quality of extended shelf life (ESL) milk in relation to HTST and UHT milk. *Int. J. Dairy Technol.* **2011**, *64*, 166–178. [CrossRef]

4. Blake, M.R.; Weimer, B.C.; Mcmahon, D.J.; Savello, P.A. Sensory and microbial quality of milk processed for extended shelf life by direct steam injection. *J. Food Prot.* **1995**, *58*, 1007–1013. [CrossRef] [PubMed]
5. Krishna, T.C.; Najda, A.; Bains, A.; Tosif, M.M.; Papliński, R.; Kapłan, M.; Chawla, P. Influence of ultra-heat treatment on properties of milk proteins. *Polymers* **2021**, *13*, 3164. [CrossRef]
6. Sun, D.W. *Thermal Food Processing: New Technologies and Quality Issues*; Taylor & Francis Group: Boca Raton, FL, USA, 2006.
7. Chavan, R.S.; Sehrawat, R.; Mishra, V.; Bhatt, S. Milk: Processing of Milk. In *Encyclopedia of Food and Health*; Elsevier Science Publishing Co., Inc.: San Diego, CA, USA, 2016; pp. 729–735.
8. Chawla, R.; Patil, G.R.; Singh, A.K. High hydrostatic pressure technology in dairy processing: A review. *J. Food Sci. Technol.* **2011**, *48*, 260–268. [CrossRef] [PubMed]
9. Mandal, R.; Kant, R. High-pressure processing and its applications in the dairy industry. *Food Sci. Technol. Int. J. FSTJ* **2017**, *1*, 33–45.
10. Srinivas, M.S.; Madhu, B.; Girijal, S. High Pressure Processing of Foods: A Review. *Andhra Agric. J.* **2018**, *65*, 467–476.
11. García-Risco, M.R.; Olano, A.; Ramos, M.; López-Fandiño, R. Micellar changes induced by high pressure. Influence in the proteolytic activity and organoleptic properties of milk. *J. Dairy Sci.* **2000**, *83*, 2184–2189. [CrossRef]
12. Mussa, D.M.; Ramaswamy, H.S. Ultra high pressure pasteurization of milk: Kinetics of microbial destruction and changes in Physico-chemical characteristics. *LWT-Food Sci. Technol.* **1997**, *30*, 551–557. [CrossRef]
13. Stratakos, A.C.; Inguglia, E.S.; Linton, M.; Tollerton, J.; Murphy, L.; Corcionivoschi, N.; Koidis, A.; Tiwari, B.K. Effect of high pressure processing on the safety, shelf life and quality of raw milk. *Innov. Food Sci. Emerg. Technol.* **2019**, *52*, 325–333. [CrossRef]
14. Tan, S.F.; Chin, N.L.; Tee, T.P.; Chooi, S.K. Physico-chemical changes, microbiological properties, and storage shelf life of cow and goat milk from industrial high-pressure processing. *Processes* **2020**, *8*, 697. [CrossRef]
15. Jackson, G.J.; Merker, R.I.; Bandler, R. Bacteriological Analytical Manual (BAM). no. January 2001. Available online: https://www.fda.gov/food/laboratory-methods-food/bacteriological-analytical-manual-bam#intro (accessed on 25 January 2022).
16. Peterkin, P.I. Compendium of methods for the microbiological examination of foods (3rd edn). *Trends Food Sci. Technol.* **1993**, *4*, 199. [CrossRef]
17. Veling, J.; van Zijderveld, F.G.; Bemmel, A.M.V.Z.-V.; Schukken, Y.H.; Barkema, H.W. Evaluation of two enzyme-linked immunosorbent assays for detecting Salmonella enterica subsp. enterica serovar Dublin antibodies in bulk milk. *Clin. Diagn. Lab. Immunol.* **2001**, *8*, 1049–1055. [CrossRef]
18. Pyne, G.T. The determination of milk-proteins by formaldehyde titration. *Biochem. J.* **1932**, *26*, 1006–1014. [CrossRef] [PubMed]
19. Horwitz, W. *AOAC: Official Methods of Analysis*; Association of Official Analytical Chemists: Washington, DC, USA, 1980; Volume 552.
20. Agostini-costa, T.S.; Teixeira-filho, J.; Scherer, R.; Prado, M.A.; Godoy, H.T. Determination of B-Group Vitamins in Enriched Flavored Milk Mixes. *Aliment. E Nutr. Araraquara* **2007**, *18*, 351–356.
21. Woollard, D.C.; Indyk, H.E.; Christiansen, S.K. The analysis of pantothenic acid in milk and infant formulas by HPLC. *Food Chem.* **2000**, *69*, 201–208. [CrossRef]
22. AOAC. AOAC Official Method 960.46. Vitamin assays, microbiological method Gaithersburg. In *The Official Methods of Analysis of AOAC International*, 18th ed.; AOAC: Arlington, TX, USA, 2006.
23. AOAC. AOAC Official Method 967.21. In *Official Methods of Analysis*, 16th ed.; AOAC: Arlington, TX, USA, 1995; Available online: https://www.scribd.com/document/176943262/AOAC-Method-Ascorbic-Ac-967-21 (accessed on 25 January 2022).
24. AOAC. AOAC Official Method 999.15. Vitamin K in Milk and Infant Formulas, Liquid Chromatographic Method. In *Official Methods of Analysis, Method 999.15*, 18th ed.; AOAC International: Gaithersburg, MD, USA, 2005.
25. Silva, M.G.; Sanders, J.K. Vitamin D in Infant Formula and Enteral Products by Liquid Chromatography: Collaborative Study. *J. AOAC Int.* **1996**, *79*, 73–80. [CrossRef]
26. Kira, C.S.; Maihara, V.A. Determination of major and minor elements in dairy products through inductively coupled plasma optical emission spectrometry after wet partial digestion and neutron activation analysis. *Food Chem.* **2007**, *100*, 390–395. [CrossRef]
27. Food Regulations. Laws of Malaysia P.U.(A) 437 Of 1985 Food Act 1983 Food Regulations 1985. 1985; p. 310. Available online: http://overseas.cas.org.tw/Data/Sites/1/media/FR1985.pdf (accessed on 21 February 2022).
28. Liepa, M.; Zagorska, J.; Galoburda, R.; Kostascuka, S. Effect of high-pressure processing on microbial quality of skimmed milk. *Proc. Latv. Acad. Sci. Sect. B Nat. Exact Appl. Sci.* **2018**, *72*, 118–122. [CrossRef]
29. Razali, M.F.; Narayanan, S.; Hazmi, N.A.M.; Shah, N.N.A.K.; Kamal, S.M.M.; Fauzi, N.A.M.; Sulaiman, A. Minimal processing for goat milk preservation: Effect of high-pressure processing on its quality. *J. Food Process. Preserv.* **2021**, *45*, e15590. [CrossRef]
30. Godič Torkar, K.; Golc Teger, S. The microbiological quality of raw milk after introducing the two day's milk collecting system. *Acta Agric. Slov.* **2008**, *92*, 61–74.
31. Martinez, B.A.; Stratton, J.; Bianchini, A. Isolation and genetic identification of spore-forming bacteria associated with concentrated-milk processing in Nebraska. *J. Dairy Sci.* **2017**, *100*, 919–932. [CrossRef]
32. Quigley, L.; O'Sullivan, O.; Stanton, C.; Beresford, T.P.; Ross, R.; Fitzgerald, G.F.; Cotter, P.D. The complex microbiota of raw milk. *FEMS Microbiol. Rev.* **2013**, *37*, 664–698. [CrossRef] [PubMed]
33. Samaržija, D.; Zamberlin, Š.; Pogačić, T. Psychrotrophic bacteria and milk and dairy products quality. *Mljekarstvo* **2012**, *62*, 77–95.
34. Xin, L.; Meng, Z.; Zhang, L.; Cui, Y.; Han, X.; Yi, H. The diversity and proteolytic properties of psychrotrophic bacteria in raw cows' milk from North China. *Int. Dairy J.* **2017**, *66*, 34–41. [CrossRef]

35. Liu, G.; Carøe, C.; Qin, Z.; Munk, D.M.; Crafack, M.; Petersen, M.A.; Ahrné, L. Comparative study on quality of whole milk processed by high hydrostatic pressure or thermal pasteurization treatment. *LWT-Food Sci. Technol.* **2020**, *127*, 109370. [CrossRef]
36. Official Journal of the European Communities. Official Journal of the European Communities. *Anal. Proc.* **1984**, *21*, 196. [CrossRef]
37. Gilbert, R.J.; De Louvois, J.; Donovan, T.; Little, C.; Nye, K.; Ribeiro, C.D.; Richards, J.; Roberts, D.; Bolton, F.J. Guidelines for the microbiological quality of some ready-to-eat foods sampled at the point of sale. PHLS Advisory Committee for Food and Dairy Products. *Commun. Dis. Public Health/PHLS* **2000**, *3*, 163–167.
38. Food Standards Australia New Zealand. *Compendium of Microbiological Criteria for Food*; Food Standards Australia New Zealand: Wellington, New Zealand, 2018.
39. Lu, M.; Shiau, Y.; Wong, J.; Lin, R.; Kravis, H.; Blackmon, T.; Pakzad, T.; Jen, T.; Cheng, A.; Chang, J.; et al. Milk Spoilage: Methods and Practices of Detecting Milk Quality. *Food Nutr. Sci.* **2013**, *04*, 113–123. [CrossRef]
40. Swartzel, K. The role of heat exchanger fouling in the formation of sediment in aseptically processed and packed milk. *J. Food Process. Prev.* **1983**, *7*, 247–251. [CrossRef]
41. Andrew, A.; Brooker, B.; Hobbs, D. Properties of aseptically packed UHT milk. Electron microscopic examination of changes occurring during storage. *J. Dairy Res.* **1977**, *44*, 283–285. [CrossRef]
42. Vankatachalm, N.; Macmahon, D. Effect of lactose concentration on age-gelation of UHT sterilized skim milk concentrate. *J. Dairy Sci.* **1991**, *74*, 101–107.
43. Benet, G.U. In Phd thesis High-Pressure Low-Temperature Processing of Foods: Impact of Metastable Phases on Process and Quality Parameters. 2005, pp. 29–30. Available online: https://depositonce.tu-berlin.de/items/ee7befa0-1511-462a-88d9-c1d84d2cb4cc (accessed on 27 January 2022).
44. Huppertz, T.; Kelly, A.L.; Fox, P.F. Effects of high pressure on constituents and properties of milk. *Int. Dairy J.* **2002**, *12*, 561–572. [CrossRef]
45. Özer, B.; Akdemir-Evrendilek, G. Non-Thermal Processing of Milk and Milk Products for Microbial Safety. In *Dairy Microbiology and Biochemistry*; CRC Press: Boca Raton, FL, USA, 2014; pp. 285–301.
46. Tran, M.; Roberts, R.; Felix, T.; Harte, F. Effect of high-pressure-jet processing on the viscosity and foaming properties of pasteurized whole milk. *J. Dairy Sci.* **2018**, *101*, 3887–3899. [CrossRef] [PubMed]
47. Dumay, E.; Lambert, C.; Funtenberger, S.; Cheftel, J.C. Effects of high pressure on the physico-chemical characteristics of dairy creams and model oil/water emulsions. *LWT-Food Sci. Technol.* **1996**, *29*, 606–625. [CrossRef]
48. Sierra, I.; Vidal-Valverde, C. Effect of high pressure on the vitamin B1 and B6 content of milk. *Milchwissenschaft* **2000**, *55*, 365–367. Available online: https://www.cabdirect.org/cabdirect/abstract/20000405287 (accessed on 16 March 2022).
49. Moltó-Puigmartí, C.; Permanyer, M.; Castellote, A.I.; López-Sabater, M.C. Effects of pasteurisation and high-pressure processing on vitamin C, tocopherols and fatty acids in mature human milk. *Food Chem.* **2011**, *124*, 697–702. [CrossRef]
50. Balci, A.T.; Wilbey, R.A. High pressure processing of milk-the first 100 years in the development of a new technology. *Int. J. Dairy Technol.* **1999**, *52*, 149–155. [CrossRef]
51. Cao, X.; Bi, X.; Huang, W.; Wu, J.; Hu, X.; Liao, X. Changes of quality of high hydrostatic pressure processed cloudy and clear strawberry juices during storage. *Innov. Food Sci. Emerg. Technol.* **2012**, *16*, 181–190. [CrossRef]
52. Terefe, N.S.; Kleintschek, T.; Gamage, T.; Fanning, K.J.; Netzel, G.; Versteeg, C.; Netzel, M. Comparative effects of thermal and high pressure processing on phenolic phytochemicals in different strawberry cultivars. *Innov. Food Sci. Emerg. Technol.* **2013**, *19*, 57–65. [CrossRef]
53. Andrés, V.; Villanueva, M.-J.; Tenorio, M.-D. Influence of high pressure processing on microbial shelf life, sensory profile, soluble sugars, organic acids, and mineral content of milk- and soy-smoothies. *LWT-Food Sci. Technol.* **2016**, *65*, 98–105. [CrossRef]

Disclaimer/Publisher's Note: The statements, opinions and data contained in all publications are solely those of the individual author(s) and contributor(s) and not of MDPI and/or the editor(s). MDPI and/or the editor(s) disclaim responsibility for any injury to people or property resulting from any ideas, methods, instructions or products referred to in the content.

Article

High Pressure and Pasteurization Effects on Dairy Cream †

Fernanda Machado [1], Ricardo V. Duarte [1], Carlos A. Pinto [1], Susana Casal [2], José A. Lopes-da-Silva [1] and Jorge A. Saraiva [1,*]

[1] Associated Laboratory for Green Chemistry of the Network of Chemistry and Technology (LAQV-REQUIMTE), Department of Chemistry, University of Aveiro, 3810-193 Aveiro, Portugal; fernandamachado@ua.pt (F.M.); ricardo.vd@ua.pt (R.V.D.); carlospinto@ua.pt (C.A.P.); jals@ua.pt (J.A.L.-d.-S.)

[2] Associated Laboratory for Green Chemistry of the Network of Chemistry and Technology (LAQV-REQUIMTE), Laboratory of Bromatology and Hydrology, Faculty of Pharmacy, Porto University, Rua Jorge Viterbo Ferreira 228, 4050-313 Porto, Portugal; sucasal@ff.up.pt

* Correspondence: jorgesaraiva@ua.pt; Tel.: +351-234-401-513

† This study is a part of the thesis of Fernanda Machado.

Abstract: Dairy cream, a common ingredient in various dishes and food products, is susceptible to rapid microbial growth due to its high water activity (≈0.97) and pH (≈6.7). Thus, it requires proper processing conditions to ensure food safety and extend shelf life. High-pressure processing (HPP) has emerged as a nonthermal food pasteurization method, offering an alternative to conventional heat-based techniques to obtain tastier, fresh-like, and safe dairy products without undesirable heat-induced alterations. This study assessed the impact of HPP (450 and 600 MPa for 5 and 15 min at 7 °C) and thermal pasteurization (75 °C for 15 s) on the microbiological and physicochemical attributes of dairy cream immediately after processing and throughout refrigerated storage (4 °C). HPP-treated samples remained microbiologically acceptable even on the 51st day of storage, unlike thermally pasteurized samples. Moreover, HPP decreased inoculated *Escherichia coli* and *Listeria innocua* counts by more than 6 log units to undetectable levels (1.00 log CFU/mL). pH, color (maximum variation of ΔE* up to 8.43), and fatty acid profiles remained relatively stable under varying processing conditions and during storage. However, viscosity exhibited higher values for HPP-treated samples (0.028 ± 0.003 Pa·s) compared to thermally processed ones (0.016 ± 0.002 Pa·s) by the 28th day of storage. Furthermore, volatile compounds (VOCs) of all treated samples presented a tendency to increase throughout storage, particularly acids and aliphatic hydrocarbons. These findings show HPP's potential to significantly extend the shelf life of highly perishable dairy cream by at least 15 days compared to thermal pasteurization.

Keywords: dairy cream; fatty acids; food safety; nonthermal and thermal processing; rheological parameters; shelf life; volatile organic compounds

1. Introduction

Dairy cream is used as an ingredient in many products, including butter, ice cream, and sour cream, among others [1]. However, it is a highly perishable product, with a pH of around 6.7 and high water activity (around 0.97), requiring adequate preservation to increase its shelf life [2]. Traditionally, most cream for retail and industrial use is thermally pasteurized [3], aiming to destroy vegetative (pathogenic and spoilage) microorganisms and inactivate enzymes, extending the cream's shelf life. Nonetheless, depending on the food matrix, heat pasteurization may not always be the ideal processing method since it may cause substantial modifications to the product's optimal quality, including the development of off-flavors and the destruction of vitamins and other minerals. Consumers place a high value on food's texture, flavor, aroma, shape, and color, and there is a growing demand for minimally processed, long-lasting products. As a result, alternative preservation techniques, particularly nonthermal ones, capable of preserving food's sensory and nutritional qualities, have been tested and developed [4].

High-pressure processing (HPP) is a common nonthermal method that utilizes elevated hydrostatic pressures (approximately 400–600 MPa) to pasteurize, denature multiple enzymes, and inactivate pathogenic and spoilage vegetative microorganisms, thereby assuring food safety [5]. Unlike thermal pasteurization, HPP does not affect covalent bonds and is able to effectively retain food quality attributes, namely sensorial and nutritional properties [4]. Additionally, as a pasteurization technique, nonthermal HPP does not target bacterial spores, such as those from *Bacillus* spp., yet it can target some spores from yeasts and molds, with the exception of those from *Byssochlamys* and *Talaromyces* spp. and some species of *Zygosaccharomyces* [6]. As such, HPP products are to be kept under refrigeration.

Only a few studies have evaluated the effects of HPP on dairy creams [7,8]. One observed that 450 MPa treatment (10 or 25 °C during 15 or 30 min) followed by refrigerated storage (4 °C) for 8 days did not affect the fat globule size distribution and other physicochemical properties of pasteurized creams [8]. Regarding microbiological changes, another study showed that it is possible to considerably reduce *Listeria innocua* load in creams (35% fat), obtaining a decimal reduction time (D) of D450 MPa/25 °C for 7.4 min [9]. Differently, Gervilla et al. (2000) obtained D400 MPa/25 °C = 4 min on ewe's milk (6% fat), showing the potential effect of fat to protect microorganisms against hydrostatic pressure [10].

Other methodologies have also been used for nonthermal pasteurization of dairy products, such as ultraviolet radiation, pulsed electric field (PEF), ultrasound, etc., yet these present lower efficacies compared to HPP, as PEF and US need to be combined with moderate temperatures to increase the inactivation rates of the target microorganisms, while ultraviolet radiation has low penetrance in opaque fluids. Other methodologies such as membrane filtration, despite its possible continuous use, require frequent cleaning and replacing the filters, which are rather expensive and do not allow a proper flow of bulky liquids through the filters [11].

To evaluate the impact and safety of this nonthermal technology and compare it with thermal pasteurization, (a) microbial load (endogenous and inoculated *Escherichia coli* and *L. innocua*), (b) fatty acid composition, (c) color parameters, (d) viscosity, and (e) volatile compounds were studied. Samples included the raw cream with no treatment (control), after the heat treatment (conventional pasteurization at 75 °C for 15 s), and after the pressure treatment (at 450 and 600 MPa for 5 min), followed by refrigerated storage (4 °C). The effect of HPP on inoculated microorganisms in dairy cream was also studied in a second set of experiments, at 600 MPa for 5 and 15 min.

2. Materials and Methods
2.1. Cream Samples

Industrially homogenized raw and thermally pasteurized (75 °C for 15 s) cream samples were kindly provided by a local cream-producing company (Portugal). Pasteurization was performed according to the commercial procedure used in the company [12].

2.2. Preparation of Cream Samples and Inoculation

Triplicated samples (20 mL each), for each storage day, were aseptically packed in UV-light sterilized low-permeability polyamide–polyethylene (PA/PE) bags and manually heat sealed prior to HPP, excluding as much air as possible.

Cultures of *E. coli* (ATCC 25922) and *L. innocua* (ATCC 33090) were grown in Tryptic Soy Broth (TSB; Liofilchem, Roseto degli Abruzzi, Italy) at 37 °C for 24 h to reach the stationary phase and then inoculated into raw cream to a final concentration of 10⁸ cells/mL.

2.3. HPP Treatment of Samples

HPP treatments were performed in a pilot scale high-pressure device (Model 55, Hiperbaric, Burgos, Spain) with 55 L of vessel capacity, 2000 mm of vessel length, and 200 mm of vessel diameter. The pressure rise time was 200 MPa/min, and the decompression time was almost instantaneous. A first cream batch was subjected to 450 MPa and 600 MPa for 5 min each, at 7 °C, to optimize the pressure level required to achieve desirable microbial inactivation levels to extend the shelf life of dairy cream. Additionally, as described in the literature, the temperature of water increases between 2 and 3 °C for each 100 MPa [13]; so, in order to have a maximum temperature of 19–25 °C while at 600 MPa, the water temperature before pressurization was 7 °C. A second cream batch was processed at 600 MPa for 5 and 15 min at 7 °C to evaluate the effects of the processing time (at the most suitable pressure obtained in the first batch) on dairy cream. After the respective processing, samples from both batches were stored at 4 °C.

2.4. Storage Conditions

Thermally pasteurized and HPP samples from the first batch were stored under refrigeration (4 °C) for 5, 9, 18, 33, and 51 days, while samples from the second batch were stored for 3, 10, 28, and 52 days to evaluate and compare the shelf life of creams processed at both conditions (thermal pasteurization and HPP).

2.5. Microbial Analyses

After each experiment, cream samples from the first batch were analyzed for total aerobic psychrophiles (TAPs), Enterobacteriaceae (ENT), and lactic acid bacteria (LAB) counts. Apart from ENT, samples from the second batch were analyzed for the same microorganisms, along with inoculated *E. coli* (ATCC 25922) and *L. innocua* (ATCC 33090). Both cultures that were used to inoculate the cream samples were stored on Trypticase Soy Agar (TSA; Liofilchem, Roseto degli Abruzzi, Italy) Petri dishes at 4 °C. Briefly, one colony of each microorganism, previously isolated in TSA plate, was collected, inoculated in 250 mL of Tryptic Soy Broth (TSB; Liofilchem, Roseto degli Abruzzi, Italy), and incubated at 37 °C, 150 rpm, for 10–12 h. The growth period was selected in order to ensure that cells reached the stationary phase to be later inoculated into raw cream, with a final concentration of about 10^8 cells/mL. Under aseptic conditions, 20 mL of each cell suspension was used to inoculate 160 mL of the second batch of cream samples. The microbiological analyses were performed as described by [14]. The results were expressed as a decimal logarithm of colony-forming units per milliliter of cream (log CFU/mL). The maximum endogenous microbial load considered in this study was 6.00 log CFU/mL [15], and the detection limit was 1.00 log CFU/mL.

The experimental design of each cream batch is reported in Table 1.

Table 1. Experimental design of each cream batch and the aim of the study.

	HPP Conditions		Storage Period (Days)	Nomenclature	To Study the Effect of HPP after Processing and during Each Storage Period On:
	Pressure (MPa)	Duration (min)			
First batch	-	-	-	Raw	General microbiology (TAP, LAB, and ENT) and physicochemical parameters
	-	-	-	Pasteurized	
	450	5	0, 5, 9, 18, 33, 51	450/5	
	600	5		600/5	
Second batch	-	-	-	Raw	• General microbiology (TAP and LAB) and physicochemical parameters
	-	-	-	Pasteurized	• Inoculated *E. coli* and *L. innocua*
	600	5	0, 3, 10, 28, 52	600/5	
	600	15		600/15	

2.6. pH and Color

The pH of all cream samples was measured at room temperature (21 ± 2 °C) in triplicate with a glass electrode (pH electrode 50 14, Crison Instruments, S.A., Barcelona, Spain).

The color was assessed using a Konica Minolta CM 2300d (Konica Minolta, Osaka, Japan) spectrophotometer on three random spots per sample, recorded according to the CIELab system, and the data were processed with the SpectraMagic™ NX software (Konica Minolta, Osaka, Japan). The obtained parameters were L*-lightness, a*-redness, and b*-yellowness. The total color difference (ΔE*) was calculated using Equation (1) [16].

$$\Delta E^* = [(L^* - L^*_0)^2 + (a^* - a^*_0)^2 + (b^* - b^*_0)^2]^{1/2} \quad (1)$$

where ΔE* is the total color change between a sample and the control (initial values identified with the subscript "0").

2.7. Apparent Viscosity Measurements

The cream's apparent viscosity was determined using a controlled-stress rheometer (AR-1000, TA Instruments, New Castle, DE, USA), equipped with a cone-and-plate geometry (acrylic cone, 6 cm diameter, and 2° angle). The bottom plate temperature was kept constant using a circulating bath (Circulating Bath 1156D, VWR International, Carnaxide, Portugal). Samples were equilibrated to 25 °C for about 15 min and then gently homogenized and placed carefully (approximately 2 mL) on the top of the bottom plate to minimize the damage to the sample structure and avoid trapping air bubbles. Flow curves were obtained by applying a continuous shear stress ramp (0 to 3 Pa) for 3 min [17].

2.8. Fatty Acid Determination

Fatty acids (FAs) were determined by gas chromatography as methyl esters (FAMEs). Briefly, fat was separated by centrifugation at 13,000 rpm for 20 min. Then, 40 μL of the upper layer (fat phase) was dissolved in hexane (2 mL), and the FAs were converted to their respective FAME by cold transmethylation (ISO 12966-2, 2011). Chromatographic separation was achieved with an Agilent J&W Select FAME column (100 m × 0.25 mm, J&W Agilent, Santa Clara, CA, USA) using a Chrompack CP 9001 gas chromatograph (Chrompack, Middelburg, The Netherlands) equipped with an FID detector. FA identification and FID calibration were accomplished with a certified reference standard mixture (TraceCert–Supelco 37 component FAME mix) and individual FAME, all from Supelco. Fatty acids were expressed in a relative percentage of their FAME.

2.9. Volatile Profile

The volatile compound (VOC) profiles were determined by headspace solid-phase microextraction (HS-SPME) followed by gas chromatography–mass spectrometry (GC-MS), as performed by [18], with modifications. Initially, 5 mL of each sample was placed in 20 mL headspace vials, and then cyclohexanone was added as an internal standard along with 28% sodium chloride (w/w). The vials were heated at 60 °C for 20 min with constant stirring (250 rpm), and the SPME fiber (DVB/CAR/PDMS; 50/30 μm; Supelco Inc., Bellefonte, PA, USA) was exposed for 30 min (60 °C). Volatiles were thermally desorbed for 5 min in the injector port (splitless mode; 250 °C). Chromatographic separation was performed on a fused-silica DB-5 MS column (30 m ×0.25 mm I.D. × 0.25 μm film thickness) from J&W Agilent (Santa Clara, CA, USA), with a temperature program going from 40 to 235 °C and a total run time of 60 min. The MS transfer line and ion source were at 280 °C and 230 °C, respectively, and the MS quadrupole temperature was at 150 °C, with an electron ionization of 70 eV set in full scan mode (m/z 40 to 650 at 1.2 scan/s). Compounds were identified by comparing their respective mass spectra with a mass spectral database (NIST v14, nist.gov, accessed on 21 September 2023). Semi-quantification was achieved as internal standard equivalents basis and expressed in μg of internal standard equivalents per 100 mL of cream.

2.10. Statistical Analysis

The experiments were performed in triplicate, each analyzed in duplicate. Statistical analysis of the results was performed using a two-way analysis of variance (ANOVA) followed by a multiple comparison post hoc test and Tukey's honest significant differences (HSD) test at a 5% level of significance.

3. Results and Discussion

3.1. Microbial Analysis

Regarding the first batch experiments, TAP, LAB, and ENT were quantified before (initial) and right after (day zero) thermal or HPP, and also on days 5, 9, 18, 33, and 51 under refrigeration (4 °C) (Figure 1).

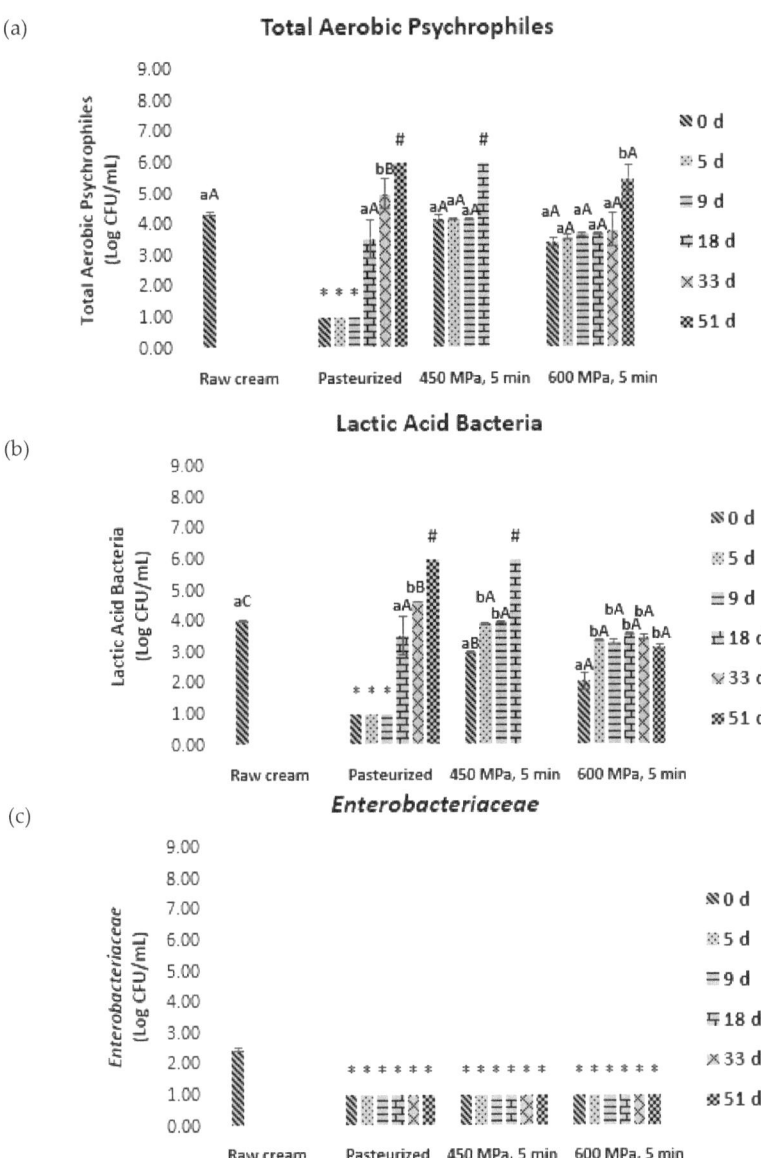

Figure 1. Microbial growth of (**a**) TAP, (**b**) LAB, and (**c**) ENT on initial raw cream after heat (75 °C, 15 s) treatment and after pressure treatment under 450 MPa and 600 MPa during 5 min (first batch). Analyses were made on the initial cream and right after processing (◧) and after 5 (⊙), 9 (−), 18 (⊥), 33 (◆), and 51 (✻) days of storage at 4 °C. Bars with * and # represent microbial loads below the detection limit (lower than 1.00 log CFU/mL) and above 6.00 log CFU/mL, respectively. Different letters denote significant differences ($p \leq 0.05$) between storage days for each condition (A,B) and treatment conditions for each storage day (a,b).

As represented in Figure 1, TAP, LAB, and ENT counts of thermally pasteurized cream samples decreased to below the detection limit (<1.00 log CFU/mL), maintaining low counts until the ninth day of storage. From the 18th day onwards, microbial growth was observed for all microorganisms, except for ENT, which remained undetected. By the 51st day, TAP and LAB counts surpassed 6.00 log CFU/mL, and, therefore, no further analyses were performed for the thermally pasteurized cream.

Right after processing at 450 MPa for 5 min, TAP loads observed were similar to those of raw dairy cream, while for samples processed at 600 MPa for 5 min, a very small (not statistically significant) decrease was observed. For ENT, regardless of the treatment, its counts were reduced to below 1.00 log CFU/mL and kept constant throughout storage. These results are in agreement with Permanyer et al. (2010), who reported a similar barosensivity of ENT when human milk was pressurized at 400, 500, and 600 MPa for 5 min at 12 °C [19]. Evert-Arriagada et al. (2014), working with starter-free fresh cheeses, also observed that after HPP (500 MPa, 5 min, 16 °C), ENT was not able to recover during all the cold storage period (21 days) [20]. From the 18th day onwards, both TAP and LAB counts were above 6.00 log CFU/mL; thus, samples treated at 450 MPa for 5 min (450/5) were considered spoiled. Samples processed at 600 MPa for 5 min (600/5) resulted in a slower recovery of TAP and LAB under refrigeration in comparison to those processed at 450/5, with TAP counts only increasing ($p \leq 0.05$) after the 51st day (Figure 1). This demonstrates the efficiency of HPP at 600 MPa to injure microorganisms, taking them additional time to recover and develop compared to the thermal pasteurization process.

To evaluate the influence of pressurization time, a second study was performed, and a new fresh cream batch was processed at 600 MPa for 15 min (600/15) instead of 5. Since in the first study, ENT exhibited high sensitivity to both high pressure and pasteurization treatments, the effect of 600/15 was evaluated only for TAP and LAB. In addition to endogenous microorganisms, the effect of HPP (600/5 and 600/15) on inoculated *L. innocua* and *E. coli* was also evaluated.

The 600/15 condition significantly reduced ($p \leq 0.05$) TAP and LAB counts by about 1.4- and 1.8-fold, respectively, compared to the initial raw cream counts (Figure 2).

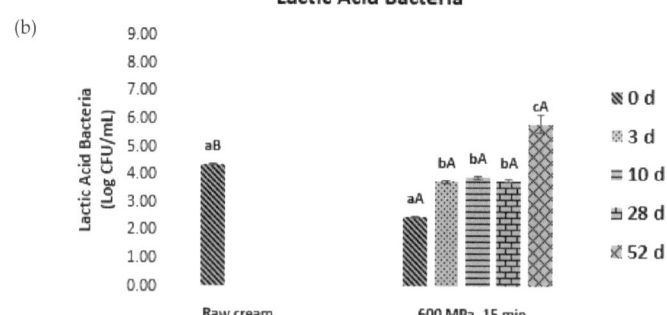

Figure 2. Cont.

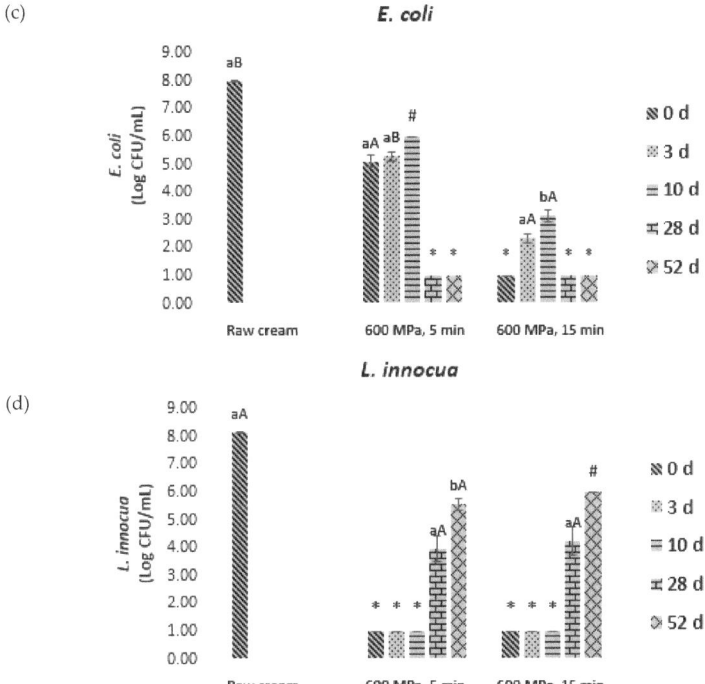

Figure 2. Microbial growth of (**a**) TAP, (**b**) LAB, (**c**) *E. coli*, and (**d**) *L. innocua* on initial raw cream after heat treatment (75 °C, 15 s) and after pressure treatment under 600 MPa for 5 min and 600 MPa during 15 min (second batch). Analyses were made on the initial cream and right after processing (◼) and after 3 (✵), 10 (═), 28 (✶), and 52 (✿) days of storage at 4 °C. Bars with * and # represent microbial loads below the detection limit (lower than 1.00 log CFU/mL) and above 6.00 log CFU/mL, respectively. Different letters denote significant differences ($p \leq 0.05$) between storage days for each condition (A,B) and treatment conditions for each storage day (a–c).

By the 28th day of storage, TAP counts on thermally pasteurized samples increased to values above 6.00 log CFU/mL, while those treated by HPP (600/15) presented counts of 4.53 ± 0.11 log CFU/mL, evidencing the efficacy of HPP in inhibiting long-term microbial development and extending shelf life. Regarding LAB, a significant increase ($p \leq 0.05$) throughout storage was observed.

Concerning inoculated microorganisms, 600/5 and 600/15 treatments were able to significantly reduce ($p \leq 0.05$) *E. coli* counts compared to the initial inoculated load (Figure 2). By the 10th day, *E. coli* counts on both 600 MPa treatments experienced a significant increase ($p \leq 0.05$), surpassing 6.00 log CFU/mL for 600/5 samples. However, on the following days, *E. coli* counts on both 600 MPa treated samples presented values below 1.00 log CFU/mL, probably due to the fact that *E. coli* is not able to survive after long exposures to low temperatures, as suggested by Arias et al. (2001) [21]. Despite the similar outcomes by the end of storage time, longer exposure to HPP at the same pressure appeared to be more effective in delaying microbial growth over time, given the lower counts registered on days 3 and 10.

Previous works revealed that gram-positive bacteria are more resistant to HPP than gram-negative [22,23]. In our study, *L. innocua* loads increased in both 600 MPa-treated samples ($p \leq 0.05$) from the 28th up to the 52nd day of storage, suggesting that cells may recover from the injuries caused by HPP and grow during cold storage [24]. A larger number of L. monocytogenes cells on milk samples, after 10 days of refrigerated storage, was also observed by Liepa et al. (2018) [25]. This is probably due to the higher-pressure resistance of gram-positive bacteria, namely regarding their metabolic repair mechanisms, in comparison with gram-negative *E. coli*.

Even though thermal pasteurization was able to reduce initial microbial loads and inhibit microbial growth on the first days of storage, it is possible to conclude that HPP has a more pronounced effect on slowing microbial growth rate over time, as evidenced by lower microbial counts on the final day of storage (52nd).

Since milk and dairy products follow very strict regulations worldwide, further research is also needed to accurately establish the safety of dairy cream processed by HPP, namely to overcome these regulatory issues. For instance, in the United States, pasteurization must inactivate *Mycobacterium tuberculosis* and *Coxiella burnettii* (which is more heat stable than the first one), while also resulting in a negative phosphatase reaction [26].

3.2. pH and Color

The initial pH of the cream used in the first and second studies was similar to the ones reported in the literature [27]: 6.74 ± 0.05 and 6.91 ± 0.14, respectively (Table S1—available in the Supplementary Materials). Regarding the first batch, all treated samples presented similar values to raw cream ($p > 0.05$), with small variations throughout storage (Table S1—available in the Supplementary Materials). Contrarily, the pH of HPP samples was higher ($p \leq 0.05$) during the first 9 days of storage, decreasing thereafter, which was probably caused by the observed microbial growth and the organic acids produced from their metabolic activity [28]. On the second batch, no significant differences ($p > 0.05$) between treatment conditions at each storage day were detected (Table S2—available in the Supplementary Materials).

Regarding color measurements, detailed results for L*, a*, and b* values are presented in Table S1 (available in the Supplementary Materials) (first batch) and Table S2 (available in the Supplementary Materials) (second batch). In general, the L* parameter on both studies remained stable at all different storage days and conditions, except on 600/5 samples, where a significant increase ($p \leq 0.05$) was observed when comparing the value obtained immediately after processing with that obtained on the 51st day of storage. For the a* parameter, it suffered some variations concerning both batches. On the first batch, compared to the initial raw cream, it was significantly higher ($p \leq 0.05$) for both HPP samples and similar ($p > 0.05$) to the thermally pasteurized samples. On the contrary, in the second study, initial a* values of all samples, treated and non-treated, were statistically different ($p \leq 0.05$) from each other, in the order from the highest a* value to the lowest: raw cream >600/15 > thermally pasteurized. These variations between the first and second studies are probably due to differences between the cream's batch. By looking at every storage period, 450/5 and 600/5 samples remained statistically similar to each other ($p > 0.05$), differing only from thermally treated samples ($p \leq 0.05$). The same happened with 600/15 and thermally treated samples on the second batch ($p \leq 0.05$); the a* parameter on HPP samples was always higher than the thermally treated samples. Despite the small variations obtained for L*, a*, and b* parameters, no statistical differences ($p > 0.05$) were observed for the total color change (ΔE^*) for all treatment conditions at each day of storage on both studies.

3.3. Viscosity

Cream's flow behavior was studied only for raw, thermally pasteurized, and 600 MPa-processed samples in the second batch. Generally, samples showed a qualitatively similar non-Newtonian flow behavior, with apparent viscosity decreasing with shear rate (shear thinning). The observed behavior was expected for an emulsion and is in accordance with Donsì et al. (2011), who evaluated the rheological behavior of milk cream under pressure (400–500 MPa for 5–10 min at 25 °C) and reported that HPP milk cream also behaved as a non-Newtonian pseudoplastic fluid [29].

The apparent viscosity of the studied samples was compared at a constant shear rate of 33 s^{-1} (Table 2).

Table 2. Apparent viscosity values determined at a particular shear rate (33 s^{-1}) for initial raw cream and cream at different treatment conditions (75 °C, 15 s, 600/5, and 600/15) right after processing and after 3, 10, 28, and 52 days of storage at 4 °C. Results are presented as mean ± standard deviation. Different superscript letters denote statistical differences ($p \leq 0.05$) between storage days for each condition (A–C) and treatment conditions for each storage day (a,b).

Storage Time (Days)	Conditions	Shear Rate (1/s)	Viscosity (Pa·s)
0	Initial		0.015 ± 0.001 ^{aA}
	Heat treated		0.017 ± 0.001 ^{aA}
	600 MPa/5 min		0.016 ± 0.001 ^{aA}
	600 MPa/15 min		0.015 ± 0.001 ^{aA}
3	Heat treated		0.018 ± 0.002 ^{aA}
	600 MPa/5 min		0.031 ± 0.003 ^{bB}
	600 MPa/15 min		0.026 ± 0.002 ^{bB}
10	Heat treated	33.19	0.017 ± 0.002 ^{aA}
	600 MPa/5 min		0.027 ± 0.003 ^{bB}
	600 MPa/15 min		0.026 ± 0.002 ^{bB}
28	Heat treated		0.016 ± 0.002 ^{aA}
	600 MPa/5 min		0.028 ± 0.003 ^{bB}
	600 MPa/15 min		0.030 ± 0.003 ^{bBC}
52	Heat treated		–
	600 MPa/5 min		0.030 ± 0.003 ^{aB}
	600 MPa/15 min		0.034 ± 0.003 ^{aC}

No major differences in the initial viscosity values were detected between the raw cream and all treated samples. Also, the apparent viscosity of the thermally pasteurized samples remained similar ($p > 0.05$) throughout the storage time. After 3 days of refrigerated storage, HPP samples presented viscosity values almost two times higher ($p \leq 0.05$) than the initial ones (immediately after processing). However, from this day forward, viscosity values of HPP samples remained statistically similar ($p > 0.05$), only increasing ($p \leq 0.05$) on the 52nd day of storage for 600/15 samples. Dumay et al. (1996) reported that after HPP, the flow behavior of the pasteurized cream samples did not show considerable changes after 7 days of storage (4 °C) [8]. In general, pressure-treated samples always presented a higher viscosity ($p \leq 0.05$) than the heat treated ones (Table 2), which can be advantageous in the development of products with different texture characteristics and consumer acceptability.

3.4. Fatty Acid Analysis

To our knowledge, this is the first study that reports the changes observed in cream's fatty acid profile after thermal pasteurization and pressurization treatments, upon storage. The GC analysis revealed the presence of twenty-nine FAs (Table S3—available in the Supplementary Materials). The cream samples were essentially rich in saturated, followed by monounsaturated, and a small percentage of polyunsaturated FAs. Raw cream's main fatty acids were palmitic (C16:0, 24.50 ± 0.12%), oleic (C18:1c, 20.03 ± 0.11%), and myristic (C14:0, 11.25 ± 0.07%), similar to what was previously reported [28,30].

In general, the different treatments did not affect ($p > 0.05$) saturated and monounsaturated FAs, while polyunsaturated decreased ($p \leq 0.05$) on the 52nd day of storage. Regarding the main FA on cream, only C16:0 was present in higher ($p \leq 0.05$) amounts in raw cream (compared to processed samples). Moltó-puigmartí et al. (2011) reported that HPP did not significantly change FA proportions compared to untreated human milk [31].

3.5. Volatile Analysis

A total of 39 different VOCs were identified in cream samples. Table 3 shows the chemical families of the VOCs and the total volatile amounts (identified and non-identified) detected. In general, there was a tendency for total volatiles to increase throughout storage. In raw cream, the most abundant families were aliphatic hydrocarbons, followed by aldehydes/ketones, acids, and lactones. Immediately after thermal and HPP treatments, a new class of compounds, alcohols, was detected on cream samples (Table 3). Alcohols can be produced by the reduction in their corresponding aldehydes and methyl ketones, through the activity of LAB dehydrogenases or by sugar fermentation, which is in accordance with the lower pH measured in these samples [32].

Table 3. Cream volatile profile (mg/100 g equivalents of cyclohexanone) at different treatment conditions (75 °C, 15 sec, 600/5, and 600/15) of the initial cream and right after processing (0 d) and after 3 (3 d), 10 (10 d), 28 (28 d), and 52 (52 d) days of storage at 4 °C. Results are presented as mean ± standard deviation. Different superscript letters denote statistical differences ($p \leq 0.05$) between storage days for each condition (A–C) and treatment conditions for each storage day (a–d).

Storage Time (Days)	Conditions	Alcohols	Acids	Aldehydes/Ketones	Aliphatic Hydrocarbons	Lactones	Total Volatiles
0	Initial	Nd	16.4 ± 2.5 [aA]	19.5 ± 1.3 [aA]	51.6 ± 1.9 [aAB]	0.2 ± 0.1 [aA]	315.1 ± 26.9 [aAB]
	Heat treated	3.2 ± 0.4 [aAB]	33.5 ± 4.7 [aAB]	36.9 ± 1.0 [aAB]	80.5 ± 0.8 [aB]	1.1 ± 0.1 [aB]	492.8 ± 36.7 [aB]
	600 MPa/5 min	4.1 ± 0.6 [aAB]	54.8 ± 3.3 [aB]	26.2 ± 4.6 [aAB]	41.6 ± 0.4 [aA]	0.4 ± 0.1 [aAB]	291.1 ± 11.4 [aA]
	600 MPa/15 min	6.6 ± 0.6 [aB]	52.6 ± 6.4 [aB]	47.5 ± 1.0 [bB]	6.9 ± 0.2 [aA]	0.4 ± 0.1 [aAB]	314.1 ± 10.7 [aAB]
3	Heat treated	3.6 ± 0.3 [aA]	34.3 ± 2.8 [aA]	49.6 ± 10.2 [abB]	94.4 ± 2.8 [abA]	0.7 ± 0.1 [aA]	427.2 ± 64.5 [aA]
	600 MPa/5 min	5.2 ± 0.3 [aA]	114.2 ± 13.7 [cB]	42.0 ± 0.7 [abAB]	152.5 ± 3.8 [bB]	0.9 ± 0.1 [aAB]	493.9 ± 21.7 [bA]
	600 MPa/15 min	4.4 ± 0.7 [aA]	53.1 ± 0.7 [aA]	23.0 ± 0.8 [aA]	79.4 ± 2.9 [bA]	1.6 ± 0.1 [bB]	479.9 ± 58.2 [aA]
10	Heat treated	4.1 ± 0.2 [aA]	26.1 ± 3.7 [aA]	60.5 ± 1.7 [bAB]	121.6 ± 3.2 [bA]	1.5 ± 0.3 [aA]	373.5 ± 35.8 [aA]
	600 MPa/5 min	6.6 ± 0.3 [aA]	84.4 ± 4.8 [bB]	70.8 ± 6.4 [bB]	220.2 ± 13.4 [cB]	2.4 ± 0.3 [bB]	703.2 ± 66.1 [cB]
	600 MPa/15 min	8.3 ± 0.2 [aA]	78.9 ± 4.12 [bB]	43.4 ± 8.0 [abA]	310.2 ± 6.8 [dC]	2.3 ± 0.2 [bAB]	782.4 ± 32.4 [bB]
28	Heat treated	20.3 ± 5.0 [bB]	62.8 ± 9.8 [bA]	74.5 ± 11.8 [bA]	69.2 ± 5.7 [aA]	2.6 ± 0.2 [bA]	687.9 ± 37.8 [bA]
	600 MPa/5 min	5.9 ± 1.2 [aA]	152.5 ± 3.8 [dB]	73.9 ± 2.6 [bA]	346.5 ± 36.4 [dB]	4.9 ± 0.3 [cB]	1007.3 ± 49.9 [dC]
	600 MPa/15 min	4.6 ± 0.4 [aA]	220.7 ± 6.7 [cC]	71.4 ± 6.4 [cA]	83.3 ± 2.2 [bA]	5.6 ± 0.1 [cB]	797.6 ± 28.6 [bB]
52	Heat treated	50.5 ± 3.7 [cB]	126.1 ± 8.9 [cA]	61.9 ± 12.7 [bB]	64.2 ± 4.2 [aA]	2.6 ± 0.2 [bA]	751.2 ± 56.4 [bA]
	600 MPa/5 min	4.1 ± 0.3 [aA]	167.1 ± 6.4 [dB]	52.2 ± 2.2 [bAB]	119.9 ± 8.1 [bB]	5.3 ± 0.5 [cB]	783.3 ± 30.2 [cA]
	600 MPa/15 min	6.1 ± 0.3 [aA]	307.0 ± 3.3 [dC]	36.2 ± 3.4 [abA]	224.2 ± 1.2 [cC]	5.7 ± 0.2 [cB]	881.1 ± 112.7 [bA]

Nd—not detected.

From the 28th day onwards, thermally treated samples presented higher ($p \leq 0.05$) amounts of alcohol compared to HPP samples. Similarly, Chugh et al. (2014) studied the effect of thermal pasteurization on skim milk's volatile composition, observing that during refrigerated storage, alcohol concentration increased as a result of the reduction in the corresponding carbonyl compounds [33].

The initial amount of acid compounds in raw cream increased ($p \leq 0.05$) immediately after thermal and HPP treatments (Table 3). Throughout the storage, the number of acids on HPP samples remained higher ($p \leq 0.05$) than on thermally treated samples. Garrido et al. (2015) observed a relevant increase of carboxylic acids in human milk after processing at 400 or 600 MPa for 6 min, which was probably due to the release of short-chain FAs from triglycerides (lipolysis) [34]. Acids can act as precursor molecules for a series of catabolic reactions, which can lead to the production of other flavor compounds such as alcohols, lactones, and methyl ketones [32].

All cream samples presented a similar ($p > 0.05$) content of aldehydes/ketones after processing compared to the raw cream, except for 600/15 samples, which presented higher levels ($p \leq 0.05$) (Table 3). Vazquez-Landaverde et al. (2006) observed that at 25 °C, ketone concentration in milk processed under 620 MPa at 1, 3, or 5 min was similar to raw milk [35]. Despite the fact that ketones are naturally present in raw milk, most of them can be formed during heat treatment by β-oxidation of saturated fatty acids or by decarboxylation of β-ketoacids. Furthermore, several authors reported that HPP enhances the oxidation of free FAs, leading to the formation of ketone VOCs [34,36,37]. Vazquez-Landaverde et al. (2006) also observed an increase in aldehyde concentration when milk was processed at 620 MPa, which was possibly due to a higher solubility of oxygen under high pressure, which could enhance the formation of hydroperoxides, resulting in more aldehydes [35].

Aliphatic hydrocarbons were the major VOCs found on cream samples, presenting no regular tendency throughout the storage days under all processing conditions; they were statistically similar ($p > 0.05$) on both thermal and HPP samples. However, their content was significantly higher on thermally treated samples ($p \leq 0.05$). Accordingly, Chugh et al.

(2014) observed a significant increase in hydrocarbon compounds after heat treatment of skim milk [33].

Lactones, detected in very low levels in all cream samples, are related to lipid degradation and are formed by the cyclization of γ- and β-hydroxy acids [32]. Lactone levels in HPP samples were similar ($p > 0.05$) to raw cream but were higher in thermally pasteurized samples ($p \leq 0.05$). Throughout the storage, lactones increased significantly on all treated samples ($p \leq 0.05$) and were always higher on HPP samples.

In summary, initially treated samples were similar to raw cream with a general tendency to increase volatile amounts throughout the storage period, without major differences between heat pasteurized and HPP samples.

4. Conclusions

The present study evaluated the feasibility of using HPP for the nonthermal pasteurization of raw dairy cream as an alternative to the conventional heat-based pasteurization processes. HPP samples were still microbiologically acceptable by the 51st day of refrigerated storage, unlike thermally processed ones, which clearly highlights the use of this nonthermal technology to extend the shelf life of dairy cream. For the effect of HPP on inoculated microorganisms, even though HPP at 600 MPa for 15 min was able to reduce microbial loads to lower counts than 600 MPa for 5 min (at the beginning of the storage experiments), a similar microbiological development pattern was observed on both processing conditions by the end of the shelf life evaluation period, indicating that the inactivation effect is less likely to be dependent on processing time. In general, pH, color (maximum variation of ΔE^* up to 8.43), and fatty acids (mainly palmitic, oleic, and myristic acids) were not considerably changed by the different processing conditions and storage, while viscosity presented higher values ($p \leq 0.05$) for HPP samples (0.034 Pa·s, at the 52nd day). Furthermore, VOCs of all treated samples presented a tendency to increase throughout storage, particularly acids and aliphatic hydrocarbons. From a practical point of view, commercial (heat pasteurized) refrigerated dairy cream usually presents a shelf life <3 weeks. This shelf life could be considerably extended by at least 30 days using HPP, without major changes in the products' quality, clearly evidencing the potential of this nonthermal technology for dairy cream pasteurization. Indeed, these results open the possibility of using HPP for the nonthermal pasteurization of dairy products, such as fresh cheeses, dairy creams, and even milk, either for retailing or using as food ingredients, as the extended shelf life can not only increase food safety but also reduce food waste.

Supplementary Materials: The following supporting information can be downloaded at https://www.mdpi.com/article/10.3390/foods12193640/s1, Table S1: Variation of pH and color throughout different storage (4 °C) days of raw cream (initial) and after submitting to the different treatment conditions (heat, 450/5 and 600/5). Results are presented as mean ± standard deviation. Different letters represent statistical differences ($p \leq 0.05$) between storage days for each condition (A,B) and treatment conditions for each storage day (a,b); Table S2: Variation of pH and color throughout different storage (4 °C) days of raw cream (initial) and after submitting to the different treatment conditions (heat and 600/15). Results are presented as mean ± standard deviation. Different letters represent statistical differences ($p \leq 0.05$) between storage days for each condition (A,B) and treatment conditions for each storage day (a,b); Table S3: Cream fatty acid composition (% of total fatty acids) throughout different storage (4 °C) days of raw cream (initial) and after submitting to the different treatment conditions (heat, 600/5 and 600/15). Results are presented as mean ± standard deviation. Different letters represent statistical differences ($p \leq 0.05$) between storage days for each condition (A,B) and treatment conditions for each storage day (a,b).

Author Contributions: Conceptualization, F.M., R.V.D., C.A.P., and J.A.S.; methodology, F.M., R.V.D., C.A.P., S.C. (volatiles and fatty acids), and J.A.L.-d.-S. (viscosity); formal analysis, F.M., R.V.D., C.A.P., S.C., and J.A.L.-d.-S.; investigation, F.M., R.V.D., and C.A.P.; writing—original draft preparation, F.M., R.V.D., and C.A.P.; writing—review and editing, J.A.S.; supervision, J.A.S. All authors have read and agreed to the published version of the manuscript.

Funding: This work received financial support from PT national funds (FCT/MCTES, Fundação para a Ciência e Tecnologia and Ministério da Ciência, Tecnologia e Ensino Superior) through the projects UIDB/50006/2020 and UIDP/50006/2020, and from FCT/MCTES through the PhD grants of Ricardo V. Duarte (SFRH/BD/121727/2016) and Carlos A. Pinto (SFRH/BD/137036/2018 and COVID/BD/153220/2023).

Data Availability Statement: Data are available upon request to the corresponding author.

Acknowledgments: Thanks are due to the University of Aveiro and FCT/MCT for the support for the LAQV and CICECO research units (UIDB/50006/2020 and UIDB/50011/2020, respectively) through national funds and, where applicable, co-financed by the FEDER, within the PT2020 Partnership Agreement and the Portuguese NMR Network.

Conflicts of Interest: The authors declare no conflict of interest.

References

1. Deosarkar, S.S.; Khedkar, C.D.; Kalyankar, S.D.; Sarode, A.R. Cream: Types of cream. In *Encyclopedia of Food and Health*; Elsevier Ltd.: Amsterdam, The Netherlands, 2016; Volume 2, pp. 331–337. ISBN 9780123849533.
2. Wang, T.; Lin, L.U.; Ou, J.I.E.; Chen, M.I.N.; Yan, W. The inhibitory effects of varying water activity, pH, and nisin content on Staphylococcus aureus growth and enterotoxin A production in whipping cream. *J. Food Saf.* **2016**, *37*, e12280. [CrossRef]
3. O'Sullivan, M. *The Technology of Dairy Products*, 2nd ed.; Early, R., Ed.; Springer: New York, NY, USA, 1993; Volume 19.
4. Hogan, E.; Kelly, A.L.; Sun, D.W. High Pressure Processing of Foods. An Overview. In *Emerging Technologies for Food Processing*; Academic Press: Cambridge, MA, USA, 2005; Volume 1, ISBN 9780126767575.
5. Elamin, W.M.; Endan, J.B.; Yosuf, Y.A.; Shamsudin, R.; Anvarjon, A. High pressure processing technology and equipment evolution: A review. *J. Eng. Sci. Technol. Rev. 8* **2015**, *8*, 75–83. [CrossRef]
6. Pinto, C.A.; Moreira, S.A.; Fidalgo, L.G.; Inácio, R.S.; Barba, F.J.; Saraiva, J.A. Effects of high-pressure processing on fungi spores: Factors affecting spore germination and inactivation and impact on ultrastructure. *Compr. Rev. Food Sci. Food Saf.* **2020**, *19*, 553–573. [CrossRef] [PubMed]
7. Ravash, N.; Peighambardoust, S.H.; Soltanzadeh, M.; Lorenzo, J.M. Impact of high-pressure treatment on casein micelles, whey proteins, fat globules and enzymes activity in dairy products: A review. *Crit. Rev. Food Sci. Nutr.* **2020**, *62*, 2888–2908. [CrossRef] [PubMed]
8. Dumay, E.; Lambert, C.; Funtenberger, S.; Cheftel, J.C. Effects of high pressure on the physico-chemical characteristics of dairy creams and model oil/water emulsions. *LWT Food Sci. Technol.* **1996**, *29*, 606–625. [CrossRef]
9. Raffalli, J.; Rosec, J.C.; Carlez, A.; Dumay, E.; Richard, N.; Cheftel, J.C. High pressure stress and inactivation of Listeria innocua in inoculated dairy cream. *Sci. Aliments* **1994**, *14*, 349–358.
10. Gervilla, R.; Ferragut, V.; Guamis, B. High pressure inactivation of microorganisms inoculated into ovine milk of different fat contents. *J. Dairy Sci.* **2000**, *83*, 674–682. [CrossRef] [PubMed]
11. Neokleous, I.; Tarapata, J.; Papademas, P. Non-thermal Processing Technologies for Dairy Products: Their Effect on Safety and Quality Characteristics. *Front. Sustain. Food Syst.* **2022**, *6*, 856199. [CrossRef]
12. *Codex Alimentarius Commission*; Food and Agriculture Organization: Rome, Italy, 2000; Volume 1, pp. 1–10.
13. Ahn, D.U.; Mendonça, A.F.; Feng, X. The Storage and Preservation of Meat: II—Nonthermal Technologies. In *Lawrie's Meat Science*; Elsevier: Amsterdam, The Netherlands, 2017; pp. 231–263. ISBN 9780081006948.
14. Pinto, C.; Moreira, S.A.; Fidalgo, L.G.; Santos, M.D.; Vidal, M.; Delgadillo, I.; Saraiva, J.A. Impact of different hyperbaric storage conditions on microbial, physicochemical and enzymatic parameters of watermelon juice. *Food Res. Int.* **2017**, *99*, 123–132. [CrossRef]
15. Institute of Medicine (US); National Research Council (US) Committee. International Microbiological Criteria for Dairy Products. In *Scientific Criteria to Ensure Safe Food*; National Academies Press: Washington, DC, USA, 2003; p. 424. ISBN 0309509203.
16. Milovanovic, B.; Tomovic, V.; Djekic, I.; Miocinovic, J.; Solowiej, B.G.; Lorenzo, J.M.; Barba, F.J.; Tomasevic, I. Colour assessment of milk and milk products using computer vision system and colorimeter. *Int. Dairy J.* **2021**, *120*, 105084. [CrossRef]
17. Duarte, R.V.; Casal, S.; Da Silva, J.A.L.; Gomes, A.; Delgadillo, I.; Saraiva, J.A. Nutritional, Physicochemical, and Endogenous Enzyme Assessment of Raw Milk Preserved under Hyperbaric Storage at Variable Room Temperature. *ACS Food Sci. Technol.* **2022**, *2*, 961–974. [CrossRef] [PubMed]
18. Shepard, L.; Miracle, R.E.; Leksrisompong, P.; Drake, M.A. Relating sensory and chemical properties of sour cream to consumer acceptance. *J. Dairy Sci.* **2013**, *96*, 5435–5454. [CrossRef] [PubMed]
19. Permanyer, M.; Castellote, C.; Audí, C.; Castell, M. Maintenance of breast milk immunoglobulin A after high-pressure processing. *J. Dairy Sci.* **2010**, *93*, 877–883. [CrossRef] [PubMed]
20. Evert-Arriagada, K.; Hernández-Herrero, M.M.; Guamis, B.; Trujillo, A.J. Commercial application of high-pressure processing for increasing starter-free fresh cheese shelf-life. *LWT Food Sci. Technol.* **2014**, *55*, 498–505. [CrossRef]
21. Arias, M.L.; Monge-Rojas, R.; Chaves, C.; Antillon, F. Effect of storage temperatures on growth and survival of Escherichia coli O157: H7 inoculated in foods from a neotropical environment. *Rev. Biol. Trop.* **2001**, *49*, 517–524. [PubMed]

22. Trujillo, A.J. Applications of high-hydrostatic pressure on milk and dairy products. *Int. J. High Press. Res.* **2002**, *22*, 619–626. [CrossRef]
23. Viazis, S.; Farkas, B.E.; Jaykus, L.A. Inactivation of bacterial pathogens in human milk by high-pressure processing. *J. Food Prot.* **2008**, *71*, 109–118. [CrossRef] [PubMed]
24. Bozoglu, F.; Alpas, H. Injury recovery of foodborne pathogens in high hydrostatic pressure treated milk during storage. *FEMS Immunol. Med. Microbiol.* **2004**, *40*, 243–247. [CrossRef]
25. Liepa, M.; Baltrukova, S.; Safety, F.; Health, A.; Bior, E.; Zagorska, J.; Galoburda, R. Survival of pathogens in high pressure processed milk. *Food Sci.* **2018**, *1*, 215–221. [CrossRef]
26. Fortin, N.D. Codex alimentarius commission. *Handb. Transnatl. Econ. Gov. Regimes* **2009**, *1*, 645–653. [CrossRef]
27. Gassi, J.; Famelart, M.; Lopez, C. Heat treatment of cream affects the physicochemical properties of sweet buttermilk. *Dairy Sci. Technol.* **2008**, *88*, 369–385. [CrossRef]
28. Decimo, M.; Ordónez, J.A.; Brasca, M.; Cabeza, M.C. Fatty acids released from cream by psychrotrophs isolated from bovine raw milk. *Int. J. Dairy Technol.* **2006**, *70*, 339–344. [CrossRef]
29. Donsì, G.; Ferrari, G.; Maresca, P. Rheological Properties of High Pressure Milk Cream. *Procedia Food Sci.* **2011**, *1*, 862–868. [CrossRef]
30. Rutkowska, J.; Bialek, M.; Adamska, A.; Zbikowska, A. Differentiation of geographical origin of cream products in Poland according to their fatty acid profile. *Food Chem.* **2015**, *178*, 26–31. [CrossRef] [PubMed]
31. Moltó-puigmartí, C.; Permanyer, M.; Isabel, A.; López-sabater, M.C. Effects of pasteurisation and high-pressure processing on vitamin C, tocopherols and fatty acids in mature human milk. *Food Chem.* **2011**, *124*, 697–702. [CrossRef]
32. Juan, B.; Barron, L.J.R.; Ferragut, V.; Trujillo, A.J. Effects of high pressure treatment on volatile profile during ripening of ewe milk cheese. *J. Dairy Sci.* **2010**, *90*, 124–135. [CrossRef] [PubMed]
33. Chugh, A.; Duizer, L.; Griffiths, M.; Walkling-Ribeiro, M.; Corredig, M.; Khanal, D. Change in color and volatile composition of skim milk processed with pulsed electric field and microfiltration treatments or heat pasteurization. *Foods* **2014**, *3*, 250–268. [CrossRef]
34. Garrido, M.; Contador, R.; García-Parra, J.; Delgado, F.J.; Delgado-Adámez, J.; Ramírez, R. Volatile profile of human milk subjected to high-pressure thermal processing. *Food Res. Int.* **2015**, *78*, 186–194. [CrossRef]
35. Vazquez-Landaverde, P.A.; Torres, J.A.; Qian, M.C. Effect of high-pressure-moderate-temperature processing on the volatile profile of milk. *J. Agric. Food Chem.* **2006**, *54*, 9184–9192. [CrossRef]
36. Oey, I.; Lille, M.; Van Loey, A.; Hendrickx, M. Effect of high-pressure processing on colour, texture and flavour of fruit- and vegetable-based food products: A review. *Trends Food Sci. Technol.* **2008**, *19*, 320–328. [CrossRef]
37. Van Der Plancken, I.; Verbeyst, L.; De Vleeschouwer, K.; Grauwet, T.; Heiniö, R.L.; Husband, F.A.; Lille, M.; Mackie, A.R.; Van Loey, A.; Viljanen, K.; et al. (Bio)chemical reactions during high pressure/high temperature processing affect safety and quality of plant-based foods. *Trends Food Sci. Technol.* **2012**, *23*, 28–38. [CrossRef]

Disclaimer/Publisher's Note: The statements, opinions and data contained in all publications are solely those of the individual author(s) and contributor(s) and not of MDPI and/or the editor(s). MDPI and/or the editor(s) disclaim responsibility for any injury to people or property resulting from any ideas, methods, instructions or products referred to in the content.

Article

Clotting and Proteolytic Activity of Freeze-Dried Crude Extracts Obtained from Wild Thistles *Cynara humilis* L. and *Onopordum platylepis* Murb.

Cindy Bande-De León [1,*,†], Laura Buendía-Moreno [2,*,†], Adela Abellán [1], Pamela Manzi [3], Bouthaina Al Mohandes Dridi [4], Ismahen Essaidi [4], Lucia Aquilanti [5] and Luis Tejada [1]

1. Department of Human Nutrition and Food Technology, Universidad Católica de Murcia-UCAM, Campus de los Jerónimos, 30107 Murcia, Spain; aabellan@ucam.edu (A.A.); ltejada@ucam.edu (L.T.)
2. Department of Food Science and Technology, Veterinary Faculty, University of Murcia, 30071 Murcia, Spain
3. CREA, Centro di Ricerca Alimenti e Nutrizione, Via Ardeatina 546, 00178 Rome, Italy; pamela.manzi@crea.gov.it
4. Research Laboratory of Agrobiodiversity and Ecotoxicology LR21AGR02, High Agronomic Institute of Chott-Mariem, University of Sousse, Sousse 4042, Tunisia; bouthaina2@yahoo.com (B.A.M.D.); saidi.ismahen@gmail.com (I.E.)
5. Departament of Agricultural, Food and Environmental Sciences, Università Politecnica delle Marche, Via Brecce Bianche, 60131 Ancona, Italy; l.aquilanti@staff.univpm.it
* Correspondence: cmbande@ucam.edu (C.B.-D.L.); laura.buendia@um.es (L.B.-M.)
† These authors contributed equally to this work.

Abstract: The rising interest in finding alternatives to animal rennet in cheese production has led to studying the technological feasibility of using and exploiting new species of herbaceous plants. In this research work, and for the first time, freeze-dried extracts from *Cynara humilis* L. (CH) and *Onopordum platylepis* Murb. (OP) were studied for mineral and protein content, and their clotting and proteolytic activity were compared to those of *Cynara cardunculus* L. (CC). The effect of extract concentration (5–40 mg extract/mL), temperature (20–85 °C), pH (5–8), and $CaCl_2$ concentration (5–70 mM) on the milk clotting activity (MCA) of CC, CH and OP extracts was evaluated. The MCA values were significantly higher in CC at the same extract concentration. The extract that showed the most significant increase in clotting activity due to increased temperature was OP, with maximum activity at 70 °C. The pH value for maximum milk clotting was 5.0 for both CC and CH, whereas, in the case of OP, the pH value was 5.5. $CaCl_2$ enhanced the clotting capacity of the extracts, particularly for OP and CH. Furthermore, proteolytic activity (PA) and the hydrolysis rate increased with increasing time and enzyme concentration, with CC being the extract that achieved the highest caseinolytic activity.

Keywords: milk clotting activity; vegetable coagulant; proteolytic activity

1. Introduction

The enzymatic clotting of milk is an essential step in most cheesemaking processes. One of the main proteases responsible for this coagulation is chymosin, which is present in ruminant rennet and has been used for centuries by the dairy industry to manufacture different types of cheese [1].

Most of the commercial animal rennet used in dairies comes from recombinant sources, while only 20–30% of it is of natural origin [2]. The worldwide increase in cheese production has led to a decrease in the supply of animal rennet, thus increasing the demand for new coagulant substitutes, such as plant-derived proteases [1].

These proteases are obtained via maceration in water of different plant sections or organs (seeds, flowers, fruits, rhizomes, etc.) and have a high potential for use as milk coagulants to manufacture cheese, replacing animal rennet [3–7].

Numerous studies have been carried out on these plant enzymes obtained from aqueous extracts, such as those derived from papaya (papain), pineapple (bromelain) [8,9], and other plants, whose enzymes are characterised by a proteolytic to clotting activity ratio that is not sufficiently high or proteolytic activity that is excessively high for the production of commercial cheeses [10]. In addition, these vegetable coagulants have certain limitations, mainly related to the texture and sensory quality of the produced cheeses [3].

However, cheeses made with plant proteases from the genus *Cynara* L. have a smooth, creamy texture and exquisite flavour, thus explaining why these perennial herbaceous plants have been used since ancient times to produce traditional cheeses [11].

The genus *Cynara* L. is native to the Mediterranean flora and belongs to the family Asteraceae; it encompasses eight species and four subspecies, including *Cynara cardunculus* L. (CC) and *Cynara humilis* L. (CH). Some of these *Cynara* species have traditionally been used as milk coagulants in cheesemaking due to their high concentration of proteolytic enzymes responsible for clotting, such as cyprosins or cardosins [12]. These enzymes, present in the characteristic violet flowers that are dried and macerated for use, have proven to be successful substitutes for animal rennet [6]. CC has been used since ancient times for the manufacture of goat and sheep cheeses in several rural areas of Spain and Portugal [13]; some examples are Los Pedroches (Córdoba), Torta del Casar (Cáceres), La Flor de Guía (Gran Canaria), Serra da Estrela (Portugal), and Serpa (Portugal), among others [14].

Various studies have shown that the use of aqueous extracts obtained via the maceration of thistle flowers increases the initial microbial count of the milk and, consequently, of the curd [15,16]. Subsequent studies have shown that freeze drying aqueous extracts does not alter the initial microbial count after addition and have recommended its use as it is soluble in water and milk, is free of viable micro-organisms, has a stable shelf life without the need for preservatives, and has a coagulating activity that does not decrease after one year of storage [7,17–19].

The available studies evaluating the milk clotting activity (MCA) of freeze-dried thistle extracts and their use in cheese production have only been assessed in CC [7,17–22].

The genus *Onopordum* L. belongs to the family Asteraceae, and some of its species are widely distributed in Europe, Northern Africa, the Canary Islands, and the Caucasus, as well as Southwest and Central Asia. Thistles within this genus are native mainly to the Mediterranean biogeographical region and have been studied for their potential as antimicrobial, haemostatic, and hypotensive agents [23,24].

Very few studies describe the use of *Onopordum* spp. in cheese production. Very recently, Mozzon et al. (2020) [25] and Foligni et al. (2022) [26] studied the milk clotting and the caseinolytic activity of a freeze-dried extract from *Onopordum tauricum* Willd. in milk of different origins. To the author's knowledge, no study has evaluated the milk clotting activity (MCA) and proteolytic activity (PA) of the species *Onopordum platylepis* Murb.

Optimal conditions for the milk clotting of aqueous extracts from flowers of *C. cardunculus* L., *C. humilis* L. and *C. scolymus* L. have already been described [27–30], as well as has the proteolytic activity of cardosins A and B on goat casein [31], the proteolytic activity of *C. humilis* L. on ovine Na-caseinate [32] and of *C. scolymus* L. flower extract on bovine casein [33]. Available studies on CH describe the effect of pH and temperature on the rheological properties of gels curdled with CH [34,35].

However, as far as the authors know, no research studies are available on freeze-dried CH and OP performance as agents with clotting and caseinolytic activity.

The main objective of this research work was to characterize the MCA (the effect of extract concentration, temperature, pH, and calcium chloride ($CaCl_2$) concentration) and PA of freeze-dried extracts obtained from CH and OP compared to those obtained from CC. In order to characterize the freeze-dried extracts, the mineral and the protein content of the extracts obtained from CH and OP were also reported.

2. Materials and Methods

2.1. Freeze-Dried Enzymatic Extract Preparation

CC (from Cáceres, Extremadura, Spain), CH (from Cáceres, Extremadura, Spain), and OP (from Sousse, Tunisia, Africa) were the thistle species used for the preparation of the freeze-dried crude extracts. These were later freeze-dried following the procedure described by Tejada and Fernández-Salguero (2003) [22]. The plant material, styles, and stigmas were macerated in distilled water for 24 h at 25 °C in a 1:10 (w/v) ratio. The aqueous extract obtained was sieved, and the filtrate was centrifuged at $3000 \times g$ for 5 min. The supernatant obtained was filtered with Whatman No. 1 paper. The filtrate obtained was frozen at -32 °C for 24 h and then lyophilised (Alpha 1-2LD plus, Christ, Osterode am Harz, Germany) at a working pressure between 4 and 13 Pa. The lyophilised powder was hermetically packed and frozen at -20 °C until its use.

2.2. Characterisation of Thistle Extracts

2.2.1. Total Protein Content

The total protein of CC, CH, and OP was determined according to the Bradford method [36] using the Sigma (Sigma-Aldrich, Milan, Italy) ready-to-use reagent. A set of bovine serum albumin (Merck, KGa, Darmstadt, Germany) solutions (0.1–1 mg/mL) was used for calibration. Absorbance readings at 595 nm were carried out using a UV-1800 Shimadzu (Kyoto, Japan) spectrophotometer. The protein content of the three enzymatic extracts studied was determined in triplicate.

2.2.2. Mineral Content

Ca, P, Na, K, Mg, Zn and Mn contents of the freeze-dried crude extracts from CH and OP were determined according to the AOAC (2002) [37] method. Briefly, 0.3 g of each sample was ashed. After mineralisation, samples were solubilised in 1 mL of HNO_3 65% and then adjusted to a final volume of 50 mL of 1% HNO_3 (v/v) with 0.1% (w/v) CsCl to avoid sodium and potassium ionisation and with 0.1% $LaCL_3$ (w/v) for Ca and Mg detection. Ca, Na, K, Mg, Zn and Mn were detected using Atomic Absorption Spectrometer A. Analyst 300 (Perkin Elmer, Norwalk, CT, USA), while phosphorus content was measured at 400 nm using a UV-1800 spectrophotometer (Shimadzu Corporation, Tokyo, Japan). NIST, SMR1570a and SMR1846 (only for Mg) were used as standard reference materials.

2.2.3. Milk Clotting Activity Assay

The milk clotting activity (MCA) of the extracts was assessed using the Berridge test according to the International Dairy Federation (IDF) [38] based on the visual evaluation of the first clotting flakes' appearance. For the clotting activity determination, 10 mL of reconstituted skimmed milk powder (0.12 kg/L) was transferred into a clean and dry test tube. A calcium chloride solution was added at the concentration established in each test (Sigma-Aldrich, Milan, Italy). The assay tube was allowed to equilibrate for 5 min at the desired temperature in an M20 thermostatic water bath (Lauda-Konigshofen, Germany) before adding the enzymatic extract (CC, CH, and OP). After reaching the temperature, 0.1 mL of the enzymatic extracts were added. The time from the addition of the enzyme to the first appearance of solid material was measured in seconds, as clotting. One Soxhlet unit (SU/mL) of clotting activity was defined as the volume of milk that can be clotted by one volume unit of the enzymatic extract in 40 min under defined temperature, pH, and $CaCl_2$ test conditions [25] and was calculated with the following equation:

$$MCA(SU/mL) = (2400 \times M) \div (T \times V) \qquad (1)$$

where M is the volume of milk (mL); T is the clotting time in seconds; and V is the volume of the enzyme (mL).

The effect of four independent variables was studied (extract concentration (5–40 mg/mL), temperature (20–85 °C), pH (5–8), and $CaCl_2$ concentration (5–70 mM)) on the milk clotting activity. The clotting activity of the three enzymatic extracts studied was determined in triplicate.

To measure the effect of extract concentration (5, 10, 20, 30 and 40 mg/mL), the conditions of temperature (32 °C), pH (6.2), and $CaCl_2$ concentration (10 mM) were set.

The effect of temperature (20, 25, 30, 35, 40, 45, 50, 60, 70, 80, and 85 °C), pH (5, 5.5, 6, 6.5, 7, 7.5 and 8) and $CaCl_2$ concentration (5, 10, 15, 20, 30, 50, 50, 60 and 70 mM) variables were measured at two extract concentrations (20 and 40 mg/mL).

2.2.4. Proteolytic Activity

The proteolytic activity of the enzymatic extracts was determined following the method employed by Silva and Malcata (2005) [39]. The substrate used was bovine milk casein, free of carbohydrates and fatty acids (Calbiochem, Darmstadt, Germany) at 1% (w/v) in a 10 mM citrate buffer (pH 6.2) (Sigma-Aldrich, St. Louis, MO, USA) with 0.03% (w/v) sodium azide (Fisher Scientific, Madrid, Spain) to avoid microbial growth, and was incubated in a bath at 32 °C. Hydrolysis was started by adding 0.12 mL of the reconstituted extract at different concentrations (5, 10, 20, 30 and 40 mg of freeze-dried extract/mL) to 3 mL of the casein solution. Subsequently, 0.5 mL aliquots were sampled at different times (5, 10, 20, 30, 40, 50, and 60 min) and put in Eppendorf tubes. The proteolytic activity was quantified via an evaluation of the peptides soluble in aqueous 5% (w/v) trichloroacetic acid (TCA) (Sigma-Aldrich, St. Louis, Missouri, USA). For this, 1 mL of 5% TCA (w/v) was added to each tube, incubated for 10 min at 25 °C, and centrifuged at $12,000 \times g$ for 10 min, while the absorbance of the supernatant was measured at 280 nm in a quartz cuvette. A proteolytic unit (U) was defined as the amount of enzymatic extract that produced a 0.001 unit increase in absorbance at 280 nm per minute under the stated test conditions. All determinations were made in triplicate.

2.3. Statistical Analysis

All experiments were conducted in triplicate, and the results were expressed with the mean and standard error. The statistical analysis of different parameters was computed using the SPSS version 21.0 software package (IBM Corporation, Armonk, NY, USA). In order to assess differences between the species, a one-way analysis of variance (ANOVA) was applied to mineral composition. Regarding the MCA analysis, a two-way ANOVA was performed to study the influence of temperature, pH, and $CaCl_2$ concentration. For PA, a two-way ANOVA was applied to analyse the effect of the species and reaction time. Tukey's HSD test ($p < 0.05$) was performed to determine significant differences between the treatment groups. Differences were considered statistically significant when p-values were equal to or below 0.05. Relationships among the studied factors are presented using appropriate curves and tables.

3. Results and Discussion

3.1. Total Protein Content

The protein content of the reconstituted freeze-dried extracts was different ($p < 0.05$) between species CC, CH, and OP, corresponding to 0.1018 ± 0.0065 (mean ± standard error), 0.1121 ± 0.0102 (mean ± standard error) and 0.0764 ± 0.0011 (mean ± standard error) mg protein/mg extract, respectively.

3.2. Mineral Content

In Table 1, the mineral contents of the freeze-dried extracts from CH and OP are shown; in more detail, CH showed a higher content of calcium, potassium, magnesium, and zinc ($p < 0.05$) than OP did, while phosphorous and manganese contents were higher in OP. No differences were seen between these two species for sodium content.

Table 1. Mineral composition (mg/100 g dry weight) of the freeze-dried extracts of *Cynara humilis* L. (CH) and *Onopordum platylepis* Murb. (OP).

Minerals	Species	
	CH	OP
Ca	346.1 ± 2.3 [a]	330.7 ± 4.1 [b]
P	638.9 ± 7.6 [b]	778.1 ± 13.3 [a]
Na	77.3 ± 3.4 [a]	73.6 ± 1.5 [a]
K	7577.7 ± 156.2 [a]	5918.6 ± 86.4 [b]
Mg	393.3 ± 0.4 [a]	311.1 ± 1.4 [b]
Zn	3.2 ± 0.1 [a]	2.5 ± 0.1 [b]
Mn	2.0 ± 0.0 [b]	2.2 ± 0.1 [a]

Data are mean ± standard deviation (n = 3). [a,b] different superscript letters in a row mean significant differences (HSD test, $p < 0.05$). CH, *Cynara humilis* L.; OP, *Onopordum platylepis* Murb.

To the authors' knowledge, very scarce data are currently available in the literature about the mineral composition of thistle extracts and, above all, about the mineral composition of CH and OP. A recent study [26] analysed the mineral composition of a freeze-dried extract prepared from *Onopordum tauricum*; the results of this investigation are consistent with those herein reported for OP, especially for P, K, and Zn content.

Some currently available data refer to *C. cardunculus* L. subsp. *scolymus* (L) and *C. cardunculus* L. var. *altilis* (DC) leaves, which have been recognised as a good source of K and Ca; nevertheless, among the micronutrients, mainly Fe and Zn, the mineral composition of thistle leaves is strictly affected by the concentration of nutrient solutions used for the treatment of thistles during their cultivation [40].

In a further investigation, the mineral content of CC flowers and seeds was also reported [41]. These vegetable parts contain considerable amounts of K, Ca, and Mg, while they are poor in Na [42]. According to Hajji Nabih et al. (2021) [43], the main microelements in stems of CC were Na, K, Ca, Mg, B, and P, along with other trace elements (including Zn and Mn).

3.3. Milk Clotting Activity Assay

The milk clotting time measured for each variable studied at the different extract concentrations is given in the Supplementary Materials (Supplementary Tables S1–S4).

3.3.1. Effect of Thistle Species and Extract Concentration

The effect of the concentration (5–40 mg/mL) of CC, CH, and OP extracts on the clotting activity in milk at 32 °C, pH 6.2 and 10 mM $CaCl_2$ is shown in Figure 1.

The milk clotting activity depends on the concentration of the enzyme. In this study, the MCA value increased with increasing extract concentration ($p < 0.001$); the concentration at which the maximum MCA was reached in all species was 40 mg/mL, and the highest MCA value (409.28 SU/mL) was seen for CC at an extract concentration of 40 mg/mL. The maximum values obtained for CH and OP were 170.64 and 63.16 (SU/mL), respectively.

The correlation between enzyme concentration and clotting time is well-known and has been studied by many authors [30,31,39,44] whose results clearly showed a decrease in clotting time as the concentration of proteases increased and are consistent with our results.

The great MCA performance of the CC species may be due to its high caseinolytic capacity and its content of chymosin-like proteases (Cardosin A and B) acting on κ-casein, more specifically on Phenylalanine$_{105}$–Methionine$_{106}$ bonds [45,46].

To the authors' knowledge, no data are currently available on the performance of *Onopordum platylepis* for MCA performance. Nevertheless, studies evaluating the milk clotting properties of other subspecies of the genus *Onopordum* [25,47] also found maximum MCA values at higher extract concentrations in bovine milk.

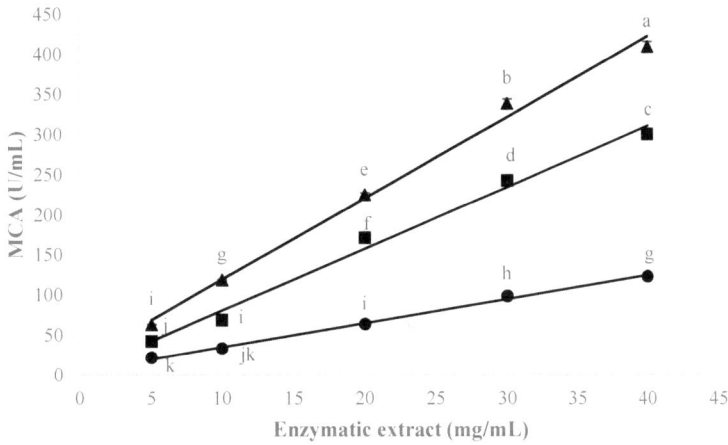

Figure 1. Effect of the enzymatic extract concentration and the species, expressed in mg extract/mL, on the clotting activity (MCA), expressed in SU/mL, of standard bovine skimmed milk at pH 6.2 and 10 mM $CaCl_2$, using CC, (▲) CH (■) and OP (•). Data are the mean of three independent experiments ($n = 3$). Error bars correspond to standard deviations. Items with different letters (a–k) are significantly different (HSD test, $p < 0.05$).

3.3.2. Effect of Thistle Species and Temperature

The effect of temperature on the clotting activity ($p < 0.001$) of the three extracts of plant origin (CC, CH, and OP) on milk at pH 6.2 and a $CaCl_2$ concentration of 10 mM was evaluated at temperatures between 20 and 85 °C and different extract concentrations (20 and 40 mg/mL).

The clotting activity of these extracts increased with temperature, with higher coagulation developing at 70 °C in all cases. Furthermore, the milk clotting activity was influenced by the concentration of extract used, as the 20 mg/mL concentration showed lower MCA values than the 40 mg/mL extract concentration did (Figure 2).

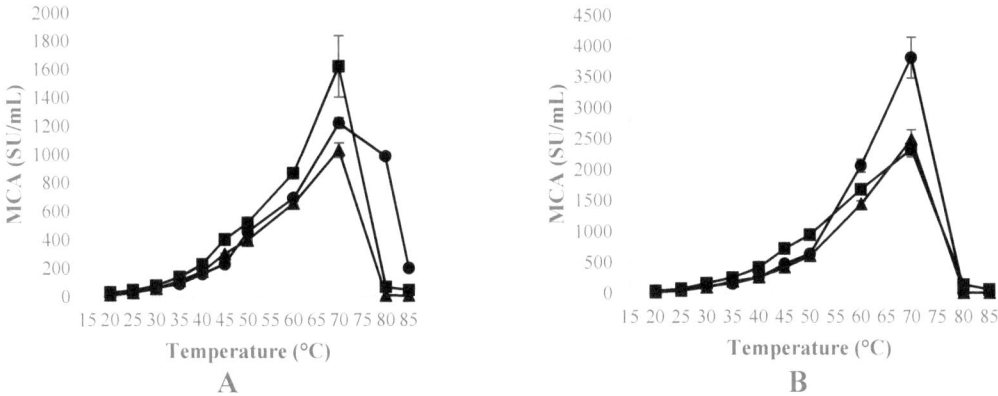

Figure 2. Effect of temperature, expressed in °C, on the clotting activity (MCA), expressed in SU/mL, of standard bovine skimmed milk at pH 6.2 with 10 mM $CaCl_2$, using CC, (▲) CH (■) and OP (•), at the concentrations of 20 mg extract/mL (**A**) and 40 mg extract/mL (**B**). Data are the means of three independent experiments ($n = 3$). Error bars correspond to standard deviations.

Comparing between species, at 70 °C and at an extract concentration of 20 mg/mL, the CH extract showed significantly higher MCA ($p < 0.05$) than the other species herein assayed did. Nevertheless, the clotting capacity of the OP extract concentration was much more highly favoured by an increase in temperature than that of the other thistle species. Therefore, at the concentration of 20 mg/mL, OP showed a MCA 14.80 and 4.5 times higher than that of CH ($p < 0.05$) at temperatures of 80 and 85 °C, respectively. Moreover, the MCA for OP at 40 mg/mL was 1.23 and 1.64 times higher than that of CH at 60 and 70 °C ($p < 0.05$).

At temperatures above 70 °C, the milk clotting activity of the extracts was found to decrease in all the species herein assayed, indicating the denaturation of the enzymes.

The milk coagulating agents of plant origin assayed in this research consist of clearly thermophilic enzymes whose clotting activity increased with temperature to relatively high values. Several studies [44,48,49] reported that the optimum activity of an aqueous crude extract obtained from CC was between 40 and 60 °C; the same authors reported a decrease in the activity at temperatures over 70 °C, thus agreeing with our results. Furthermore, Ref. [50] confirmed the thermal stability of the aqueous extracts obtained from CH flowers. More recently, Mozzon et al. (2020) [25] investigated the clotting properties of a freeze-dried extract from *O. tauricum* L., observing that its optimum coagulation temperature was 55 °C, the highest in the range tested (35–55 °C), thus leading to the conclusion that temperature positively affects the MCA of the freeze-dried extract.

3.3.3. Effect of Thistle Species and pH

Figure 3 shows the influence of pH on the clotting activity of CC, CH, and OP freeze-dried crude extracts, at concentrations of 20 and 40 mg extract/mL, on milk at 32 °C with 10 mM CaCL$_2$.

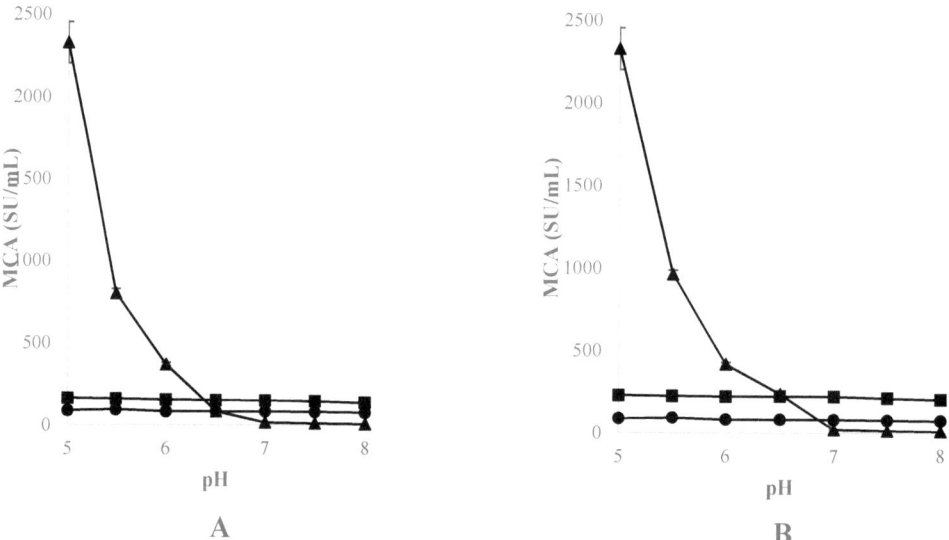

Figure 3. Effect of pH on the clotting activity (MCA), expressed in SU/mL, of standard bovine skimmed milk at 32 °C with 10 mM of CaCl$_2$, using CC, (▲) CH (■) and OP (●), at the concentrations of 20 mg extract/mL (**A**) and 40 mg extract/mL (**B**). Data are mean of three independent experiments ($n = 3$). Error bars correspond to standard deviations.

The clotting activity of these extracts was observed to respond to a wide range of pH values (5.0–8.0), with the maximum MCA value reaching between 5.0 and 5.5. This evidence was expected since the aspartic proteinases from the Cardueae tribe have been shown to have higher milk clotting and caseinolytic activity in acidic pH ranges [39,47]. More specifically, CC and CH presented a maximum MCA at an extract concentration of 40 mg/mL and a pH of 5.0, while OP presented this at a pH of 5.5. As the pH of the milk was increased, the clotting activity of all extracts was observed to decrease drastically at both extract concentrations studied. The extract concentration used influenced the MCA, which was more significant at higher concentrations.

These results indicate that an increase in pH has a greater negative effect on the MCA of CC than that of CH and OP, and CH is the species whose MCA is the least affected by the increase in pH.

Other studies concluded that aqueous extracts obtained from flowers of *Cynara* species are more active at a slightly acidic pH [51]. Campos et al. (1990) [48] and Heimgartner et al. (1990) [27] previously demonstrated that the proteases, mainly endopeptidases, have a maximum clotting activity at pH values in the range of 5.1–5.7; these data are in accordance with our results. According to Sousa-Gallagher and Malcata (1996) [52] and Chen et al. (2003) [53], the maximum activity of an aqueous crude extract obtained from CC was seen at pH 5.9–6.0. However, Martínez and Esteban (1980) [30] reported that CH shows its highest clotting activity at pH 7.

To date, no data are available on OP; however, for a freeze-dried extract from *Onopordum tauricum*, a higher MCA was recorded at pH values ranging from 4.9 to 5.7 [25,26].

3.3.4. Effect of Thistle Species and $CaCl_2$

The effect of adding $CaCl_2$ at different concentrations (5, 10, 15, 20, 30, 40, 50, 60 and 70 mM) on milk at a temperature of 32 °C and a pH of 6.2 is shown in Figure 4.

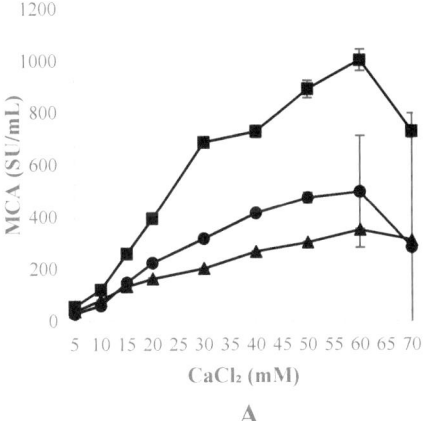
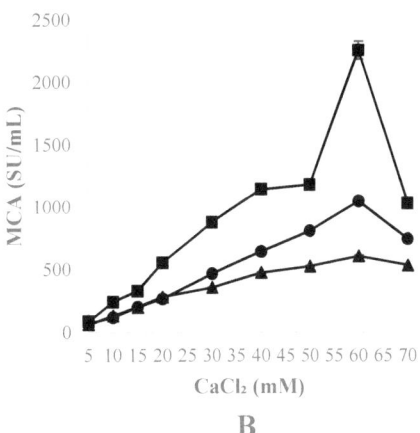

Figure 4. Effect of the concentration of $CaCl_2$, expressed in mM, on the clotting activity (MCA), expressed in U/mL, of standard bovine skimmed milk at 32 °C and pH 6.2, using CC, (▲) CH (■) and OP (●), at the concentrations of 20 mg extract/mL (**A**) and 40 mg extract/mL (**B**). Data are mean of three independent experiments (n = 3). Error bars correspond to standard deviations.

For this purpose, three freeze-dried crude extracts (CC, CH, and OP) were assayed at concentrations of 20 and 40 mg/mL. As a general trend, MCA increased at higher $CaCl_2$ concentrations for all thistle extracts assayed ($p < 0.001$). As far as the extract concentration is concerned, all the assayed species showed higher MCA values at 40 mg/mL than at 20 mg/mL.

All the species presented a maximum MCA value at 60 mM $CaCl_2$ and at an extract concentration of 40 mg/mL. By comparing the three thistle species, the MCA of CH, at both extract concentrations, was significantly higher than that of CC and OP ($p < 0.05$) at all $CaCl_2$ concentrations tested. The MCA of OP at 20 mg/mL was higher ($p < 0.05$) than that of CC at $CaCl_2$ concentrations between 20–60 mM. At a 40 mg/mL extract concentration, the MCA of OP was significantly higher than the MCA of CC except at $CaCl_2$ concentrations between 30 and 70 mM.

These results indicate that the addition of $CaCl_2$ significantly improved the clotting activity of these freeze-dried crude extracts. Similar evidence emerges from other studies on aqueous extracts from CC and CH, showing that an increase in $CaCl_2$ concentration leads to an improvement in clotting activity [30,44,54]. Nevertheless, based on the data herein collected, the MCA of the freeze-dried extract from CC was enhanced the least by an increase in $CaCl_2$ concentration, compared to that of CH and OP; in contrast, the performance of these latter two species was most positively affected by the addition of high concentrations of $CaCl_2$.

In the first phase of the coagulation process, once most of the Phenylalanine$_{105}$–Methionine$_{106}$ bonds have been cleaved, Ca^{2+} ions combine with para-kappa casein fractions to form firm clots. For this reason, the addition of $CaCl_2$ to milk reduces the coagulation time and allows the aggregation of casein micelles [25,55–57].

3.4. Proteolytic Activity and MCA/PA Ratio

The proteolytic activity (U) in bovine casein (32 °C and pH 6.2) at the reaction times (5, 10, 20, 30, 40, 50 and 60 min) for each of the extract concentrations assayed (5, 10, 20, 30 and 40 mg extract/mL) are shown in Figure 5. For the calculation of the MCA/PA ratio, the MCA values obtained at 32 °C, pH 6.2 and a concentration of 10 mM of $CaCl_2$, and the PA values obtained at 60 min were considered (Table 2).

Table 2. Milk clotting activity and proteolytic activity ratio (MCA/PA) of the species at different extract concentrations.

[Extract] (mg/mL)	MCA/PA [1]		
	CC	CH	OP
5	3.2740	2.7359	1.8765
10	3.3861	3.2477	2.5854
20	4.3226	6.0250	3.4263
30	7.3651	7.0204	4.6564
40	7.7977	7.2768	3.5884

[1] MCA expressed in SU/mL; PA expressed in U. [Extract], extract concentration; CC, *Cynara cardunculus* L.; CH, *Cynara humilis* L.; OP, *Onopordum platylepis* Murb.

Comparison data referring to proteolytic activity between the different species, concentrations of extracts, and reaction times are given in the (Supporting Information Supplementary Tables S5 and S6).

The effects of the species on the proteolytic activity were compared at the same extract concentrations previously evaluated for the MCA. As shown in Figure 5, significant differences emerged by comparing species and the reaction time at a specific extract concentration ($p < 0.001$).

The CC and CH species reached a maximum PA value at the 40 mg/mL extract concentration and at 60 min of the reaction, respectively, these values being 52.49 U and 41.24 U. Regarding OP, the highest PA value (34.18 U) was recorded at an extract concentration of 40 mg/mL and at 50 min of hydrolysis.

As previously suggested, the increase in reaction time and extract concentration favourably affects the hydrolysis rate [31,33]. This is consistent with the results herein obtained.

It would be crucial to find the right extract concentration as an excess of proteolytic enzymes can increase secondary proteolysis related to bitter flavours, and an insufficient amount of these enzymes can affect the texture of the cheese by decreasing its consistency [58].

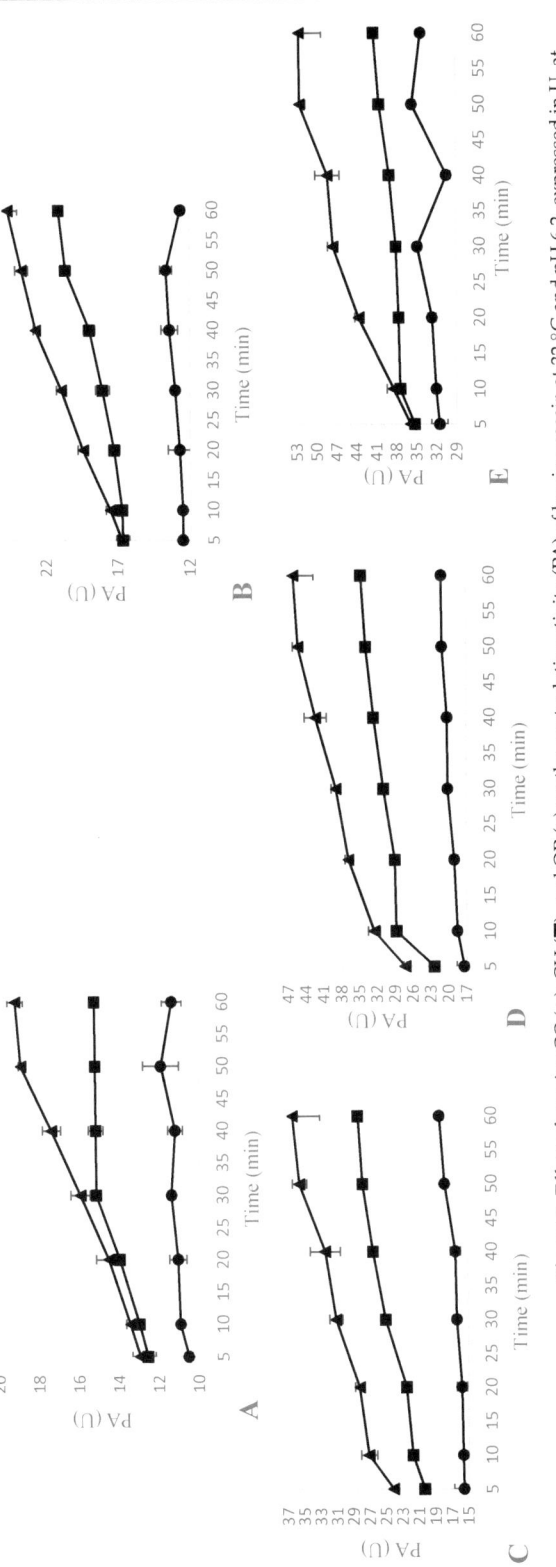

Figure 5. Effect of species CC (▲), CH (■), and OP (●) on the proteolytic activity (PA) of bovine casein at 32 °C and pH 6.2, expressed in U, at the concentrations of 5 mg extract/mL (**A**), 10 mg extract/mL (**B**), 20 mg extract/mL (**C**), 30 mg extract/mL (**D**) and 40 mg extract/mL (**E**). Data are the means of three independent experiments (n = 3). Error bars correspond to standard deviations. To the authors' knowledge, no data are currently available on the performance of *Onopordum platylepis* for casein hydrolysis. Its enzymatic composition and the specificity of its component proteases are also unknown; nevertheless, the proteolytic activity of other species belonging to the genus *Onopordum*, such as *Onopordum acanthium* [47] and *Onopordum tauricum* [26], has previously been evaluated, showing a higher specificity for β- and α-caseins and lower specificity for κ-caseins. More specifically, the *Onopordum acanthium* enzymatic extract was shown to have a lower PA than chymosin and the *Cynara cardunculus* extract was.

As far as the increase in hydrolysis rate is concerned, the extract obtained from CC showed the highest rate of hydrolysis at the different extract concentrations assayed, although in some cases, CH and CC extracts showed comparable activities. The slight difference that emerged in the proteolytic activity of CC and CH might be due to the fact that both species have in common the occurrence of an aspartic protease known as cardosin A; however, CC has a second protease called cardosin B, which is even more proteolytic [59].

As a general rule, a balanced degradation of caseins is necessary to develop favourable organoleptic characteristics in cheese. An excess of proteolytic activity during cheesemaking is associated with a low yield and an intense bitter taste due to the accumulation of small low-molecular-weight peptides and hydrophobic peptides responsible for the bitter taste [47,60].

Comparing the values obtained by the different species, the CC and CH species showed a better MCA/PA relation than the OP species did under the conditions tested in this study. Brutti et al. (2012) [47] found higher MCA and PA in the *Cynara cardunculus* extract than in *Onopordum acanthium*; however, the MCA/PA ratio of onopordosin was higher.

The MCA/PA ratio is an important measure related to higher cheese yield and quality. Therefore, the species with the highest MCA/PA ratio is the most suitable for use in cheesemaking. However, it would be advisable to explore the effect of various factors on the MCA/PA ratio, and the optimal conditions of extract concentration, temperature, pH and $CaCl_2$ at which the performance of the extracts is increased should be taken into account.

4. Conclusions

The performance of the freeze-dried thistle extracts against variations in the factors involved in MCA and PA was as expected. The effect of the parameters (temperature, pH, and $CaCl_2$ concentration) on the MCA of the extracts was similar at two extract concentrations (20 and 40 mg/mL). The results clearly showed the stability of these extracts at elevated temperatures, showing clotting activity up to a maximum temperature of 70 °C. An increase in pH adversely affects milk coagulation, but an increase in the extract concentration and the addition of $CaCl_2$ improve this activity. Furthermore, proteolytic activity in bovine casein increased at higher extract concentrations and longer hydrolysis times.

To summarize, the milk clotting conditions to achieve maximum values for CH and OP are [extract] = 40 mg/mL, T = 70 °C, pH 5, and 60 mM $CaCl_2$, and [extract] = 40 mg/mL, T = 70 °C, pH 5.5, and 60 mM $CaCl_2$, respectively. The extract from CH showed similar behaviour to that obtained from CC in the hydrolysis of bovine casein, while OP was the species with the lowest caseinolytic capacity.

Given these points, CH and OP proved to be good milk coagulants, at least at the laboratory level, for cheese production. Nevertheless, further research on the use of these thistle-based coagulants in cheesemaking is ongoing to evaluate the changes they generate in proteolysis and the sensory characteristics of final cheeses.

Supplementary Materials: The following supporting information can be downloaded at: https://www.mdpi.com/article/10.3390/foods12122325/s1, Table S1: Milk clotting time (seconds) for milk coagulation of vegetable coagulants from Cynara cardunculus, Cynara humilis and Onopordum platylepis at different extract concentrations; Table S2: Effect of temperature (°C) on the MCT (seconds) of the vegetable coagulants from Cynara cardunculus, Cynara humilis, and Onopordum platylepis at different extract concentrations (mg/mL); Table S3: Effect of the pH on the MCT (seconds) of the vegetable coagulants from Cynara cardunculus, Cynara humilis, and Onopordum platylepis at different extract concentrations (mg/mL); Table S4: Effect of calcium chloride concentration (CaCl2) on the MCT (seconds) of the vegetable coagulants from Cynara cardunculus, Cynara humilis, and Onopordum platylepis at different extract concentrations (mg/mL); Table S5: Effect of species and reaction times (minutes) on the proteolytic activity (Uabs) of Cynara cardunculus, Cynara humilis, and Onopordum platylepis plant extracts at different extract concentrations (mg/mL); Table S6: Effect of species and extract concentrations (mg/mL) on the proteolytic activity (Uabs) of Cynara cardunculus, Cynara humilis, and Onopordum platylepis plant extracts at different reaction times (minutes).

Author Contributions: Conceptualization, L.T.; methodology, L.T. and P.M.; formal analysis, C.B.-D.L., L.B.-M. and P.M.; investigation, A.A., C.B.-D.L. and L.B.-M.; resources, A.A., B.A.M.D. and I.E.; visualization, C.B.-D.L.; writing—original draft preparation, C.B.-D.L. and L.B.-M.; writing—review and editing, C.B.-D.L., L.B.-M. and L.T.; supervisión, L.T. and L.A.; project administration, L.A.; funding acquisition, L.T. and L.A. All authors have read and agreed to the published version of the manuscript.

Funding: This work is supported by the Ministry of Economy, Industry and Competitiveness-through the Agencia Estatal de Investigación (AEI-MINECO, PCI2019-103551, España), Italian Ministry of Education, University and Research (MIUR), and Ministry of Higher Education and Scientific Research (MHESR) and is part of the PRIMA programme supported by the European Union, project title "Valorisation of thistle-curdled CHEESEs in MEDiterranean marginal areas" (https://veggiemedcheeses.com/).

Data Availability Statement: Data is contained within the article or supplementary material.

Conflicts of Interest: The authors have not stated any conflict of interest.

References

1. Mazorra-Manzano, M.A.; Perea-Gutiérrez, T.C.; Lugo-Sánchez, M.E.; Ramirez-Suarez, J.C.; Torres-Llanez, M.J.; González-Córdova, A.F.; Vallejo-Cordoba, B. Comparison of the milk-clotting properties of three plant extracts. *Food Chem.* **2013**, *141*, 1902–1907. [CrossRef] [PubMed]
2. Jacob, M.; Jaros, D.; Rohm, H. Recent advances in milk clotting enzymes. *Int. J. Dairy Technol.* **2011**, *64*, 14–33. [CrossRef]
3. Shah, M.A.; Mir, S.A.; Paray, M.A. Plant proteases as milk-clotting enzymes in cheesemaking: A review. *Dairy Sci. Technol.* **2014**, *94*, 5–16. [CrossRef]
4. Galán, E.; Cabezas, L.; Fernández-Salguero, J. Proteolysis, microbiology and sensory properties of ewes' milk cheese produced with plant coagulant from cardoon *Cynara cardunculus*, calf rennet or a mixture thereof. *Int. Dairy J.* **2012**, *25*, 92–96. [CrossRef]
5. Khaldi, S.; Sonnante, G.; el Gazzah, M. Analysis of molecular genetic diversity of cardoon (*Cynara cardunculus* L.) in Tunisia. *C. R. Biol.* **2012**, *335*, 389–397. [CrossRef]
6. Roseiro, L.B.; Barbosa, M.; Ames, J.M.; Wilbey, R.A. Cheesemaking with vegetable coagulants—The use of *Cynara* L. for the production of ovine milk cheeses. *Int. J. Dairy Technol.* **2003**, *56*, 76–85. [CrossRef]
7. Tejada, L.; Abellán, A.; Prados, F.; Cayuela, J.M. Compositional characteristics of Murcia al Vino goat's cheese made with calf rennet and plant coagulant. *Int. J. Dairy Technol.* **2008**, *61*, 119–125. [CrossRef]
8. Getu Derso, A.; Gashaw Dagnew, G. Isolate and extract for milk clotting enzymes from the leaves of *Moringa oleifera*, *Carica papaya* and *Mangifera indica* and use in cheese making: The case of Western Hararage Region, Ethiopia. *J. Food Nutr. Res.* **2019**, *7*, 244–254. [CrossRef]
9. Beltrán-Espinoza, J.A.; Domínguez-Lujan, B.; Guitierrez-Méndez, N.; Chávez-Garay, D.R.; Nájera-Domínguez, C.; Leal-Ramos, M.Y. The impact of chymosin and plant-derived proteases on the acid-induced gelation of milk. *Int. J. Dairy Technol.* **2021**, *74*, 297–306. [CrossRef]
10. Feijoo-Siota, L.; Villa, T.G. Native and biotechnologically engineered plant proteases with industrial applications. *Food Bioproc Technol.* **2011**, *4*, 1066–1088. [CrossRef]
11. Gomes, S.; Belo, A.T.; Alvarenga, N.; Dias, J.; Lage, P.; Pinheiro, C.; Pinto-Cruz, C.; Brás, T.; Duarte, M.F.; Martins, A.P.L. Characterization of *Cynara cardunculus* L. flower from Alentejo as a coagulant agent for cheesemaking. *Int. Dairy J.* **2019**, *91*, 178–184. [CrossRef]
12. Sarmento, A.C.; Lopes, H.; Oliveira, C.S.; Vitorino, R.; Samyn, B.; Sergeant, K.; Debyser, G.; Van Beeumen, J.; Domingues, P.; Amado, F.; et al. Multiplicity of aspartic proteinases from *Cynara cardunculus* L. *Planta* **2009**, *230*, 429–439. [CrossRef] [PubMed]
13. Reis, P.M.; Lourenço, P.L.; Domingos, A.; Clemente, A.F.; Pais, M.S.; Malcata, F.X. Applicability of extracts from *Centaurea calcitrapa* in ripening of bovine cheese. *Int. Dairy J.* **2000**, *10*, 775–780. [CrossRef]
14. Silva, S.V.; Barros, R.M.; Malcata, F.X. Hydrolysis of caseins by extracts of *Cynara cardunculus* precipitated by ammonium sulfate. *J. Food Sci.* **2002**, *67*, 1746–1751. [CrossRef]
15. Fernández-Salguero, J.; Sánchez, E.; Gómez, R.; Mata, C.; Vioque, M.; Tejada, L. A preliminary study of microbiological quality of cardoons of genus *Cynara* L. used in manufacture of traditional cheese. *Milchwissenschaft* **1999**, *54*, 688–690.
16. Vioque, M.; Gómez, R.; Sánchez, E.; Mata, C.; Tejada, L.; Fernández-Salguero, J. Chemical and microbiological characteristics of ewes' milk cheese manufactured with extracts from flowers of *Cynara cardunculus* and *Cynara humilis* as coagulants. *J. Agric. Food Chem.* **2000**, *48*, 451–456. [CrossRef]

17. Fernández-Salguero, J.; Tejada, L.; Gómez, R. Use of powdered vegetable coagulant in the manufacture of ewe's milk cheeses. *J. Sci. Food Agric.* **2002**, *82*, 464–468. [CrossRef]
18. Tejada, L.; Gómez, R.; Fernández-Salguero, J. Sensory characteristics of ewe milk cheese made with three types of coagulant: Calf rennet, powdered vegetable coagulant and crude aqueous extract from *Cynara cardunculus*. *J. Food Qual.* **2007**, *30*, 91–103. [CrossRef]
19. Tejada, L.; Vioque, M.; Gómez, R.; Fernández-Salguero, J. Effect of lyophilisation, refrigerated storage and frozen storage on the coagulant activity and microbiological quality of *Cynara cardunculus* L. extracts. *J. Sci. Food Agric.* **2008**, *88*, 1301–1306. [CrossRef]
20. Abellán, A.; Cayuela, J.M.; Pino, A.; Martínez-Cachá, A.; Salazar, E.; Tejada, L. Free amino acid content of goat's milk cheese made with animal rennet and plant coagulant. *J. Sci. Food Agric.* **2012**, *92*, 1657–1664. [CrossRef]
21. Tejada, L.; Abellán, A.; Cayuela, J.M.; Martínez-Cachá, A.; Fernández-Salguero, J. Proteolysis in goats' milk cheese made with calf rennet and plant coagulant. *Int. Dairy J.* **2008**, *18*, 139–146. [CrossRef]
22. Tejada, L.; Fernández-Salguero, J. Chemical and microbiological characteristics of ewe milk cheese (Los Pedroches) made with a powdered vegetable coagulant or calf rennet. *Ital. J. Food Sci.* **2003**, *15*, 125–131.
23. Bouazzi, S.; El Mokni, R.; Nakbi, H.; Dhaouadi, H.; Joshi, R.K.; Hammami, S. Chemical composition and antioxidant activity of essential oils and hexane extract of *Onopordum arenarium* from Tunisia. *J. Chromatogr. Sci.* **2020**, *58*, 287–293. [CrossRef] [PubMed]
24. Zenobi, S.; Fiorentini, M.; Aquilanti, L.; Foligni, R.; Mannozzi, C.; Mozzon, M.; Zitti, S.; Casavecchia, S.; Al Mohandes Dridi, B.; Orsini, R. Effect of planting density in two thistle species used for vegetable rennet production in marginal Mediterranean areas. *Agronomy* **2021**, *11*, 135. [CrossRef]
25. Mozzon, M.; Foligni, R.; Mannozzi, C.; Zamporlini, F.; Raffaelli, N.; Aquilanti, L. Clotting properties of *Onopordum tauricum* (Willd.) aqueous extract in milk of different species. *Foods* **2020**, *9*, 692. [CrossRef]
26. Foligni, R.; Mannozzi, C.; Gasparrini, M.; Raffaelli, N.; Zamporlini, F.; Tejada, L.; Bande-De León, C.; Orsini, R.; Manzi, P.; Di Costanzo, M.G.; et al. Potentialities of aqueous extract from cultivated *Onopordum tauricum* (Willd.) as milk clotting agent for cheesemaking. *Food Res. Int.* **2022**, *158*, 111592. [CrossRef] [PubMed]
27. Heimgartner, U.; Pietrzak, M.; Geertsen, R.; Brodelius, P.; da Silva Figueiredo, A.C.; Pais, M.S.S. Purification and partial characterization of milk clotting proteases from flowers of *Cynara cardunculus*. *Phytochemistry* **1990**, *29*, 1405–1410. [CrossRef]
28. Llorente, B.E.; Brutti, C.B.; Caffini, N.O. Purification and characterization of a milk-clotting aspartic proteinase from globe artichoke (*Cynara scolymus* L.). *J. Agric. Food Chem.* **2004**, *52*, 8182–8189. [CrossRef]
29. Llorente, B.E.; Obregón, W.D.; Avilés, F.X.; Caffini, N.O.; Vairo-Cavalli, S. Use of artichoke (*Cynara scolymus*) flower extract as a substitute for bovine rennet in the manufacture of Gouda-type cheese: Characterization of aspartic proteases. *Food Chem.* **2014**, *159*, 55–63. [CrossRef]
30. Martínez, E.; Esteban, M.A. Clotting activity of an extract from the flowers of the cardon *Cynara humilis*. *Arch. Zootec.* **1980**, *29*, 107–116.
31. Silva, S.V.; Malcata, F.X. Influence of the coagulant level on early proteolysis in ovine cheese-like systems made with sterilized milk and *Cynara cardunculus*. *J. Food Sci.* **2004**, *69*, C579–C584. [CrossRef]
32. Serrano, E.; Marcos, A. Proteolytic activity of the extract from the flowers of *Cynara humilis* L. *Arch. Zootec.* **1980**, *29*, 11.
33. Bueno-Gavilá, E.; Abellán, A.; Bermejo, M.S.; Salazar, E.; Cayuela, J.M.; Prieto-merino, D.; Tejada, L. Characterization of proteolytic activity of artichoke (*Cynara scolymus* L.) flower extracts on bovine casein to obtain bioactive peptides. *Animals* **2020**, *10*, 914. [CrossRef] [PubMed]
34. Esteves, C.L.C.; Lucey, J.A.; Hyslop, D.B.; Pires, E.M.V. Effect of gelation temperature on the properties of skim milk gels made from plant coagulants and chymosin. *Int. Dairy J.* **2003**, *13*, 877–885. [CrossRef]
35. Esteves, C.L.C.; Lucey, J.A.; Wang, T.; Pires, E.M.V. Effect of pH on the gelation properties of skim milk gels made from plant coagulants and chymosin. *J. Dairy Sci.* **2003**, *86*, 2558–2567. [CrossRef]
36. Bradford, M.M. A rapid and sensitive method for the quantitation of microgram quantities of protein utilizing the principle of protein-dye binding. *Anal. Biochem.* **1976**, *72*, 248–254. [CrossRef] [PubMed]
37. AOAC. AOAC Official Method 991.25. Calcium, Magnesium, and Phosphorus in Cheese, Atomic Absorption Spectrophotometric and Colorimetric Method. In *Official Methods of Analysis*, 17th ed.; AOAC International: Arlington, VA, USA, 2002.
38. *IDF Standard 110A*; Determination of the Clotting Time of Milk to Which a Milk Clotting Enzyme Solution has Been Added. International Dairy Federation: Brussels, Belgium, 1987.
39. Silva, S.V.; Malcata, F.X. Studies pertaining to coagulant and proteolytic activities of plant proteases from *Cynara cardunculus*. *Food Chem.* **2005**, *89*, 19–26. [CrossRef]
40. Rouphael, Y.; Cardarelli, M.; Lucini, L.; Rea, E.; Colla, G. Nutrient solution concentration affects growth, mineral composition, phenolic acids, and flavonoids in leaves of artichoke and cardoon. *HortScience* **2012**, *47*, 1424–1429. [CrossRef]
41. Petropoulos, S.; Fernandes, Â.; Pereira, C.; Tzortzakis, N.; Vaz, J.; Soković, M.; Barros, L.; Ferreira, I.C.F.R. Bioactivities, chemical composition and nutritional value of *Cynara cardunculus* L. seeds. *Food Chem.* **2019**, *289*, 404–412. [CrossRef]
42. Silva, L.R.; Jacinto, T.A.; Coutinho, P. Bioactive Compounds from Cardoon as Health Promoters in Metabolic Disorders. *Foods* **2022**, *11*, 336. [CrossRef]
43. Hajji Nabih, M.; El Hajam, M.; Boulika, H.; Hassan, M.M.; Idrissi Kandri, N.; Hedfi, A.; Zerouale, A.; Boufahja, F. Physicochemical Characterization of Cardoon "*Cynara cardunculus*" Wastes (Leaves and Stems): A Comparative Study. *Sustainability* **2021**, *13*, 13905. [CrossRef]

44. Vieira de Sá, F.; Barbosa, M. Cheese-making with a vegetable rennet from cardo (*Cynara cardunculus*). *J. Dairy Res.* **1972**, *39*, 335–343. [CrossRef]
45. Galán, E.; Prados, F.; Pino, A.; Tejada, L.; Fernández-Salguero, J. Influence of different amounts of vegetable coagulant from cardoon Cynara cardunculus and calf rennet on the proteolysis and sensory characteristics of cheeses made with sheep milk. *Int. Dairy J.* **2008**, *18*, 93–98. [CrossRef]
46. Chazarra, S.; Sidrach, L.; López-Molina, D.; Rodríguez-López, J.N. Characterization of the milk-clotting properties of extracts from artichoke (*Cynara scolymus*, L.) flowers. *Int. Dairy J.* **2007**, *17*, 1393–1400. [CrossRef]
47. Brutti, C.B.; Pardo, M.F.; Caffini, N.O.; Natalucci, C.L. *Onopordum acanthium* L. (Asteraceae) flowers as coagulating agent for cheesemaking. *LWT-Food Sci. Technol.* **2012**, *45*, 172–179. [CrossRef]
48. Campos, R.; Guerra, R.; Aguilar, M.; Ventura, O.; Camacho, L. Chemical characterization of proteases extracted from wild thistle (*Cynara cardunculus*). *Food Chem.* **1990**, *35*, 89–97. [CrossRef]
49. Christen, C.; Virasoro, E. Présures végétales. Extraction et propriétés. *Le Lait* **1935**, *15*, 354–363. [CrossRef]
50. Marcos, A.; Esteban, M.A.; Martínez, E.; Alcala, M.; Fernández-Salguero, J. Thermal inactivation of proteinases from the flowers of the cardoon *Cynara humilis* L.: Kinetic and thermodynamic parameters. *Arch. Zootec.* **1980**, *29*, 283–294.
51. Claverie-Martín, F.; Vega-Hernández, M.C. Aspartic proteases used in cheese making. In *Industrial Enzymes*; Polaina, J., MacCabe, A.P., Eds.; Springer: Dordrecht, The Netherlands, 2007; pp. 207–219. [CrossRef]
52. Sousa-Gallagher, M.J.; Malcata, F. Effects of processing conditions on the caseinolytic activity of crude extracts of *Cynara cardunculus* L. *Food Sci. Technol. Int.* **1996**, *2*, 255–263. [CrossRef]
53. Chen, S.J.; Zhao, J.; Agboola, S. Isolation and partial characterization of rennet-like proteases from Australian cardoon (*Cynara cardunculus* L.). *J. Agric. Food Chem.* **2003**, *51*, 3127–3134. [CrossRef]
54. Zikiou, A.; Zidoune, M.N. Enzymatic extract from flowers of Algerian spontaneous *Cynara cardunculus*: Milk-clotting properties and use in the manufacture of a Camembert-type cheese. *Int. J. Dairy Technol.* **2019**, *72*, 89–99. [CrossRef]
55. Bencini, R. Factors affecting the clotting properties of sheep milk. *J. Sci. Food Agric.* **2002**, *82*, 705–719. [CrossRef]
56. Lagaude, A.; Fernandez, L.; Cuq, J.L.; Marchesseau, S. Characterization of curd formation during the rennet coagulation of milk by an optical microscopic method. *Int. Dairy J.* **2004**, *14*, 1033–1039. [CrossRef]
57. Nájera, A.I.; de Renobales, M.; Barron, L.J.R. Effects of pH, temperature, $CaCl_2$ and enzyme concentrations on the rennet-clotting properties of milk: A multifactorial study. *Food Chem.* **2003**, *80*, 345–352. [CrossRef]
58. Liu, X.; Wu, Y.; Guan, R.; Jia, G.; Ma, Y.; Zhang, Y. Advances in Research on Calf Rennet Substitutes and Their Effects on Cheese Quality. *Food Res. Int.* **2021**, *149*, 110704. [CrossRef]
59. Esteves, C.L.C.; Lucey, J.A.; Pires, E.M.V. Rheological properties of milk gels made with coagulants of plant origin and chymosin. *Int. Dairy J.* **2002**, *12*, 427–434. [CrossRef]
60. Pires, E.; Faro, C.; Macedo, I.; Esteves, C.; Morgado, J.; Veríssimo, P.; Pereira, C.; Gomes, D. Flor do cardo versus quimosina no fabrico de queijos artesanais. *Bol. Soc. Port. Química* **1994**, *54*, 66–68. [CrossRef]

Disclaimer/Publisher's Note: The statements, opinions and data contained in all publications are solely those of the individual author(s) and contributor(s) and not of MDPI and/or the editor(s). MDPI and/or the editor(s) disclaim responsibility for any injury to people or property resulting from any ideas, methods, instructions or products referred to in the content.

Article

An Easy and Cheap Kiwi-Based Preparation as Vegetable Milk Coagulant: Preliminary Study at the Laboratory Scale

Fabrizio Domenico Nicosia [1], Ivana Puglisi [1], Alessandra Pino [1,2], Andrea Baglieri [1], Rosita La Cava [3], Cinzia Caggia [1,2,4], Antonio Fernandes de Carvalho [5] and Cinzia Lucia Randazzo [1,2,4,*]

1. Department of Agricultural, Food and Environment, University of Catania, 95123 Catania, Italy; fabrizio.nicosia@phd.unict.it (F.D.N.); ipuglisi@unict.it (I.P.); alessandra.pino@unict.it (A.P.); abaglier@unict.it (A.B.); ccaggia@unict.it (C.C.)
2. ProBioEtna SRL, Spin off of University of Catania, Via Santa Sofia, 100, 95123 Catania, Italy
3. Caseificio La Cava, 95036 Randazzo, Italy; rosita.lacava@caseificiolacava.com
4. CERNUT, Interdepartmental Research Centre in Nutraceuticals and Health Products, University of Catania, 14, viale A. Doria 6, 95125 Catania, Italy
5. InovaLeite—Laboratório de Pesquisa em Leite e Derivados, Departamento de Tecnologia de Alimentos, Universidade Federal de Viçosa, Viçosa 36570 900, MG, Brazil; antoniofernandes@ufv.br
* Correspondence: cinzia.randazzo@unict.it; Tel.: +39-0957580218

Abstract: In the present study, a kiwifruit aqueous extract was developed and used as a coagulant enzyme in cheesemaking. In detail, polyacrylamide gel electrophoresis (SDS-PAGE) was used to investigate the presence of actinidin, the kiwifruit enzyme involved in κ-casein hydrolysis, in different tissues (pulp, peel, and whole fruit) of ripe and unripe kiwifruits. Data revealed the presence of the enzyme both in the peel and in the pulp of the fruit. Although the aqueous extract obtained from the kiwifruit peel was able to hydrolyze semi-skimmed milk, it did not break down κ-casein. The aqueous extract obtained from the pulp showed a hydrolytic activity toward both κ-casein and semi-skimmed milk. The values for milk-clotting and proteolytic activity of the kiwifruit pulp extract were evaluated at different temperatures and pH parameters in order to obtain a high value of the MCA/PA ratio; we found that a temperature of 40 °C in combination with a pH value of 5.5 allowed us to obtain the best performance. In addition, the data revealed a higher hydrolytic activity of the enzymatic preparation from ripe kiwifruits than that from unripe ones, suggesting the use of the extract from pulp of ripe kiwifruits in the laboratory-scale cheesemaking. The data showed that 3% (v/v) of the ripe kiwifruit pulp extract determined a curd yield of 20.27%, comparable to chymosin yield. In conclusion, the extraction procedure for kiwifruit aqueous extract proposed in the present study was shown to be a fast, cheap, chemical-free, and ecofriendly technology as a plant coagulant for cheese manufacturing.

Keywords: vegetable coagulant; plant proteases; actinidin; milk-clotting activity; *Actinidia deliciosa*

1. Introduction

Milk-clotting proteases are essential enzymes for cheesemaking, and among them, animal rennet, which is extracted from the abomasum of the newborn ruminants, is the most widely used. It contains a high amount of chymosin (EC 3.4.23.4), an aspartic protease that can hydrolyze a specific κ-casein bond (Phe$_{105}$-Met$_{106}$), thus causing the coagulation of milk during cheesemaking [1]. The worldwide increase in cheese production, combined with the reduction in supply and the increasing prices of calf rennet [2], as well as religious components (in Islam and Judaism) and factors related to vegetarianism, have led to the search for alternative enzymes for coagulation of milk as appropriate substitutes for animal rennet [3]. Microbial coagulants are the most commonly substituted enzymes available; they are aspartic proteases produced by *Rhizomucor miehei* and *Rhizomucor pusillus* used to obtain various types of cheeses [4]. These enzymes have a low production cost, but

also have some defects such as heat resistance, which involves a high proteolytic activity and therefore defects in the cheese (bitterness, low cheese yield) [5]. The development of recombinant DNA technology allowed the creation of fermentation-produced chymosin (FPC) as an innovative substitute for animal rennet. This low-cost technology clones the bovine chymosin gene into a host microorganism that will produce it through fermentation [6], which obtains chymosin with the same characteristics of the animal one. By using EPS, it is possible to obtain cheese with excellent quality [6]. However, technologies involving recombinant DNA are the subject of great debate for ethical reasons, and many countries, such as France, Germany and the Netherlands, have restricted the use of FPC [7,8]. Nowadays, a completely natural alternative is represented by plant-derived milk-clotting enzymes, which are commercialized and used as valid substitutes for animal rennet [3]. They have become of growing interest in the cheese industry due to their easy availability and simple purification processes. Among them, proteases are present in tissues of various plants, such as cardoon flower [9], *Cynara scolymus* artichoke [10], ginger (*Zingiber officinale*) rhizomes [11], *Dregea sinensis* stems [12], and *Citrus aurantium* flowers [13]. Many of these enzymes have been widely used for the production of the Portuguese and Spanish soft cheese types. However, due the excessive proteolytic activity, plant-derived enzymes generate bitter flavors, limiting their industrial use. Among plant-derived enzymes, kiwifruit extract was revealed to have promising milk-clotting properties. Kiwifruit (*Actinidia deliciosa*) contains high amounts of actinidin (EC. 3.4.22.14), a cysteine protease, which showed a high potential for its use as a milk-clotting agent in cheesemaking [14]. Actinidin is composed of 220 amino acid residues with a molecular mass of approximately 23.5 kDa [15]. The optimal parameters for actinidin activity, such as pH and temperature, are compatible with those used during cheesemaking. Actinidin and hydrolyze β-casein, followed by κ-casein at several points (Arg_{97}–His_{98} or Lys_{111}–Lys_{112} bonds), possess a higher specificity of hydrolysis against caseins than other proteases (such as papain) [16]. Mazorra-Manzano et al. (2013) [17] demonstrated that cheese made with kiwifruit extract showed a higher dairy yield, chewiness, springiness, and gumminess than cheeses produced with melon (*Cucumis melo*) or ginger rhizomes (*Zingiber officinale*). In addition, kiwifruit extract revealed the highest milk-clotting activity (MCA)/proteolytic activity (PA) ratio, thus representing the most effective alternative to calf rennet. However, the purification of kiwifruit extracts is a complex and time-consuming process, and often requires expensive equipment [14,16–18].

In the present study, a cheap, quick, and easy kiwifruit aqueous preparation was developed and used as a coagulant enzyme at the laboratory scale. The kiwifruit extract was obtained from both the pulp and peel of the fruit at different ripening times, and different extract concentrations of ripe fruit pulp were tested in a cheesemaking trial at the laboratory scale to establish the effect of the vegetable coagulant on the yield of the cheeses.

2. Materials and Methods

2.1. Raw Materials

The kiwifruits (*Actinidia deliciosa* cv. Hayward) used to obtain the aqueous extracts were purchased at a local market in Catania, Italy. The fruits were preliminarily evaluated for their ripeness degree using a digital refractometer (Atago, RX-5000, Tokyo, Japan): unripe (9.5° Brix) and ripe (14° Brix). Cow's milk was kindly provided by the La Cava dairy farm (Randazzo, Italy), and was used for cheesemaking at the laboratory scale.

2.2. Kiwifruit Aqueous Extracts Preparation

Unripe and ripe kiwifruits were washed, peeled with a knife, weighted, and pulped. The aqueous extract of the kiwifruit pulp was obtained by pressure using a manual stainless-steel press. The obtained juice was filtered twice through sterile gauze (50 grade, 20 × 20 cm) to separate all the seeds and the coarse content from the juice. The residual peel was blended (30 s) and then filtered through two layers of sterile gauze to obtain the peel's aqueous extract.

Whole fruits were separately weighted, washed, blended (30 s), pressed (as above), and filtered to obtain whole-fruit aqueous extract following the same extraction procedure used for pulp. The kiwifruits were processed to obtain different aqueous extract preparations as follows: (i) ripe fruit pulp, (ii) ripe fruit peel, (iii) ripe whole fruit, (iv) unripe fruit pulp, (v) unripe fruit peel, and (vi) unripe whole fruit.

Aqueous extract yields of kiwifruits were expressed as mL of obtained juice per kg of whole kiwifruits processed. Finally, samples of 10 mL of each extract were placed in an convective oven at $105 \pm 2\ °C$ until reaching a constant weight (24 h) to determine the dry matter (DM) of the extracts.

2.3. Electrophoretic Analyses

An aliquot of all kiwifruit aqueous extract preparations was centrifuged for 10 min at 5000 rpm at 4 °C; then the supernatant was recovered and used for electrophoretic analyses. The protein content of extract samples was assessed using a Pierce™ BCA Protein Assay Kit (Thermo Fisher Scientific, Waltham, MA, USA). Extract samples (2 µg in protein) were prepared for SDS-PAGE by adding an equal volume of loading buffer (62.5 mM Tris-HCl, pH 6.8, 2% SDS, 25% glycerol, 0.01% bromophenol blue, and 5% 100 mM DTT).

The hydrolytic action of the aqueous extracts from the different tissues of ripe and unripe fruits toward κ-casein (Sigma, Darmstadt, Germany) and semi-skimmed cow milk was also evaluated. The hydrolysis was performed according to Puglisi et al. (2014) [14]. Partially skimmed milk and κ-casein (10 mg in protein) dissolved in 67 mM NaH_2PO_4 pH 7.2 buffer (final volume 300 µL) were used as substrates, and were incubated with each aqueous kiwifruit extracts (10 µg protein) for 20 min at 55 °C. Aliquots of 10 µL samples were loaded into gel wells.

SDS-PAGE patterns (4–20% slab gels) were determined according to the method of Laemmli (1970) [19]. After the electrophoresis run was over (24 mA for 6 h), the gels were immersed for 8 h in a dye solution (50% methanol, 7.5% acetic acid, and 0.2% Comassie Blue R-250). The excess dye was removed via several washings (for about 4 h) with a bleaching solution (15% methanol and 7.5% acetic acid).

2.4. Milk-Clotting Activity Determination

Based on the results obtained from the electrophoretic analysis, further tests were conducted only on the extract obtained from ripe kiwifruit pulp. The milk-clotting activity (MCA) was determined as described by Arima et al. (1970) [20] with slight modifications. In detail, the aqueous extract obtained from pulp of ripe fruits was stabilized by adding an equal volume (v/v) of 20 mM sodium phosphate buffer (pH 7.0), then 1 mL of coagulant was added to 10 mL of low-fat (1%) pasteurized milk (containing 0.02% $CaCl_2$). The period elapsing between inoculation with the coagulant and the appearance of the first clot was calculated and expressed as the clotting time (seconds). The MCA was defined in terms of the Soxhlet unit (SU), representing the amount of protein in 1 mL of coagulant able to clot 1 mL of low-fat milk in 40 min (2400 s), and was expressed as: MCA (SU) = $2400/t \times S/E$, where t = clotting time (sec), S = volume of milk (mL), and E = volume of extract (mL).

The MCA was tested at different temperatures (35 °C, 40 °C, 45 °C, 50 °C, and 55 °C) and pH values (5.2, 5.5, 6.0, 6.5, 7.0, and 7.5). The temperatures were settled using a thermostatic bath (WB-M50, Falc Instruments, Treviglio, BG, Italy), and the MCA was assayed at a constant pH of 7.0. The pH effect was monitored at the optimal temperature of 40 °C by adjusting milk samples at the different pH values and monitoring with a digital pH-meter (MettlerDL25, Mettler-Toledo International Inc., Columbus, OH, USA). The assay was performed in triplicate.

2.5. Proteolytic Activity Determination

Proteolytic activity (PA) was determined using the method of Kunitz (1947) [21] with low-fat milk powder as the substrate. Briefly, 50 µL of kiwifruit aqueous extract from the pulp of ripe fruits was added to 450 µL of 1% substrate solution (0.1 M phosphate buffer, pH 7.0) and incubated at 40 °C for 60 min. After incubation, the reaction was stopped by the addition of 500 µL of 5% (w/v) trichloroacetic acid (TCA); for the control sample, the TCA was added immediately before incubation, and then the sample was placed on ice. The mixture was vortexed (ZX3, Velp Scientifica, Usmate Velate, MB, Italy), left to stand on ice for 30 min, and then centrifuged at 13,000 rpm for 20 min. The optical density (OD) of the supernatant was then measured at 280 nm using a Perkin-Elmer Lambda 1 spectrophotometer. One unit of enzyme activity (U) was defined as the amount of protein that increased the absorbance by one unit at 280 nm under the conditions described above. PA was tested at different pH values (5.2, 5.5, 6.0, 6.5, 7.0, and 7.5) and at the optimum temperature of MCA (40 °C) using 0.1 M sodium phosphate buffer at the different pHs. The pH value was monitored using a digital pH meter (MettlerDL25, Mettler-Toledo International Inc., Columbus, OH, USA). The assay was performed in triplicate.

2.6. Cheesemaking at the Laboratory Scale

The cheesemaking was carried out at the laboratory scale following the method of Cologna et al. (2009) [22] with slight modifications, by placing Pyrex beakers in a thermostatic bath using 500 mL of cow milk at the optimal MCA temperature (40° ± 1). Once the temperature of 40 °C was reached, lyophilized commercial starter cultures, provided by the La Cava farm, were added for milk acidification until reaching a pH of 5.5 ± 0.2, the optimal pH for the highest MCA/PA ratio. Different amounts of aqueous extract, corresponding to 1.6, 2.0, 3.0, and 4.0% v/v, obtained from the pulp of ripe kiwifruits were added to the milk samples. When coagulation was completed, the curd was broken into very small irregular granules (4–6 mm) with a thorn, then pressed for 15 min in a cheese mold. The curd was turned and pressed for 30 min to facilitate the purging of the whey. Curd yields were calculated as curd weight/milk weight × 100. Chymosin and microbial coagulant (supplied by Caglificio Clerici, Como, Italy) were used as a control. The cheesemaking at the laboratory scale was conducted in triplicate.

2.7. Statistical Analysis

One-way ANOVA analysis with Tukey's post hoc test was applied in the MCA determination, PA determination, and cheesemaking test in three replicates using Statistica software (TIBCO Software, Palo Alto, CA, USA) to evaluate the statistical differences between the samples. Differences were statistically significant at $p < 0.05$.

3. Results

3.1. Kiwifruit Aqueous Extracts Preparation

The aqueous kiwifruit extract preparation was obtained by using the procedure described in Section 2, and showed extract yields of 364 mL/kg and 457 mL/kg from unripe and ripe kiwifruit pulp, respectively. Similarly, yields of the aqueous extracts obtained from the whole ripe and unripe fruits were 469 mL/kg and 371 mL/kg, respectively, as shown in Table 1. Yields from the peels of both ripe and unripe fruits were low (15 mL/kg and 11 mL/kg, respectively).

Table 1. Aqueous extract yields from ripe and unripe kiwifruit tissues expressed as mL of obtained juice/kg of kiwifruit and as percentage of dry matter (DM) of the extract.

Tissue	Ripe		Unripe	
	Extract (mL/kg)	DM (%)	Extract (mL/kg)	DM (%)
Pulp	457 ± 2.5	17.5 ± 0.3	364 ± 3.1	14.5 ± 0.6
Peel	15 ± 2.1	12.4 ± 1.1	11 ± 1.8	10.8 ± 0.8
Whole fruits	469 ± 1.5	16.8 ± 0.4	371 ± 1.1	15.7 ± 0.9

3.2. SDS-PAGE Electrophoretic Profile

The presence of actinidin in the fruit tissues at different ripeness degrees was determined using electrophoresis; Figure 1 shows the pattern profiles of extracts from ripe fruit pulp (lane B), ripe fruit peel (lane C), ripe whole fruit (lane D), unripe fruit pulp (lane E), unripe peel (lane F), and unripe whole fruit (lane G). The results showed the presence of two main bands of approximately 20 kDa and 23 kDa in all tissue samples. The band of approximately 23 kDa corresponded to that of actinidin's molecular weight, as previously reported [23]. The hydrolysis electrophoretic patterns of κ-casein subjected to the enzymatic action of kiwifruit aqueous extracts are reported in Figure 2. The band with a molecular weight of 19 kDa, corresponding to κ-casein as reported by the producer (Sigma), was completely hydrolyzed by the aqueous extracts from the pulp of ripe fruits (Figure 2, lane D). Similarly, extracts from the pulp of unripe fruits (Figure 2, lane H) hydrolyzed κ-casein, but to a lesser extent than the extract from pulp of ripe ones. On the contrary, the aqueous extract from the peels of both ripe and unripe kiwifruits (Figure 2, lanes E and I) did not hydrolyze κ-casein, which remained intact after treatment, as indicated in Figure 2 (lanes E and I) by arrows. Finally, a partial hydrolysis of κ-casein due to the action of extracts from both ripe and unripe whole fruits was detected (Figure 2, lanes F and L). Figure 3 shows the patterns of semi-skimmed milk treated with the kiwifruit aqueous preparation. In all samples, hydrolysis of the milk proteins produced peptides showing an apparent molecular weight ranging from 8 to 9.6 kDa. These bands were less evident in milk treated with extracts from peel samples (Figure 3, lanes E and H), suggesting that a different degree of hydrolysis of the milk occurred. Furthermore, a band with a molecular weight of 16.9 kDa was clearly detected in all samples.

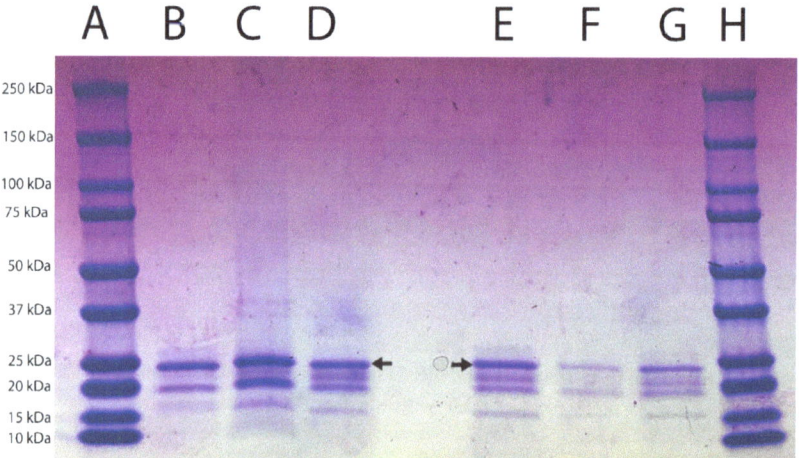

Figure 1. SDS-PAGE patterns of aqueous extract of ripe and unripe kiwifruits. Lanes A, H: molecular markers; lane B: ripe fruit pulp; lane C: ripe fruit peel; lane D: ripe whole fruit; lane E: unripe fruit pulp; lane F: unripe fruit peel; lane G: whole unripe fruit. The arrows indicate actinidin bands corresponding to approximately 23 kDa.

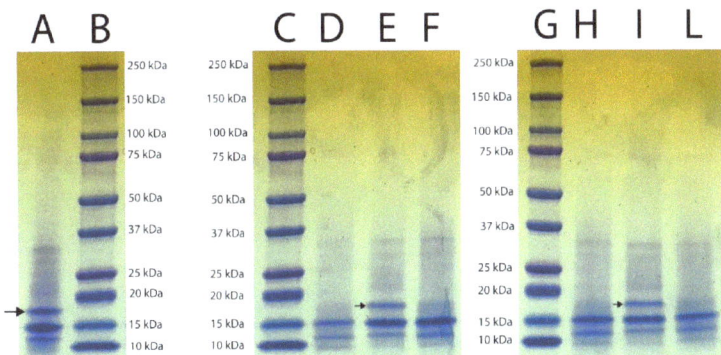

Figure 2. SDS-PAGE patterns of κ-casein subjected to treatment with aqueous extracts of kiwifruit tissues. Lane A: control κ-casein; lanes B, C, G: molecular markers; lane D: ripe fruit pulp; lane E: ripe fruit peel; lane F: ripe whole fruit; Lane H: unripe fruit pulp; Lane I: unripe fruit peel; lane L: whole unripe fruit. The arrows indicate the band corresponding to κ-casein (19 kDa, Sigma).

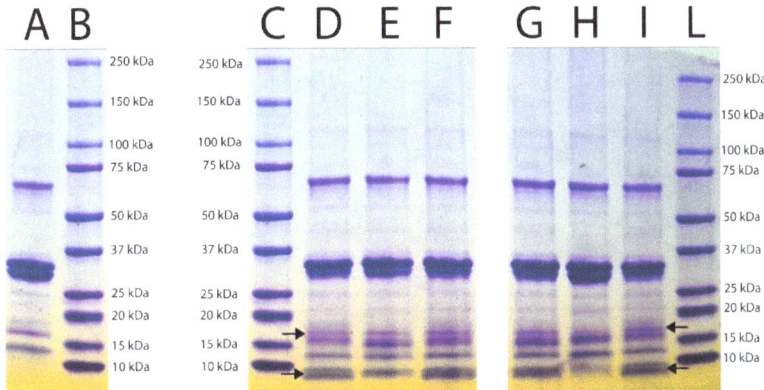

Figure 3. SDS-PAGE patterns of semi-skimmed milk treated with aqueous extract of kiwifruit tissues. Lane A: control semi-skimmed milk; lanes B, C, L: molecular marker; lane D: ripe fruit pulp; lane E: ripe fruit peel; lane F: ripe whole fruit; lane G: unripe fruit pulp; Lane H: unripe fruit peel; lane I: whole unripe fruit. Arrows indicate the peptides formed after milk hydrolysis.

3.3. Temperature and pH Effects on Milk-Clotting Activity (MCA)

Based on previous results, the aqueous extract from the pulp of ripe kiwifruits was used for further experiments. The effect of temperature on MCA is shown in Figure 4. The maximum (100%) MCA value (3.87 SU/mL) was reported at a temperature of 40 °C. The MCA slowly decreased when the temperature reached 45 °C (3.41 SU/mL), showing a drastic decrease at 55 °C (0.73 SU/mL). Furthermore, the effect of pH (ranging from 5.2 to 7.5) on MCA was evaluated, and the results are shown in Figure 5. The maximum (100%) value of MCA was detected at pH 5.5 (5.43 SU/mL), which decreased at pH 6.0 and above.

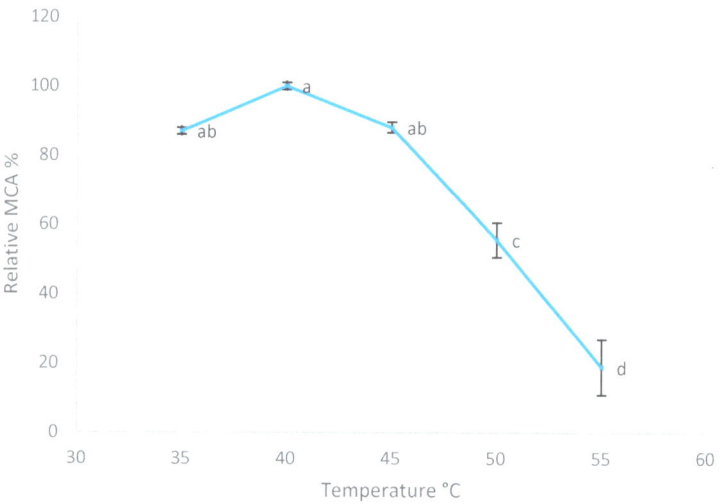

Figure 4. Temperature's effect on MCA of aqueous extract of pulp ripe kiwifruits. The percentage (%) of relative MCA represents the mean of three independent determinations performed in triplicate. The maximum value of MCA was 100%. Error bars represent standard deviations. Different lowercase letters (a, b, c and d) indicate a significant difference among samples at $p < 0.05$ (ANOVA with Tukey's post hoc test).

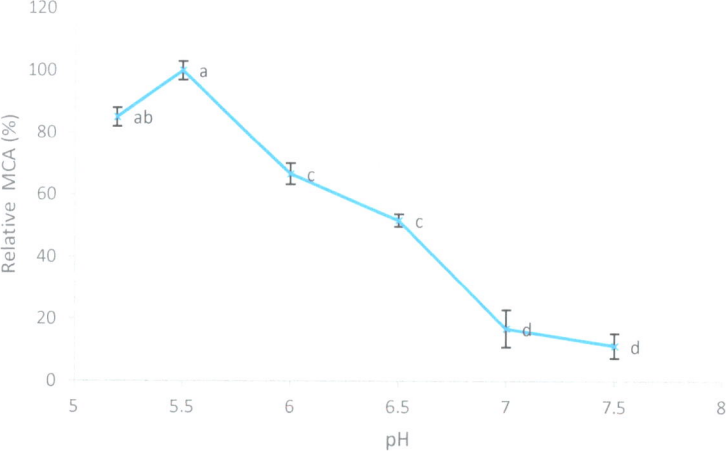

Figure 5. Effect of different pHs on milk-clotting activity of aqueous extract from ripe pulp kiwifruits. The percentage (%) of relative MCA represents the mean of three independent determinations performed in triplicate. The maximum value of MCA was 100%. Error bars represent standard deviations. Different lowercase letters (a, b, c and d) indicate a significant difference among samples at $p < 0.05$ (ANOVA with Tukey's post hoc test).

3.4. Effects of Different pHs on Proteolytic Activity (PA) and MCA/PA Ratio

The proteolytic activity was tested at different pH values (from 5.2 to 7.5) at a constant temperature of 40 °C (which was the best temperature previously determined for MCA), and results are shown in Figure 6. Proteolytic activity (PA) showed the maximum value (0.754 U/mL) at pH 6.5, then a slight decrease at a more acid or alkaline pH. At pH 5.5 (the best pH for MCA), a lower PA value was recorded (0.596 U/mL), suggesting that this pH may be a suitable value for an optimal condition for cheesemaking. Moreover, Table 2 shows the MCA/PA ratio at different pH values.

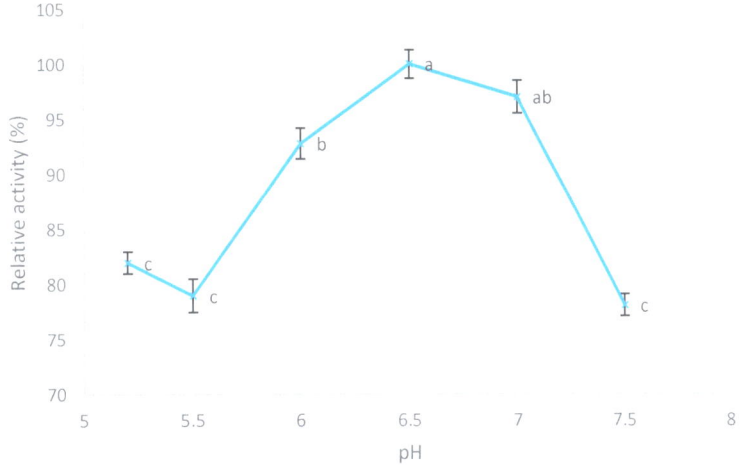

Figure 6. Effect of different pHs on proteolytic activity (PA) of aqueous extract from ripe pulp kiwifruit. The percentage (%) of relative PA represents the mean of three independent determinations performed in triplicate. The maximum value of PA was 100%. Error bars represent standard deviations. Different lowercase letters (a, b and c) indicate a significant difference among samples at $p < 0.05$ (ANOVA with Tukey's post hoc test).

Table 2. Effect of pH on MCA/PA ratio at 40 °C. MCA was expressed in SU/mL and PA was expressed in U.

	pH					
	5.2	5.5	6.0	6.5	7.0	7.5
MCA/PA	7.43 ± 0.4 [b]	9.1 ± 0.5 [a]	5.1 ± 0.5 [c]	3.7 ± 0.9 [c]	1.3 ± 0.8 [d]	1.1 ± 0.5 [d]

Different lowercase letters (a, b, c and d) indicate a significant difference among samples at $p < 0.05$ (ANOVA with Tukey's post hoc test).

3.5. Cheesemaking Test

A cheesemaking trial at the laboratory scale was conducted using different amounts of the aqueous extract from the pulp of ripe kiwifruits. Table 3 reports the results for yields (%) of curds obtained from different percentages of the extract. In addition, the data were compared to those obtained from chymosin and microbial coagulant. The data revealed that a percentage (20%) of 3 (v/v) and 4 (v/v) of the kiwifruit extracts showed a comparable yield to those of chymosin and microbial coagulants.

Table 3. Comparison between yield of curds obtained with chymosin, microbial coagulant, and different aliquots of aqueous extract from ripe kiwifruit.

	Ripe Aqueous Extract				Chymosin	Microbial
Inoculum (v/v)	4%	3%	2%	1.6%	0.04%	0.1%
Yield (%)	20.03 ± 1.84 [a]	20.27 ± 1.16 [a]	11.43 ± 1.96 [b]	10.90 ± 0.79 [b]	20.93 ± 0.90 [a]	20.06 ± 0.70 [a]

The values are the means of data from three replications. Data are reported as percentage value, and the standard deviation was calculated using three replications. Different lowercase letters (a and b) indicate a significant difference among samples at $p < 0.05$ (ANOVA with Tukey's post hoc test).

4. Discussion

Plant-derived enzymes have become of growing interest in dairy technology. Among them, actinidin, a cysteine protease from kiwifruit, is a promising substitute for chymosin due to its ability to form a good milk clot; moreover, the enzyme is fully compatible with technological parameters used during cheese manufacturing. Currently, several kiwifruit-extract preparation methods are already available in the literature [14,16–18,24]; however, they are time-consuming and high-cost procedures.

In the present study, a fast and cheap preparation of the aqueous extract of kiwifruit was developed and used for laboratory-scale cheesemaking. A higher yield in juice extract may be obtained by using ripe kiwifruits as a starting material, thus allowing the use of waste fruits with a high ripeness degree that are discarded by the food industry in the preparation of different kiwi-based foods and drinks. The data revealed that the kiwifruit aqueous extract from ripe fruit showed a higher clotting yield than that of unripe ones, suggesting a higher actinidin concentration in ripe fruit. It was noteworthy that the concentration of protein in kiwifruit is related to the growth stage, cultivar types, and treatment of the fruit during postharvest storage [25,26]. Our data from the SDS PAGE profile suggested that the quantity of actinidin in the fruits' tissues was strongly dependent on the ripeness degree. These data were supported by other studies that demonstrated that plant coagulants, such as those prepared using Noni (*Morinda citrifolia* L.) fruits [27] and berries of *S. elaeagnifolium* [28], were obtained from ripe fruits. The hypothesis that the ripeness degree of the fruit positively influenced the presence and the activity of the actinidin enzyme was also confirmed by Karki et al. (2018) [29], who found that the amount of actinidin was greater in ripe kiwifruits, which showed a higher level of activity (299 U/mg) against total casein with respect to unripe fruits.

In the present study, the presence of actinidin in the peel samples was also revealed, confirming the results of the study conducted by Nieuwenhuzen et al. (2007) [30], in which the amount of actinidin was determined both in the pulp and in the peel of the fruits. However, although actinidin was detected in the peel samples, the hydrolysis electrophoretic pattern of the kiwifruit aqueous extracts toward κ-casein showed that extracts obtained from the peel of the fruits, both ripe and unripe, seemed unable to hydrolyze κ-casein (corresponding to the band at 19 kDa). The factors affecting hydrolysis activity toward κ-casein could be the presence of actinidin in the peel of the fruit, which was in a minimal quantity compared to the content present in the pulp, as reported by Lewis and Luh (1988) [31]; or an insufficient extraction of the target components from the peel during the procedure. Moreover, a lower hydrolysis pattern on κ-casein treated with the extract of whole kiwifruits was shown with respect to that obtained using pulp extract. This could be explained by the possible presence of substances in the peel of kiwifruits, which can determine a putative inhibition of κ-casein hydrolysis. The patterns of semi-skimmed milk treated with the aqueous extract of kiwifruits showed, in all the samples, bands of about 8–9.6 kDa, which were similar to those generated by the hydrolysis of semi-skimmed milk after the treatment with pure actinidin (Lo Piero et al., 2011) [16]. Moreover, according to Chalabi et al. (2014) [32], bands corresponding to 16.9 kDa may be the result of hydrolysis produced by actinidin upon α-casein. These results were also comparable to those obtained by Puglisi et al. (2014) [14], which showed a digestion product with a molecular weight of 16.9 kDa generated by the hydrolytic action of an aqueous kiwifruit extract toward semi-skimmed milk. The aqueous extract from the pulp of ripe fruits showed an optimum temperature for MCA at 40 °C, in agreement with Mazorra-Manzano et al. (2013) [17]. This optimum temperature revealed that this preparation may be compatible with industrial cheesemaking. The MCA slowly decreased when the temperature reached 45 °C and 55 °C, probably due to the protease denaturation after the heat treatment. This result was comparable with that reported by Lo Piero et al. (2011) [16], who demonstrated that when 55 °C was reached, the caseinolytic activity of actinidin was reduced by 30%. Finally, the maximum value of MCA detected at pH 5.5 was in agreement with results obtained by Grozdanovic et al. (2013) [23] and Chalabi et al. (2014) [32], who demonstrated that actinidin showed better performances at acid pHs ranging between 5.0 and 5.5.

The proteolytic activity (PA) of a clotting enzyme is an important parameter for cheese ripening; high values of proteolysis are often associated with the formation of off flavor and bitter taste due to the production of short peptides [7,23]. The maximum value of proteolytic activity of the aqueous extract from the pulp of kiwifruits was reached at pH 6.5, and decreased at more acidic or alkaline pHs (Figure 6), in agreement with Lo Piero et al. (2011) [16]. Dehkordi et al. (2021) [33] found that, although kiwifruit extract was a good substitute for animal rennet during cheesemaking, the use of this extract could lead to defects in the cheese (bitterness and soft texture) caused by its high proteolytic activity. Katsaros et al. (2009) [34] inactivated the actinidin after coagulation of the cheese using high hydrostatic pressure (HHP) to reduce the defects caused by the high proteolytic activity of the kiwifruit extract. Another method that allows researchers to obtain a better cheese is to vary the parameters of pH and temperature in order to manage the MCA/PA ratio, as high values of the MCA/PA ratio correspond to a coagulant capable of providing a cheese free of defects such as bitter flavors [35]. In the present work, the highest calculated MCA/PA ratio was 9.1, and it was reached at 40 °C and a pH of 5.5. The aqueous extract at pH 6.0 reached an MCA/PA ratio of 5.2, which was similar to the ratio of 5.0 calculated by Mazorra-Manzano (2013) [17].

The search for the suitable amount of coagulant is a critical point in cheesemaking, as an excess of proteases can influence secondary proteolysis (developing defects such as bitter flavors), while insufficient quantities lead to a loss in consistency of the cheese [5]. Moreover, a long exposition to protease action can determine a proteolytic degradation of the casein network (especially α- and β-casein), thus reducing approximately 0.3–0.7% the curd yield [36]. In this work, curd made with an inoculum of 3% (v/v) of aqueous extract from pulp of ripe fruits provided a yield of 20.27%, and similar results were achieved when 4% of the coagulant was used (20.03%). These results suggested that an inoculum of about 3% (v/v) is enough to obtain a maximum curd yield that is comparable to the yields obtained using chymosin (20.93%) and microbial coagulant (20.06%). Lower yield values were found in similar studies. Mazzorra-Manzano et al. (2013) [17] used an extract of kiwifruit prepared from slices of peeled fruit stabilized by adding one equal part (w/v) of 20 mM sodium phosphate buffer (pH 7.2) and then homogenizing in a blender, obtaining a yield of 17.8% coagulated bovine low-fat milk. Ojha et al. (2021) [37] reached a yield of 16.69% (goat milk) using a kiwifruit extract obtained after pressing the pulp and centrifuging it at 3000 rpm for 10 min as a coagulant. A statistical analysis showed that only the samples inoculated with 2% and 1.6% of the extract produced significantly different results, which meant that both amounts were not enough to reach an optimal yield.

The curd yield produced using 3% of the extract from the pulp of ripe kiwifruits was also comparable to those obtained from other plant extracts: the yields (bovine milk) obtained using latex from the plants of *Euphorbiaceae* family as a coagulant ranged from 20.73% (for *E. tirucalli*) to 21.30% (for *E. nerifolia*) [38]. Furthermore, enzymes from sunflower used to coagulate cow milk showed a curd yield of 20.78% [39]. Finally, lower yields than that obtained using the preparation from pulp of ripe kiwifruits occurred when using several vegetable coagulants from berries of *Solanum elaeagnifolium* (17.77%) [40], melon, and ginger (15.1% and 15.4%, respectively) [17].

The high curd yield achieved in the present study may also have been related to the use of ripe kiwifruits and to the extraction method proposed, which was based on a fast and cheap procedure that was free of any added chemicals.

5. Conclusions

The extraction procedure for the kiwifruit aqueous extract proposed in the present study is a fast, cheap, -free, and ecofriendly technology to obtain an aqueous kiwifruit extract. The extract obtained from both the pulp and peel of the fruit at different ripening degrees exhibited different hydrolytic actions on κ-casein, suggesting that the actinidin concentration was influenced by the fruit ripening. Further studies on cheesemaking are ongoing to develop cheese types with desirable organoleptic and textual characteristics for industrial-scale cheese production.

Author Contributions: Conceptualization, C.L.R. and I.P.; methodology, F.D.N. and A.P.; data curation, C.L.R., A.B., C.C. and A.F.d.C.; writing—original draft preparation, F.D.N., I.P. and A.P., writing—review and editing, C.L.R., C.C., A.F.d.C and A.B.; funding acquisition, R.L.C. and C.L.R. All authors have read and agreed to the published version of the manuscript.

Funding: This research was partially funded by the "Innovazioni tecnologiche di prodotto e di processo per lo sviluppo della filiera lattiero casearia Halal (Cheesehal)", -cod. domanda: 94250045039, cod. CUP: G66D20000640009, sottomisura 16.1 del PSR-Sicilia 2014/2020.

Institutional Review Board Statement: Not applicable.

Informed Consent Statement: Not applicable.

Data Availability Statement: The data presented is contained within the article.

Acknowledgments: The authors thank the La Cava Farm (Randazzo, Catania, Italy) for providing the pasteurized cow milk.

Conflicts of Interest: A.P., C.C. and C.L.R. declare that they are members of ProBioEtna, a spinoff of the University of Catania, Italy. In addition, the authors declare that they do not have any personal, financial, professional, political, or legal interests with a significant chance of interfering with the performance of their ethical or legal duties. R.L.C., as quality control manager of La Cava srl, declares that she does not have any personal, financial, professional, political, or legal interests with a significant chance of interfering with the performance of her ethical or legal duties.

References

1. Kumar, A.; Grover, S.; Sharma, J.; Batish, V.K. Chymosin and Other Milk Coagulants: Sources and Biotechnological Interventions. *Crit. Rev. Biotechnol.* **2010**, *30*, 243–258. [CrossRef]
2. García-Gómez, B.; Vázquez-Odériz, L.; Muñoz-Ferreiro, N.; Romero-Rodríguez, Á.; Vázquez, M. Rennet Type and Microbial Transglutaminase in Cheese: Effect on Sensory Properties. *Eur. Food Res. Technol.* **2020**, *246*, 513–526. [CrossRef]
3. Colombo, M.L.; Cimino, C.V.; Bruno, M.A.; Hugo, A.; Liggieri, C.; Fernández, A.; Vairo-Cavalli, S. Artichoke cv. Francés flower extract as a rennet substitute: Effect on textural, microstructural, microbiological, antioxidant properties, and sensory acceptance of miniature cheeses. *J. Sci. Food Agric.* **2021**, *101*, 1382–1388. [CrossRef] [PubMed]
4. Mamo, J.; Assefa, F. The Role of Microbial Aspartic Protease Enzyme in Food and Beverage Industries. *J. Food Qual.* **2018**, *2018*, 7957269. [CrossRef]
5. Liu, X.; Wu, Y.; Guan, R.; Jia, G.; Ma, Y.; Zhang, Y. Advances in Research on Calf Rennet Substitutes and Their Effects on Cheese Quality. *Food Res. Int.* **2021**, *149*, 110704. [CrossRef] [PubMed]
6. Dekker, P. Dairy Enzymes. In *Industrial Enzyme Applications*; John Wiley & Sons, Ltd.: Chichester, UK, 2019; pp. 143–166.
7. Nicosia, F.D.; Puglisi, I.; Pino, A.; Caggia, C.; Randazzo, C.L. Plant Milk-Clotting Enzymes for Cheesemaking. *Foods* **2022**, *11*, 871. [CrossRef]
8. Roseiro, L.B.; Barbosa, M.; Ames, J.M.; Wilbey, R.A. Cheesemaking with Vegetable Coagulants-the Use of *Cynara* L. for the Production of Ovine Milk Cheeses. *Int. J. Dairy Technol.* **2003**, *56*, 76–85. [CrossRef]
9. Zikiou, A.; Esteves, A.C.; Esteves, E.; Rosa, N.; Gomes, S.; Louro Martins, A.P.; Zidoune, M.N.; Barros, M. Algerian Cardoon Flowers Express a Large Spectrum of Coagulant Enzymes with Potential Applications in Cheesemaking. *Int. Dairy J.* **2020**, *105*, 104689. [CrossRef]
10. Bueno-Gavilá, E.; Abellán, A.; Bermejo, M.S.; Salazar, E.; Cayuela, J.M.; Prieto-Merino, D.; Tejada, L. Characterization of Proteolytic Activity of Artichoke (*Cynara scolymus* L.) Flower Extracts on Bovine Casein to Obtain Bioactive Peptides. *Animals* **2020**, *10*, 914. [CrossRef]

11. Hashim, M.M.; Mingsheng, D.; Iqbal, M.F.; Xiaohong, C. Ginger Rhizome as a Potential Source of Milk Coagulating Cysteine Protease. *Phytochemistry* **2011**, *72*, 458–464. [CrossRef]
12. Zhao, Q.; Zhao, C.; Shi, Y.; Wei, G.; Yang, K.; Wang, X.; Huang, A. Proteomics Analysis of the Bio-Functions of *Dregea Sinensis* Stems Provides Insights Regarding Milk-Clotting Enzyme. *Food Res. Int.* **2021**, *144*, 110340. [CrossRef] [PubMed]
13. Mazorra-Manzano, M.A.; Moreno-Hernández, J.M.; Ramírez-Suarez, J.C.; de Jesús Torres-Llanez, M.; González-Córdova, A.F.; Vallejo-Córdoba, B. Sour Orange *Citrus aurantium* L. Flowers: A New Vegetable Source of Milk-Clotting Proteases. *LWT—Food Sci. Technol.* **2013**, *54*, 325–330. [CrossRef]
14. Puglisi, I.; Petrone, G.; Lo Piero, A.R. A Kiwi Juice Aqueous Solution as Coagulant of Bovine Milk and Its Potential in Mozzarella Cheese Manufacture. *Food Bioprod. Process.* **2014**, *92*, 67–72. [CrossRef]
15. Carne, A.; Moore, C.H. The Amino Acid Sequence of the Tryptic Peptides from Actinidin, a Proteolytic Enzyme from the Fruit of *Actinidia chinensis*. *Biochem. J.* **1978**, *173*, 73–83. [CrossRef]
16. Lo Piero, A.R.; Puglisi, I.; Petrone, G. Characterization of the Purified Actinidin as a Plant Coagulant of Bovine Milk. *Eur. Food Res. Technol.* **2011**, *233*, 517–524. [CrossRef]
17. Mazorra-Manzano, M.A.; Perea-Gutiérrez, T.C.; Lugo-Sánchez, M.E.; Ramirez-Suarez, J.C.; Torres-Llanez, M.J.; González-Córdova, A.F.; Vallejo-Cordoba, B. Comparison of the Milk-Clotting Properties of Three Plant Extracts. *Food Chem.* **2013**, *141*, 1902–1907. [CrossRef]
18. Serra, A.; Conte, G.; Corrales-Retana, L.; Casarosa, L.; Ciucci, F.; Mele, M. Nutraceutical and Technological Properties of Buffalo and Sheep Cheese Produced by the Addition of Kiwi Juice as a Coagulant. *Foods* **2020**, *9*, 637. [CrossRef] [PubMed]
19. Laemmli, U.K. Cleavage of structural proteins during the assembly of the head of bacteriophage T4. *Nature* **1970**, *227*, 680–685. [CrossRef]
20. Arima, K.; Yu, J.; Iwasaki, S. Milk-clotting enzyme from *Mucor pusillus* var. Lindt. In *Methods in Enzymology*; Perlmann, G., Lorand, L., Eds.; Academic Press: New York, NY, USA, 1970; Volume 19, pp. 446–459. [CrossRef]
21. Kunitz, M. Crystalline Soybean Trypsin Inhibitor. *J. Gen. Physiol.* **1947**, *30*, 291–310. [CrossRef]
22. Cologna, N.; Dal Zotto, R.; Penasa, M.; Gallo, L.; Bittante, G. A Laboratory Micro-Manufacturing Method for Assessing Individual Cheese Yield. *Ital. J. Animal Sci.* **2009**, *8*, 393–395. [CrossRef]
23. Grozdanovic, M.M.; Burazer, L.; Gavrovic-Jankulovic, M. Kiwifruit (*Actinidia deliciosa*) Extract Shows Potential as a Low-Cost and Efficient Milk-Clotting Agent. *Int. Dairy J.* **2013**, *32*, 46–52. [CrossRef]
24. Raquib, M.; Borpuzar, T.; Hazarika, M.; Kumar Laskar, S.; Saikia, G.K.; Ahmed Hazarika, R. Effect of Coagulating Enzymes and Types of Milk on the Physico-Chemical, Proximate Composition and Sensory Attributes of Iron Fortified Mozzarella Cheese. *Emir. J. Food Agric.* **2022**, *34*, 180–187. [CrossRef]
25. Nishiyama, I.; Oota, T. Varietal Difference in Actinidin Concentration and Protease Activity in the Kiwifruit Juice. *J. Jpn. Soc. Food Sci. Technol.* **2002**, *49*, 401–408. [CrossRef]
26. Salzano, A.M.; Renzone, G.; Sobolev, A.P.; Carbone, V.; Petriccione, M.; Capitani, D.; Vitale, M.; Novi, G.; Zambrano, N.; Pasquariello, M.S.; et al. Unveiling Kiwifruit Metabolite and Protein Changes in the Course of Postharvest Cold Storage. *Front. Plant Sci.* **2019**, *10*, 71. [CrossRef] [PubMed]
27. De Farias, V.A.; da Rocha Lima, A.D.; Santos Costa, A.; de Freitas, C.D.T.; da Silva Araújo, I.M.; dos Santos Garruti, D.; de Figueiredo, E.A.T.; de Oliveira, H.D. Noni (*Morinda citrifolia* L.) Fruit as a New Source of Milk-Clotting Cysteine Proteases. *Food Res. Int.* **2020**, *127*, 108689. [CrossRef] [PubMed]
28. Gutiérrez-Méndez, N.; Balderrama-Carmona, A.; García-Sandoval, S.; Ramírez-Vigil, P.; Leal-Ramos, M.; García-Triana, A. Proteolysis and Rheological Properties of Cream Cheese Made with a Plant-Derived Coagulant from *Solanum elaeagnifolium*. *Foods* **2019**, *8*, 44. [CrossRef] [PubMed]
29. Karki, A.; Ojha, P. Quality Evaluation of Kiwi Juice Coagulated Mozzarella Cheese. *J. Food Sci. Technol. Nepal* **2018**, *10*, 7–10. [CrossRef]
30. Nieuwenhuizen, N.J.; Beuning, L.L.; Sutherland, P.W.; Sharma, N.N.; Cooney, J.M.; Bieleski, L.R.F.; Schröder, R.; MacRae, E.A.; Atkinson, R.G. Identification and Characterisation of Acidic and Novel Basic Forms of Actinidin, the Highly Abundant Cysteine Protease from Kiwifruit. *Funct. Plant Biol.* **2007**, *34*, 946. [CrossRef]
31. Lewis, D.A.; Luh, B.S. Application OF Actinidin From Kiwifruit to Meat Tenderization and Characterization of Beef Muscle Protein Hydrolysis. *J. Food Biochem.* **1988**, *12*, 147–158. [CrossRef]
32. Chalabi, M.; Khademi, F.; Yarani, R.; Mostafaie, A. Proteolytic Activities of Kiwifruit Actinidin (*Actinidia deliciosa* cv. Hayward) on Different Fibrous and Globular Proteins: A Comparative Study of Actinidin with Papain. *Appl. Biochem. Biotechnol.* **2014**, *172*, 4025–4037. [CrossRef]
33. Mahdian Dehkordi, A.; Rezazadeh Bari, M.; Babaie, G.; Amiri, S. Application of Actinidin as Coagulants to Produce Iranian White Brined Cheese: Investigating the Technological, Textural, and Sensory Properties. *J. Food Meas. Charact.* **2022**, *16*, 957–963. [CrossRef]
34. Katsaros, G.I.; Tavantzis, G.; Taoukis, P.S. Production of Novel Dairy Products Using Actinidin and High Pressure as Enzyme Activity Regulator. *Innov. Food Sci. Emerg. Technol.* **2010**, *11*, 47–51. [CrossRef]
35. Amira, A.; Besbes, S.; Attia, H.; Blecker, C. Milk-Clotting Properties of Plant Rennets and Their Enzymatic, Rheological, and Sensory Role in Cheese Making: A Review. *Int. J. Food Prop.* **2017**, *20*, S76–S93. [CrossRef]
36. Jacob, M.; Jaros, D.; Rohm, H. Recent Advances in Milk Clotting Enzymes. *Int. J. Dairy Technol.* **2011**, *64*, 14–33. [CrossRef]

37. Ojha, P.; Dhakal, S.; Subedi, U. Comparative Quality Evaluation of Goat Cheese Made from Kiwi Juice and Rennet. *Natl. Workshop Livest. Fish. Res. Nepal* **2021**, *3*, 183.
38. Mahajan, R.T.; Chaudhari, G.M. Plant latex as vegetable source for milk clotting enzymes and their use in cheese preparation. *Int. J. Adv. Res.* **2014**, *2*, 1173–1181.
39. Nasr, A.I.A.M.; Mohamed Ahmed, I.A.; Hamid, O.I.A. Characterization of Partially Purified Milk-clotting Enzyme from Sunflower (*Helianthus annuus*) Seeds. *Food Sci. Nutr.* **2016**, *4*, 733–741. [CrossRef] [PubMed]
40. Chávez-Garay, D.R.; Gutiérrez-Méndez, N.; Valenzuela-Soto, M.E.; García-Triana, A. Partial Characterization of a Plant Coagulant Obtained from the Berries of *Solanum elaeagnifolium*. *CyTA J. Food.* **2016**, *14*, 200–205. [CrossRef]

Article

Impact of Lavender Flower Powder as a Flavoring Ingredient on Volatile Composition and Quality Characteristics of Gouda-Type Cheese during Ripening

Cristina Anamaria Semeniuc [1,2], Mara Mandrioli [3], Matilde Tura [3,*], Beatrice Sabrina Socaci [1,2], Maria-Ioana Socaciu [1,2], Melinda Fogarasi [1,2], Delia Michiu [1,2], Anamaria Mirela Jimborean [1,2], Vlad Mureșan [1,2], Simona Raluca Ionescu [1,2], Mihaela Ancuța Rotar [1,2] and Tullia Gallina Toschi [3]

[1] Faculty of Food Science and Technology, University of Agricultural Sciences and Veterinary Medicine of Cluj-Napoca, 3-5 Calea Mănăștur, 400372 Cluj-Napoca, Romania; cristina.semeniuc@usamvcluj.ro (C.A.S.); socaciubeatricesabrina@gmail.com (B.S.S.); maria-ioana.socaciu@usamvcluj.ro (M.-I.S.); melinda.fogarasi@usamvcluj.ro (M.F.); delia.michiu@usamvcluj.ro (D.M.); mirela.jimborean@usamvcluj.ro (A.M.J.); vlad.muresan@usamvcluj.ro (V.M.); rallucab@yahoo.com (S.R.I.); anca.rotar@usamvcluj.ro (M.A.R.)

[2] Centre for Technology Transfer-BioTech, 64 Calea Florești, 400509 Cluj-Napoca, Romania

[3] Department of Agricultural and Food Sciences, Alma Mater Studiorum-Università di Bologna, Viale Giuseppe Fanin 40, 40127 Bologna, Italy; mara.mandrioli@unibo.it (M.M.); tullia.gallinatoschi@unibo.it (T.G.T.)

* Correspondence: matilde.tura2@unibo.it; Tel.: +39-051-20-96-015

Abstract: This study aimed to formulate a Gouda-type cheese from cow's milk, flavored with lavender flower powder (0.5 g/L matured milk), ripened for 30 days at 14 °C and 85% relative humidity. Physicochemical, microbiological, and textural characteristics, as well as the volatile composition of the control (CC—cheese without lavender) and lavender cheese (LC), were assessed at 10-day intervals of ripening. Consumers' perception, acceptance, and purchase intention were only evaluated for ripened cheeses. Moisture and carbohydrate contents, the pH, cohesiveness, indexes of springiness and chewiness decreased during ripening in both CC and LC; however, protein, ash, and sodium chloride contents, titratable acidity, hardness, lactobacilli, streptococci, and volatiles increased. Fat and fat in dry matter contents, respectively, the energy value did not vary with ripening time in LC and increased in CC; gumminess decreased in CC and did not change in LC. Lavender flower powder significantly affected the cheese's microbiological and sensory characteristics and volatile composition but did not considerably impact physicochemical and textural ones. Populations of lactobacilli and streptococci were substantially higher in LC compared to CC. The volatile profile of LC was dominated by terpene and terpenoids, and that of CC by haloalkanes. Sensory scores were slightly lower for LC than CC, even if it did not considerably affect consumers' acceptance and purchase intention.

Keywords: lavender-flavored Gouda-type cheese; physicochemical properties; textural properties; microbiological properties; volatile compounds; sensory properties

1. Introduction

Gouda is a ripened firm/semi-hard cheese with a body color ranging from near-white or ivory to light yellow or yellow and a firm texture. Its interior has a few to plenty of gas holes uniformly distributed. This type of cheese has a smooth and dry rind, with few openings and splits accepted [1]. It is ordinarily made from pasteurized cow's milk and acidified with a mesophilic starter culture containing miscellaneous lactic acid bacteria [2]. Gouda cheese generally has an inside diameter of approximately 25.4 cm and a thickness of 16.5 cm. The percentage of water varies from 41.25 to 45.43%, with an average of 43.5% [3].

Furthermore, since several brands of Gouda-type cheese are commercially available, the sensory properties, mainly the cheese flavor, are the key factors affecting consumers' acceptance and are decisive for purchase intention [4,5]. Thus, dairy manufacturers have started producing differentiated cheeses by incorporating some atypical ingredients, such as lavender [6], cumin [7], red chili pepper [8], fenugreek [9], mustard [10], and garlic [10]. However, despite the potential appeal of these cheeses to consumers, there needs to be more knowledge regarding the effects of adding these flavoring ingredients on the development of texture and flavor in ripened cheese [6].

It is well known that microorganisms, especially those in the starter culture, play an essential role in cheese making; the enzymes produced by them break down cheese constituents such as lipids, proteins, and, to a lesser extent, carbohydrates, improving the product texture and flavor during ripening [11]. The starter culture used in Gouda-type cheese manufacturing includes *Lactococcus*, *Leuconostoc*, *Lactobacillus*, and *Streptococcus* strains [12]. Members of the *Leuconostoc* genus and lactococcal variant *Lactococcus lactis* subsp. *lactis* biovar. *diacetylactis* ferment milk citrate into various compounds, the most significant being carbon dioxide (necessary for eye formation in Gouda cheese) and the diacetyl and acetic acid flavors [13].

The antimicrobial properties of some spices/herbs used as ingredients are widely known, and, thus, their possible effects on starter organisms in cheese. On the other hand, the microbiome of these flavoring ingredients also affects the starter culture during ripening, and hence the development of the cheese's texture and flavor [11]. However, to our knowledge, work has yet to be published regarding the impact of lavender flower powder on the Gouda-type cheese's ripening process and implicitly on its properties. Therefore, this study proposes a manufacturing process for a Gouda-type cheese flavored by adding lavender flower powder into matured milk to evaluate the effect of lavender flower powder and the ripening time on the cheese's volatile compounds and the physicochemical, textural, and microbiological characteristics.

2. Materials and Methods

Raw milk (3.8% fat, 3.4% protein, 4.5% lactose, 87.5% moisture, and a pH of 6.5) from Bălțată Românească cattle was used to manufacture the Gouda-type cheeses (LC—with lavender flower powder and CC—without the flavoring ingredient). Cow's milk was received (as a donation) from a milk farm (P.F.A. Socaci L. Maria) in Chirileu, Mureș County, Romania, with a herd of 30 cows (1–6 lactation cycles). Cows were kept in free stabulation during the daytime to graze on green pastures. As additional roughage, they were fed with a mixture of alfalfa, hay, orchard grass, and silage or a blend of corn, wheat, sunflower, and barley flour.

Lavender (dried bunches of *Lavandula angustifolia* Mill.) was purchased from a lavender farm (Lavanda Lola) in Bonțida, Cluj County, Romania. The flavoring ingredient, lavender flower powder, was prepared by grinding lavender flowers (manually detached from stems and separated from impurities) to a fine powder using a mortar grinder (RM 200; Retsch GmbH, Haan, Germany); the powder thus obtained was then sealed into a glass jar with a lid and kept in a cool, dry place until use.

2.1. Manufacturing of Gouda-Type Cheese

Gouda-type cheese was made as described in our previous paper [6]. An average quantity of 3.85 kg CC and 3.95 kg LC was obtained from processing 30 L cow's milk. The flowchart in Figure 1 shows the manufacturing process steps for LC (see Figure 2a); CC (see Figure 2b) was manufactured similarly but without lavender flower powder. Both treated (LC) and untreated (CC) cheeses were produced in two batches. The amount of lavender flower powder (15 g) added into matured milk (30 L) to flavor the cheese was selected from a series of tested concentrations (30, 25, 20, and 15 g lavender powder per 30 L of milk) based on a sensory evaluation of cheese (performed using an internal, trained panel of 6 assessors).

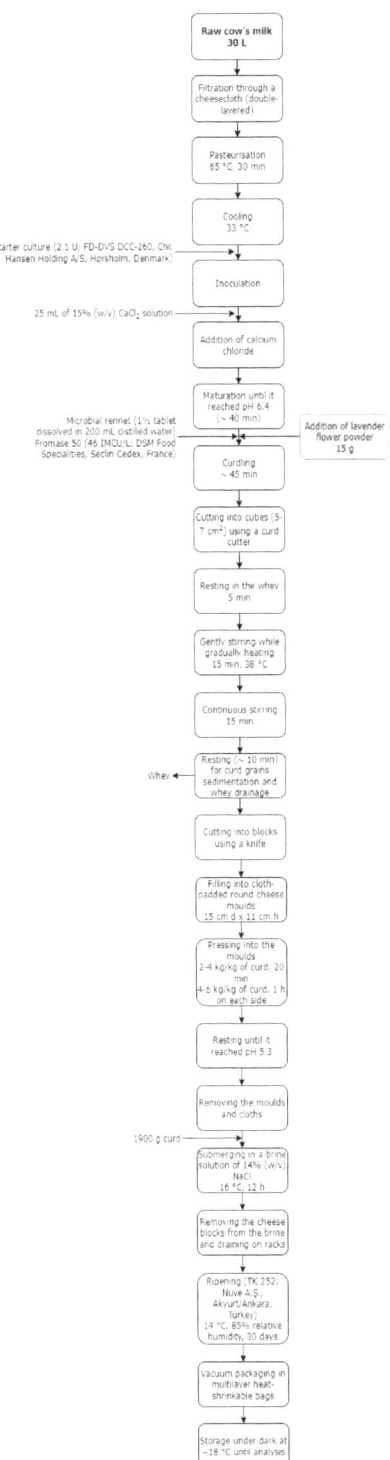

Figure 1. Flowchart of lavender Gouda-type cheese manufacturing process.

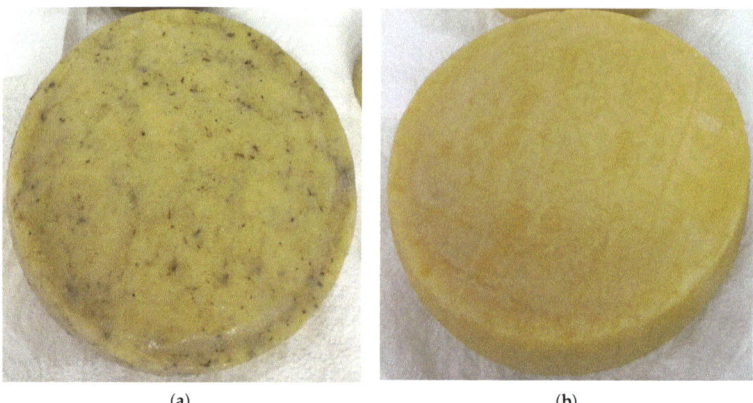

Figure 2. Gouda-type cheeses: (**a**) LC—lavender cheese; (**b**) CC—control cheese.

All analyses were performed both on the lavender (LC) and control cheese (CC) at ten (T1), twenty (T2), and thirty days (T3) of ripening, as mentioned in Sections 2.2–2.6. In addition, consumers' perception, acceptance, and purchase intention were also determined for ripened LC and CC, as described in Section 2.7.

2.2. Proximate Composition Analysis of Cheese and Calculation of Total Carbohydrate Content, Fat in Dry Matter Content, and Energy Value

Moisture content was determined by drying the sample to a constant weight in an electric oven (Digitheat; J.P. Selecta S.A., Barcelona, Spain), following instructions provided by ISO 5534:2004 | IDF 4:2004 [14].

Determination of the nitrogen content and calculation of crude protein was carried out as described in ISO 8968-1:2014 | IDF 20-1:2014 [15] using the Kjeldahl method. It involved acid digestion of the sample (DK6 Heating Digester; VELP Scientifica SRL, Usmate Velate, Italy), followed by alkalization and steam distillation of the acid digest (UDK 129 Distillation Unit; VELP Scientifica S.R.L., Usmate Velate, Italy), and finally by quantification of the trapped ammonia through titration. The protein content in cheese was estimated by multiplying the total nitrogen by 6.25, a nitrogen-to-protein conversion factor.

Fat content was determined directly using the Van Gulik method from ISO 3433:2008 | IDF 222:2008 [16], while fat in dry matter (FDM) content was calculated by the following Formula (1):

$$FDM\ (\%) = \frac{F}{DM} \times 100 \quad (1)$$

where F is the fat content (%) of the cheese, and DM is the dry matter content (%) of the cheese (computed by subtracting moisture content (%) from 100).

Ash content was determined by incineration of the sample in a muffle furnace (L3/11/B170; Nabertherm GmbH, Bremen, Germany), as detailed by Nagy et al. [17]. First, approximately 1.0 g of grated cheese was weighed to the nearest 1 mg (ABJ-220-4NM; Kern & Sohn GmbH, Balingen, Germany) into a porcelain crucible, and it was then heated at 600 °C for 12 h in the muffle furnace, cooled in a desiccator, and weighed again. Finally, the percentage of ash content was calculated with the following Formula (2):

$$Ash\ (\%) = \frac{w_a}{w_s} \times 100 \quad (2)$$

where w_a is the weight (g) of ash, and w_s is the weight (g) of the sample.

All samples were analyzed in triplicate, and results were expressed in percentages.

Total carbohydrate content (%) was calculated from Formula (3) used by Nagy et al. [17]:

$$\text{Total carbohydrate } (\%) = 100 - (\% \text{ moisture} + \% \text{ protein} + \% \text{ fat} + \% \text{ ash}) \quad (3)$$

The energy value of cheese was calculated according to Nagy et al. [17] using the following Formula (4):

$$\text{Energy value } (\text{kcal}/100 \text{ g}) = 4 \times (\text{g protein} + \text{g carbohydrate}) + 9 \times (\text{g fat}) \quad (4)$$

2.3. Determination of Sodium Chloride Content, pH, and Titratable Acidity of the Cheese

Sodium chloride content, expressed as a percentage, was determined following the potentiometric titration method specified in ISO 5943:2006 | IDF 88:2006 [18]. The pH measurement of cheese was performed with a portable pH meter (HI 99161; Hanna Instruments, Limena, Italy). Titratable acidity, expressed in Thörner degrees (°T), was determined using the titrimetric method from STAS 6353-85 [19], a Romanian standard. All samples were analyzed in triplicate.

2.4. Texture Profile Analysis of Cheese

This analysis included measurement of the hardness (N), cohesiveness, springiness index, gumminess (N), and chewiness index (N) of cheese sample using a texture analyzer (CT3; Brookfield Engineering Laboratories Inc., Middleboro, MA, USA), according to the method described by Ong et al. [20]. Samples, in 1.5 cm cubes, were taken from the central part of the cheese using a knife and kept at 20 °C for 1 h in a closed container (to prevent moisture loss) until analysis. The test consisted of sample deformation to 50% of its height (7.5 mm), at a speed of 2 mm/s, with a TA25/1000 cylindrical probe attached to a 10 kg compression cell. Measurements were carried out on each cheese batch in triplicate (six in total for each treatment) using the TexturePro CT software.

2.5. Microbiological Analysis of Cheese

A portion of 5.0 g grated cheese was aseptically weighed into a sterile stomacher bag and homogenized in 45 mL of 0.85% (w/v) sodium chloride solution (27810.295P; VWR Chemicals, Leuven, Belgium) for 1 min using a laboratory blender (MiniMix 100 P CC; Interscience, Saint-Nom-la-Bretèche, France), as described by Socaciu et al. [21]. Seven ten-fold serial dilutions (10^{-2}, 10^{-3}, 10^{-4}, 10^{-5}, 10^{-6}, 10^{-7}, and 10^{-8}) were prepared from this stock solution (10^{-1}).

Lactobacilli and streptococci counts were determined in cheese according to the method described by the ISO 7889:2003 | IDF 117:2003 standard [22] using MRS broth (110611; Merck KGaA, Darmstadt, Germany) and M17 agar (CM0785; Oxoid Ltd., Basingstoke, England), respectively. Enumeration of lactobacilli and streptococci colonies was performed after incubating inoculated plates at 37 °C for 48 h under anaerobic conditions. All samples were analyzed in triplicate, and results were reported as log cfu/g. In addition, the total lactic acid bacteria count (log cfu/g) was calculated as the sum of the lactobacilli and streptococci counts.

2.6. Analysis of Volatile Compounds in Cheese

This was performed following the method described by Cozzolino et al. [23], with minor modifications. Into a 20-mL SPME crimp neck vial (22.5 × 75.5 mm; VWR International s.r.l., Milano, Italy), approximately 3.5 g of grated cheese was weighed to the nearest 0.1 mg using an analytical balance (E42; Gibertini Elettronica S.R.L., Milano, Italy); 3 mL aqueous solution of hydrochloric acid (0.03% 6 M HCl; Merk KGaA, Darmstadt, Germany) was added to cheese sample and mixed, followed by 5 µL methanolic solution of methyl nonanoate (1175 µg/mL; Sigma-Aldrich Chemie GmbH, Steinheim, Switzerland), used as an internal standard. First, the vial's content was stirred and then kept for 30 min at 40 °C in the autosampler thermostat (HT2850T autosampler; HTA S.r.l., Brescia, Italy) to reach equilibrium. Next, an SPME fiber (50/30 µm DVB/CAR/PDMS;

Supelco Inc., Bellefonte, PA, USA) was inserted into the vial's septum and exposed to the sample headspace at 40 °C for 30 min to adsorb the volatile compounds. Before its first use, the fiber was conditioned at 270 °C for 60 min, according to the manufacturer's recommendations. After sampling, it was retracted and automatically injected into the gas chromatograph injection port of a GCMS-QP2010 Plus instrument (Shimadzu Corp., Kyoto, Japan) equipped with an Rtx-Wax capillary column (Crossbond® Carbowax® polyethylene glycol; 30 mL × 0.25 mm ID × 0.25 µm film thickness; Restek Corp., Bellefonte, PA, USA), where volatile compounds were desorbed for 10 min in spitless mode. Helium was the carrier gas at 1 mL/min constant flow. The injector temperature was set to 250 °C, and the oven one was programmed initially at 40 °C and held at this temperature for 2 min. Subsequently, it was increased by 5 °C/min up to 65 °C and kept at 65 °C for 2 min, and then by 10 °C/min up to 240 °C and held at 240 °C for 9 min. Interface and ion source temperatures were set to 210 and 230 °C, respectively, and the filament voltage to 70 eV (electronic impact). Chromatographic analysis was carried out in triplicate for each cheese sample. Volatile compounds were identified by comparing their recorded mass spectra with those found in the NIST27 and NIST147 libraries. The relative content of each volatile compound was calculated as the ratio of its total ion current (TIC) area to the TIC area of the internal standard using the following Formula (5):

$$Conc\ A\ (\mu g/kg\ cheese) = \left(\frac{Peak\ area_A}{Peak\ area_{IS}} \times \frac{a_{IS}}{w_c}\right) \times 1000 \qquad (5)$$

where $Conc\ A$ is the analyte concentration, $Peak\ area_A$ is the analyte peak area, $Peak\ area_{IS}$ is the internal standard peak area, a_{IS} is the amount of internal standard added to the sample (µg), and w_c is the cheese weight (g).

2.7. Sensory Analysis of Cheese and Determination of Consumers' Acceptance and Purchase Intention

Cheese samples were assessed for appearance and color, consistency and texture, odor and taste, aftertaste, and overall liking using a nine-point hedonic scale (1—dislike extremely; 2—dislike very much; 3—dislike moderately; 4—dislike slightly; 5—neither like nor dislike; 6—like slightly; 7—like moderately; 8—like very much; 9—like extremely). They were coded with 3-digit random numbers and presented to panelists (50 women and 30 men aged 20–43 years) on white ceramic plates. The sensory attributes of CC and LC were rated at 20 °C (air conditioning) under white light. Panelists were asked not to eat, drink, or smoke for at least 1 h before the evaluation session (conducted in individual booths).

Consumers' purchase intention was rated on a 5-point scale (1—definitely will buy; 2—probably will buy; 3—might or might not buy; 4—probably will not buy; 5—definitely will not buy) [24]. The acceptance rate (AR) of each cheese type was calculated as described previously by dos Reis Santos et al. [25] using the following Formula (6):

$$AR\ (\%) = \frac{X}{N} \times 100 \qquad (6)$$

where X is the mean sensory score of the cheese, and N is the maximum sensory score given by panelists to the cheese.

2.8. Statistical Analysis

The Minitab statistical software (version 19.1.1; LEAD Technologies, Inc., Charlotte, NC, USA) was used for data analysis. The effects of the lavender flower powder and ripening time on the Gouda-type cheese's characteristics and volatile compounds were analyzed by one-way ANOVA with a post-hoc Tukey's test at a 95% confidence level ($p < 0.05$). In addition, hierarchical cluster analysis (HCA) was performed using the MetaboAnalyst software (version 5.0; Xia Lab at McGill University, Montreal, QC, Canada).

3. Results and Discussion

3.1. Nutritional Properties of Gouda-Type Cheese

Changes in the proximate composition, sodium chloride content, pH, titratable acidity, and energy value of the control (CC) and lavender Gouda-type cheese (LC) at 10-day intervals during 30 days of ripening are shown in Table 1. Moisture content significantly decreased with ripening time, from 36.84 to 34.97% in CC and from 37.87 to 35.48% in LC, causing an increase in protein (from 21.13 to 23.17% in CC; from 21.55 to 24.90% in LC), ash (from 4.28 to 5.45% in CC; from 4.25 to 5.06% in LC), and sodium chloride contents (from 4.28 to 5.45% in CC; from 4.25 to 5.06% in LC). On the other hand, the fat content of the Gouda-type cheese increased in CC (from 28.25 to 30.75%) while this matured but did not vary significantly in LC (from 29.25 to 29.75%); the same trends were noticed for the fat in dry matter content and energy value of CC and LC, respectively (see Table 1). Regarding carbohydrates, their content significantly declined in both CC and LC during ripening, as lactic acid bacteria consumed them, causing a fall in pH (from 4.87 to 4.60 in CC; from 4.99 to 4.65 in LC) and an increase in titratable acidity (from 85.2 to 95.0 °T in CC; from 79.8 to 92.0 °T in LC). It is well known that lactic acid bacteria use carbohydrates as a primary carbon source [26], lactic acid being the major end-product of milk lactose fermentation [27]. These results are corroborated by microbiological findings showing the multiplication of lactic acid bacteria in CC and LC with ripening. Furthermore, at the ripening period's end, protein and sodium chloride contents in LC were significantly higher than in CC, while the ash content and titratable acidity were lower. Nevertheless, both CC and LC met the quality requirements in the Codex standard for Gouda, namely CXS 266-1966 [1].

Table 1. Proximate chemical composition, sodium chloride content, pH, titratable acidity, and energy value of control and lavender Gouda-type cheese at different ripening times (T1, T2, and T3).

Parameter/Energy Value	Control Cheese			Lavender Cheese		
	T1	T2	T3	T1	T2	T3
Moisture (%)	36.84 ± 0.270^{aA}	35.80 ± 0.060^{abA}	34.97 ± 0.064^{bA}	37.87 ± 0.282^{aA}	36.57 ± 0.630^{abA}	35.48 ± 0.165^{bA}
Protein (%)	21.13 ± 0.191^{cA}	22.11 ± 0.014^{bA}	23.17 ± 0.212^{aB}	21.55 ± 0.078^{cA}	22.39 ± 0.134^{bA}	24.90 ± 0.113^{aA}
Fat (%)	28.25 ± 0.354^{cA}	29.50 ± 0.0^{bA}	30.75 ± 0.354^{aA}	29.25 ± 0.354^{aA}	29.25 ± 0.354^{aA}	29.75 ± 0.354^{aA}
Fat in dry matter (%)	44.73 ± 0.057^{cB}	45.95 ± 0.042^{bA}	47.29 ± 0.495^{aA}	47.08 ± 0.354^{aA}	46.11 ± 0.099^{aA}	46.11 ± 0.431^{aA}
Ash (%)	4.28 ± 0.014^{bA}	4.49 ± 0.156^{bA}	5.45 ± 0.014^{aA}	4.25 ± 0.014^{aA}	4.31 ± 0.035^{bA}	5.06 ± 0.028^{aB}
Total carbohydrate (%)	9.51 ± 0.700^{aA}	8.10 ± 0.205^{bA}	5.66 ± 0.488^{bA}	7.09 ± 0.0^{aB}	7.50 ± 0.177^{aA}	4.82 ± 0.106^{bA}
Sodium chloride (%)	2.37 ± 0.028^{cA}	2.60 ± 0.021^{bB}	2.71 ± 0.007^{aB}	2.26 ± 0.028^{cA}	2.91 ± 0.028^{bA}	3.28 ± 0.071^{aA}
pH	4.87 ± 0.014^{aB}	4.73 ± 0.007^{bA}	4.60 ± 0.0^{cA}	4.99 ± 0.007^{aA}	4.83 ± 0.092^{abA}	4.65 ± 0.042^{bA}
Titratable acidity (°T)	85.2 ± 0.566^{cA}	90.6 ± 0.849^{bA}	95.0 ± 0.283^{aA}	79.8 ± 0.283^{cB}	83.8 ± 0.283^{bB}	92.0 ± 0.0^{aB}
Energy value (kcal/100 g)	377 ± 4.950^{bA}	387 ± 0.707^{abA}	393 ± 2.121^{aA}	378 ± 2.828^{aA}	383 ± 4.243^{aA}	387 ± 2.121^{aA}

T1—10 days of ripening; T2—20 days of ripening; T3—30 days of ripening. Data are expressed as mean ± standard deviation values of all measurements. Different lowercase letters in the same row indicate significant differences between ripening times ($p < 0.05$, Tukey's test), and different uppercase letters show significant differences between cheeses ($p < 0.05$).

3.2. Textural Properties of Gouda-Type Cheese

Instrumental measurements of texture attributes in CC and LC are presented in Table 2. Consistent with the findings of Kanawjia et al. [28] on Gouda cheese, the hardness, springiness index, gumminess, and chewiness decreased in CC and LC during ripening, except for gumminess in LC, which did not change significantly. However, no significant differences between the texture attribute values of CC and LC were found at the final stage of ripening.

Hardness indicates the maximum force required to compress cheese between the molar teeth. In CC, it significantly increased from a value of 35.43 N on day 10 of ripening to 53.07 N on day 30, and in LC, it raised from 24.72 to 41.51 N, most likely due to a loss of moisture.

Table 2. Texture attribute values for control and lavender Gouda-type cheese at different ripening times (T1, T2, and T3).

Texture Attribute	Control Cheese			Lavender Cheese		
	T1	T2	T3	T1	T2	T3
Hardness (N)	35.43 ± 2.456[bA]	36.52 ± 5.547[bA]	53.07 ± 10.105[aA]	24.72 ± 3.534[bB]	40.56 ± 10.605[aA]	41.51 ± 9.777[aA]
Cohesiveness	0.46 ± 0.066[aA]	0.37 ± 0.047[bA]	0.24 ± 0.028[cA]	0.44 ± 0.061[aA]	0.29 ± 0.036[bB]	0.24 ± 0.030[bA]
Springiness index	0.83 ± 0.026[aA]	0.79 ± 0.061[aA]	0.54 ± 0.076[bA]	0.82 ± 0.088[aA]	0.80 ± 0.048[aA]	0.68 ± 0.019[bA]
Gumminess (N)	16.09 ± 2.530[aA]	19.90 ± 6.244[aA]	8.94 ± 2.153[bA]	10.72 ± 0.987[aB]	11.72 ± 2.580[aB]	10.17 ± 3.366[aA]
Chewiness index (N)	13.35 ± 2.229[aA]	15.71 ± 5.299[aA]	4.90 ± 1.587[bA]	8.57 ± 1.121[aB]	9.43 ± 2.914[aB]	6.96 ± 1.060[bA]

T1—10 days of ripening; T2—20 days of ripening; T3—30 days of ripening. Data are expressed as mean ± standard deviation values of all measurements. Different lowercase letters in the same row indicate significant differences between ripening times ($p < 0.05$, Tukey's test), and different uppercase letters show significant differences between cheeses ($p < 0.05$).

Cohesiveness, also known as consistency, shows the strength of the internal bonds making up a cheese's body. Surprisingly, and contrary to the findings of Kanawjia et al. [28], we noticed a downward trend with ripening time for both CC (from 0.46 to 0.24) and LC (from 0.44 to 0.24). Nevertheless, Ivanov et al. [29] reported changes in the cohesiveness of Kashkaval cheese during ripening, like those we observed, being attributed to proteolysis.

The springiness index is a texture attribute that shows the viscoelastic properties of cheese and ranges from 0 (completely viscous material) to 1 (completely elastic material). As can be seen in Table 2 below, it significantly decreased during ripening, from 0.73 to 0.54 in CC and from 0.82 to 0.68 in LC. Our results are in accordance with those reported by Zheng et al. [30] and reveal that the low moisture content in Gouda-type cheese is associated with high firmness but low springiness and cohesiveness.

Gumminess is the energy required to disintegrate cheese into a state ready for swallowing. Its level significantly decreased in CC as it ripened (from 16.09 to 8.94 N), as in the study of Pinto et al. [31], while in LC (from 10.72 to 10.17 N), it did not vary considerably.

Chewiness indicates the energy required to chew cheese to a state whereby it is ready for swallowing. The chewiness index is estimated as hardness × cohesiveness × springiness index. It decreased with ripening time, from 13.35 N in CC and 8.57 N in LC to 8.57 and 6.96 N, respectively. In another study, Pinto et al. [31] also noticed a decreasing change in cheese chewiness during ripening.

3.3. Microbiological Properties of Gouda-Type Cheese

The starter culture used for Gouda-type cheese-making contains a mixture of thermophilic and mesophilic bacteria such as *Lactococcus lactis* ssp. *cremoris*, *Lactococcus lactis* subsp. *lactis*, *Lactococcus lactis* subsp. *lactis* biovar. *diacetylactis*, *Lactobacillus helveticus*, *Lactobacillus paracasei*, *Leuconostoc* species, and *Streptococcus thermophilus*. Therefore, the effect of lavender flower powder on lactic acid bacteria growth in Gouda-type cheese was evaluated by determining the lactobacilli and streptococci count, also, the total lactic acid bacteria count (see Table 3).

Table 3. Counts of lactobacilli, streptococci, and total lactic acid bacteria in control and lavender Gouda-type cheese at different ripening times (T1, T2, and T3).

Parameter	Control Cheese			Lavender Cheese		
	T1	T2	T3	T1	T2	T3
Lactobacilli count	9.5 ± 0.007[cB]	9.7 ± 0.009[bB]	9.8 ± 0.004[aB]	9.7 ± 0.005[cA]	9.8 ± 0.004[bA]	9.9 ± 0.006[aA]
Streptococci count	8.7 ± 0.011[cB]	9.2 ± 0.011[bB]	9.3 ± 0.016[aB]	9.0 ± 0.024[cA]	9.5 ± 0.012[bA]	9.6 ± 0.013[aA]
Total lactic acid bacteria count	18.2 ± 0.004[cB]	18.9 ± 0.002[bB]	19.1 ± 0.020[aB]	18.7 ± 0.019[cA]	19.3 ± 0.008[bA]	19.5 ± 0.020[aA]

T1—10 days of ripening; T2—20 days of ripening; T3—30 days of ripening. Data are expressed as mean ± standard deviation values of all enumerations. Different lowercase letters in the same row indicate significant differences between ripening times ($p < 0.05$, Tukey's test), and different uppercase letters show significant differences between cheeses ($p < 0.05$).

The antibacterial effect of lavender against *Escherichia coli*, responsible for early cheese blowing, and *Clostridium tyrobutiricum*, responsible for late cheese blowing, was reported by Librán et al. [32] in a previous study. Therefore, we assumed that lavender flower powder, used as a flavoring ingredient in our Gouda-type cheese, could inhibit the growth of starter culture microorganisms during ripening. Nevertheless, contrary to our expectation, it stimulated the development of lactic acid bacteria since the counts of lactobacilli, streptococci, and total lactic acid bacteria were significantly higher in LC at all ripening times.

However, in line with the study of Öztürk et al. [33], the flavoring ingredient reported herein caused a significant increase in the lactobacilli and streptococci count in CC and LC with ripening time (in CC, from 9.5 to 9.8 log cfu/g, and in LC, from 9.7 to 9.9 log cfu/g for lactobacilli count; in CC, from 8.7 to 9.3 log cfu/g, and in LC, from 9.0 to 9.6 log cfu/g for streptococci count; in CC, from 18.2 to 19.1 log cfu/g, and in LC, from 18.7 to 19.5 log cfu/g for total lactic acid bacteria count). These results also explain the below-discussed accumulation of volatile compounds during cheese ripening in both CC and LC.

3.4. Volatile Compounds in Gouda-Type Cheese

Results of headspace-SPME-GC/MS analysis for quantifying volatile compounds in the Gouda-type cheeses are shown in Table 4. Thirty-seven volatile compounds were detected in CC at T1, classified under six different identified compounds: haloalkane (32.0%—one compound) was the most dominant group, followed by alcohols (23.2%—nine compounds), carboxylic acids (15.4%—six compounds), cyanoalkanes (8.3%—one compound), esters (8.1%—six compounds), and ketones (5.9%—three compounds). Moreover, five compounds were classified within the chemical class of terpenes and terpenoids (3.3%), three in aldehydes (2.2%), two in aromatic hydrocarbons (1.2%), and one in pyridines (0.4%). To better visualize the similarities and differences between cheese samples at different ripening stages regarding the volatile composition, hierarchical cluster analysis (HCA) was run (see Figure 3).

Control cheese. Chloroform (32.0%; No. **5** in Table 4) was the most dominant compound present in CC at T1, followed by acetonitrile (8.3%; No. **4**), isopropanol (6.6%; No. **2**), 1-butanol (6.0%; No. **12**), isoamyl alcohol (5.8%; No. **19**), caproic acid (5.6%; No. **59**), acetoin (5.6%; No. **28**), and methyl hexanoate (4.2%; No. **17**); of these, chloroform, toluene, acetonitrile, and isopropanol were found in the Gouda-type cheese for the first time. Previous findings have revealed the presence of chloroform in semi-hard goat cheese (Kınık et al. [34]), chloroform and toluene in some varieties of Turkish cheese (Hayaloglu and Karabulut [35]), and acetonitrile and isopropanol in Manchego cheese (Gómez-Ruiz et al. [36]). It is well known that chloroform and acetonitrile derive from the breakdown of milk carotene [35,36]. Chloroform may also derive from the chlorine-containing cleaning products used for cheese-processing equipment disinfection [37]. Alcohols (23.2%), the second most dominant class in CC at T1, would be the consequence of the abovementioned compounds' abundance, together with the presence of optical (0.9%) and meso isomers (2.4%) of 2,3-butanediol. Meanwhile, 1-butanol, isoamyl alcohol, and 2,3-butanediol were also discovered in Gouda-type cheese by Van Leuven et al. [38] and Van Hoorde et al. [39]. Some authors reported that they arise from butanal (resulting from fatty acid or amino acid metabolism), leucine, and acetoin, respectively [36,40,41]. Carboxylic acids (15.4%), the third class in abundance in CC at T1, were represented by acetic acid (2.8%), butyric acid (2.4%), and caproic acid (10.2%), all previously reported in Gouda-type cheese by other studies [38,39,42–46]. They can originate from milk fat lipolysis or lactose and lactic acid fermentation [35]. Cyanoalkanes (8.3%), esters (8.1%), ketones (5.9%), terpene and terpenoids (3.3%), aldehydes (2.2%), aromatic hydrocarbons (1.2%), and pyridines (0.4%) were the subsequent classes of volatile compounds grouped in CC at T1. Acetoin, listed among the significant volatile constituents in CC at T1, is a volatile compound from the ketone class. It was identified as a derivative of milk citrates [36] and, therefore, detected in Gouda-type cheese by other researchers [38,45,46]. Methyl hexanoate (4.2%), from the ester class, another notable volatile compound in CC at T1, was also reported in some commercial Gouda cheeses [2]. Its occurrence is probably related to the esterase activity of lactic acid bacteria [36]. As for aldehydes in cheese, they are produced by the catabolism of fatty acids or amino acids via decarboxylation or deamination [35,47]; only benzaldehyde (1.5%), hexanal (0.2%), and heptanal (0.5%) were detected in CC at T1, but it seems that they are commonly present in Gouda-type cheeses [38,39,45]. Both hexanal and heptanal disappeared during ripening; heptanal was no longer detected in CC on the 20th day of maturation and hexanal on the 30th—hence the lower number of volatile compounds found in CC at T2 (thirty-six) and CC at T3 (thirty-five). o-Cymene (1.7%), β-ocimene, trans-β-ocimene (0.6%), γ-terpinene (0.2%), and linalool (0.3%), volatile compounds of the terpene and terpenoid class, were also present in CC at T1; since this is the first time that they have been detected in a Gouda-type cheese, they most likely arise from animal feed, considering that the cows had pasture access in the afternoon and evening.

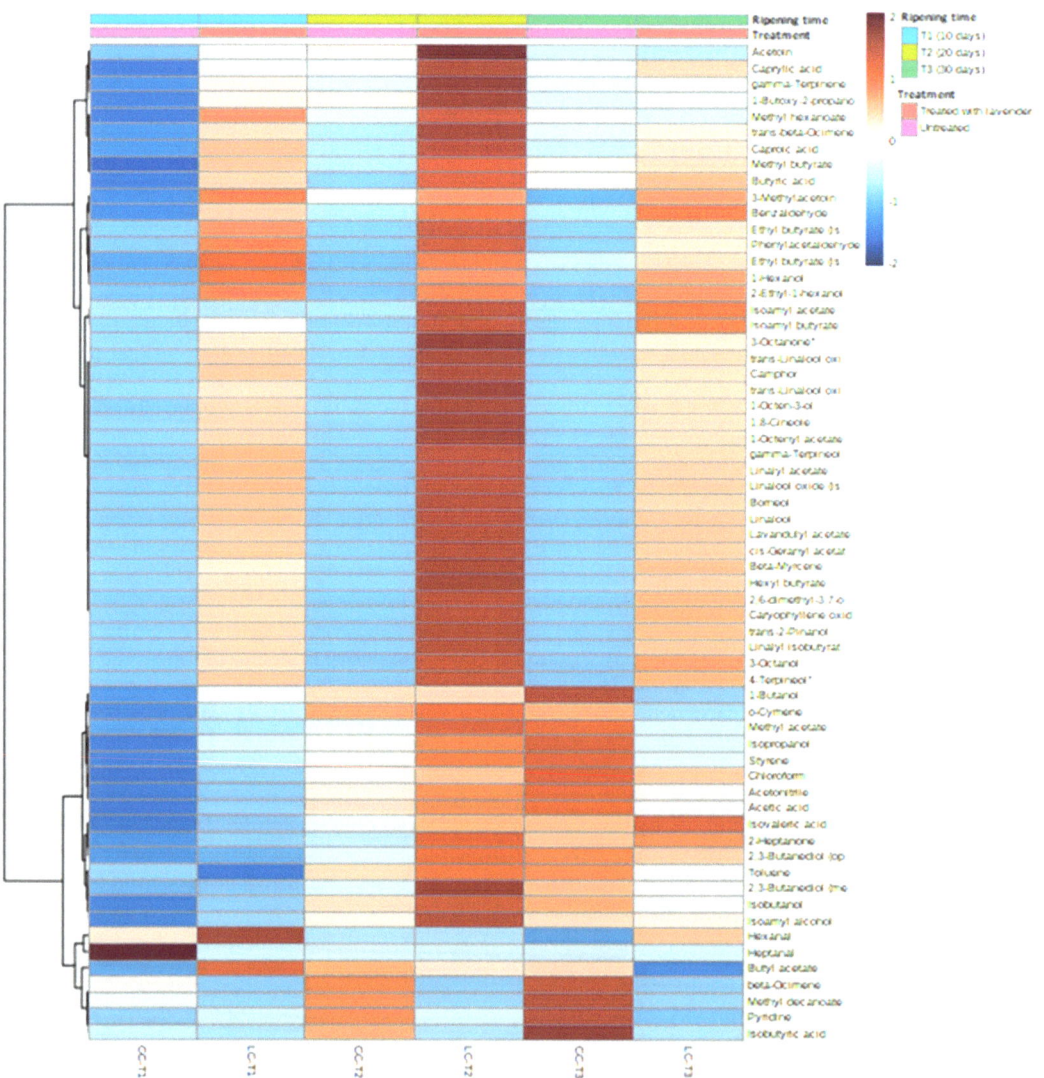

Figure 3. HCA heatmap of Gouda-type cheese based on volatile constituents.

Table 4. Volatile compound concentrations in control and lavender Gouda-type cheese at different ripening times (T1, T2, and T3).

Crt. No.	Volatile Compound	MSS (%)	Odor Descriptor	Reference	CC	Control Cheese T1	Control Cheese T2	Control Cheese T3	Lavender Cheese T1	Lavender Cheese T2	Lavender Cheese T3
						μg/kg Cheese					
1	Methyl acetate	96	Fruity, solvent, blackcurrant-like	[48]	Es	38.56 ± 4.199^{bB}	67.26 ± 2.948^{aB}	111.20 ± 26.935^{aA}	53.50 ± 1.607^{cA}	115.42 ± 25.312^{aA}	61.89 ± 5.974^{bB}
2	Isopropanol	97	Sharp musty	[49]	Alc	221.75 ± 46.558^{bB}	722.34 ± 89.846^{aB}	1155.17 ± 222.182^{aA}	599.38 ± 75.037^{bA}	1047.99 ± 120.237^{aA}	590.10 ± 7.424^{bB}
3	Methyl butyrate	96	Fruity, apple-like	[50]	Es	54.21 ± 11.983^{bB}	114.20 ± 17.312^{aB}	153.96 ± 31.418^{aA}	180.06 ± 25.014^{aA}	222.23 ± 34.620^{aA}	170.39 ± 1.721^{aA}
4	Acetonitrile	98	Choking alcohol	-	CyAlka	275.86 ± 51.660^{cB}	1083.19 ± 169.607^{bA}	1636.99 ± 279.534^{aA}	575.97 ± 89.883^{cA}	1495.49 ± 197.115^{aA}	1022.22 ± 52.569^{bB}
5	Chloroform	77	Ether-like	[49]	HoAlka	1068.81 ± 206.553^{cB}	3527.63 ± 402.152^{bA}	5590.37 ± 1139.680^{aA}	2176.24 ± 339.353^{bA}	4329.89 ± 565.593^{aA}	4329.89 ± 565.593^{aB}
6	Toluene	94	Nutty, bitter, almonds, paint, fruity	[51,52]	AH	23.03 ± 3.945^b	51.99 ± 9.667^{abA}	65.66 ± 18.515^{aA}	n.d.	68.95 ± 22.883^{aA}	41.11 ± 5.903^{aA}
7	Ethyl butyrate (isomer)	95	Fruity, pineapple, acetone, caramel	[48]	Es	n.d.	n.d.	n.d.	39.08 ± 2.587^a	50.16 ± 15.106^a	28.71 ± 12.886^a
8	Butyl acetate	94	Fruity	[49]	Es	15.34 ± 3.106^{aB}	25.07 ± 3.967^A	23.29 ± 5.877^{aA}	27.71 ± 2.459^{bA}	22.75 ± 2.323^{bA}	13.80 ± 0.133^{cB}
9	Hexanal	95	Cut grass	[53]	Ald	16.51 ± 5.051^{aA}	6.68 ± 2.494^{bA}	n.d.	31.52 ± 20.601^{aA}	6.86 ± 2.523^b	19.44 ± 0.720^a
10	Isobutanol	91	Choking alcohol	[49]	Alc	6.31 ± 1.424^{aB}	24.40 ± 3.487^{aB}	28.14 ± 8.926^{aA}	12.67 ± 3.009^{bA}	34.31 ± 1.756^{aA}	20.10 ± 3.050^{bA}
11	Isoamyl acetate	97	Banana, chewing gum	[48]	Es	n.d.	n.d.	n.d.	n.d.	62.98 ± 10.744^a	52.95 ± 4.624^a
12	1-Butanol	97	Medicinal, fruity	[51]	Alc	201.21 ± 41.648^{cB}	640.06 ± 83.138^{bA}	950.89 ± 166.505^{aA}	505.86 ± 71.458^{bA}	638.14 ± 80.605^{aA}	319.35 ± 28.604^{bB}
13	β-Myrcene	93	Balsamic, spice	[54]	Te&Ts	12.38 ± 3.015^{bA}	50.35 ± 9.850^{aA}	63.41 ± 11.766^{aA}	29.92 ± 9.055^b	62.30 ± 6.066^a	38.28 ± 2.828^b
14	Pyridine	92	Fruity	-	Pyr	6.17 ± 1.257^{cA}	13.62 ± 2.483^{bB}	23.78 ± 1.680^{aA}	21.12 ± 6.264^{aA}	21.45 ± 2.673^{aB}	10.31 ± 1.399^{bB}
15	2-Heptanone	88	Sour milk/dairy	[52]	Ket	5.94 ± 1.317	n.d.	n.d.	10.71 ± 2.732^{bA}	30.92 ± 3.245^{aA}	27.73 ± 2.546^{aA}
16	Heptanal	80	Fruity, ester-like	[52]	Ald	141.11 ± 25.302^{bB}	288.30 ± 40.965^{aB}	321.42 ± 53.508^{aA}	428.97 ± 58.989^{aA}	496.81 ± 69.633^{bA}	279.30 ± 29.445^{bA}
17	Methyl hexanoate	96	Floral, minty, fruity	[50]	Es	n.d.	n.d.	n.d.	370.96 ± 56.128^b	678.05 ± 61.261^a	356.96 ± 17.757^b
18	1,8-Cineole	97	Banana	[54]	Te&Ts	194.44 ± 43.626^{bB}	750.88 ± 94.873^{aB}	831.51 ± 172.253^{aA}	433.41 ± 95.289^{cA}	1284.59 ± 151.738^{aA}	758.69 ± 54.400^{bA}
19	Isoamyl alcohol	97	Pineapple-like	[55]	Alc	8.62 ± 2.144^{bB}	14.59 ± 2.407^{aB}	29.48 ± 7.633^{aA}	79.59 ± 8.303^{aA}	74.89 ± 2.553^{aA}	54.02 ± 9.024^{bB}
20	Ethyl butyrate (isomer)	96	Mushroom-like	[49]	Es	21.33 ± 4.752^{cA}	50.26 ± 4.803^{aB}	66.13 ± 7.807^{aB}	97.95 ± 13.079^{bB}	158.27 ± 8.496^{aA}	90.65 ± 9.332^{bA}
21	trans-β-Ocimene	82	Unpleasant	[56]	Te&Ts	5.57 ± 0.261^{cB}	13.84 ± 1.000^{aB}	14.45 ± 1.618^{aA}	17.75 ± 2.660^{bA}	30.85 ± 2.920^{aA}	16.42 ± 1.305^{bA}
22	γ-Terpinene	83	Sweet smell	[57]	Te&Ts	15.76 ± 1.719^{cB}	42.52 ± 7.501^{aB}	66.04 ± 17.135^{aA}	32.90 ± 3.271^{bB}	61.74 ± 5.802^{aA}	35.94 ± 4.774^{bB}
23	Styrene	87	Mushroom-like/buttery	[49]	AH	n.d.	n.d.	n.d.	192.64 ± 49.231^b	400.72 ± 37.208^a	181.33 ± 18.574^b
24	3-Octanone *	89	Sweet, herb	[58]	Ket	n.d.	34.83 ± 9.189^a	42.49 ± 4.247^a	n.d.	n.d.	n.d.
25	β-Ocimene	87	Gasoline, citrus	[54]	Te&Ts	18.66 ± 2.322^b	153.78 ± 18.436^{aA}	151.50 ± 28.686^{aA}	94.46 ± 15.536^{bA}	172.38 ± 21.329^{aA}	86.03 ± 15.595^{bB}
26	o-Cymene	96	Green, fruity	[54,59]	Te&Ts	56.15 ± 11.023^{bB}	n.d.	n.d.	25.73 ± 5.723^b	66.30 ± 16.486^a	55.30 ± 5.393^a
27	Isoamyl butyrate	90	Sour milk	[60]	Es	n.d.	n.d.	n.d.	n.d.	n.d.	n.d.
28	Acetoin	96	-	[51]	Ket	185.42 ± 44.259^{cB}	484.17 ± 63.701^{aB}	331.24 ± 59.406^{bA}	439.72 ± 24.064^{bA}	1026.19 ± 125.392^{aA}	283.79 ± 6.800^{bA}
29	3-Methylacetoin	87	-	-	Ket	6.87 ± 0.932^{bB}	13.14 ± 0.796^{aA}	8.24 ± 1.433^{bB}	18.18 ± 2.239^{aA}	17.12 ± 4.537^{aA}	17.12 ± 4.537^{aA}
30	1-Butoxy-2-propanol	94	Sweet, green	[49]	Alc	7.00 ± 1.595^{bB}	33.38 ± 5.554^{aB}	25.65 ± 4.623^{aA}	33.53 ± 4.275^{bA}	58.20 ± 7.734^{aA}	28.39 ± 3.179^{bA}
31	1-Hexanol	96	Mushroom, cheese	[49]	Alc	28.91 ± 5.329^{bB}	62.36 ± 46.499^{aB}	95.27 ± 14.049^{aB}	623.90 ± 309.213^{aA}	552.70 ± 159.944^{aA}	528.37 ± 238.769^{aA}
32	1-Octenyl acetate	95	Fruity, green	[54]	Es	n.d.	n.d.	n.d.	426.19 ± 57.704^b	775.69 ± 80.136^a	402.63 ± 25.727^b
33	3-Octanol	84	Vinegar, sour, sharp, peppery, green	[54]	Alc	n.d.	n.d.	n.d.	52.82 ± 5.691^b	89.54 ± 9.653^a	67.24 ± 10.446^b
34	Hexyl butyrate	93	-	[61]	Es	n.d.	n.d.	n.d.	17.50 ± 2.985^b	34.34 ± 1.539^a	19.27 ± 3.282^b
35	Acetic acid	95	Mushroom, metallic	[51]	CA	92.92 ± 21.149^{bB}	445.64 ± 79.887^{aA}	605.66 ± 92.910^{aA}	226.10 ± 45.177^{cA}	562.83 ± 68.060^{bA}	363.00 ± 28.421^{bB}
36	trans-Linalool oxide (isomer)	95	-	-	Te&Ts	n.d.	n.d.	n.d.	638.94 ± 108.438^b	1247.76 ± 186.259^a	633.61 ± 60.692^b
37	1-Octen-3-ol	97	-	[54]	Alc	5.90 ± 1.212^{bB}	10.13 ± 0.379^{aB}	11.35 ± 2.807^{aB}	84.88 ± 9.979^{bA}	160.27 ± 25.469^{aA}	82.59 ± 2.777^{bA}
38	trans-2-Pinanol	86	-	-	Te&Ts	n.d.	n.d.	n.d.	17.53 ± 1.531^b	30.69 ± 3.142^a	19.49 ± 3.055^b

Table 4. Cont.

Crt. No.	Volatile Compound	MSS (%)	Odor Descriptor	Reference	CC	Control Cheese T1	Control Cheese T2	Control Cheese T3	Lavender Cheese T1	Lavender Cheese T2	Lavender Cheese T3
								μg/kg Cheese			
39	trans-Linalool oxide (isomer)	97	-	-	Te&Ts	n.d.	n.d.	n.d.	327.46 ± 49.616[b]	558.54 ± 25.266[a]	305.77 ± 29.301[b]
40	Linalyl acetate	72	Flower, fruit, lavender	[54]	Te&Ts	n.d.	n.d.	n.d.	6.97 ± 0.964[b]	10.38 ± 0.517[a]	6.33 ± 0.437[b]
41	2-Ethyl-1-hexanol	90	Mild, oily, sweet, slightly floral odor reminiscent of rose	[49]	Alc	n.d.	n.d.	n.d.	11.11 ± 1.098[a]	11.17 ± 1.211[a]	10.57 ± 1.624[a]
42	Camphor	93	Camphor	[54]	Te&Ts	n.d.	n.d.	n.d.	61.08 ± 7.291[b]	101.92 ± 6.717[a]	54.25 ± 4.126[b]
43	Benzaldehyde	97	Almonds, sugar, burnt	[51]	Ald	49.32 ± 11.585[bB]	116.58 ± 14.396[aB]	115.88 ± 24.301[aB]	211.96 ± 12.459[aA]	271.88 ± 66.353[aA]	266.70 ± 52.937[aA]
44	2,3-Butanediol (optical isomer)	95	Fruity, onions	[51]	Alc	28.42 ± 9.483[cA]	77.80 ± 11.120[bB]	145.40 ± 24.605[aA]	39.73 ± 5.638[bA]	156.05 ± 25.139[aA]	119.33 ± 23.194[aA]
45	Linalool	86	Flower, lavender	[54,62]	Te&Ts	9.49 ± 1.481[a]	8.77 ± 0.928[bB]	15.56 ± 3.930[aB]	2416.24 ± 313.723[bA]	4130.24 ± 295.527[aA]	2371.14 ± 188.355[bA]
46	Isobutyric acid	94	Cheesy, butter, rancid	[51]	CA	12.31 ± 1.076[c]	81.03 ± 16.303[b]	142.94 ± 26.383[a]	n.d.	n.d.	n.d.
47	Linalyl isobutyrate **	87	Cheesy, butter, rancid	[51]	Te&Ts	n.d.	n.d.	n.d.	2143.98 ± 264.564[b]	3817.29 ± 361.877[a]	2297.54 ± 183.074[b]
48	2,3-Butanediol (meso isomer)	97	Fruity, onions	[51]	Alc	81.21 ± 31.384[cA]	175.49 ± 32.577[bB]	307.94 ± 34.889[aA]	98.55 ± 13.951[bA]	478.83 ± 157.897[aA]	215.29 ± 86.648[bA]
49	4-Terpineol ***	86	Green, liquorice, moldy	[48]	Te&Ts	n.d.	n.d.	n.d.	237.05 ± 30.071[b]	373.97 ± 35.340[a]	258.37 ± 31.058[b]
50	Methyl decanoate	89	Camphor	-	Es	13.02 ± 1.438[c]	30.64 ± 4.324[b]	40.91 ± 4.701[a]	n.d.	n.d.	n.d.
51	Lavandulyl acetate	93	Floral odor reminiscent of lavender	[63]	Te&Ts	n.d.	n.d.	n.d.	211.16 ± 29.910[b]	355.30 ± 35.686[a]	216.28 ± 19.026[b]
52	Butyric acid	93	Parmesan cheese, vomit, butanoic acid	[48]	CA	81.43 ± 21.084[cB]	212.27 ± 56.676[bB]	389.59 ± 64.874[aA]	469.11 ± 26.717[nA]	643.62 ± 144.232[aA]	506.19 ± 81.424[aA]
53	Phenylacetaldehyde	94	Floral, rosy, sweet, hyacinth-like	[62]	Ald	n.d.	n.d.	n.d.	20.02 ± 1.922[a]	23.38 ± 1.934[a]	12.98 ± 3.641[b]
54	Isovaleric acid	94	Sweet, rancid, rotten, cheesy, Swiss cheese	[51]	CA	64.68 ± 6.667[aB]	373.42 ± 17.242[bB]	537.57 ± 30.138[aB]	207.08 ± 13.678[cA]	553.87 ± 37.967[bA]	702.86 ± 33.921[aA]
55	α-Terpineol	83	Oil, anise, mint	[54]	Te&Ts	n.d.	n.d.	n.d.	20.99 ± 2.956[b]	31.51 ± 2.595[a]	18.49 ± 3.808[b]
56	Borneol	95	Camphor	[54]	Te&Ts	n.d.	n.d.	n.d.	155.76 ± 21.932[b]	250.67 ± 27.919[a]	145.19 ± 32.109[b]
57	cis-Geranyl acetate	86	Rose, fruity, flower	[54]	Te&Ts	n.d.	n.d.	n.d.	34.31 ± 3.226[b]	56.58 ± 6.103[a]	34.09 ± 6.532[b]
58	Linalool oxide (isomer)	77	Floral, honey-like	[61]	Te&Ts	n.d.	n.d.	n.d.	10.48 ± 2.264[b]	15.63 ± 1.521[a]	9.73 ± 2.250[b]
59	Caproic acid	93	Bad breath, popcorn, goaty	[51]	CA	185.93 ± 46.042[bB]	403.05 ± 90.032[aB]	451.28 ± 36.396[aB]	807.79 ± 45.073[bA]	1133.32 ± 206.390[aA]	722.20 ± 16.136[bA]
60	2,6-dimethyl-3,7-octadien-2,6-diol	83	-	-	Alc	n.d.	n.d.	n.d.	37.95 ± 6.169[b]	65.88 ± 5.583[a]	43.95 ± 2.542[b]
61	Caryophyllene oxide	87	Terpene notes, weak woody-spicy	[64]	Te&Ts	n.d.	n.d.	n.d.	26.76 ± 3.983[b]	47.81 ± 1.866[a]	31.78 ± 4.108[b]
62	Caprylic acid	97	Sweaty, rancid	[51]	CA	76.51 ± 7.885[bB]	188.50 ± 28.549[aB]	192.79 ± 32.975[aA]	193.80 ± 21.998[bA]	342.19 ± 29.252[aA]	240.40 ± 41.430[bA]
					TOTAL, of which:	3337.08	10,392.16	14,743.17	16,366.71	29,619.78	19,695.86
					Alc	775.14	2496.84	3551.32	2533.80	4577.67	2783.98
					Ald	71.77	123.26	115.88	263.50	302.12	299.12
					AH	38.79	94.52	131.71	32.90	130.69	77.05
					CA	513.79	1703.91	2319.84	1903.87	3235.83	2534.65
					CyAlka	275.86	1083.19	1636.99	575.97	1495.49	1022.22

Table 4. Cont.

Crt. No.	Volatile Compound	Odor Descriptor	Reference	CC	Control Cheese μg/kg Cheese			Lavender Cheese		
					T1	T2	T3	T1	T2	T3
				Es	270.86	540.06	680.26	1278.34	1921.57	1138.26
				HoAlka	1068.81	3527.63	5590.37	2176.24	4329.89	4329.89
				Ket	198.46	510.93	363.26	661.25	1474.94	509.98
				Pyr	12.38	50.35	63.41	21.12	21.45	10.31
				Te&Ts	111.21	261.48	290.13	6919.72	12130.13	6990.40

MSS (%)

* 3-octanone was identified along with β-ocimene in LC; ** linalyl isobutyrate was identified along with methyl decanoate in LC; *** 4-terpineol was identified along with isobutyric acid in LC; MSS—mass spectra similarity; T1—10 days of ripening; T2—20 days of ripening; T3—30 days of ripening; CC—chemical class; Es—esters; Alc—alcohols; CyAlka—cyanoalkanes; HoAlka—haloalkanes; AH—aromatic hydrocarbons; Ald—aldehydes; Te&Ts—terpene and terpenoids; Pyr—pyridines; Ket—ketones; CA—carboxylic acids; n.d.—not detected. Data are expressed as mean ± standard deviation values of all measurements. Different lowercase letters in the same row indicate significant differences between ripening times ($p < 0.05$, Tukey's test), and different uppercase letters show significant differences between cheeses ($p < 0.05$).

Lavender cheese. The volatile profile of LC at T1 (Table 4), however, was dominated by terpenes and terpenoids (Te&Ts; 42.3%), followed by alcohols (Alc; 15.5%), haloalkanes (HoAlka; 13.3%), carboxylic acids (CA; 11.6%), esters (Es; 7.8%), ketones (Ket; 4%), cyanoalkanes (CyAlka; 3.5%), aldehydes (Ald; 1.6%), aromatic hydrocarbons (AH; 0.2%), and pyridines (Pyr; 0.1%). Sixty-two volatile compounds were detected in this sample, including 12 alcohols, 3 aldehydes, 1 aromatic hydrocarbon, 5 carboxylic acids, 1 cyanoalkane, 10 esters, 1 haloalkane, 4 ketones, 1 pyridine, and 19 compounds from the terpene and terpenoid class, with three more alcohols than in CC at T1 (3-octanol; 2-ethyl-1-hexanol; 2,6-dimethyl-3,7-octadien-2,6-diol), 1 aldehyde (phenylacetaldehyde), 5 esters (an isomer of ethyl butyrate; isoamyl acetate; isoamyl butyrate; 1-octenyl acetate; hexyl butyrate), 1 ketone (3-octanone), and 15 compounds of the terpene and terpenoid class (β-myrcene; 1,8-cineole; 3 isomers of linalool oxide; trans-2-pinanol; linalyl acetate; camphor; linalyl isobutyrate; 4-terpineol; lavandulyl acetate; α-terpineol; borneol; cis-geranyl acetate; caryophyllene oxide). Moreover, isobutyric acid (CA) was identified along with linalyl isobutyrate (Te&Ts), methyl decanoate (Es) with 4-terpineol (Te&Ts), and β-ocimene (Te&Ts) with 3-octanone (Ket). Different from CC at T1, linalool (14.8%; Te&Ts; No. **45** in Table 4) was the main volatile compound in LC at T1, followed by chloroform (13.3%; HoAlka; No. **5**) and linalyl isobutyrate (13.1%; Te&Ts; No. **47**), and then by caproic acid (4.9%; CA; No. **59**), an isomer of trans-linalool oxide (3.9%; Te&Ts; No. **36**), 1-hexanol (3.8%; Alc; No. **31**), isopropanol (3.7%; Alc; No. **2**), acetonitrile (3.5%; CyAlka; No. **4**), and 1-butanol (3.1%; Alc; No. **12**). The number of volatile constituents in LC did not change with ripening time, but their amount increased from 16,366.71 µg/kg cheese at T1 to 29,619.78 µg/kg cheese at T2 and decreased to 19,695.86 µg/kg cheese at T3. As for CC, the total amount of volatile compounds increased during ripening from 3337.08 µg/kg cheese at T1 to 10,392.16 µg/kg cheese at T2 and further to 14,743.17 µg/kg cheese at T3. The content of total volatile compounds in the Gouda-type cheese was 79.6% higher in LC than in CC at T1, with 64.9% at T2 and 25.1% at T3. As can be noticed in Figure 3, the concentration of volatile compounds either decreased with ripening time, varied in a $\wedge$-pattern, or remained unchanged (cluster 1); in other cases, it increased (cluster 2).

3.5. Sensory Properties of Gouda-Type Cheese and Consumers' Acceptance and Purchase Intention

Consumers' perception of LC compared to CC was evaluated based on hedonic scores of their sensory attributes (Figure 4). The use of lavender flower powder as a flavoring ingredient in Gouda-type cheese-making has significantly affected its sensory properties by reducing its rating for appearance and color (0.6 points), consistency and texture (0.7 points), odor and taste (1.2 points), aftertaste (1.1 points), and overall liking (0.7 points). The overall score for LC was 7.2, and that for CC was 8.1. However, the calculation resulted in an acceptance rate of 80.2% for LC and 89.5% for CC. Furthermore, when asked about their willingness to buy the Gouda-type cheese, 34% of respondents answered that they "definitely will buy" LC, and 42% responded in this way for CC (see Figure 5). As regards the undecided subjects, 24% answered "might or might not buy" and 8% "probably will buy" the LC, which shows greater indecision about the flavored cheese. This outcome highlights the lack of familiarity but a curiosity towards this new product. For LC, 8% of survey participants responded with "definitely will not buy". Overall, these results show that there would be customers for LC if this were available on the market.

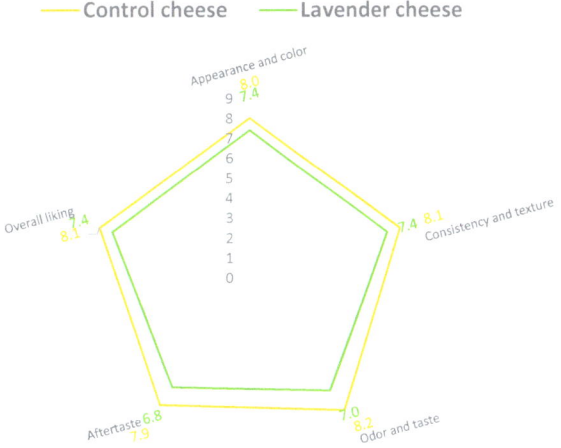

Figure 4. Hedonic scores for sensory attributes of control and lavender Gouda-type cheese. Data are expressed as mean values of all responses.

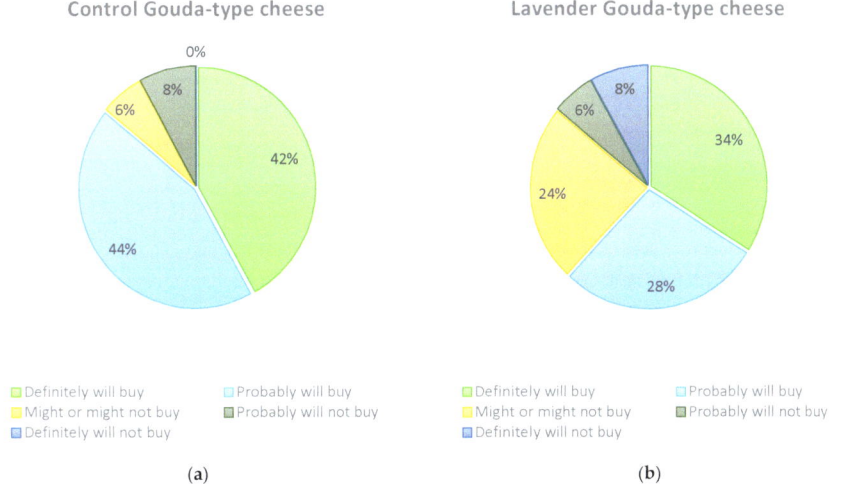

Figure 5. Response rates (%) for purchase intention of (**a**) control Gouda-type cheese; (**b**) lavender Gouda-type cheese.

4. Conclusions

The manufacturing process proposed in this study resulted in a Gouda-type cheese formulation with a lavender aroma. Using lavender flower powder as a flavoring ingredient at a concentration of 0.5 g/L matured milk in Gouda-type cheese manufacturing conferred upon the cheese a terpenic volatile profile and stimulated the growth of lactic acid bacteria from the starter culture. During ripening, a concentration of volatile compounds and lactic acid bacteria, both in the control and lavender cheese, and an improvement in their nutritional and textural properties was noticed. However, it should be underlined that the flavoring ingredient did not significantly impact the Gouda-type cheese's gross composition and textural properties. Regarding the sensory perception of the lavender cheese, it should be noted that its overall score was slightly lower but very close to that received by the unflavored cheese, definitively more familiar. Although the acceptance rate of the lavender-flavored cheese was high, consumers were less willing to buy it, being a gourmet product. However, with effective communication of this new product, perhaps accompanied by a food pairing study, the lavender-flavored Gouda-type cheese could be a successful novelty in the worldwide dairy market.

Author Contributions: Conceptualization, C.A.S.; methodology, C.A.S. and M.A.R.; formal analysis and investigation, M.M., M.T., B.S.S., M.-I.S., M.F., D.M., A.M.J., V.M. and S.R.I.; resources, C.A.S. and T.G.T.; data curation, C.A.S., M.M. and V.M.; writing—original draft preparation, C.A.S. and M.M.; writing—review and editing, M.T. and T.G.T.; visualization, M.A.R., M.M. and M.-I.S. were equal contributors to this work and are designated as co-first authors. All authors have read and agreed to the published version of the manuscript.

Funding: This work was supported by a grant from the Romanian Ministry of Research, Innovation and Digitization CNCS/CCCDI—UEFISCDI, project number PN-III-P1-1.1-TE-2019-2212, within PNCDI III, and an internal research grant (No. 26021/22.11.2021), type SOLUTIONS USAVM-CN: Development of ripened cheeses in the purpose of BIO milk superior valorization at SDE Cojocna

Institutional Review Board Statement: Not applicable.

Informed Consent Statement: Not applicable.

Data Availability Statement: All data generated or analyzed during this study are included in this published article.

Acknowledgments: We are grateful for the administrative and financial support of the Alma Mater Studiorum-Università di Bologna, Italy.

Conflicts of Interest: The authors declare no conflict of interest.

References

1. Codex Alimentarius Commission. Standard for Gouda CXS 266-1966. Available online: https://www.fao.org/fao-who-codexalimentarius/sh-proxy/en/?lnk=1&url=https%253A%252F%252Fworkspace.fao.org%252Fsites%252Fcodex%252FStandards%252FCXS%2B266-1966%252FCXS_266e.pdf (accessed on 11 February 2023).
2. Chen, C.; Tian, T.; Yu, H.; Yuan, H.; Wang, B.; Xu, Z.; Tian, H. Characterisation of the key volatile compounds of commercial Gouda cheeses and their contribution to aromas according to Chinese consumers' preferences. *Food Chem. X* **2022**, *15*, 100416. [CrossRef] [PubMed]
3. Choi, H.Y.; Yang, C.J.; Choi, K.S.; Bae, I. Characteristics of Gouda cheese supplemented with fruit liquors. *J. Anim. Sci. Technol.* **2015**, *57*, 15. [CrossRef]
4. Semeniuc, C.A.; Zăpârțan, L.; Stan, L.; Pop, C.R.; Borș, M.D.; Rotar, A.M. Physicochemical and sensory properties of whey cheese with pine nuts. *Bull. UASVM Food Sci. Technol.* **2015**, *72*, 177–181. [CrossRef]
5. Zhang, X.Y.; Guo, H.Y.; Zhao, L.; Sun, W.F.; Zeng, S.S.; Lu, X.M.; Cao, X.; Ren, F.Z. Sensory profile and Beijing youth preference of seven cheese varieties. *Food Qual. Prefer.* **2011**, *22*, 101–109. [CrossRef]
6. Semeniuc, C.A.; Mandrioli, M.; Socaci, B.S.; Socaciu, M.-I.; Fogarasi, M.; Podar, A.S.; Michiu, D.; Jimborean, A.M.; Mureșan, V.; Ionescu, S.R.; et al. Changes in lipid composition and oxidative status during ripening of Gouda-type cheese as influenced by addition of lavender flower powder. *Int. Dairy J.* **2022**, *133*, 105427. [CrossRef]
7. Taherkhani, P.; Noori, N.; Akhondzadeh Basti, A.; Gandomi, H.; Alimohammadi, M. Antimicrobial effects of Kermanian black cumin (*Bunium persicum* Boiss.) essential oil in Gouda cheese matrix. *J. Med. Plants* **2015**, *14*, 76–85.
8. Kim, Y.K.; Nam, M.S.; Bae, H.C. Characteristics of Gouda cheese supplemented with chili pepper extract microcapsules. *Korean J. Food Sci. Anim. Resour.* **2017**, *37*, 833–839. [PubMed]
9. Düsterhöft, E.M.; Engels, W.; Huppertz, T. Dutch-type cheeses. In *Global Cheesemaking Technology: Cheese Quality and Characteristics*, 1st ed.; Papademas, P., Bintsis, T., Eds.; John Wiley & Sons, Ltd.: Chichester, UK, 2018; pp. 326–335.
10. Majdik, A. Experimental Studies on Changes Made to Pre Press for Gouda Cheese with Spices. In Proceedings of the International Multidisciplinary Scientific GeoConference: SGEM, Varna, Bulgaria, 16–22 June 2013.
11. Agboola, S.O.; Radovanovic-Tesic, M. Influence of Australian native herbs on the maturation of vacuum-packed cheese. *LWT-Food Sci. Technol.* **2002**, *35*, 575–583. [CrossRef]
12. Park, W.; Yoo, J.; Oh, S.; Ham, J.S.; Jeong, S.G.; Kim, Y. Microbiological characteristics of Gouda cheese manufactured with pasteurized and raw milk during ripening using next generation sequencing. *Food Sci. Anim. Resour.* **2019**, *39*, 585. [CrossRef]

13. Open Universiteit; Thames Polytechnic. Starter cultures for cheese production. In *CIP: Biotechnological Innovations in Food Processing*; Open Universiteit, Thames Polytechnic, Eds.; Butterworth-Heinemann: Oxford, UK, 1991; pp. 77–112.
14. *ISO 5534:2004 | IDF 4:2004*; Cheese and Processed Cheese—Determination of the Total Solids Content (Reference Method). ISO: Geneva, Switzerland, 2004.
15. *ISO 8968-1:2014 | IDF 20-1:2014*; Milk and Milk Products—Determination of Nitrogen Content—Part 1: Kjeldahl Principle and Crude Protein Calculation. ISO: Geneva, Switzerland, 2014.
16. *ISO 3433:2008 | IDF 222:2008*; Cheese—Determination of Fat Content—Van Gulik Method. ISO: Geneva, Switzerland, 2008.
17. Nagy, M.; Semeniuc, C.A.; Socaci, S.A.; Pop, C.R.; Rotar, A.M.; Sălăgean, C.D.; Tofană, M. Utilization of brewer's spent grain and mushrooms in fortification of smoked sausages. *Food Sci. Technol.* **2017**, *37*, 315–320. [CrossRef]
18. *ISO 5943:2006 | IDF 88:2006*; Cheese and Processed Cheese Product—Determination of Chloride Conten—Potentiometric Titration Method. ISO: Geneva, Switzerland, 2006.
19. *STAS 6353-85*; Milk and Milk Products—Determination of Acidity. ASRO: Bucharest, Romania, 1985.
20. Ong, L.; Dagastine, R.R.; Kentish, S.E.; Gras, S.L. The effect of pH at rennetting on the microstructure, composition and texture of Cheddar cheese. *Food Res. Int.* **2012**, *48*, 119–130. [CrossRef]
21. Socaciu, M.I.; Fogarasi, M.; Simon, E.L.; Semeniuc, C.A.; Socaci, S.A.; Podar, A.S.; Vodnar, D.C. Effects of whey protein isolate-based film incorporated with tarragon essential oil on the quality and shelf-life of refrigerated brook trout. *Foods* **2021**, *10*, 401. [CrossRef]
22. *ISO 7889:2003 | IDF 117:2003*; Yogurt—Enumeration of Characteristic Microorganisms—Colony-Count Technique at 37 °C. ISO: Geneva, Switzerland, 2003.
23. Cozzolino, R.; Martignetti, A.; De Giulio, B.; Malorni, L.; Addeo, F.; Picariello, G. SPME GC-MS monitoring of volatile organic compounds to assess typicity of Pecorino di Carmasciano ewe-milk cheese. *Int. J. Dairy Technol.* **2021**, *74*, 383–392. [CrossRef]
24. Lignou, S.; Oloyede, O.O. Consumer acceptability and sensory profile of sustainable paper-based packaging. *Foods* **2021**, *10*, 990. [CrossRef]
25. dos Reis Santos, J.; Gomes Hafemann, S.P.; Giani Pieretti, G.; Antigo, J.L.; Soares dos Santos Pozza, M.; da Silva Scapim, M.R.; Scaramal Madrona, G. Sensorial, microbiological, and physico-chemical analysis of Minas frescal cheese with oregano essential oil (*Origanum vulgare*) addition. *Int. J. Food Sci. Nutr. Eng.* **2014**, *4*, 86–90.
26. Wang, Y.; Wu, J.; Lv, M.; Shao, Z.; Hungwe, M.; Wang, J.; Bai, X.; Xie, J.; Wang, Y.; Geng, W. Metabolism characteristics of lactic acid bacteria and the expanding applications in food industry. *Front. Bioeng. Biotechnol.* **2021**, *9*, 612285. [CrossRef]
27. Mureşan, C.C.; Marc Vlaic, R.A.; Semeniuc, C.A.; Socaci, A.S.; Fărcaş, A.; Francisc, D.; Pop, C.R.; Rotar, A.; Dodan, A.; Mureşan, V.; et al. Changes in physicochemical and microbiological properties, fatty acid and volatile compound profiles of Apuseni cheese during ripening. *Foods* **2021**, *10*, 258. [CrossRef]
28. Kanawjia, S.K.; Rajesh, P.; Sabikhi, L.; Singh, S. Flavour, chemical and textural profile changes in accelerated ripened Gouda cheese. *LWT-Food Sci. Technol.* **1995**, *28*, 577–583. [CrossRef]
29. Ivanov, G.; Bogdanova, A.; Zsivanovits, G. Effect of ripening temperature on the texture of cow milk Kashkaval cheese. *Prog. Agri. Eng. Sci.* **2018**, *14*, 69–78. [CrossRef]
30. Zheng, Y.; Liu, Z.; Mo, B. Texture profile analysis of sliced cheese in relation to chemical composition and storage temperature. *J. Chem.* **2016**, *2016*, 8690380. [CrossRef]
31. Pinho, O.; Mendes, E.; Alves, M.M.; Ferreira, I.M.P.L.V.O. Chemical, physical, and sensorial characteristics of "Terrincho" ewe cheese: Changes during ripening and intravarietal comparison. *J. Dairy Sci.* **2004**, *87*, 249–257. [CrossRef]
32. Librán, C.M.; Moro, A.; Zalacain, A.; Molina, A.; Carmona, M.; Berruga, M.I. Potential application of aromatic plant extracts to prevent cheese blowing. *World J. Microbiol. Biotechnol.* **2013**, *29*, 1179–1188. [CrossRef] [PubMed]
33. Öztürk, H.İ.; Aydın, S.; Akın, N. Effect of lavender powder on microbial, physicochemical, sensory and functional properties of yoghurt. *Int. J. Second. Metab.* **2017**, *4*, 94–102. [CrossRef]
34. Kınık, O.; Kesenkaş, H.; Ergönül, P.G.; Akan, E. The effect of using pro and prebiotics on the aromatic compounds, textural and sensorial properties of symbiotic goat cheese. *Mljekarstvo* **2017**, *67*, 71–85.
35. Hayaloglu, A.A.; Karabulut, I. SPME/GC-MS characterization and comparison of volatiles of eleven varieties of Turkish cheeses. *Int. J. Food Prop.* **2013**, *16*, 1630–1653. [CrossRef]
36. Gómez-Ruiz, J.Á.; Ballesteros, C.; González Viñas, M.Á.; Cabezas, L.; Martínez-Castro, I. Relationships between volatile compounds and odour in Manchego cheese: Comparison between artisanal and industrial cheeses at different ripening times. *Lait* **2002**, *82*, 613–628. [CrossRef]
37. Resch, P.; Guthy, K. Chloroform in milk and dairy products. Part A: Analysis of chloroform using static headspace gaschromatography. *Dtsch. Leb. Rundsch.* **1999**, *95*, 418–423.
38. Van Leuven, I.; Van Caelenberg, T.; Dirinck, P. Aroma characterisation of Gouda-type cheeses. *Int. Dairy J.* **2008**, *18*, 790–800. [CrossRef]
39. Van Hoorde, K.; Van Leuven, I.; Dirinck, P.; Heyndrickx, M.; Coudijzer, K.; Vandamme, P.; Huys, G. Selection, application and monitoring of *Lactobacillus paracasei* strains as adjunct cultures in the production of Gouda-type cheeses. *Int. J. Food Microbiol.* **2010**, *144*, 226–235. [CrossRef] [PubMed]
40. Ordiales, E.; Martín, A.; Benito, M.J.; Hernández, A.; Ruiz-Moyano, S.; de Guía Córdoba, M. Role of the microbial population on the flavor of the soft-bodied cheese Torta del Casar. *J. Dairy Sci.* **2013**, *96*, 5477–5486. [CrossRef]

41. Aminifar, M.; Hamedi, M.; Emam-Djomeh, Z.; Mehdinia, A. Investigation on proteolysis and formation of volatile compounds of Lighvan cheese during ripening. *J. Food Sci. Technol.* **2014**, *51*, 2454–2462. [CrossRef] [PubMed]
42. Jung, H.J.; Ganesan, P.; Lee, S.J.; Kwak, H.S. Comparative study of flavor in cholesterol-removed Gouda cheese and Gouda cheese during ripening. *J. Dairy Sci.* **2013**, *96*, 1972–1983. [CrossRef] [PubMed]
43. Ruyssen, T.; Janssens, M.; Van Gasse, B.; Van Laere, D.; Van der Eecken, N.; De Meerleer, M.; Vermeiren, L.; Van Hoorde, K.; Martins, J.C.; Uyttendaele, M.; et al. Characterisation of Gouda cheeses based on sensory, analytical and high-field 1H nuclear magnetic resonance spectroscopy determinations: Effect of adjunct cultures and brine composition on sodium-reduced Gouda cheese. *Int. Dairy J.* **2013**, *33*, 142–152. [CrossRef]
44. Shiota, M.; Iwasawa, A.; Suzuki-Iwashima, A.; Iida, F. Effects of flavor and texture on the sensory perception of Gouda-type cheese varieties during ripening using multivariate analysis. *J. Food Sci.* **2015**, *80*, C2740–C2750. [CrossRef]
45. Jo, Y.; Benoist, D.M.; Ameerally, A.; Drake, M.A. Sensory and chemical properties of Gouda cheese. *J. Dairy Sci.* **2017**, *101*, 1967–1989. [CrossRef] [PubMed]
46. Sýkora, M.; Vítová, E.; Jeleń, H.H. Application of vacuum solid-phase microextraction for the analysis of semi-hard cheese volatiles. *Eur. Food Res. Technol.* **2020**, *246*, 573–580. [CrossRef]
47. Semeniuc, C.A.; Mandrioli, M.; Rodriguez-Estrada, M.T.; Muste, S.; Lercker, G. Thiobarbituric acid reactive substances in flavored phytosterol-enriched drinking yogurts during storage: Formation and matrix interferences. *Eur. Food Res. Technol.* **2016**, *242*, 431–439. [CrossRef]
48. Varming, C.; Petersen, M.A.; Poll, L. Comparison of isolation methods for the determination of important aroma compounds in black currant (*Ribes nigrum* L.) juice, using nasal impact frequency profiling. *J. Agric. Food Chem.* **2004**, *52*, 1647–1652. [CrossRef]
49. PubChem. Available online: https://pubchem.ncbi.nlm.nih.gov/ (accessed on 12 February 2023).
50. Zheng, L.-Y.; Sun, G.-M.; Liu, Y.-G.; Lv, L.-L.; Yang, W.-X.; Zhao, W.-F.; Wei, C.-B. Aroma volatile compounds from two fresh pineapple varieties in China. *Int. J. Mol. Sci.* **2012**, *13*, 7383–7392. [CrossRef]
51. Lacroix, N.; St-Gelais, D.; Champagne, C.P.; Fortin, J.; Vuillemard, J.-C. Characterization of aromatic properties of old-style cheese starters. *J. Dairy Sci.* **2010**, *93*, 3427–3441. [CrossRef]
52. O'Riordan, P.J.; Delahunty, C.M. Comparison of volatile compounds released during the consumption of Cheddar cheese with compounds extracted by vacuum distillation using gas chromatography–olfactometry. *Flavour. Fragr. J.* **2001**, *16*, 425–434. [CrossRef]
53. Karagul Yuceer, Y.; Tuncel, B.; Guneser, O.; Engin, B.; Isleten, M.; Yasar, K.; Mendes, M. Characterization of aroma-active compounds, sensory properties, and proteolysis in Ezine cheese. *J. Dairy Sci.* **2009**, *92*, 4146–4157. [CrossRef] [PubMed]
54. Xiao, Z.; Li, Q.; Niu, Y.; Zhou, X.; Liu, J.; Xu, Y.; Xu, Z. Odor-active compounds of different lavender essential oils and their correlation with sensory attributes. *Ind. Crop. Prod.* **2017**, *108*, 748–755. [CrossRef]
55. Gómez-Míguez, M.J.; Cacho, J.F.; Ferreira, V.; Vicario, I.M.; Heredia, F.J. Volatile components of Zalema white wines. *Food Chem.* **2007**, *100*, 1464–1473. [CrossRef]
56. Song, H.S.; Sawamura, M.; Ito, T.; Ido, A.; Ukeda, H. Quantitative determination and characteristic flavour of daidai (*Citrus aurantium* L. var. *cyathifera* Y. Tanaka) peel oil. *Flavour. Fragr. J.* **2000**, *15*, 323–328. [CrossRef]
57. Senoussi, A.; Rapisarda, T.; Schadt, I.; Chenchouni, H.; Saoudi, Z.; Senoussi, S.; Zidoune, O.A.; Zidoune, M.N.; Carpino, S. Formation and dynamics of aroma compounds during manufacturing-ripening of Bouhezza goat cheese. *Int. Dairy J.* **2022**, *129*, 105349. [CrossRef]
58. Cho, I.H.; Namgung, H.-J.; Choi, H.-K.; Kim, Y.-S. Volatiles and key odorants in the pileus and stipe of pine-mushroom (*Tricholoma matsutake* Sing.). *Food Chem.* **2008**, *106*, 71–76. [CrossRef]
59. Semeniuc, C.A.; Rotar, A.; Stan, L.; Pop, C.R.; Socaci, S.; Mireșan, V.; Muste, S. Characterization of pine bud syrup and its effect on physicochemical and sensory properties of kefir. *CyTA-J. Food* **2016**, *14*, 213–218. [CrossRef]
60. Tsouli Sarhir, S.; Amanpour, A.; Bouseta, A.; Selli, S. Potent odorants and sensory characteristics of the soft white cheese "Jben": Effect of salt content. *Flavour. Fragr. J.* **2022**, *37*, 243–253. [CrossRef]
61. Guo, X.; Ho, C.-T.; Wan, X.; Zhu, H.; Liu, Q.; Wen, Z. Changes of volatile compounds and odor profiles in Wuyi rock tea during processing. *Food Chem.* **2021**, *341*, 128230. [CrossRef]
62. Zhu, M.; Sun, J.; Zhao, H.; Wu, F.; Xue, X.; Wu, L.; Cao, W. Volatile compounds of five types of unifloral honey in Northwest China: Correlation with aroma and floral origin based on HS-SPME/GC–MS combined with chemometrics. *Food Chem.* **2022**, *384*, 132461. [CrossRef] [PubMed]
63. Guo, X.; Wang, P. Aroma characteristics of lavender extract and essential oil from *Lavandula angustifolia* Mill. *Molecules* **2020**, *25*, 5541. [CrossRef] [PubMed]
64. Bail, S.; Buchbauer, G.; Schmidt, E.; Wanner, J.; Slavchev, A.; Stoyanova, A.; Denkova, Z.; Geissler, M.; Jirovetz, L. GC-MS-analysis, antimicrobial activities and olfactory evaluation of essential davana (*Artemisia pallens* Wall. ex DC) oil from India. *Nat. Prod. Commun.* **2008**, *3*, 1057–1062. [CrossRef]

Disclaimer/Publisher's Note: The statements, opinions and data contained in all publications are solely those of the individual author(s) and contributor(s) and not of MDPI and/or the editor(s). MDPI and/or the editor(s) disclaim responsibility for any injury to people or property resulting from any ideas, methods, instructions or products referred to in the content.

Review

Combination of Milk and Plant Proteins to Develop Novel Food Systems: What Are the Limits?

Luis Gustavo Lima Nascimento [1,2,†], Davide Odelli [1,†], Antônio Fernandes de Carvalho [1,*], Evandro Martins [1], Guillaume Delaplace [2], Paulo Peres de sá Peixoto Júnior [2], Naaman Francisco Nogueira Silva [3] and Federico Casanova [4,*]

[1] Departamento de Tecnologia de Alimentos, Universidade Federal de Viçosa (UFV), Viçosa 36570-900, MG, Brazil; luisgusta.ln@gmail.com (L.G.L.N.); davide.odelli@ufv.br (D.O.); evandromartins@ufv.br (E.M.)

[2] Laboratoire de Processus aux Interfaces et Hygiène des Matériaux, INRAE, 59009 Lille, France; guillaume.delaplace@inrae.fr (G.D.); paulo.peres-de-sa-peixoto-junior@inrae.fr (P.P.d.s.P.J.)

[3] Center of Natural Sciences, Federal University of São Carlos (UFSCar), Campus de Lagoa do Sino, Buri 18290-000, SP, Brazil; naaman.nogueira@ufscar.br

[4] Research Group for Food Production Engineering, National Food Institute, Technical University of Denmark, Søltofts Plads, 2800 Kongens Lyngby, Denmark

* Correspondence: antoniofernandes@ufv.br (A.F.d.C.); feca@food.dtu.dk (F.C.)

† These authors contributed equally to this work.

Abstract: In the context of a diet transition from animal protein to plant protein, both for sustainable and healthy scopes, innovative plant-based foods are being developing. A combination with milk proteins has been proposed as a strategy to overcome the scarce functional and sensorial properties of plant proteins. Based on this mixture were designed several colloidal systems such as suspensions, gels, emulsions, and foams which can be found in many food products. This review aims to give profound scientific insights on the challenges and opportunities of developing such binary systems which could soon open a new market category in the food industry. The recent trends in the formulation of each colloidal system, as well as their limits and advantages are here considered. Lastly, new approaches to improve the coexistence of both milk and plant proteins and how they affect the sensorial profile of food products are discussed.

Keywords: milk proteins; plant proteins; mixed systems; colloidal properties; innovative foods; sensorial properties

1. Introduction

The human population is continuously growing, and it is estimated to reach 9.7 billion people in 2050, which will naturally increase the demand for animal protein for human nourishment (United Nations, 2015). A report conducted by Poore and Nemecek (2018) [1] considered the environmental footprint of the production of 90% of global proteins based on land use, freshwater usage, GHG emissions, and chemical emissions in soil and water. The authors showed that proteins from animal sources (meat, dairy, eggs, and aquaculture) use ~83% of the world's available farmland and are responsible for 56–58% of general emissions, providing, in the end, only 37% of food protein supply [1]. Thus, considering the crescent demand for proteins and the deployment of the Glasgow Climate Pact (UNFCCC, 2021), the development of sustainable production systems to obtain alternative protein sources is required.

Inside this scenario, plant proteins are good candidates to partially substitute animal proteins in food since their production process has been associated with low cost and low greenhouse effect [2]. Beyond this, plant proteins are less allergenic than animal counterparts [3]. The consumer's increasing awareness of healthy and sustainable food

products has recently enhanced the demand for plant-based proteins as food ingredients worldwide, and only in the United States, 83% of North American consumers are adding plant-based foods into their diets to improve health (NDC, 2019). Proteins from various vegetables have already been studied and employed as animal protein replacers in meat and dairy analog products [4–6].

Many researchers have highlighted the positive nutritional aspects of this kind of protein which, among others, include a reduced glycemic index, reduced incidence/probability of developing cardiovascular diseases, obesity, and metabolic syndromes; conditions that reduce the overall all-cause mortality [7–9]. Therefore, the incorporation of plant proteins is not only a necessity but also a consumer tendency to maintain its well-being and healthy condition [10].

Despite the advantages in the use of plant proteins for human nutrition, their pronounced taste and poor solubility limit their applicability in the food industry [11]. To overcome this techno-functional drawback, association with animal proteins, such as milk proteins, can be an interesting strategy to increase the use of plant proteins with low compromising of food sensorial aspects. Among the potential animal proteins that can be combined with the plant ones, milk proteins stand out due to high productions, easy isolation and purification by membrane filtration systems, stability in the dry form, tecno-functionality in dairy and non-dairy products and good acceptability for consumers [12–14]. The milk and plant proteins association can improve the sensorial and nutritional aspects of foods, increase the intake of plant proteins in processed foods, reduce costs of ingredients, decrease phase separation and/or syneresis in dairy gels [15,16]. It is desired that plant proteins addition in milk-based foods can improve some properties of the system, however, this addition can alter significatively the characteristics of products, which could result in consumer rejection.

Thus, the impact of this association as well as the optimization of protein interactions must be better understood for the development of innovative products with sensory characteristics suited to the needs of consumers. In recent years, consistent research has been delivered to study these associations in different colloidal states such as dispersions, foams, gels and emulsions [17]. Indeed, these interactions depend on several aspects such as type of proteins, protein ratio, pH, ionic strength, presence of salts; additionally, industrial processes such as temperature, acids and enzymes can cause protein modification [18].

In this context, this review aims to describe the scientific advances regarding how the mixing of milk and plant protein change the features of protein systems and how these new characteristics can be useful in the formulation of foods with new textural and sensorial aspects. Moreover, innovative approaches to modify protein techno-functionality will be discussed here as a possible way to improve this combined system, limiting their drawbacks and promoting their application in the food industry.

2. Milk Proteins

The main milk components are listed in Table 1 [19]. The raw fluid milk can be transformed into a variety of food products such as ice cream, concentrated milk, milk powders, yogurt, cheese, etc. These transformations come mainly from manipulating the structure and organization of milk proteins, which influence taste, appearance, texture, color, and stability of these products [20]. The milk protein fraction can be grouped into two main classes: caseins, which are thermal resistant and have an isoelectric point in pH around 4.6, and whey proteins, which are soluble at their isoelectric point (~pH 4.8–5.0) but are precipitated by increasing the ionic strength and temperature [21,22].

Table 1. Components of raw milk.

Component	Proportion (%)
Water	85–87
Lipids	3.8–5.5
Lactose	4.8–5.0
Proteins	2.9–3.5

The structures and functional properties of these two main groups of milk protein will be better discussed in the following paragraphs.

2.1. Caseins

Caseins compose about 80% of total milk proteins and are represented by four main fractions: αs1-, αs2-, β-, and k- caseins in a molar ratio of 11:3:10:4, respectively [20]. In natural milk conditions, these fractions interact with each other by hydrophobic and electrostatic interactions, and calcium phosphate nanoclusters forming supramolecular structures named casein micelles (CMs) (Figure 1). κ-casein fraction contributes to the electrostatic and steric repulsion among CMs and is the main casein responsible for stabilizing and maintaining CMs in suspensions [23].

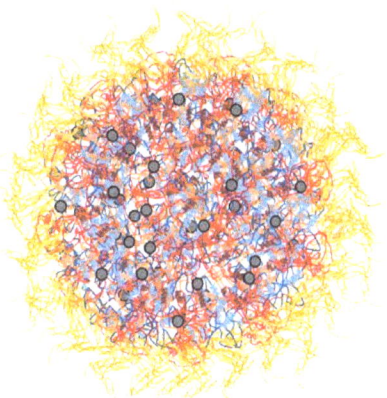

Figure 1. Casein micelle structure and its components: κ-casein (**yellow**); α-casein (**blue**); β-casein (**red**); calcium phosphate nanoclusters (**grey dots**).

Commercially, the separation of caseins from other constituents of milk, occurs by isoelectric precipitation, ultra and microfiltration, and rennet coagulation [24,25]. Acid caseins can be obtained by adjusting milk pH to 4.6, after that, a centrifugation step can separate the fractions. Due to their spherical structure, with large particle sizes, caseins obtained by acidification are insoluble in water and generally requires neutralization for their solubilization [26]. To overcome this problem, in food formulation caseins are typically applied in the form of sodium or calcium caseinates. They are produced from CMs by the addition of NaOH or CaOH to skimmed milk. The resulting caseinates are more soluble and have better water holding capacity (WHC) compared to native CMs thanks to their non-spherical shape and improved hydration of the particles, which confer small particle sizes of about ~20 nm [27]. Caseins in different configurations, i.e., CMs, caseinates, acid caseins, and rennet casein (caseins enzymatically precipitated), can be incorporated as food ingredients in a variety of food products such as waffles, cake mixtures, bread, cream liqueurs, coffee whiteners, processed meat and fish products and also dairy products such as cheese analogs and ice cream yogurt, among others [25]. Despite their nutritional features, the main reasons for caseins applications are their suitable functional properties. Indeed, thanks to their exceptional surface activity, emulsifying and self-assembly properties, and gelation and water binding capacities, caseins and caseinates are largely employed in food products as emulsifiers and foaming agents, fat replacers, and texture and thickening improvers [28]. These properties derive from the ability of caseins to be modified and form different colloidal systems such as dispersions, emulsions, foams, and gels [29].

2.2. Whey Proteins

In the past, whey was considered a waste created by the cheese and caseins production but the panorama has changed since then, mainly due to the discovery of its nutritional and techno-functional properties, which boost whey applications in the food industry [30]. Whey proteins account for approximately 20% of milk proteins and are composed mainly of β-lactoglobulin (60% w/w) and α-lactalbumin (20% w/w) with lower contents of immunoglobulins (10% w/w), bovine serum albumin (3% w/w), and lactoferrin (<0.1% w/w) [31]. Contrarily to the caseins, whey proteins are globular proteins with well-defined secondary, tertiary, and quaternary structures that depend on medium conditions such as pH, ionic strength, and temperature, but can also be modified by different treatments such as pressure, ultrasounds, pulsed electric field, and enzymatic reactions [22]. When the whey proteins are heated above their denaturation temperature, the molecular structure is unfolded and the formation of new hydrophobic interactions, hydrogen bonds, and disulfide bounds is favored [32].

Whey proteins are usually obtained by ultra and microfiltration. This technology allows the use of low temperatures and the absence of chemicals or enzymes added to the milk, which results in the purification of whey proteins with similar structure to their natural conformation, thus without interfering with their physicochemical properties. In the food industry, the main products obtained from whey processing are whey protein concentrates (WPCs) and whey protein isolates (WPIs). Those products can be used as ingredients in food formulations due to their ability to strengthen food gels and/or stabilize emulsions and foams. Additionally, WPCs and WPIs can be directly consumed by the final consumer after powder resuspension, giving protein solubility a paramount role for consumer acceptance [30].

3. Plant Proteins

3.1. Sources

Plant proteins are characterized by a different structure and morphology than animal proteins, which highly influences their functionality [33]. During their evolution history, plants have developed the ability to biosynthesize a large number of proteins for different purposes and can be generally classified into two different groups: "metabolic" and "storage" proteins. The first ones represent crucial proteins for the development of the plant, while the second ones consist of the reservoir of vital amino acids to sustain plant life [34]. These groups represent an important nutritional source for both humans and livestock or animal feed thanks to the presence of essential amino acids which can satisfy their nutritional requirements [35,36]. Plant proteins are generally obtained by dry or wet extraction methods as co- or by-products from various starting materials of the oil and starch extraction industries. More than 30 plant protein sources are currently used in food formulation, and overall, they can be organized into three general groups: legumes, cereals, and oilseeds [33] (Figure 2). Among the legumes, soybean and green peas are the most employed nowadays, but also proteins from other beans such as fava beans, chickpeas, and lentils are commonly requested by the food industry [37,38]. Regarding the cereals group, the main sources of proteins are provided by wheat gluten, corn zein, and rice, while proteins from oilseeds are separated from the oil, starch, and fibers of products such as canola, sunflower, peanut, rapeseed, and flaxseed [39–42].

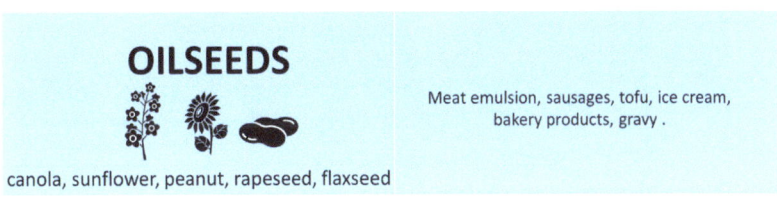

Figure 2. Main plant-based sources and some of their applications. Adapted from Akharume et al., (2021) [43].

3.2. Structure and Functionality

Proteins' polypeptide composition, in terms of amino acids and functional groups, and their rearranged spatial structure greatly influence their physicochemical properties and functionalities [44]. Plant proteins display a specific morphology when they are biosynthesized, which allows them to express their biological functions. The natural 3D structure is obtained through the folding and interaction of the protein-peptide chains, driven by several forces such as van der Waals and hydrophobic attractions [34]. Hydrogen bonds, disulfide bonds, electrostatic and steric attraction/repulsion, torsional angles, and solvent interactions also participate in the morphology of the amino acid chains within a protein, but the same interactions can also occur within different protein molecules. For this reason, it is reasonable to believe that proteins physiologically exist in different states, which can range from monomers to oligomers and, at a certain concentration, to assemblies and aggregates, all characterized by this kind of natural forces [15,45]. It is good to know that any kind of process applied to the raw vegetable material is possibly able to interfere with these forces and thus influence biomolecules' native structures and functionalities.

For example, the employed extraction methodology, purification, and any other processing method can largely modify protein three-dimensional organization. It has been proven that proteins extracted from the same source with different methodology may present a greater functionality variation than proteins extracted from different sources with the same method [46]. For instance, with a dry extraction, proteins tend to maintain their native organization, while with a wet extraction, different solvents are adopted such as water or an alkali, acid, or a salt solution which interact with the native proteins causing a potential disruption and rearrangement of their structures [15]. Therefore, it is fundamental to adapt all of these processes to obtain the desired characteristics of isolated proteins and be able to design specific food products.

When added as functional ingredients, proteins exhibit many roles in food matrices influencing for example their texture and structure but also their organoleptic properties such as flavor, color, odor, and appearance. Indeed, thanks to their amphipathic nature they can interact with other macronutrients such as carbohydrates and fats but also with water and air, working as gelling and thickening agents, stabilizers of foams and emulsions, film-forming polymers, and binding agents for fat and water, which all together represent colloidal properties [43]. Moreover, they could also have biological properties exhibiting antimicrobial and antioxidant effects [11]. Some examples of plant protein sources and their functional utilization in food formulation are resumed in Table 2.

Table 2. Plant proteins and their functionalities based on native physicochemical properties.

Plant Source	Physicochemical Properties	Functionality
Soy, almond, rice	Hydrophilicity; surface charge; hydrogen bonding	Solubility
Soy, pea, lentils, beans	Aggregative behavior after thermal and pH denaturation; electrostatic and hydrophobic interactions; disulfide bonding	Gelling
Soy, pea, faba, sunflower, pumpkin	Surface tension; interfacial film forming ability; amphipathic behavior	Emulsifying
Potato, pea, lentils, chickpea	Surface tension; interfacial film forming ability; amphipathic behavior	Foaming

4. Protein-Protein Interactions to Modify Food Techno-Functional Properties and Colloidal Properties

Protein techno-functionality can be described as the protein behavior during food processing and in a food system, a behavior which is strictly based on protein physicochemical properties, without necessarily including its biological and nutritional activities [47]. For example, caseins' biological function is not to make dairy products, but their colloidal properties are responsible for several interactions that play a fundamental role in cheese and yogurt manufacturing. Physicochemical interactions such as electrostatic attraction and repulsion, hydrogen bonds, hydrophobic interactions, and disulfide bonds affect protein colloidal properties. Thanks to these properties, proteins interact with each other and other ingredients in food formulations determining their overall structure and colloidal state. Colloidal systems such as dispersions, gels, emulsions, and foams containing proteins are, thus, extremely influenced by their techno-functional properties, which can further be modulated during food processing, when other ingredients are added into the formulation and/or physical, chemical, or enzymatic treatments are employed. As complex systems, foods are usually composed of more than one colloidal state; therefore, the knowledge of how proteins behave in each of these colloidal states is precious to design any food formulation and can be used as a tool to predict and tailor the final product features.

The following paragraphs will specifically focus on the protein-protein interactions between milk and plant proteins, describing their role and characteristic in any type of colloidal state, in order to obtain a general consciousness of their relationship for the development of innovative food systems.

4.1. Milk: Plant Proteins Dispersions

A dispersion is a colloidal system where a solid material is dispersed into a liquid, where the solid is the dispersed phase, and the liquid is the continuous phase [48]. Thus, the formulation of beverages arises as to the direct application of the knowledge gained in these studies. Additionally, dispersions must be made before other systems, i.e., gels, emulsions, and foams, and the type of interactions, as well as the dispersion properties as viscosity, particle sizes, and solubility, affect the final product [49]. By the dispersion's definition, solubility is the most important factor and can be understood as the resultant of the protein-protein and protein-solvent interactions [50]. The challenge increases when a high percentage of protein dispersed is required, as observed in high-protein beverages, mainly designed for the market of sports drinks [51]. The solubility of plant proteins, in general, is lower than milk proteins and can even be worse when high temperatures are used for the protein extraction. In mixed systems, the presence of a different protein can impact the overall system solubility. Ben-Harb et al., 2018 (Ben-Harb et al., 2018), observed an antagonistic effect in the solubility of mixed pea/milk proteins, where the mixture of pea and milk proteins was less soluble compared to each protein individually. However, other treatments can improve the solubility of mixed systems as demonstrated by Wang et al., (2019) [52], with the application of a pH-cycle technique. By variation of the dispersion pH from 12.0 to 7.0, the authors observed an increase in proximately 30 times of rice protein solubility when it was associated with WPI (1:1) compared to the pure rice suspension. The main reason for the observed phenomenon was attributed to proteins complexation, driven mainly by the formation of hydrogen bonds. Using the same method, Wang et al., (2018) [53], observed an increase in proximately 52 times in the solubility of rice protein when combined with sodium caseinate in the ratio 1:0.01. In addition, the increase in sodium caseinate content up to a 1:1 ratio did not significantly change the solubility of the systems, while when it was reduced, a lower solubility of rice proteins was detected.

The viscosity and the particle size of the proteins in dispersion also change regarding protein combination ratios, which directly impact the process parameters. It was observed by Singh et al., (2019) [54], that the mixture of milk protein concentrate (MPC) and soy protein hydrolytes (SPH) resulted in dispersions with higher viscosities when compared with the systems formed only by one type of protein at equal protein concentrations. The coagulation time of the systems also was impacted, SPH does not coagulate when exposed to 145 °C for 15 min, and the MPC took 14 min to present the first sign of coagulation. After the mix, depending on the ratio, coagulation time decreased to less than 2 min. These results, which are summarized in Table 3, show that the general processing carried out in the food industry for systems with only one protein source cannot be directly applied to mixed systems.

Table 3. Summary of the key results obtained for mixed protein dispersions.

Mixed Dispersions	Functionalities	Reference
Milk/Pea	Antagonistic effect on protein solubility	Ben-Harb et al., (2018) [18]
Whey/Rice	Improved solubility with pH shift	Wang et al., (2019) [52]
Caseinates/Rice	Improved solubility with pH shift	Wang et al., (2018) [53]
Milk/Soy	Higher viscosities of the dispersions with decreased coagulation	Singh et al., (2019) [54]

Therefore, it can be said that, in general, the presence of a mixture of proteins in a dispersion may negatively affect their solubility due to complex intermolecular interaction between the proteins and the solvent, causing higher molecular aggregates and precipitates. However, different treatments could be employed in order to modify the component structure and improve its overall solubility. For example, physical treatments such as homogenization, ultrafiltration, or ultrasounds have been applied to affect plant proteins particle sizes and second structures, generally enhancing their solubility [55]. Saricaoglu

(2020) [56], used high-pressure homogenization to significantly reduce particle size of lentil proteins and increase their solubility, as well as influence their rheological properties. In the same context, Wang et al., (2020) [57], employed an ultrasound-assisted extraction method for pea protein isolate, which resulted in a partial protein unfolding and smaller particle size that significantly improved their dispersion. Additionally, chemical modification, such as phosphorylation, enzymatic hydrolysis, and biopolymer complexation, have been applied to modify protein functional groups, structure and viscosity in order to improve the interaction with the solvent [58–60].

Thus, solubility, viscosity and particle size, and how they are affected by different treatments have to be considered when a mixed protein system is designed in order to evaluate its stability and sensorial characteristics.

4.2. Milk: Plant Proteins Gels

A gel can be defined as a colloidal system where long thread-like molecules cross-link, chemically or physically, and/or entangle to such an extent that a continuous three-dimensional network is formed [61]. In rheological studies, a system where the elastic modulus (G') is higher than its viscous modulus (G'') is defined as a gel, thus resembling a solid-like material [62]. In protein systems, gelation properties depend on intrinsic and extrinsic factors, such as amino acid composition, presence of disulfide bonds, hydrophobic and electrostatic interactions, protein concentration, ionic strength, temperature, pressure, and pH [32]. Particularly in a mixed system, the type of proteins, their concentration, and ratios affect the final gel properties. For example, the minimal protein concentration to achieve thermal and acid gelation was determined for mixed pea and β-lactoglobulin systems [63]. In the pure systems, the minimal concentration required for thermal gelation was 7 and 5% for pea proteins and β-lactoglobulin, respectively. The mixed systems minimal thermal gelation varied according to the protein ratio, 5% being the least concentration between the mixed samples for 1:4 pea: β-lactoglobulin protein ratio. Smaller values were found by Wong et al., (2013) [64], where the least gelation concentrations diminished when different protein rates were mixed, i.e., 3% of total protein concentration is required to form whey and pea gels. However, because of the synergistic enhancement of 2:8 pea/whey, 2% of total protein was necessary for gelation to occur. Additionally, the methodology used to obtain the gel is responsible for protein structures modification and their intermolecular interactions which influence the final features of the gel product.

4.2.1. Heat-Induced Milk: Plant Proteins Gels

Protein gelation can occur when a sufficient amount of energy, in the form of heat, is applied to a system. Generally, at high temperatures, globular proteins unfold, exposing their hydrophobic residues that are hidden in the natural conformation. Once exposed, the amino acids can associate by hydrophobic interaction, Van-der-walls forces, and hydrogen bonds or can associate more strongly with disulfide bonds [46]. These new interactions between the protein chains lead to aggregation and a three-dimensional structure starts to form. In milk processing, heat treatment is used to promote aggregation between whey proteins and CMs, which in turn leads to stiffer gels after acidification. Thus, when plant proteins are added to milk, the first question that appears is if plant proteins can aggregate with CMs as whey proteins. Some authors have investigated the interactions between CMs and pea and soy proteins after heat treatment [65–68].

The common approach to access this information is using small-amplitude oscillation shear (SAOS) rheology technique to follow G' during heat application. Silva et al., (2018) [65] studied the gelation profile of suspensions composed of CMs alone or in the presence of whey, pea, or soy protein at pH 5.8. As expected, a reinforcement of CMs gels was observed in the presence of whey protein, which was attributed to their co-aggregation. However, no reinforcement of the gel was observed even at high temperatures for both pea and soy protein, which suggests the absence of co-aggregation. Additionally, the protein ratios, i.e., the proportion between CMs and plant proteins, or protein concentrations, did not lead to their co-aggregation. In mixed systems where both CMs and plant proteins can form gels, the gel features are driven mainly by the protein that is in high concentration and the presence of two independent three-dimensional structures leads to less stiff gels [66]. Indeed, Message et al., (2017) [68], studied the aggregation patterns of CMs and two fractions of pea protein, i.e., legumin, and vicilin, at pH 7.2 using reducing and non-reducing electrophoresis, DSC, and liquid chromatography. They concluded that during heat treatment, denaturation of both pea protein fractions took place, followed by the formation of protein aggregates. This aggregation occurs differently in each protein fraction, with the formation of disulfide bonds for legumin and non-covalent interactions for vicilin. However, the CMs did not participate in aggregation.

Despite the absence of co-aggregation between CMs and pea and soy proteins, the presence of the plant proteins impacts the availability of free calcium in the mixed systems, which seems to increase the CMs gelation temperature (Tgel) [65]. Tgel is defined as the temperature where the sol-gel transition occurs, and in the case of CMs suspensions, it is affected by free calcium concentration in the medium [69]. As the temperature increases the calcium solubility decreases, which leads to calcium precipitation on the CMs surface. As consequence, CMs destabilization occurs and ultimately leads to aggregation [70]. Thus, the less calcium available to precipitate, the harder it will be for aggregation to occur. As observed by Silva et al., (2018), pea and soy protein can bind calcium from the medium, where soy proteins bind more calcium than pea proteins, which resulted in higher gelation temperature of CMs in the systems where soy proteins were present. Thus, the authors argued that these plant proteins work as a chelating agent in mixed systems, increasing the heat stability of mixed systems in comparison to the suspensions of pure/isolated/native CMs.

The studies of how the plant protein specifically interacts with CMs are important to understand the potential application of mixed systems in the food industry. Ben-Harb et al., (2018) [18], studied heat-induced gelation in mixed milk/pea suspensions at pH 6.33. They found that 14.8% (w/w) mixed systems gel in protein ratios of 1:1 showed G' as high as pea protein alone, while the sample containing solely milk fractions formed a weak gel at 14.8% (w/w) and did not gel at 7.4% (w/w) concentration. The data indicate that pea proteins were responsible for gel structuration, since CMs do not form gels when heated at a pH as high as 6.33 [69]. Nevertheless, pea proteins could not be the unique responsible for gel structure since the mixed systems only with 7.4% (w/w) of pea protein showed gel stiffness as high as the 14.8% (w/w) pea gels. Thus, interactions between whey and pea proteins may take place. Indeed, Wong et al., (2013) [64], studied the gel formation achieved by heating pea and whey protein in different rations, concentrations, and pH values. The best synergistic enhancement in G' was achieved by 16% (w/w) total protein concentration, 2:8 pea/whey ratio at pH 6.0. In general, small amounts of pea protein increased the gel stiffness, but it varies depending on pH and protein concentration. Each protein has its isoelectric point and solubility, thus a pH value that promotes a similar aggregation profile of both proteins leads to the formation of a more homogeneous network. Additionally, the decrease in the electrostatic repulsion caused by pHs close to the protein isoelectric point leads to an increase in protein-protein interaction [64].

The mechanism of the interaction between β-lactoglobulin and pea after heat treatment at pH 7.2 was hypothesized by Chihi et al., (2016) [71]. The authors suggested that the unfolding of both protein types after heating exposed thiol groups and previously buried hydrophobic groups. In this way, the proteins started to self-aggregate, and aggregations between β-lactoglobulin and legumin potentially occur by disulfide bonds. Then, those small protein aggregates interact mainly by hydrophobic and/or electrostatic interactions, which increase their sizes. Despite the differences between soy and pea proteins, it is reasonable to think that the interactions with whey proteins for both plant proteins are similar. For instance, the formation of disulfide bonds after heat treatment of 6% (w/w) soy-whey protein mixed system has been proposed by Roesh and Corredig, (2005) [72]. The authors showed that when high amounts of whey are present, the incorporation of soy proteins occurs, and the formed aggregate is composed of both proteins. However, the presence of low amounts of plant proteins also led to the formation of aggregates formed solely by whey proteins. These diverse profiles led to differences in the gel network, where the gels formed with higher amount of whey protein showed a more homogeneous network and higher G'. The same feature of mixed soy-whey protein gels was observed, even at 12 and 16% (w/w) total protein concentration. Thus, in mixed systems, the whey protein is responsible for gel formation, while soy proteins appear as filler material within the gel structure [73]. To resume, the incorporation of soy protein in whey gels decreases the G' and changes the network structure. Additionally, the modeling the soy/whey protein ratio allows the creation of 16% (w/w) protein gels with the same strength of 6% (w/w) [73].

In conclusion, it can be said that mixed-system heat-induced gel properties mainly depend on intrinsic factors such as type of proteins, their concentration, and ratio, and on environmental factors such as pH, ionic strength, and temperature applied for gel formation. Thus, at an industrial level, the strict control of all of these parameters can allow the design of products with desired techno-functional properties. In particular, the employment of different protein fractions may allow to obtain heat induced gel with the advantage of using less quantity of dairy protein as well as increasing the employment and consumption of plant-based proteins, obtaining similar gel structures to heat induced gel made out of animal proteins only.

4.2.2. Acid Induced Milk: Plant Proteins Gels

The acid gelation is induced by pH modification toward the isoelectric point of the proteins. During the pH decreasing of protein suspensions, the electrostatic and steric repulsion between the proteins is reduced, which causes approximation between them, formation of new interactions, aggregation, and ultimately the formation of a continuous three-dimensional network [74]. In milk, the solubilization of calcium phosphate cannot be neglected once it causes protein rearrangement of the gel matrix. Acid gelation is widely applied in the dairy industry, mainly in the production of fermented milks and cheeses to develop desirable textural properties [75].

In mixed protein systems, the difference in protein origins and properties interfere in gel formation during acidification. For example, the pH, where the gelation starts for each protein, impacts directly the structure of the gel network [76]. In the acid gelation of pea and milk proteins, Ben-Harb et al., (2018) [18], observed that pea proteins play a major role in the first stages of gel formation because they reach the isoelectric point at higher rates, due to their lower buffer capacity. Chihi et al., (2018) [63], showed that the rates of acidification were equal for single and mixed systems composed of β-lactoglobulin and pea protein at 4% protein concentration. According to the authors, pea protein gelation occurred after 24 min of acidification in pH 6.6, while β-lactoglobulin gelation occurred 58 min after acidification at pH 5.7 in single systems. Thus, the increase in β-lactoglobulin in the mixed systems resulted in a decrease in gelation pH. The same was observed for an acid gel formed by mixing soy and cow's milk at 4.5% total protein [76]. Soy gelation pH is around 6.0, while milk did not form gels by acidification in pH higher than 5.6. Thus, a gel network formed by the mixture of soy and milk in pH above 5.6 will be composed only of soy protein. In addition, the presence of milk proteins interfered in the soy network formation. If a rennet treatment is applied, the milk gelation pH rises to around 6.1; in this way, the formed gel network counts either with soy protein or milk protein contributions. The gelation of both proteins occurring at the same time increased G' and formed a more homogeneous network compared to cow's milk not treated with chymosin. However, the gels formed with only cow's milk or soy milk presented higher G' compared to the mixed systems. It indicates that there is no co-aggregation of the proteins and a network formation interferes in the other [76].

The presence of plant proteins in dairy products requires the evaluation of the changes during the production process and the interferences caused by the presence of lactic acid bacteria (LABs). Yousseef et al., (2016) [77], developed pea-milk yogurts with several

LABs. In those systems, altering the pea/milk protein ratio from 0:100 to 40:60 at 4.5% total protein led to faster gel formation and increased the product acidity. The same occurred with the addition of lentil flour [37]. This phenomenon was explained by the lower buffer capacity in the systems with less casein content. Another effect after increasing pea protein amount was the increase in gels syneresis, which was related to the differences in gel network formation. The presence of pea protein decreased the firmness of the mixed gels compared to milk gels. It was suggested that the pea proteins prevented the formation of most homogeneous casein networks, thus weakening the resulting gel. This behavior highlights the possibility to develop gelled products of similar firmness with higher protein content using vegetable proteins.

The supplementation of milk with milk protein powder to increase the solid content, aiming at the development of a more elastic gel, is usual in the dairy industry. The substitution of milk protein powder for lentil flour as a source of solids was evaluated by Zare et al., (2011) [78]. The syneresis of yogurts supplemented with 3% lentil flower was similar to the samples with 3% milk powder. However, the syneresis increased when a lower quantity of lentil flour was added (1 and 2%). The increase in protein content in the samples, with the addition of more solids, lead to more water retention in the gel matrix compared to control samples. After 28 days of storage, the samples containing lentil flour presented G′ comparable to the samples supplemented with milk proteins, showing the potential of replacement of milk proteins for plant proteins.

The formulation of an acid-induced gel system does not exclude the application of a pre-treatment before gelling. Indeed, thermal treatment of milk is generally applied before fermentation in yogurt production to increase the stiffness of the final product. Pre-treatments, such as heat, are useful in modifying the proteins and the types of interactions between them, changing the building blocks of the acid gel. These building blocks are the foundation of the gel and their size and organization can be modulated by modifying the processing parameters such as pH, protein ratio, and the order of heat treatment, i.e., heating proteins separated with posterior mixing or mix the proteins with posterior heating [68]. The effect of the pre-heat treatment in the gel composed of sodium caseinate (CasNa), an important milk ingredient used in the dairy industry with several applications, and soy proteins were studied by Martin et al., (2016) [79]. The pH of the suspension during heat impacted the acid gel structure. In general, the heat treatment in lower pHs lead to more fragile and coarse gels. Additionally, the addition of soy protein without any heat treatment resulted in a gel with a coarser microstructure. Concerning the processing order, the heat treatment of only soy protein with posterior mixing with CasNa lowered the mechanical properties of the gel in comparison with CasNa alone. However, mixing the proteins before heat treatment increased the gel's mechanical properties to a value close to CasNa alone. Similar results were observed by Chihi et al., (2018) [63], studying mixed β-lactoglobulin: pea protein gels. The authors showed that when the proteins were heated separately and then mixed, the gels show a more open and disordered structure with lower WHC compared to the sample in which the proteins were heated together before acidification. However, in a study of casein-pea gels, Message et al., (2017) [68], observed that heating the proteins separately with their post-treatment mixture produced more elastic gels, explaining that the type of pea protein fraction utilized represents a critical factor for the final gel stiffness.

Thus, since each plant protein seems to have different acid gelation properties influenced by the presence of several fractions, by the specific isoelectric point, and by the characteristic structure and functional groups of the molecule, a general rule for all mixed milk-plant protein systems cannot be established. In other words, the nature of the proteins involved changes completely the characteristics of the acid gel. Even though in fermented products, plant proteins addition fastened the gelation and increased the acidity, when pea, soy and lentils proteins were added into acid dairy gels, their presence sterically inhibited the formation of a strong network, resulting in an increased syneresis and decreased gel firmness. However, the employment of heat treatment before acidification has shown relatively positive effects on gelling properties, opening the possibility of the application of a preliminary treatment to enhance proteins functionality. In this context, up to date, there is a lack of comprehensive information on the effects of different protein modification approaches to improve acid composite milk: plant protein gel, which needs to be carefully addressed for future perspectives in the food industry.

4.2.3. Gelation Induced by Other Methods

Other physical methods employed to modify protein conformation, and thus structuring food, is the use of enzymatic reactions and/or ultrasounds. While enzymatic gelation has been known for many years in the dairy industry, i.e., the use of rennet in cheese-making processes, the utilization of ultrasound treatment is increasing in the food industry as a way to develop products with new features. Opposing the results reported by McCann et al., (2018) [73], who used heat treatment, Cui et al., (2020) [80], developed a whey-soy-based gel with higher hardness compared to the gels produced from the sole protein sources. However, the authors used a combination of ultrasound treatment and transglutaminase enzyme (Tgase). While ultrasounds treatment is known to promote the exposure of hidden amino acid residues, Tgase can promote a cross-link reaction between them. The higher hardness was recorded when the system was sonicated for 45 min. The ultrasound treatment also influenced the water holding capacity (WHC) of all different systems, with or without protein combination, and, in particular, the maximum WHC was recorded at 30 min of ultrasound treatment, without differences between the mixed and separated systems. In this specific enzymatic gelation, the caseins are the main product responsible for gel formation since they are more susceptible to Tgase action. However, the mixed gels had a lower store modulus (G′) compared to pure milk gels, perhaps because the presence of a different protein fraction inhibited the action of the enzyme [80]. A similar study was reported by Ma et al., (2022) [81], where a combined treatment of ultrasounds and enzymatic hydrolysis was applied to develop soy protein isolate (SPI) gels cross-linked by transglutaminase. In this case, papain-mediated hydrolysis was also added as a pre-treatment, in order to obtain a pool of different and modified proteins and peptides, which, associated with the ultrasounds, facilitated the cross-linking action of transglutaminase. The treated SPI gel showed a more uniform and dense structure, with significantly improved gel strength and water-holding capacity when compared to the untreated SPI gel. The results obtained by these studies highlighted the possible synergistic effects of these treatments, which could thus represent an effective way of improving gelling properties also of combined protein systems in which dairy and plant proteins coexist.

Another effective method to improve the gelation process is to combine the proteins with biopolymers with gelling capacity. In particular, protein amyloid fibrils have recently gained popularity for their ability to reinforce hydrogels thanks to their specific structure and availability of functional groups [82]. Protein amyloid fibrils can be obtained from a wide range of food proteins, including dairy and plant proteins, by hydrolysis and unfolding mediated by thermal treatments in acidic environments [83]. For example, Khalesi et al., (2021) [84] designed a gel composed of whey protein isolate and their amyloid fibrils and discovered that it was a brilliant strategy to improve the gelling properties of proteins. Indeed, the newly composed gel showed an enhanced elastic modulus by approximately 11 times compared to the control gel. This method has proven to be useful also for plant proteins where Wu et al., (2022) [85], created amyloid from pea proteins to form an enhanced gel for lutein encapsulation with better stability against environmental stresses. Additionally, the study conducted by Ge et al., (2022) [86], used amyloid fibrils from panda bean to reinforce the gel structure of pea protein isolate gel. Even though the water holding capacity and secondary structure were not modified, the gel strength was significantly enhanced and intermolecular interactions such as hydrogen bonds and hydrophobic interactions increased with increased fibrils concentration. Thus, this approach could also be used in mixed dairy and plant proteins gel to modify their textural and rheological properties, opening the area for innovative research that could finally be employed by the food industry.

As a conclusion, Table 4 summarizes all of the features and characteristics that have been studied so far about mixed milk-plant gels. From all of these insights, it can be stated that knowing the characteristic techno-functional properties of all the protein fractions involved in a mixed system and how they behave in different environmental conditions (pH; temperature; ionic strength) may allow the development of innovative gel products such as yogurt, cheese analogues, and beverages characterized by appropriate textural and sensorial properties.

Table 4. Summary of mixed dairy-plant gel.

Mixed Heat-Induced Gel	Functionalities	Reference
Caseins/Pea/soy	No interaction in the gel formation and presence of distinct phases	Silva et. al., 2018 [66]
Whey/Pea	Increased gelation temperature; modulation of gel structure and rheological properties; increased gel stiffness; increased gel homogeneity at isoelectric pH	Wong et. al., 2013 [64]
Whey/Soy	Formation of aggregates with disulfide bonds; reduced gel strength	Corredig et. al., 2015 [87]
Mixed Acid-induced gel		
Whey/Pea	Decreased gelation pH; decreased gel stiffness; no interactions	Chihi et. al., 2018 [63]
Milk/Pea	Faster gelation; increased acidity; increased syneresis; decreased gel firmness	Yousseef et. al., 2016 [77]
Milk/Lentil	Similar syneresis and rheological behavior to milk control	Zare et. al., 2011 [78]
Enzymatic gel		
Whey/Soy	Increased gel hardness; optimal water holding capacity; decreased rheological properties	Cui et. al., 2020 [80]

4.3. Mixed Milk: Plant Proteins Emulsions

Emulsions are colloidal systems formed by two immiscible liquids, where one liquid is scattered in small droplets, the dispersed phase, into the other liquid, the continuous phase. Naturally, these systems are unstable, and require molecules able to adsorb in the interphases to decrease the interfacial tension and increase their stability [88]. In foods, emulsion systems are usually represented by water dispersed in oil (w/o emulsion) or oil dispersed in water (o/w emulsion). Margarine and butters are practical examples of the former, while mayonnaise and creams of the latter [48]. Milk by itself is an emulsion, where the lipids are finely dispersed in the continuous water phase, and stabilized by phospholipids, CMs, and whey proteins [89].

The combination of sodium caseinate and soy proteins with 5% oil fraction at 2% protein concentration in a 1:1 ratio was performed by Ji et al., (2015) [90]. The emulsions showed an average droplet size of 250 nm and a zeta potential of −45 mV at pH of 6.8. This high zeta potential value associated with the small droplet's sizes conferred remarkable stability to the emulsion. The long-term stability of the emulsions stabilized by mixed proteins was higher than that of single proteins. After two weeks at room temperature, the droplet's sizes grow from 250 nm to more than 1100 nm for the single protein emulsion, while it did not change for mixed system. Similar results were found by Hinderink et al., (2019) [17], where emulsions stabilized by combination of pea/WPI and pea/CasNa presented better stability after 14 days storage compared to emulsions where only one kind of protein was present, showing the synergic effect of the protein blends in the emulsion stability. The mixed emulsions layer was denser than the single proteins, and it may be a reason for better emulsion stabilization, where the systems were mainly stabilized by steric repulsion [90]. In the mixed systems, both proteins are absorbed at the interfacial layer with low concentration of proteins in the aqueous phase. However, during the storage time, a displacement of interfacial proteins can occur, as observed by Hinderink et al., (2019) [17], where whey proteins could substitute pea proteins in the interface, as well as pea protein displaced CasNa but without stability loss.

Liang et al., (2016) [91], studied emulsions formed by mixing CMs, pea, soy, and whey protein with a protein total concentration of 10% w/w, which is high if compared to the concentration of emulsifiers generally used. CMs mixed with plant proteins showed lower droplet size compared to a combination of CM-whey. As a general consideration, the higher the amount of whey, the higher the droplet size. Concerning heat stability, the systems containing soy protein presented better results in comparison to the systems formed by pea and whey.

Le Roux et al., (2020) [92], tried to produce infant formulas with a partial substitution of dairy proteins by pea and faba proteins and compared their functional properties with a traditional reference made entirely by dairy proteins. They found that the plant-based products showed, in general, very similar physicochemical and functional properties to the fully dairy infant formula reference. In particular, when the powders were mixed with an oil component to produce an emulsion, all of the samples presented similar emulsion stability with equivalent free fat release, independently from the protein source. However, it was also seen that pea and faba proteins were difficult to disperse and created larger aggregates with higher particle sizes when the powders were reconstituted. Further analyses are therefore necessary to elucidate the protein functions in such emulsion system, as well as to find a solution for particle size reduction.

The emulsions can also work as a delivery system for sensible hydrophobic bioactive molecules, which can be applied in the fortification of foods. The mixed system CasNa/ Soy proteins showed better protection properties compared to single protein systems, showing retention of vitamin A around 93% after three months of storage [90]. This protection over Vitamin A is due mainly to two factors: i. Proteins light deviation which diminishes Vitamin A light exposure and ii. ability of protein to bind metals in the aqueous phase. Milk: plant protein blends were also used as emulsifiers in lycopene emulsions [93]. The blends containing whey-soy and whey-pea presented better emulsion stability than the proteins alone. However, an antagonistic effect was observed in the blends of CasNa and the plant fractions, which cause emulsion destabilization after 7 days of production, probably caused by competitive absorption at the oil-drop surface between CasNa and pea [93].

The process that milk undergoes to develop milk products changes the protein structure and interaction. The understanding of the different processes employed in the dairy and beverage industries for mixed systems is relevant to give a more concrete idea of their potential uses. In addition to temperature, homogenization plays an important role during milk and plant beverage processing. The impact of the homogenization order, i.e., homogenize cow's milk with cream followed by soy milk addition, or homogenize soy milk with cream followed by cow's milk addition or homogenize both kinds of milk together was studied by Grygorczyk et al., (2014) [87]. The homogenization order modulates which protein will be predominant in the fat globule interfaces. When soy milk is homogenized with milk cream in absence of cows' milk, soy proteins are the major constituents in the fat globule interface. The same occurs when milk is homogenized in absence of soy milk. However, when both milks and cream are homogenized together, the fat globule interface is composed mainly of milk proteins. The homogenization process did not have an impact on the fat droplet's sizes.

Based on the results of all of the studies here considered, which key aspects are summarized in Table 5, it can be definitely said that mixed systems of milk and plant proteins may represent a very useful tool to improve stability and techno-functionalities of many food emulsions such as beverages, salad dressings, desserts, and cheese analogues. Their synergic effect at the oil droplet interface indeed forms a dense protein layer and maintain a steady droplet size for long periods of time; a feature that allows the inhibition of creaming and sedimentation which cause emulsion breakdown and instability. Moreover, dairy and plant protein mix also manifested interesting results as bioactive compounds carrier. Their employment in emulsion food formulations therefore could not only reduce the use of additives and emulsifiers but could also be used to improve the transport of several bioactive compounds, producing foods characterized by high protein content and healthy claims. However, sensorial properties of the composed foods need to be evaluated in order to also promote consumer acceptance.

Table 5. Summary of mixed dairy: plant emulsions.

Mixed Emulsions	Functionalities	Reference
Caseins/Soy	Improved stability; small droplet sizes	Ji et. al., 2015 [90]
Milk/Pea	High stability; small droplet sizes; dense interfacial layer	Hinderink et. al., 2019 [17]
Milk/Pea/soy	Reduced droplet sizes and improved heat-stability	Liang et. al., 2016 [91]
Milk/Faba	Overall optimal stability; decreased solubility; larger particle sizes	Le Roux et. al., 2020 [92]
Caseins/Soy	Enhanced encapsulation ability for vitamin a and lycopene	Ho et. al., 2018 [93]

4.4. Mixed Milk: Plant Proteins Foams

Foams are described as mixed systems in which gas bubbles are uniformly dispersed in a continuous liquid or solid phase. In food products such as cake, meringue, bread, and whipped toppings, they are essential parts contributing to properties such as texture and palatability [94]. Foams are thermodynamically unstable since gravitational forces and colloidal activities can be responsible for bubble coalescence and disproportionation, destabilizing the overall system. To prevent their collapse, surface-active substances are needed to reduce the surface tension around each air bubble and inhibit their burst, enhancing foam stability. Thanks to their amphipathic properties, proteins can be adsorbed at the interface of the phases and form a viscoelastic film which physically entraps air bubbles. Because of their high efficiency in these stabilization mechanisms, proteins from both animal and plant sources are being employed in many food-grade foams [95,96]. Alves et al., (2022) [97], evaluated the structural and foaming properties of mixing whey (WPI) and soy protein (SPI) isolates in different ratios before and after heat treatment. They found out that foam capacity (FC) values were similar for all of the samples despite their ratio and the submitted heat treatment. However, the blends of the two proteins negatively affected the foam stability (FS) even at moderate blends, with further antagonistic effect after heat treatment. They hypothesized that, even in small amounts, the more hydrated SP aggregates sterically prevented the formation of a strong and compact viscoelastic protein film at the air-water interface. Moreover, the high temperature contributed to the formation of insoluble aggregates, mainly stabilized by hydrophobic interactions, of both WPI and SP which further reduced the flexibility of the interfacial film. Both of these phenomena contributed to the reduction in FS values for the mixed proteins samples. Similar results were obtained by the study of Krentz et al., 2022 [98], where the authors evaluated the foaming properties of a mixture of casein micelles (CMs) and pea protein isolate (PPI). The blends were compared to skim milk and pea protein isolate slurries which, respectively, exhibited the highest and the lowest values for both FC and FS. The incorporation of CMs in the blends enhanced the foaming properties of PPI control, while PPI presence did not improve values of CMs control. It is then reasonable to say that in this study, pea protein aggregates behaved similarly to the soy protein aggregates, sterically preventing the formation of a strong viscoelastic film at the bubble interfaces and destabilizing the overall system.

Therefore, it can be concluded that, even though producers may be able to find a useful ratio of dairy and plant proteins to obtain a required functionality, their mix does not manifest a synergic effect in stabilizing a foam system; a property which is instead given by the sole characteristics of the proteins employed.

However, proteins techno-functionality can be modified using different and nonconventional physicochemical treatments in order to stimulate protein-protein interaction and improve the overall stability of the system. Up to date, there are not many studies regarding the possibility of treatment applications on dairy and plant proteins in mixed system to increase their colloidal properties. More knowledge of these systems and their possible manipulation is therefore required in the next years.

5. Possible Approaches to Improve Dairy-Plant Proteins Interaction and Techno-Functionalities

To try to overcome dairy and plant proteins limits in forming and stabilizing any different colloidal system, lately new approaches have been explored. For example, the design of innovative food products based on the combination of milk and plant proteins have also been attempted exploiting microorganisms. While microbial fermentation is well known to modify milk protein behavior, it has also been employed to reduce many off-flavors linked to beans presence as well to improve plant proteins techno-functional properties on different plant-based food systems [99,100]. Canon et al., (2022) [101], manufactured plant-based yogurt alternatives by emulsifying milk and lupin protein and fermenting it with a coculture of several species of lactic acid bacteria. The addition of the fermentation process presented encouraging results; indeed, some cocultures developed a more firm and viscous structure with a higher water holding capacity, in particular, when the milk/lupin protein ratio was 67:33. Moreover, these yogurt alternatives were sensorially discriminated on the sole protein ratio and fat type, not from the different starters employed. These findings could thus lead to a wide variety of formulations with several interesting features that could also promote the consumption of such innovative plant-based products. Additionally, the fermentation with different starting cultures could represent a strategic tool to manufacture newly mixed foods with the desired techno-functional and sensorial properties.

Other alternative technologies have also been studied recently to modify dairy and plant protein behavior. Pulsed electric field (PEF), for example, are being used for their ability to change the protein structure and, consequently, their physicochemical properties [102,103]. Indeed, PEF treatments can improve protein interactions by promoting the unfolding of the molecules and the polarization of the amino acids by exposing hydrophobic regions as well as sulfhydryl groups and, thus, enhancing protein aggregation. Several studies applied PEF treatments on dairy, and plant proteins and the main effects are summarized in Table 6.

Table 6. Adapted with permission from Taha et al., (2022) [102].

Protein	Effects on Functionality	References
Whey Protein Isolate (WPI)	Improved gelling properties of WPI when treated with an intensity lower than 45 kV/cm. However, weaker gel strengths compared to heat-treated gels.	Jin et al., 2013; Rodrigues et al., 2015 [104,105]
Caseins and WPI	Increased rate of unfolding proteins and their surface hydrophobicity	Sharma et al., 2016; Subasi et al., 2021 [106,107]
Soy protein isolate (SPI)	Decreased solubility and surface hydrophobicity	Li et al., 2007 [108]
Canola protein	Improved solubility, foaming and emulsifying properties	Zhang et al., 2017 [109]
Sunflower protein	Reduced interfacial tension at protein/water interface	Subasi et al., 2021 [107]
Pea protein isolate (PPI)	Increased surface hydrophobicity and gelling properties	Chen et al., 2022 [110]

The effectiveness of PEF technology to improve protein functionality highly depends on the specific conditions used (intensity, time, and temperature of the treatment), which could thus be tailored for each type of protein to obtain the desired features in every colloidal system. However, up to date, PEF has not yet been applied in a system where dairy and plant proteins coexist. Thus, more studies are needed to elucidate if this technology could promote their interaction and affinity.

High pressure processing (HPP) is a nonthermal processing technology used in the food industry to extend food products shelf-life [111], but it has been also employed to modify structure and functionality of food proteins in colloidal systems. While HPP for animal and dairy proteins have been largely studied, little knowledge exists nowadays on the effects of this processing method on alternative proteins. Queirós et al., (2018) [112], reviewed the recent applications on a large variety of plant proteins and concluded that HPP can tailor their functionalities by inducing unfolding paths and better exposition of functional groups promoting their aggregation, solubility, emulsifying, gelling, and foaming properties. Sim et al., (2020) [113], applied HPP treatment to develop plant protein (mung bean, chickpea, pea, lentil, and faba bean) gels and emulsions and compared the results to commercial dairy yogurts. The study revealed that HPP developed viscoelastic gels and emulsions with comparable gel strength and viscosities to the controls, proposing a new methodology to develop plant-based yogurt alternatives. However, how HPP affects techo-functionalities of the proteins highly depends on a complex series of relationships between the intrinsic characteristics and type of protein, the environmental conditions, and the high-pressure parameters. Thus, in this context, when a system is composed of two or more different biopolymers, such as dairy and plant proteins, the HPP conditions to improve properties of both are most likely to be incompatible, thus, if this technology is to be used to combine dairy and plant proteins in a mixture, perhaps it would be wise to treat them separately, tailoring specific characteristics and then proceed with their mixture.

Nowadays, many other innovative processes are being investigated to try to improve plant-based proteins' physicochemical properties, such as partial hydrolysis, ohmic heating and freeze-thaw cycle, which are capable of modifying proteins' structure and intermolecular forces [114–116]. However, currently, little is known about these modifications when milk and plant proteins are both present in the same medium. It will therefore be a task and trend for the future food industry research to investigate how these processes can affect a single or both proteins and if it could be useful to improve their interactions for the development of innovations in the food industry.

6. Sensory Attributes of Mixed Systems

An important feature of any food is its sensorial attributes; required consumer intent, desirable texture, taste, flavor, among others. However, studies regarding the sensory evaluation of mixed proteins systems are still scarce. Zare et al., (2011) [78], compared the smoothness, graininess, flavor, color, and overall acceptance of two supplemented yogurts. The replacement of skim milk powder with lentil flour was evaluated sensorially. The yogurt supplemented with 1.2 and 3% of lentil flour showed no significant difference in smoothness, graininess, flavor, and overall acceptance when compared to yogurts supplemented with 1.2 and 3% of skim milk powder. However, in the color parameter, yogurts added with 2 and 3% of lentil flour were different from 2 and 3% skim milk yogurt. Thus, the impact of the addition of vegetable protein was not perceived by the consumers in the concentration studied, indicating that lentil flour can be used to fortify yogurts without sensorial loss. The concentration of plant protein added in dairy products must be high enough to cause desirable changes in the functional properties and, at the same time, cause minimum interference in the sensorial attributes.

The sensorial impact caused by increasing concentration of pea protein in yogurts produced using several starter cultures was evaluated by Yousseef et al., (2016) [77]. As pea concentration increases, the intensity of the terms pea, earth, smoked, and vinegar increased, which are considered negative sensory characteristics, while the positive terms dairy and creamy decreased in intensity. Among the pea concentrations studied, yogurts containing 20 to 40% of pea protein were characterized as products with undesirable features, while 10% pea protein concentration was considered the closest to the control yogurt sample. In addition, the fermentation process also showed the potentiality of decreasing the undesirable beany flavor of mixed milk-pea gels [117]. However, the type of metabolites, as well as the microorganism growth, depend on the composition of the gel matrix [118]. Canon et al., (2022) [101], also manufactured mixed dairy and plant protein yogurt alternatives, mixing, in particular, skim milk powder or whey proteins with lupin protein isolate and milk fat or coconut oil. They evaluated not only the protein type proportion but also the fermentation with several lactic acid bacteria strains. The sensorial results obtained showed that yogurt alternatives were discriminated only on the basis of protein ratios and fat components but not of starters. In particular, the milk/lupin ratio of 67:33 was more accepted than the 50:50 ratio; however, the employment of cocultures of lactic acid bacteria produced different aroma compounds, which increased 50:50 acceptance. Thus, the sensory changes promoted by the addition of plant proteins cannot be underestimated, and studies regarding the maximum quantity of protein addition are highly required.

Lastly, Grygorczyk et al., (2013) [76], using napping methodology investigated the effect of the order of homogenization in the texture of systems formed by soy milk and cow's milk. The homogenization of milk with cream in the presence or absence of soymilk leads to yogurts with high thickness, roughness, and mouthcoating. When the cream was homogenized in soymilk with posterior addition of skim milk, the formed gel exhibits thinner and watery features. The perception of fatty attributes was also influenced by the homogenization order, since the fat content of all samples was the same. The samples, where the aggregation of milk proteins started first, had more fatty-related attributes, while the opposite happened when the aggregation of the protein occurred at the same time, showing that, how the fat globules are disposed in the matrix, can influence the perception of fats in the product.

7. Conclusions and Perspectives

The studies address evidence of concept, indicating that combination of plant and milk proteins can be used to module colloidal systems with direct application in the food industry. Some studies pointed out that the addition of plant protein can solve some technological problems such as the production of high-protein dairy beverages which tend to form a gel in the packaging. Moreover, the presence of dairy proteins can increase the solubility of some plant proteins in mixed protein dispersions. However, plant protein solubility remains the greatest challenge to overcome in the designing and manufacturing of such products. The low solubility can in fact lead to complex intermolecular interactions between the proteins and the solvent, causing higher molecular aggregates and precipitates.

On the other hand, this complex behavior could be useful for the development of a gel system. Even though there is no general rule for the manufacturing of mixed dairy and plant proteins gels, knowing the specific techno-functional properties of all the protein fractions involved and how they are influenced by environmental conditions such as pH and temperature, may allow the development of innovative gel products such as yogurt, cheese analogues, and beverages characterized by appropriate textural and sensorial properties. Moreover, different pre-treatments could be applied to the proteins such as ultrasounds and enzymatic reactions that can improve their intermolecular interactions and impact the overall structure of the final gel.

Regarding the emulsion colloidal system, mixed systems of dairy and plant proteins manifested a better synergy for stabilizing the system. Indeed, the presence of both types of protein at the interfaces of the emulsions (o/w or w/o) stabilized the droplets for long periods of time and created a denser interfacial layer, inhibiting at the same time creaming and sedimentation. The results here presented could be of paramount importance for the design of enhanced emulsion food products such as salad dressings, sauces, and cheese analogues.

Finally, foam systems with the presence of both proteins were taken into consideration, reaching the conclusion that their mixture does not manifest a synergic effect in stabilizing a foam system. Nonetheless, both proteins were able to create a viscoelastic film around each air bubble, they did not interact with themself, and the stabilization mechanisms were only given via the characteristics of the single proteins employed, where dairy sources presented higher foaming properties than plant ones.

However, to try to enhance these colloidal systems and thus promote their application in the food industry, emergent and innovative approaches are being evaluated to modify the techno-functional properties of both dairy and plant proteins. For example, precise fermentation, pulsed electric field, and high hydrostatic pressure are able to modify protein structures and physicochemical properties and can therefore be employed in these binary systems to obtain the desired characteristics and promote their application in the food system. Moreover, the sensorial profile of such foods needs always to be taken into account since it can also be responsible for consumer rejection. In this context, a balanced proportion of milk and plant proteins needs to be achieved depending on the desired characteristics of the final product.

Thus, new studies are needed in the near future to evaluate the applicability of pretreatments to promote dairy and plant proteins' coexistence, to improve their functional and sensorial properties, and, consequently, to open a new market category of innovative food products.

Author Contributions: Conceptualization, L.G.L.N. and D.O.; data curation, L.G.L.N. and D.O.; methodology, N.F.N.S., P.P.d.s.P.J., G.D., E.M. and A.F.d.C.; validation, A.F.d.C., F.C., G.D. and N.F.N.S.; formal analysis, L.G.L.N. and D.O.; investigation, P.P.d.s.P.J., L.G.L.N. and D.O.; resources, L.G.L.N., D.O., E.M., N.F.N.S. and P.P.d.s.P.J.; writing—original draft preparation, L.G.L.N. and D.O.; writing—review and editing, D.O.; visualization, D.O., L.G.L.N. and E.M.; supervision, F.C. and A.F.d.C.; project administration, A.F.d.C., G.D. and F.C.; funding acquisition, F.C. All authors have read and agreed to the published version of the manuscript.

Funding: This study was financed in part by the Coordenação de Aperfeiçoamento de Pessoal de Nível Superior—Brasil (CAPES)—Finance Code 001.

Data Availability Statement: No new data were created for the production of this manuscript. All of the data here discussed and presented are available in the relative references here cited and listed.

Conflicts of Interest: The authors declare no conflict of interest.

References

1. Poore, J.; Nemecek, T. Reducing Food's Environmental Impacts through Producers and Consumers. *Science* **2018**, *360*, 987–992. [CrossRef] [PubMed]
2. Clune, S.; Crossin, E.; Verghese, K. Systematic Review of Greenhouse Gas Emissions for Different Fresh Food Categories. *J. Clean. Prod.* **2017**, *140*, 766–783. [CrossRef]
3. Matsumiya, K.; Murray, B.S. Soybean Protein Isolate Gel Particles as Foaming and Emulsifying Agents. *Food Hydrocoll.* **2016**, *60*, 206–215. [CrossRef]
4. Bergsma, J. Vegan Cheese Analogue. Patent WO2017150973A1, 1 March 2017.
5. Lipan, L.; Rusu, B.; Sendra, E.; Hernández, F.; Vázquez-Araújo, L.; Vodnar, D.C.; Carbonell-Barrachina, Á.A. Spray Drying and Storage of Probiotic-Enriched Almond Milk: Probiotic Survival and Physicochemical Properties. *J. Sci. Food Agric.* **2020**, *100*, 3697–3708. [CrossRef]
6. Schreuders, F.K.G.; Dekkers, B.L.; Bodnár, I.; Erni, P.; Boom, R.M.; van der Goot, A.J. Comparing Structuring Potential of Pea and Soy Protein with Gluten for Meat Analogue Preparation. *J. Food Eng.* **2019**, *261*, 32–39. [CrossRef]
7. Lonnie, M.; Johnstone, A.M. The Public Health Rationale for Promoting Plant Protein as an Important Part of a Sustainable and Healthy Diet. *Nutr. Bull.* **2020**, *45*, 281–293. [CrossRef]
8. Budhathoki, S.; Sawada, N.; Iwasaki, M.; Yamaji, T.; Goto, A.; Kotemori, A.; Ishihara, J.; Takachi, R.; Charvat, H.; Mizoue, T.; et al. Association of Animal and Plant Protein Intake with All-Cause and Cause-Specific Mortality in a Japanese Cohort. *JAMA Intern. Med.* **2019**, *179*, 1509–1518. [CrossRef]
9. Qian, F.; Liu, G.; Hu, F.B.; Bhupathiraju, S.N.; Sun, Q. Association between Plant-Based Dietary Patterns and Risk of Type 2 Diabetes: A Systematic Review and Meta-Analysis. *JAMA Intern. Med.* **2019**, *179*, 1335–1344. [CrossRef]
10. Sá, A.G.A.; Moreno, Y.M.F.; Carciofi, B.A.M. Plant Proteins as High-Quality Nutritional Source for Human Diet. *Trends Food Sci. Technol.* **2020**, *97*, 170–184. [CrossRef]
11. Nikbakht Nasrabadi, M.; Sedaghat Doost, A.; Mezzenga, R. Modification Approaches of Plant-Based Proteins to Improve Their Techno-Functionality and Use in Food Products. *Food Hydrocoll.* **2021**, *118*, 106789. [CrossRef]

12. Goulding, D.A.; Fox, P.F.; O'Mahony, J.A. Milk Proteins: An Overview. In *Milk Proteins: From Expression to Food*; Academic Press: Cambridge, MA, USA, 2019.
13. Uluko, H.; Liu, L.; Lv, J.P.; Zhang, S.W. Functional Characteristics of Milk Protein Concentrates and Their Modification. *Crit. Rev. Food Sci. Nutr.* **2016**, *56*, 1193–1208. [CrossRef] [PubMed]
14. Pouliot, Y. Membrane Processes in Dairy Technology-From a Simple Idea to Worldwide Panacea. *Int. Dairy J.* **2008**, *18*, 735–740. [CrossRef]
15. Schmitt, C.; Bovetto, L.; Buczkowski, J.; De Oliveira Reis, G.; Pibarot, P.; Amagliani, L.; Dombrowski, J. Plant Proteins and Their Colloidal State. *Curr. Opin. Colloid Interface Sci.* **2021**, *56*, 101510. [CrossRef]
16. Guyomarc'h, F.; Arvisenet, G.; Bouhallab, S.; Canon, F.; Deutsch, S.M.; Drigon, V.; Dupont, D.; Famelart, M.H.; Garric, G.; Guédon, E.; et al. Mixing Milk, Egg and Plant Resources to Obtain Safe and Tasty Foods with Environmental and Health Benefits. *Trends Food Sci. Technol.* **2021**, *108*, 119–132. [CrossRef]
17. Hinderink, E.B.A.; Münch, K.; Sagis, L.; Schroën, K.; Berton-Carabin, C.C. Synergistic Stabilisation of Emulsions by Blends of Dairy and Soluble Pea Proteins: Contribution of the Interfacial Composition. *Food Hydrocoll.* **2019**, *97*, 105206. [CrossRef]
18. Ben-Harb, S.; Panouillé, M.; Huc-Mathis, D.; Moulin, G.; Saint-Eve, A.; Irlinger, F.; Bonnarme, P.; Michon, C.; Souchon, I. The Rheological and Microstructural Properties of Pea, Milk, Mixed Pea/Milk Gels and Gelled Emulsions Designed by Thermal, Acid, and Enzyme Treatments. *Food Hydrocoll.* **2018**, *77*, 75–84. [CrossRef]
19. Foroutan, A.; Guo, A.C.; Vazquez-Fresno, R.; Lipfert, M.; Zhang, L.; Zheng, J.; Badran, H.; Budinski, Z.; Mandal, R.; Ametaj, B.N.; et al. Chemical Composition of Commercial Cow's Milk. *J. Agric. Food Chem.* **2019**, *67*, 4897–4914. [CrossRef]
20. Walstra, P.; Wouters, J.T.M.; Geurts, T.J. *Dairy Science and Technology*, 2nd ed.; Taylor & Francis: Abingdon, UK, 2006; Volume 4, ISBN 9780824727635.
21. Horne, D.S. Casein Micelle Structure and Stability. In *Milk Proteins: From Expression to Food*; Academic Press: Cambridge, MA, USA, 2019.
22. Edwards, P.J.B.; Jameson, G.B. Structure and Stability of Whey Proteins. In *Milk Proteins: From Expression to Food*; Academic Press: Cambridge, MA, USA, 2019.
23. Holt, C.; Carver, J.A.; Ecroyd, H.; Thorn, D.C. Invited Review: Caseins and the Casein Micelle: Their Biological Functions, Structures, and Behavior in Foods1. *J. Dairy Sci.* **2013**, *96*, 6127–6146. [CrossRef]
24. Carter, B.G.; Cheng, N.; Kapoor, R.; Meletharayil, G.H.; Drake, M.A. Invited Review: Microfiltration-Derived Casein and Whey Proteins from Milk. *J. Dairy Sci.* **2021**, *104*, 2465–2479. [CrossRef]
25. Carr, A.; Golding, M. Functional Milk Proteins Production and Utilization: Casein-Based Ingredients. In *Advanced Dairy Chemistry: Volume 1B: Proteins: Applied Aspects*, 4th ed.; Springer: New York, NY, USA, 2016.
26. Huppertz, T.; Fox, P.F.; Kelly, A.L. The Caseins: Structure, Stability, and Functionality. In *Proteins in Food Processing*, 2nd ed.; Woodhead Publishing: Sawston, UK, 2017.
27. Huppertz, T.; Gazi, I.; Luyten, H.; Nieuwenhuijse, H.; Alting, A.; Schokker, E. Hydration of Casein Micelles and Caseinates: Implications for Casein Micelle Structure. *Int. Dairy J.* **2017**, *74*, 1–11. [CrossRef]
28. Hammam, A.R.A.; Martínez-Monteagudo, S.I.; Metzger, L.E. Progress in Micellar Casein Concentrate: Production and Applications. *Compr. Rev. Food Sci. Food Saf.* **2021**, *20*, 4426–4449. [CrossRef] [PubMed]
29. Oliveira, I.C.; de Paula Ferreira, I.E.; Casanova, F.; Cavallieri, A.L.F.; Lima Nascimento, L.G.; de Carvalho, A.F.; Nogueira Silva, N.F. Colloidal and Acid Gelling Properties of Mixed Milk and Pea Protein Suspensions. *Foods* **2022**, *11*, 1383. [CrossRef] [PubMed]
30. Smithers, G.W. Whey and Whey Proteins-From "Gutter-to-Gold". *Int. Dairy J.* **2008**, *18*, 695–704. [CrossRef]
31. Farrell, H.M.; Jimenez-Flores, R.; Bleck, G.T.; Brown, E.M.; Butler, J.E.; Creamer, L.K.; Hicks, C.L.; Hollar, C.M.; Ng-Kwai-Hang, K.F.; Swaisgood, H.E. Nomenclature of the Proteins of Cows' Milk-Sixth Revision. *J. Dairy Sci.* **2004**, *87*, 1641–1674. [CrossRef]
32. Guyomarc'h, F.; Famelart, M.H.; Henry, G.; Gulzar, M.; Leonil, J.; Hamon, P.; Bouhallab, S.; Croguennec, T. Current Ways to Modify the Structure of Whey Proteins for Specific Functionalities—A Review. *Dairy Sci. Technol.* **2015**, *95*, 795–814. [CrossRef]
33. Loveday, S.M. Plant Protein Ingredients with Food Functionality Potential. *Nutr. Bull.* **2020**, *45*, 321–327. [CrossRef]
34. Tan, M.; Nawaz, M.A.; Buckow, R. Functional and Food Application of Plant Proteins–A Review. *Food Rev. Int.* **2021**. [CrossRef]
35. Bessada, S.M.F.; Barreira, J.C.M.; Oliveira, M.B.P.P. Pulses and Food Security: Dietary Protein, Digestibility, Bioactive and Functional Properties. *Trends Food Sci. Technol.* **2019**, *93*, 53–68. [CrossRef]
36. Hara-Nishimura, I.; Shimada, T.; Hatano, K.; Takeuchi, Y.; Nishimura, M. Transport of Storage Proteins to Protein Storage Vacuoles Is Mediated by Large Precursor-Accumulating Vesicles. *Plant Cell* **1998**, *10*, 825–836. [CrossRef]
37. Boye, J.; Zare, F.; Pletch, A. Pulse Proteins: Processing, Characterization, Functional Properties and Applications in Food and Feed. *Food Res. Int.* **2010**, *43*, 414–431. [CrossRef]
38. Gumus, C.E.; Decker, E.A.; McClements, D.J. Impact of Legume Protein Type and Location on Lipid Oxidation in Fish Oil-in-Water Emulsions: Lentil, Pea, and Faba Bean Proteins. *Food Res. Int.* **2017**, *100*, 175–185. [CrossRef] [PubMed]
39. Aachary, A.A.; Thiyam-Hollander, U.; Eskin, M.N.A. Canola/Rapeseed Proteins and Peptides. In *Applied Food Protein Chemistry*; John Wiley & Sons, Ltd.: Hoboken, NJ, USA, 2014; ISBN 978-1-119-94449-2.
40. Loveday, S.M. Food Proteins: Technological, Nutritional, and Sustainability Attributes of Traditional and Emerging Proteins. *Annu. Rev. Food Sci. Technol.* **2019**, *10*, 311–339. [CrossRef] [PubMed]
41. Mohammed, K.; Obadi, M.; Omedi, J.O.; Letsididi, K.S.; Koko, M.; Zaaboul, F.; Siddeeg, A.; Liu, Y. Effect of Sunflower Meal Protein Isolate (SMPI) Addition on Wheat Bread Quality. *J. Acad. Ind. Res.* **2018**, *6*, 159–163.

42. Pereira, J.; Zhou, G.; Zhang, W. Effects of Rice Flour on Emulsion Stability, Organoleptic Characteristics and Thermal Rheology of Emulsified Sausage. *J. Food Nutr. Res.* **2020**, *4*, 216–222. [CrossRef]
43. Akharume, F.U.; Aluko, R.E.; Adedeji, A.A. Modification of Plant Proteins for Improved Functionality: A Review. *Compr. Rev. Food Sci. Food Saf.* **2021**, *20*, 198–224. [CrossRef]
44. Nwachukwu, I.D.; Aluko, R.E. Chapter 1: Food Protein Structures, Functionality and Product Development. In *Food Chemistry, Function and Analysis*; Royal Society of Chemistry: London, UK, 2021; Volume 2021.
45. Roberts, C.J. Protein Aggregation and Its Impact on Product Quality. *Curr. Opin. Biotechnol.* **2014**, *30*, 211–217. [CrossRef]
46. Nicolai, T.; Chassenieux, C. Heat-Induced Gelation of Plant Globulins. *Curr. Opin. Food Sci.* **2019**, *27*, 18–22. [CrossRef]
47. Foegeding, E.A. Food Protein Functionality-A New Model. *J. Food Sci.* **2015**, *80*, C2670–C2677. [CrossRef]
48. Milani, J.M.; Golkar, A. Introductory Chapter: Some New Aspects of Colloidal Systems in Foods. In *Some New Aspects of Colloidal Systems in Foods*; IntechOpen: London, UK, 2019.
49. Alrosan, M.; Tan, T.C.; Easa, A.M.; Gammoh, S.; Alu'datt, M.H. Molecular Forces Governing Protein-Protein Interaction: Structure-Function Relationship of Complexes Protein in the Food Industry. *Crit. Rev. Food Sci. Nutr.* **2022**, *62*, 4036–4052. [CrossRef]
50. Pace, C.N.; Treviño, S.; Prabhakaran, E.; Scholtz, J.M.; Franks, F.; Wilson, K.; Daniel, R.M.; Halling, P.J.; Clark, D.S.; Purkiss, A. Protein Structure, Stability and Solubility in Water and Other Solvents. *Philos. Trans. R. Soc. B Biol. Sci.* **2004**, *359*, 1225–1235. [CrossRef]
51. Shire, S.J.; Shahrokh, Z.; Liu, J. Challenges in the Development of High Protein Concentration Formulations. *J. Pharm. Sci.* **2004**, *93*, 1390–1402. [CrossRef] [PubMed]
52. Wang, R.; Xu, P.; Chen, Z.; Zhou, X.; Wang, T. Complexation of Rice Proteins and Whey Protein Isolates by Structural Interactions to Prepare Soluble Protein Composites. *LWT* **2019**, *101*, 207–213. [CrossRef]
53. Wang, T.; Yue, M.; Xu, P.; Wang, R.; Chen, Z. Toward Water-Solvation of Rice Proteins via Backbone Hybridization by Casein. *Food Chem.* **2018**, *258*, 278–283. [CrossRef] [PubMed]
54. Singh, J.; Prakash, S.; Bhandari, B.; Bansal, N. Ultra High Temperature (UHT) Stability of Casein-Whey Protein Mixtures at High Protein Content: Heat Induced Protein Interactions. *Food Res. Int.* **2019**, *116*, 103–113. [CrossRef]
55. Yousefi, N.; Abbasi, S. Food Proteins: Solubility & Thermal Stability Improvement Techniques. *Food Chem. Adv.* **2022**, *1*, 100090.
56. Saricaoglu, F.T. Application of High-Pressure Homogenization (HPH) to Modify Functional, Structural and Rheological Properties of Lentil (Lens Culinaris) Proteins. *Int. J. Biol. Macromol.* **2020**, *144*, 760–769. [CrossRef]
57. Wang, F.; Zhang, Y.; Xu, L.; Ma, H. An Efficient Ultrasound-Assisted Extraction Method of Pea Protein and Its Effect on Protein Functional Properties and Biological Activities. *LWT* **2020**, *127*, 109348. [CrossRef]
58. Sánchez-Reséndiz, A.; Rodríguez-Barrientos, S.; Rodríguez-Rodríguez, J.; Barba-Dávila, B.; Serna-Saldívar, S.O.; Chuck-Hernández, C. Phosphoesterification of Soybean and Peanut Proteins with Sodium Trimetaphosphate (STMP): Changes in Structure to Improve Functionality for Food Applications. *Food Chem.* **2018**, *260*, 299–305. [CrossRef]
59. Eckert, E.; Han, J.; Swallow, K.; Tian, Z.; Jarpa-Parra, M.; Chen, L. Effects of Enzymatic Hydrolysis and Ultrafiltration on Physicochemical and Functional Properties of Faba Bean Protein. *Cereal Chem.* **2019**, *96*, 725–741. [CrossRef]
60. Yildiz, G.; Ding, J.; Andrade, J.; Engeseth, N.J.; Feng, H. Effect of Plant Protein-Polysaccharide Complexes Produced by Mano-Thermo-Sonication and PH-Shifting on the Structure and Stability of Oil-in-Water Emulsions. *Innov. Food Sci. Emerg. Technol.* **2018**, *47*, 317–325. [CrossRef]
61. Kontogeorgis, G.M.; Kiil, S. Introduction to Colloid and Surface Chemistry. In *Introduction to Applied Colloid and Surface Chemistry*; John Wiley & Sons, Ltd.: Hoboken, NJ, USA, 2016.
62. Gunasekaran, S.; Ak, M.M. Dynamic Oscillatory Shear Testing of Foods-Selected Applications. *Trends Food Sci. Technol.* **2000**, *11*, 115–127. [CrossRef]
63. Chihi, M.L.; Sok, N.; Saurel, R. Acid Gelation of Mixed Thermal Aggregates of Pea Globulins and β-Lactoglobulin. *Food Hydrocoll.* **2018**, *85*, 120–128. [CrossRef]
64. Wong, D.; Vasanthan, T.; Ozimek, L. Synergistic Enhancement in the Co-Gelation of Salt-Soluble Pea Proteins and Whey Proteins. *Food Chem.* **2013**, *141*, 3913–3919. [CrossRef]
65. Silva, J.V.C.; Balakrishnan, G.; Schmitt, C.; Chassenieux, C.; Nicolai, T. Heat-Induced Gelation of Aqueous Micellar Casein Suspensions as Affected by Globular Protein Addition. *Food Hydrocoll.* **2018**, *82*, 258–267. [CrossRef]
66. Silva, J.V.C.; Cochereau, R.; Schmitt, C.; Chassenieux, C.; Nicolai, T. Heat-Induced Gelation of Mixtures of Micellar Caseins and Plant Proteins in Aqueous Solution. *Food Res. Int.* **2019**, *116*, 1135–1143. [CrossRef] [PubMed]
67. Silva, J.V.C.; Jacquette, B.; Amagliani, L.; Schmitt, C.; Nicolai, T.; Chassenieux, C. Heat-Induced Gelation of Micellar Casein/Plant Protein Oil-in-Water Emulsions. *Colloids Surf. A Physicochem. Eng. Asp.* **2019**, *569*, 85–92. [CrossRef]
68. Mession, J.L.; Roustel, S.; Saurel, R. Interactions in Casein Micelle–Pea Protein System (Part I): Heat-Induced Denaturation and Aggregation. *Food Hydrocoll.* **2017**, *67*, 229–242. [CrossRef]
69. Nicolai, T.; Chassenieux, C. Heat-Induced Gelation of Casein Micelles. *Food Hydrocoll.* **2021**, *118*, 106755. [CrossRef]
70. Huppertz, T.; Nieuwenhuijse, H. Constituent Fouling during Heat Treatment of Milk: A Review. *Int. Dairy J.* **2022**, *126*, 105236. [CrossRef]
71. Chihi, M.L.; Mession, J.L.; Sok, N.; Saurel, R. Heat-Induced Soluble Protein Aggregates from Mixed Pea Globulins and β-Lactoglobulin. *J. Agric. Food Chem.* **2016**, *64*, 2780–2791. [CrossRef]

72. Roesch, R.R.; Corredig, M. Heat-Induced Soy-Whey Proteins Interactions: Formation of Soluble and Insoluble Protein Complexes. *J. Agric. Food Chem.* **2005**, *53*, 3476–3482. [CrossRef] [PubMed]
73. McCann, T.H.; Guyon, L.; Fischer, P.; Day, L. Rheological Properties and Microstructure of Soy-Whey Protein. *Food Hydrocoll.* **2018**, *82*, 434–441. [CrossRef]
74. Totosaus, A.; Montejano, J.G.; Salazar, J.A.; Guerrero, I. A Review of Physical and Chemical Protein-Gel Induction. *Int. J. Food Sci. Technol.* **2002**, *37*, 589–601. [CrossRef]
75. Lucey, J.A. Milk Protein Gels. In *Milk Proteins: From Expression to Food*; Academic Press: Cambridge, MA, USA, 2020; pp. 599–632. [CrossRef]
76. Grygorczyk, A.; Alexander, M.; Corredig, M. Combined Acid- and Rennet-Induced Gelation of a Mixed Soya Milk-Cow's Milk System. *Int. J. Food Sci. Technol.* **2013**, *48*, 2306–2314. [CrossRef]
77. Yousseef, M.; Lafarge, C.; Valentin, D.; Lubbers, S.; Husson, F. Fermentation of Cow Milk and/or Pea Milk Mixtures by Different Starter Cultures: Physico-Chemical and Sensorial Properties. *LWT* **2016**, *69*, 430–437. [CrossRef]
78. Zare, F.; Boye, J.I.; Orsat, V.; Champagne, C.; Simpson, B.K. Microbial, Physical and Sensory Properties of Yogurt Supplemented with Lentil Flour. *Food Res. Int.* **2011**, *44*, 2482–2488. [CrossRef]
79. Martin, A.H.; De Los Reyes Jiménez, M.L.; Pouvreau, L. Modulating the Aggregation Behaviour to Restore the Mechanical Response of Acid Induced Mixed Gels of Sodium Caseinate and Soy Proteins. *Food Hydrocoll.* **2016**, *58*, 215–223. [CrossRef]
80. Cui, Q.; Wang, G.; Gao, D.; Wang, L.; Zhang, A.; Wang, X.; Xu, N.; Jiang, L. Improving the Gel Properties of Transgenic Microbial Transglutaminase Cross-Linked Soybean-Whey Mixed Protein by Ultrasonic Pretreatment. *Process Biochem.* **2020**, *91*, 104–112. [CrossRef]
81. Ma, Z.; Li, L.; Wu, C.; Huang, Y.; Teng, F.; Li, Y. Effects of Combined Enzymatic and Ultrasonic Treatments on the Structure and Gel Properties of Soybean Protein Isolate. *LWT* **2022**, *158*, 113123. [CrossRef]
82. Means, A.K.; Grunlan, M.A. Modern Strategies to Achieve Tissue-Mimetic, Mechanically Robust Hydrogels. *ACS Macro Lett.* **2019**, *8*, 705–713. [CrossRef]
83. Jansens, K.J.A.; Rombouts, I.; Grootaert, C.; Brijs, K.; Van Camp, J.; Van der Meeren, P.; Rousseau, F.; Schymkowitz, J.; Delcour, J.A. Rational Design of Amyloid-Like Fibrillary Structures for Tailoring Food Protein Techno-Functionality and Their Potential Health Implications. *Compr. Rev. Food Sci. Food Saf.* **2019**, *18*, 84–105. [CrossRef]
84. Khalesi, H.; Sun, C.; He, J.; Lu, W.; Fang, Y. The Role of Amyloid Fibrils in the Modification of Whey Protein Isolate Gels with the Form of Stranded and Particulate Microstructures. *Food Res. Int.* **2021**, *140*, 109856. [CrossRef]
85. Wu, W.; Nian, Y.; Liu, Y.; Zhang, Y.; Hu, B. Formation of Pea Protein Amyloid Fibrils to Stabilize High Internal Phase Emulsions for Encapsulation of Lutein. *J. Funct. Foods* **2022**, *94*, 105110. [CrossRef]
86. Ge, J.; Sun, C.; Chang, Y.; Sun, M.; Zhang, Y.; Fang, Y. Heat-Induced Pea Protein Isolate Gels Reinforced by Panda Bean Protein Amyloid Fibrils: Gelling Properties and Formation Mechanism. *Food Res. Int.* **2022**, *162*, 112053. [CrossRef] [PubMed]
87. Grygorczyk, A.; Duizer, L.; Lesschaeve, I.; Corredig, M. Gelation of Recombined Soymilk and Cow's Milk Gels: Effect Ofhomogenization Order and Mode of Gelation on Microstructure Andtexture of the Final Matrix. *Food Hydrocoll.* **2014**, *35*, 69–77. [CrossRef]
88. Derkach, S.R. Rheology of Emulsions. *Adv. Colloid Interface Sci.* **2009**, *151*, 1–23. [CrossRef] [PubMed]
89. Singh, H.; Gallier, S. Nature's Complex Emulsion: The Fat Globules of Milk. *Food Hydrocoll.* **2017**, *68*, 81–89. [CrossRef]
90. Ji, J.; Zhang, J.; Chen, J.; Wang, Y.; Dong, N.; Hu, C.; Chen, H.; Li, G.; Pan, X.; Wu, C. Preparation and Stabilization of Emulsions Stabilized by Mixed Sodium Caseinate and Soy Protein Isolate. *Food Hydrocoll.* **2015**, *51*, 156–165. [CrossRef]
91. Liang, Y.; Wong, S.S.; Pham, S.Q.; Tan, J.J. Effects of Globular Protein Type and Concentration on the Physical Properties and Flow Behaviors of Oil-in-Water Emulsions Stabilized by Micellar Casein-Globular Protein Mixtures. *Food Hydrocoll.* **2016**, *54*, 89–98. [CrossRef]
92. Le Roux, L.; Mejean, S.; Chacon, R.; Lopez, C.; Dupont, D.; Deglaire, A.; Nau, F.; Jeantet, R. Plant Proteins Partially Replacing Dairy Proteins Greatly Influence Infant Formula Functionalities. *LWT* **2020**, *120*, 108891. [CrossRef]
93. Ho, K.K.H.Y.; Schroën, K.; San Martín-González, M.F.; Berton-Carabin, C.C. Synergistic and Antagonistic Effects of Plant and Dairy Protein Blends on the Physicochemical Stability of Lycopene-Loaded Emulsions. *Food Hydrocoll.* **2018**, *81*, 180–190. [CrossRef]
94. Patel, S.G.; Siddaiah, M. Formulation and Evaluation of Effervescent Tablets: A Review. *J. Drug Deliv. Ther.* **2018**, *8*, 296–303. [CrossRef]
95. Mohanan, A.; Nickerson, M.T.; Ghosh, S. Utilization of Pulse Protein-Xanthan Gum Complexes for Foam Stabilization: The Effect of Protein Concentrate and Isolate at Various PH. *Food Chem.* **2020**, *316*, 126282. [CrossRef] [PubMed]
96. Odelli, D.; Sarigiannidou, K.; Soliani, A.; Marie, R.; Mohammadifar, M.A.; Jessen, F.; Spigno, G.; Vall-llosera, M.; De Carvalho, A.F.; Verni, M.; et al. Interaction between Fish Skin Gelatin and Pea Protein at Air-Water Interface after Ultrasound Treatment. *Foods* **2022**, *11*, 659. [CrossRef] [PubMed]
97. Alves, A.C.; Martha, L.; Casanova, F.; Tavares, G.M. Structural and Foaming Properties of Whey and Soy Protein Isolates in Mixed Systems before and after Heat Treatment. *Food Sci. Technol. Int.* **2022**, *28*, 545–553. [CrossRef] [PubMed]
98. Krentz, A.; García-Cano, I.; Jiménez-Flores, R. Functional, Textural, and Rheological Properties of Mixed Casein Micelle and Pea Protein Isolate Co-Dispersions. *JDS Commun.* **2022**, *3*, 85–90. [CrossRef]
99. Fischer, E.; Cayot, N.; Cachon, R. Potential of Microorganisms to Decrease the "Beany" Off-Flavor: A Review. *J. Agric. Food Chem.* **2022**, *70*, 4493–4508. [CrossRef]

100. Clark, A.J.; Soni, B.K.; Sharkey, B.; Acree, T.; Lavin, E.; Bailey, H.M.; Stein, H.H.; Han, A.; Elie, M.; Nadal, M. Shiitake Mycelium Fermentation Improves Digestibility, Nutritional Value, Flavor and Functionality of Plant Proteins. *LWT* **2022**, *156*, 113065. [CrossRef]
101. Canon, F.; Maillard, M.B.; Famelart, M.H.; Thierry, A.; Gagnaire, V. Mixed Dairy and Plant-Based Yogurt Alternatives: Improving Their Physical and Sensorial Properties through Formulation and Lactic Acid Bacteria Cocultures. *Curr. Res. Food Sci.* **2022**, *5*, 665–676. [CrossRef]
102. Taha, A.; Casanova, F.; Šimonis, P.; Stankevič, V.; Gomaa, M.A.E.; Stirkė, A. Pulsed Electric Field: Fundamentals and Effects on the Structural and Techno-Functional Properties of Dairy and Plant Proteins. *Foods* **2022**, *11*, 1556. [CrossRef]
103. Giteru, S.G.; Oey, I.; Ali, M.A. Feasibility of Using Pulsed Electric Fields to Modify Biomacromolecules: A Review. *Trends Food Sci. Technol.* **2018**, *72*, 91–113. [CrossRef]
104. Jin, S.; Yin, Y.; Wang, Y. Effects of Combined Pulsed Electric Field and Heat Treatment on Texture Characteristics of Whey Protein Gels. *Nongye Jixie Xuebao/Trans. Chin. Soc. Agric. Mach.* **2013**, *44*, 142–146. [CrossRef]
105. Rodrigues, R.M.; Martins, A.J.; Ramos, O.L.; Malcata, F.X.; Teixeira, J.A.; Vicente, A.A.; Pereira, R.N. Influence of Moderate Electric Fields on Gelation of Whey Protein Isolate. *Food Hydrocoll.* **2015**, *43*, 329–339. [CrossRef]
106. Sharma, P.; Oey, I.; Everett, D.W. Thermal Properties of Milk Fat, Xanthine Oxidase, Caseins and Whey Proteins in Pulsed Electric Field-Treated Bovine Whole Milk. *Food Chem.* **2016**, *207*, 34–42. [CrossRef]
107. Subaşı, B.G.; Jahromi, M.; Casanova, F.; Capanoglu, E.; Ajalloueian, F.; Mohammadifar, M.A. Effect of Moderate Electric Field on Structural and Thermo-Physical Properties of Sunflower Protein and Sodium Caseinate. *Innov. Food Sci. Emerg. Technol.* **2021**, *67*, 102593. [CrossRef]
108. Li, Y.; Chen, Z.; Mo, H. Effects of Pulsed Electric Fields on Physicochemical Properties of Soybean Protein Isolates. *LWT* **2007**, *40*, 1167–1175. [CrossRef]
109. Zhang, L.; Wang, L.J.; Jiang, W.; Qian, J.Y. Effect of Pulsed Electric Field on Functional and Structural Properties of Canola Protein by Pretreating Seeds to Elevate Oil Yield. *LWT* **2017**, *84*, 73–81. [CrossRef]
110. Chen, Y.; Wang, T.; Zhang, Y.; Yang, X.; Du, J.; Yu, D.; Xie, F. Effect of Moderate Electric Fields on the Structural and Gelation Properties of Pea Protein Isolate. *Innov. Food Sci. Emerg. Technol.* **2022**, *77*, 102959. [CrossRef]
111. Balasubramaniam, V.M.; Barbosa-Canovas, G.V.; Lelieveld, H.L. *High Pressure Processing of Food: Principles, Technology and Application*; Springer: Berlin/Heidelberg, Germany, 2016.
112. Queirós, R.P.; Saraiva, J.A.; da Silva, J.A.L. Tailoring Structure and Technological Properties of Plant Proteins Using High Hydrostatic Pressure. *Crit. Rev. Food Sci. Nutr.* **2018**, *58*, 1538–1556. [CrossRef]
113. Sim, S.Y.J.; Hua, X.Y.; Henry, C.J. A Novel Approach to Structure Plant-Based Yogurts Using High Pressure Processing. *Foods* **2020**, *9*, 1126. [CrossRef]
114. Kumar, P.K.; Sivabalan, S.; Parhi, A.; Sablani, S.S. Modification of Pea Protein Isolate Functionality by Freeze–Thaw Cycling. *J. Food Meas. Charact.* **2022**, *16*, 162–170. [CrossRef]
115. Li, X.; Ye, C.; Tian, Y.; Pan, S.; Wang, L. Effect of Ohmic Heating on Fundamental Properties of Protein in Soybean Milk. *J. Food Process Eng.* **2018**, *41*, e12660. [CrossRef]
116. Sarigiannidou, K.; Odelli, D.; Jessen, F.; Mohammadifar, M.A.; Ajalloueian, F.; Vall-llosera, M.; Carvalho, A.F.; Casanova, F. Interfacial Properties of Pea Protein Hydrolysate: The Effect of Ionic Strength. *SSRN Electron. J.* **2022**, *6*, 76. [CrossRef]
117. Pua, A.; Tang, V.C.Y.; Goh, R.M.V.; Sun, J.; Lassabliere, B.; Liu, S.Q. Ingredients, Processing, and Fermentation: Addressing the Organoleptic Boundaries of Plant-Based Dairy Analogues. *Foods* **2022**, *11*, 875. [CrossRef]
118. Harper, A.R.; Dobson, R.C.J.; Morris, V.K.; Moggré, G.J. Fermentation of Plant-Based Dairy Alternatives by Lactic Acid Bacteria. *Microb. Biotechnol.* **2022**, *15*, 1404–1421. [CrossRef] [PubMed]

Disclaimer/Publisher's Note: The statements, opinions and data contained in all publications are solely those of the individual author(s) and contributor(s) and not of MDPI and/or the editor(s). MDPI and/or the editor(s) disclaim responsibility for any injury to people or property resulting from any ideas, methods, instructions or products referred to in the content.

Article

Microproteomic-Based Analysis of the Goat Milk Protein Synthesis Network and Casein Production Evaluation

Li Chen [1,2,*], Hiroaki Taniguchi [3,4] and Emilia Bagnicka [1,*]

1. Department of Biotechnology and Nutrigenomics, Institute of Genetics and Animal Biotechnology, Polish Academy of Sciences, 05-552 Jastrzębiec, Poland
2. College of Food Engineering and Nutritional Science, Shaanxi Normal University, Xi'an 710119, China
3. Department of Experimental Embryology, Institute of Genetics and Animal Biotechnology, Polish Academy of Sciences, 05-552 Jastrzębiec, Poland; h.taniguchi@igbzpan.pl
4. African Genome Center, University Mohammed VI Polytechnic (UM6P), Lot 660, Hay Moulay Rachid, Ben Guerir 43150, Morocco

* Correspondence: l.chen@igbzpan.pl (L.C.); e.bagnicka@igbzpan.pl (E.B.); Tel.: +48-(22)-7367000 (E.B.)

Abstract: Goat milk has been consumed by humans since ancient times and is highly nutritious. Its quality is mainly determined by its casein content. Milk protein synthesis is controlled by a complex network with many signal pathways. Therefore, the aim of our study is to clearly depict the signal pathways involved in milk protein synthesis in goat mammary epithelial cells (GMECs) using state-of-the-art microproteomic techniques and to identify the key genes involved in the signal pathway. The microproteomic analysis identified more than 2253 proteins, with 323 pathways annotated from the identified proteins. Knockdown of *IRS1* expression significantly influenced goat casein composition (α, β, and κ); therefore, this study also examined the insulin receptor substrate 1 (IRS1) gene more closely. A total of 12 differential expression proteins (DEPs) were characterized as upregulated or downregulated in the *IRS1*-silenced sample compared to the negative control. The enrichment and signal pathways of these DEPs in GMECs were identified using GO annotation and KEGG, as well as KOG analysis. Our findings expand our understanding of the functional genes involved in milk protein synthesis in goats, paving the way for new approaches for modifying casein content for the dairy goat industry and milk product development.

Keywords: goat milk protein synthesis network; microproteomic analysis; *IRS1* gene; milk protein content evaluation

Citation: Chen, L.; Taniguchi, H.; Bagnicka, E. Microproteomic-Based Analysis of the Goat Milk Protein Synthesis Network and Casein Production Evaluation. *Foods* **2024**, *13*, 619. https://doi.org/10.3390/foods13040619

Academic Editors: Paolo Formaggioni and Piero Franceschi

Received: 23 January 2024
Revised: 9 February 2024
Accepted: 13 February 2024
Published: 19 February 2024

Copyright: © 2024 by the authors. Licensee MDPI, Basel, Switzerland. This article is an open access article distributed under the terms and conditions of the Creative Commons Attribution (CC BY) license (https://creativecommons.org/licenses/by/4.0/).

1. Introduction

Goat milk has greater digestibility and alkalinity, as well as a higher buffering capacity, than cow's milk. Therefore, it is highly praised for its unique nutritional and functional properties. It also has better emulsifying and foaming properties and is favored by manufacturers in developing new food products. Goat milk proteins also contain higher levels of certain amino acids, such as tryptophan and cysteine, compared to cow milk proteins and are believed to possess immunomodulatory, allergy management, anti-inflammatory, and antioxidant effects, as well as antimicrobial and anticancer properties [1,2]. Furthermore, people who are allergic to cow milk may feel comfortable with goat milk because of its lower lactose content and the different forms of proteins found therein [3–6].

Initial information on milk secretion was obtained from goats' milk, and this has provided an insight into the processes occurring in mammary glands and cows' udders. Milk protein is secreted by mammary epithelial cells (MECs), in which milk quality is strongly influenced by casein production [7]. Milk protein, consisting of approximately 80% casein and 20% whey, plays an important role in the production of cheese and other dairy products. Promoting milk production is a priority for food science in general, and the dairy goat sector is particularly in need of a way to increase casein content to ensure its development.

Due to the high kinase activity of insulin receptors, the mammary gland remains highly sensitive to insulin throughout pregnancy and the lactation period [8]. Milk protein synthesis requires the activity of insulin, amino acids, and amino acid transporters, as well as the mammalian target of rapamycin (mTOR) pathway [9–11].

To better understand the pathways of milk protein synthesis, proteomic techniques have been used to investigate the functional proteins in animal tissues [12–14]. Although standard (macro)proteomic application is suitable for large samples with protein losses measured in micrograms or milligrams, it is not sensitive enough for small numbers of cell samples. Moreover, sample preparation consists of several steps that can lead to protein loss, thus reducing the levels of low-abundance proteins and preventing their accurate identification. Fortunately, microproteomic (μP) approaches have been developed for the analysis of samples with attomolar protein concentrations, where even proteins present in sub-microgram levels can be analyzed while retrieving useful proteome data [15,16].

To date, no μP pipeline analyses have been performed on milk protein synthesis pathways in goat mammary epithelial cells (GMECs). Therefore, the present study examines the pathways of milk protein synthesis in GMECs using μP pipelines with the aid of state-of-the-art mass spectrometers and Orbitrap instruments. The results will shed greater light on the key genes taking part in milk protein synthesis networks and provide a novel insight into milk protein synthesis mechanisms in GMECs.

2. Materials and Methods

2.1. Cell Culture

The purified primary GMECs were donated by Prof. Jernej Ogorevc from the University of Ljubljana, Slovenia. Mammary tissue was obtained from slaughtered lactating Saanen goats (*Capra hircus*), which were approximately three years old at the peak of lactation [17]. The purified fourth-passage cells were selected with the basal medium, including 90% DMEM/F12 (11320033, Gibco, Thermo Fisher Scientific, Waltham, MA, USA), 10% fetal bovine serum (E5050, EURX, Gdańsk, Poland), 1% penicillin–streptomycin supplemented with 1 μg/mL of hydrocortisone (H6909, BioXtra, Sigma-Aldrich, Darmstadt, Germany), 10 ng/mL of epidermal growth factor (PHG0311, Gibco, Thermo Fisher Scientific, Waltham, MA, USA), 5 μg/mL of insulin solution from bovine pancreas (I0516, BioReagent, Sigma-Aldrich, Darmstadt, Germany), and L-glutamine (G7513, BioXtra, Sigma-Aldrich, Darmstadt, Germany) at a final concentration of 4.5 mM [17].

2.2. Microproteomic Analysis

A high-resolution mass spectrometer (MS) was used to analyze the microsample data. The microsample data of MS were processed with MaxQuant's integrated Andromeda engine and the "match between runs" mode [15]. The analysis was based on peptide peak intensity, peak area, and LC retention time related to MS1, as well as other information. The data were subjected to statistical analysis and quality control before the GO, KOG, pathway, and other functional annotation analyses.

2.2.1. Microsample Protein Extraction and Enzymolysis

Protein extraction and enzymolysis were performed by BGI Genomics Co., Ltd. (Shenzhen, China). The cell sample was mixed with 10 μL 50 mM ammonium bicarbonate, subjected to ultrasonic lysis for 10 min, and incubated with DL-Dithiothreitol (DTT) at a final concentration of 10 mM in a water bath at 37 °C for 30 min. Following this, iodoacetamide solution (IAM) was added to a final concentration of 55 mM and left to react for 45 min in the dark. Finally, 1 μg trypsin was added for enzymatic hydrolysis at 37 °C for two hours [15].

2.2.2. Microsample MS Analysis

Protein separation was performed using a Thermo UltiMate 3000 UHPLC through a trap column and a self-packed C18 column at a flow rate of 500 nL/min. Peptide separation for DDA (data-dependent acquisition) mode was performed using a combined nanoESI source and Orbitrap Fusion™ Lumos™ Tribrid™ (Thermo Fisher Scientific, San Jose, CA, USA). The identification data were selected at PSM-level FDR $\leq 1\%$, and the significant identification was collected at protein-level FDR $\leq 1\%$ [15].

2.3. Differential Quantification Analysis

The proteins identified in each sample were quantified using MaxQuant to determine their levels in each sample [18]. The data were subjected to Welch's t-test to test the preset comparison group and calculate the multiple of differences. Significant differences were indicated by a fold change > 1.5 and p value < 0.05.

2.4. Bioinformatics Analysis

In all samples, proteins were identified using Gene Ontology (GO) functional annotation analysis [19]. The GO analysis was based on three ontologies (cellular component, biological process, and molecular function); the IDs and the number of proteins of all the corresponding proteins were listed. The identified proteins were classified into functional divisions using eukaryotic orthologous group (KOG) annotation according to the KOG database. The Kyoto Encyclopedia of Genes and Genomes (KEGG) pathway database was used to help further understand their biological functions.

2.5. RNAi

The siRNA used in this study was synthesized by Merck Life Science (Poznań, Poland) (Table 1). RNAi was performed by Lipofectamine™ RNAiMAX Transfection Reagent (13778075, Invitrogen, Thermo Fisher, Waltham, MA, USA) with the Opti-MEM® I Reduced Serum Medium (31985070, Gibco, Thermo Fisher, Waltham, MA, USA) [20]. MISSION® siRNA Universal Negative Control #1 (SIC001, Sigma-Aldrich, Darmstadt, Germany) was used as a negative control at a concentration of 20 µM. The cells were combined with a transfection mixture at a concentration of 0.15×10^6/mL and then incubated for 48 h at 37 °C and 5% CO_2. RNA was isolated to determine silencing efficiency.

Table 1. The design of siRNA.

Target Name	Sense/Antisense	siRNA Design	Start on Target	Target Sequence
IRS1	s	CUACCAUUUCCCACCAGAAdTdT	1595	CTACCATTTCCCACCAGAA
IRS1	a	UUCUGGUGGGAAAUGGUAGdTdT	1595	TTCTGGTGGGAAATGGTAG
IRS1	s	CACUUUACCUCGGGCCCGAdTdT	2607	CACTTTACCTCGGGCCCGA
IRS1	a	UCGGGCCCGAGGUAAAGUGdTdT	2607	TCGGGCCCGAGGTAAAGTG
IRS1	s	CAUUGAGGAAUAUACUGAAdTdT	1641	CATTGAGGAATATACTGAA
IRS1	a	UUCAGUAUAUUCCUCAAUGdTdT	1641	TTCAGTATATTCCTCAATG
IRS1	s	CAAAGAACCUGAUUGGCAUdTdT	527	CAAAGAACCTGATTGGCAT
IRS1	a	AUGCCAAUCAGGUUCUUUGdTdT	527	ATGCCAATCAGGTTCTTTG

2.6. RNA Isolation and Reverse Transcription Quantitative PCR (RT-qPCR)

Total RNA was isolated and purified with a NucleoSpin RNA Mini kit for RNA purification (Macherey-Nagel GmbH & Co. KG, Düren, Germany). RNA quantity and purity were determined using a Nanodrop 1000 (Thermo Scientific, Waltham, MA, USA) and a Bioanalyzer (Agilent 2100, Santa Clara, CA, USA). RNA was reverse transcribed with a Transcriptor First Strand cDNA synthesis kit (Roche, LifeScience Solutions, Basel, Switzerland) with random hexamer primers, according to the manufacturer's instructions [20]. The final concentration of total RNA was approximately 595 ng/μL in all samples for cDNA synthesis. The relative expression of genes was determined by RT-qPCR. Glyceraldehyde-3-phosphate dehydrogenase (GAPDH) was adopted as a reference. The primers were copied from previous studies on goat cells (Table 2).

Table 2. Primers for RT-qPCRs.

Gene	Sequence	Reference
GAPDH	For 5′-CATGTTTGTGATGGGCGTGAACCA-3′, Rev 5′-TAAGTCCCTCCACGATGCCAAAGT-3	[17]
IRS1	IRS1-F GTAGTGGCAAACTCCTGTCTTGT, IRS1-R GAGTAGTAGGAGAGGACGGGCT	[21]

Expression analysis was performed using a LightCycler 480 SYBR Green I Master (Roche, Basel, Switzerland) using at least three technical replicates for each sample [20]. The amplification reactions contained 2× Master Mix, 10× each PCR primer (0.4 μM), and water, to a total volume of 20 μL. The following sequence was performed: pre-incubation at 95 °C for 10 min, followed by amplification for 45 cycles at 95 °C for 10 s, 60 °C for 10 s, and 72 °C for 10 s. The melting curve was 95 °C for 5 s and 65 °C and 97 °C for 1 min. The $2^{-\Delta\Delta Ct}$ method was adopted to calculate relative gene expression.

2.7. Milk Protein Secretion Determination

The protein content was determined by the goat casein alpha (CSN1) ELISA kit, goat beta-casein (Csn2) ELISA kit, and goat kappa casein (κ-CN) ELISA kit. The absorbance was measured at OD450 nm with a microplate reader (TECAN F039300, Männedorf, Switzerland). All calculations were performed using CurveExpert Professional 2.6.5 software. All reagents were obtained from the Wuhan Xinqidi Biological technology Co., Ltd. (Wuhan, China).

2.8. Statistical Analysis

All the experiments were repeated three times with three replicates. All results were analyzed using Duncan's multiple-range tests ($p < 0.05$) by SAS 9.0 software (Cary, NC, USA).

3. Results

3.1. Pathway Annotation Analysis of GMEC Proteins

Protein function is associated with its biological behavior, which is related to many complex signal transduction pathways. The most important biochemical metabolic pathways and the signal transduction pathways formed by a protein can be determined by pathway analysis. Our present findings indicate the presence of more than 2253 proteins and about 337 pathway annotations among the quantified key proteins in GMECs (Table S1). About 42 of the identified proteins have been recorded in the insulin signaling pathway, and 44 in the mTOR signaling pathway. Milk protein synthesis is known to involve many complex factors [22]. The insulin–mTOR signal pathway merited particular attention because insulin has been reported to directly stimulate mTOR protein activity through phosphorylation [23].

3.2. Relative Quantitation of IRS1 Expression during Silence

The proteins belonging to the Insulin Receptor Substrate (IRS) family, IRS1, IRS2, IRS3, and IRS4, play a vital role in insulin signal transduction [24,25]. All four are phosphorylated on multiple tyrosine residues following insulin receptor kinase activation [26]. Previous studies found IRS1 to remarkably affect insulin-like growth factor and stimulate growth [27]. IRS1-deficient mice have mild glucose intolerance and insulin resistance [28]. IRS1 has also been found to be downregulated and to play a key role in cell proliferation and survival in breast cancer [29,30].

The present study used four pairs of synthetic siRNAs to silence the *IRS1* gene and then measure the relative expression of mRNA in all samples using RT-qPCR. It was found that mRNA expression was significantly reduced in all four siRNA samples compared to the negative control (NC), indicating successful blockage by the four synthetic siRNAs (Figure 1). The samples treated with the four siRNAs demonstrated similar mRNA expression, with no statistically significant difference between them.

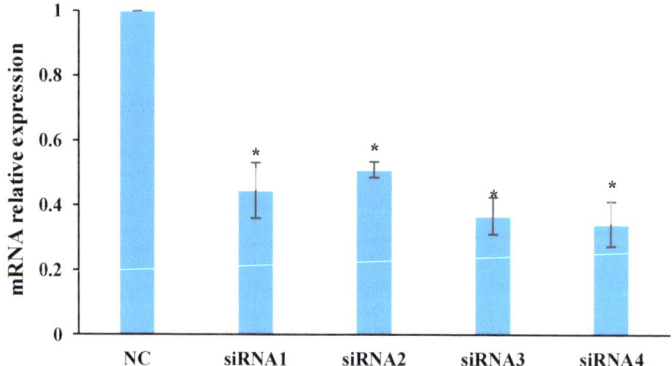

Figure 1. RT-qPCR analysis of *IRS1* expression in silenced GMECs. Relative gene expression was determined after transfection with negative control (NC), Lipo (lipofectamine™ RNAiMAX), siRNA1, siRNA2, siRNA3, and siRNA4 in GMEC for 48 h. The results are shown as mean ± SD, and the statistically significant analysis was calculated by Duncan's multiple-range tests. The asterisks indicate statistically significant differences, $p < 0.05$.

3.3. Casein Production Detection of GMECs

Goat milk protein consists of approximately 80% casein and 20% whey. The two have unique properties that can support the conversion of milk into yogurt and cheese. In turn, goat milk casein consists of four principal proteins: α_{s1}-casein (α_{s1}-CN), α_{s2}-casein (α_{s2}-CN), β-casein (β-CN), and κ-casein (κ-CN) [1]. Of these, β-casein is the most abundant in goat milk, and the allergen α_{s1}-casein is the most abundant in cow milk. As shown in Figure 2, the content of κ-, β-, and α-casein differed significantly between *IRS1*-silenced cells and controls: κ- and β-casein contents were higher, while α-casein content was lower. As the samples treated with the four siRNAs demonstrated similar casein contents, siRNA1 was selected for further study.

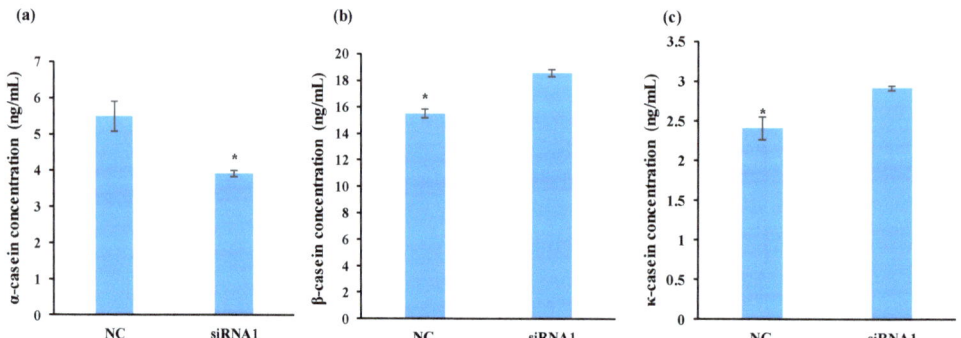

Figure 2. Determination of the content of (**a**) αs-1, (**b**) β-casein, and (**c**) κ-casein in GMECs. The casein content was determined after transfection with negative control (NC), Lipo (lipofectamine™ RNAiMAX), and siRNA1 in GMEC for 48 h. The column represents the mean ± SD; statistically significant differences were calculated by Duncan's multiple-range tests; the asterisks indicate statistically significant differences, $p < 0.05$.

3.4. Identification of Differential Expression Proteins (DEPs) by Microproteomic Analysis

The DEPs in the test samples are depicted in volcano plots in Figure 3. Nine DEPs in the siRNA1 sample were found to be upregulated, and three were downregulated compared to the NC samples in GMECs (Table S2). The upregulated DEPs were identified as Keratin, MAP7 domain, Syntaxin, KIAA1217 ortholog, Phosphodiesterase, Heme binding protein, Rhophilin Rho GTPase binding protein, and Myosin XVIIIA. The downregulated proteins were Protein arginine N-methyltransferase, Glutaredoxin, and Protein MAK16.

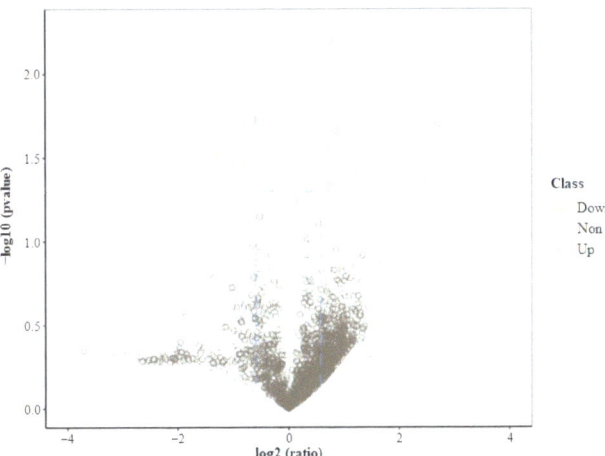

Figure 3. Volcano plot of screened DEPs. The x-axis indicates the protein difference multiple, while the y-axis is the $-\log 10$ (p value). A gray dot indicates a non-significantly altered protein (following silencing), a red dot indicates an upregulated protein, and a green dot indicates a downregulated protein.

3.5. KOG Analysis of the DEPs in GMECs

KOG analysis was used to classify DEPs in the NC vs. siRNA1 samples into three divisions: cellular process and signaling, information storage and processing, and poorly characterized. In Figure 4, it can be seen that most DEPs belong to the cellular process and signaling division: post-translational modification, protein turnover, and chaperones. Others were classified as information storage and processing, with the most common function being transcription. Finally, in the poorly characterized division, general function prediction only was noted.

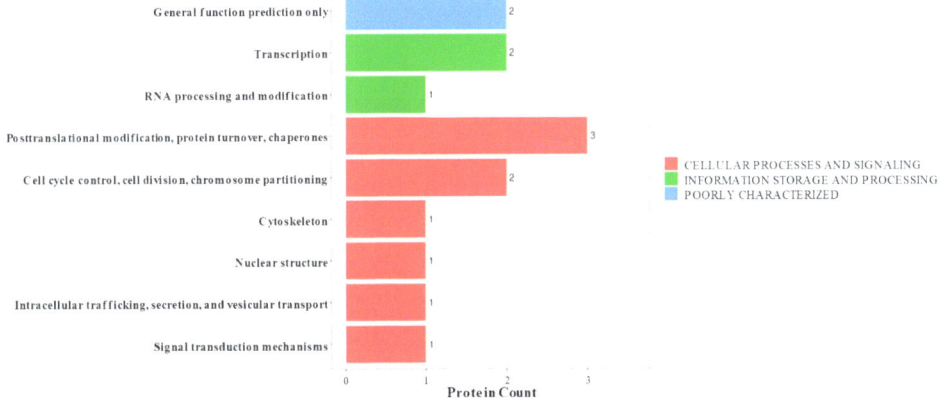

Figure 4. DEPs' KOG annotation. The x-axis represents protein count, and the y-axis represents KOG terms.

3.6. GO Analysis of DEPs

The DEPs in GMECs in NC vs. siRNA1 were classified into the cellular composition, biological processes, and molecular function groups according to GO annotation (Table S3). The GO function up and down chart of the DEPs is given in Figure 5. In the *biological process* division, the most upregulated proteins belong to biological regulation, cellular process, and regulation of biological processes, while the major downregulated proteins belong to cellular process and metabolic process. In the *cellular component* division, both the most upregulated and downregulated proteins belonged to the cell, cell part, and organelle groups. Finally, in *molecular function*, the most upregulated components belonged to binding, while the most downregulated proteins belonged to binding and catalytic activity.

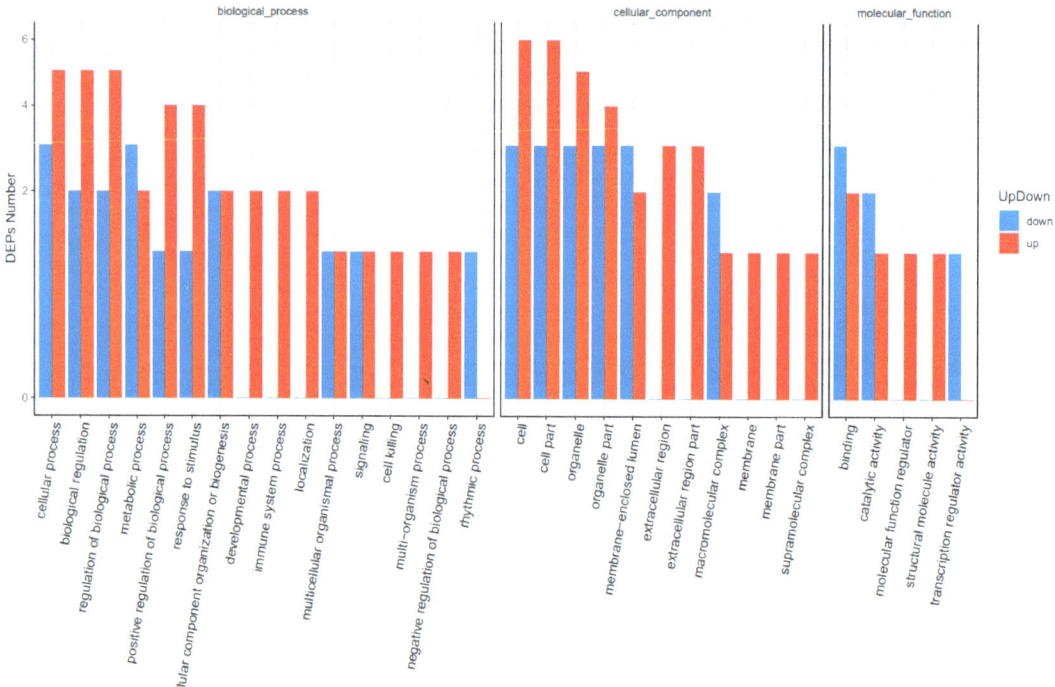

Figure 5. DEP GO function classification up and down chart. The *x*-axis is GO annotation, and the *y*-axis is DEP number.

A GO term relationship network was established to describe DEP enrichment (Figure 6). In the diagram, a node indicates a GO term. Green indicates cellular components, red biological processes, and blue molecular functions. *Biological processes* included two positive regulations (cellular process and response to stimulus) and nine negative regulations (RNA metabolic process, cellular metabolic process, macromolecule metabolic process, and nucleobase-containing compound metabolic process). No GO term regulation was observed in the *cellular component* or *molecular function* divisions.

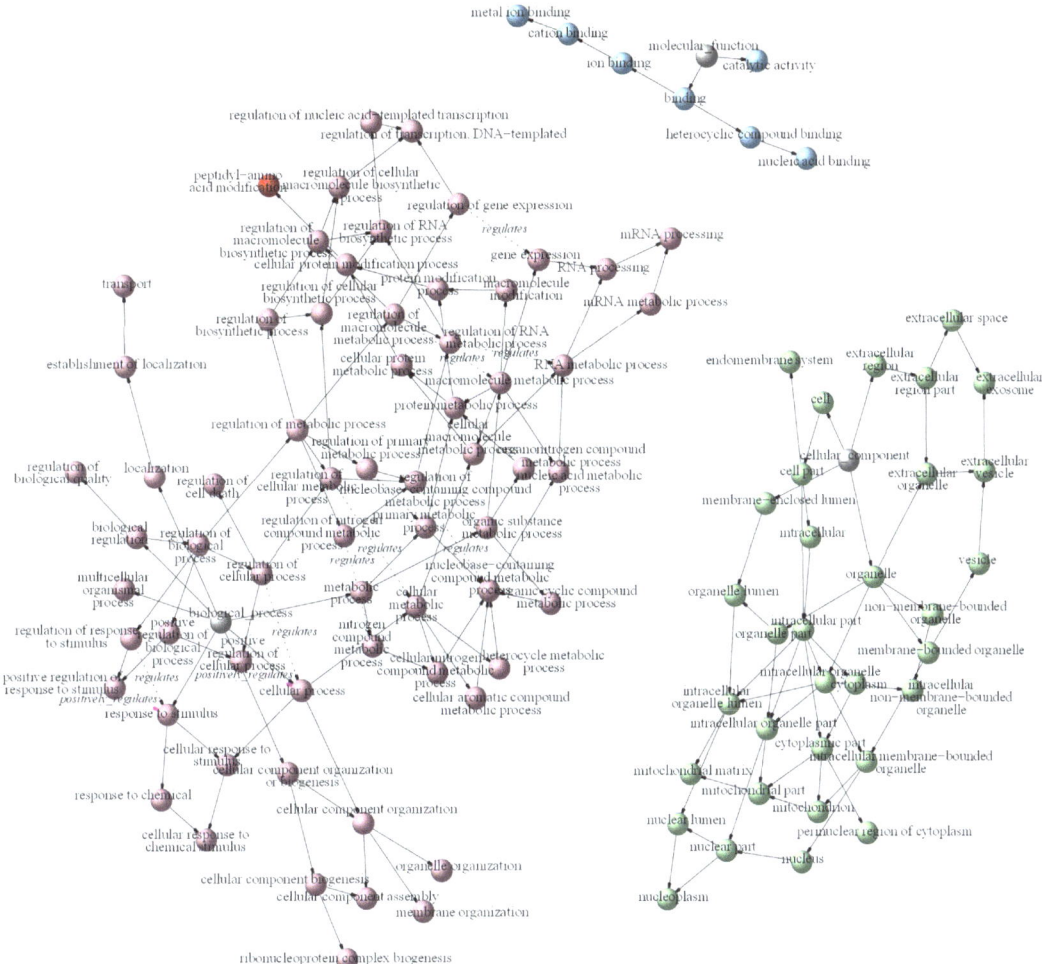

Figure 6. GO term relationship network diagram. A node is a GO term. The colors indicate different functional categories. Red indicates biological processes, green cellular components, and blue molecular functions. Dark colors indicate significantly enriched GO terms, light colors indicate insignificant GO terms, and grays indicate no enriched GO terms. A solid arrow indicates an inclusion relationship between GO terms, while a dotted arrow indicates a regulation relationship. A red dotted line suggests positive regulation, and a green dotted line negative regulation.

3.7. KEGG Pathway Analysis of DEPs

The KEGG pathway analyses classified the DEPs into cellular process, genetic information processing, and metabolism divisions (Figure 7). The main pathways involved were folding, sorting and degradation, translation, global and overview maps, and metabolism of cofactors and vitamins.

3.8. Subcellular Localization Prediction of DEPs

Subcellular localization prediction refers to the computational task of determining the specific location of a protein within a cell. Proteins perform their functions within specific compartments or organelles within the cell. Understanding the subcellular localization of molecules and organelles is essential for studying cellular processes, signaling pathways, and the mechanisms underlying health and disease. Subcellular localization prediction can be crucial for understanding the functions of DEPs within cells, as it can provide insights into their functions and roles within cellular processes. The DEPs were classified into six divisions: plasma membrane (plas), cytosol (cyto), mitochondrion (mito), nucleus (nucl), cytosol and nucleus (cyto_nucl), and endoplasmic reticulum (E.R.) (Table S4). The main subcellular locations of DEPs were cyto, nucl, and mito (Figure 8).

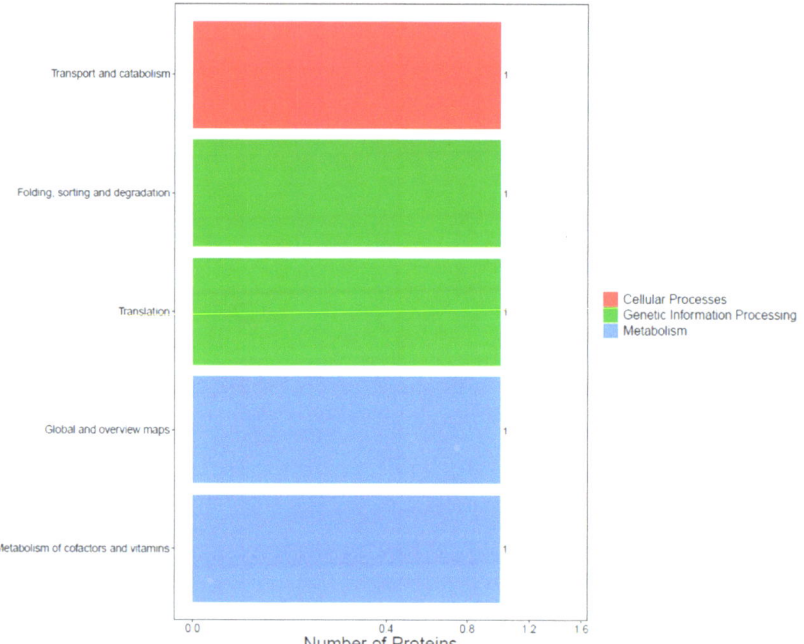

Figure 7. DEP pathway classification by KEGG enrichment. The x-axis represents protein number, and the y-axis represents KEGG pathway enrichment.

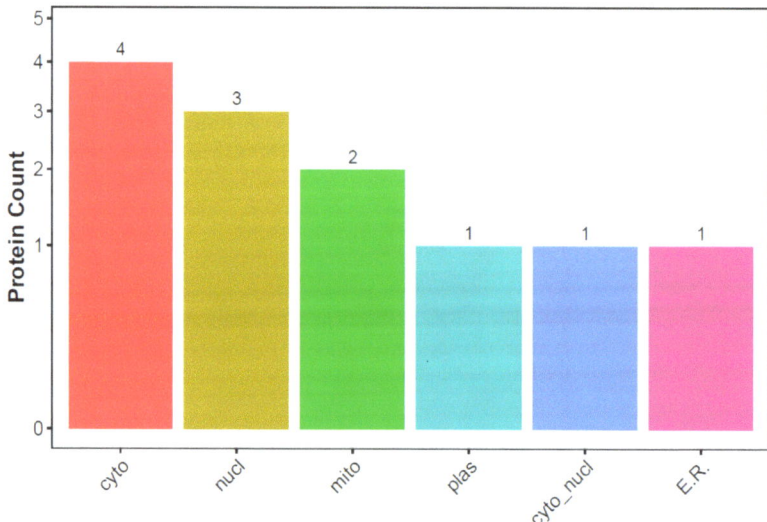

Figure 8. Subcellular localization prediction. The x-axis represents subcellular structure, and the y-axis represents protein number.

4. Discussion

Milk protein is secreted by mammary epithelial cells (MECs), and casein content is a key determinant of milk quality [31]. Recent years have seen a number of studies aimed at increasing milk protein secretion in MECs based on molecular mechanisms and signal pathways [32,33]. However, milk protein synthesis is a complex process. AMP-activated protein kinase (AMPK) and tumor suppressor LKB1 are located upstream of mTOR [34]. AMPK activates ATP-generating pathways and inhibits ATP consumption. The inhibition of mTORC1 mediated by LKB1 relies on AMPK and TSC2. Milk protein synthesis also involves the insulin–mTOR pathway. All these signaling pathways have been confirmed by our μP approaches.

As shown in Table S1, about 66 of the proteins identified in GMECs belong to the PI3K-Akt signaling pathway, which has been shown to play an important role in camel milk protein networks [35]. Interference in the PI3K-Akt signal may lead to insulin resistance, resulting in the creation of a vicious circle [36]. Additionally, many of the identified proteins were found to be associated with more than 330 pathways, including MAPK signaling, insulin signaling, necroptosis, apoptosis, biosynthesis of amino acids, AMPK signaling, mTOR signaling, and TNF signaling; some of these are closely linked with milk protein synthesis (Table S1). Previous studies indicate that the insulin–mTOR pathway plays a role in regulating milk protein synthesis, with insulin directly stimulating the mTOR protein.

The IRS-family proteins are closely associated with the insulin signal pathway [37,38]. Indeed, IRS1 can be found in the central part of a signaling pathway network diagram of camel milk proteins designed by Han (2022) [35]. Our study explored the role of *IRS1* in goat milk synthesis and casein composition. Casein plays an important role in cheese making as it dictates how well, and how rapidly, the milk clots and forms a curd. Any changes in the amount of α-CN or β-CN would alter the properties of the milk and the resulting cheese [39]. Our findings indicate that *IRS1* silencing significantly influenced the content of κ-casein, β-casein, and α-casein in GMECs (Figure 2).

Previous studies found goat milk with altered α_{S1}-CN contents to be allergenic in guinea pigs [40]. Goat milk lacking α_{S1}-CN was less allergenic than other goat milks, probably due to its modified ratio of β-LG to α_S-CN. In the present study, the *IRS1*-silenced sample demonstrated higher levels of β-casein and lower levels of α-casein. Unfortunately, as little is currently known about casein synthesis, it is hard to explain these changes. Nevertheless, these findings encourage further research in the area.

In the present study, microproteomic analysis identified about 12 DEPs among more than 2253 proteins (Figure 3 and Table S2). Among these, the upregulated DEPs were Keratin, MAP7 domain, Syntaxin, KI-AA1217 ortholog, Phosphodiesterase, Heme binding protein, Rhophilin Rho GTPase binding protein, and Myosin XVIIIA. The downregulated DEPs were Protein arginine N-methyltransferase, Glutaredoxin, and Protein MAK16.

Syntaxin is involved in vesicle trafficking and membrane fusion events within cells, particularly during exocytosis [41], the cellular process in which substances are released from vesicles into the extracellular space. Although syntaxin itself is not directly implicated in the synthesis of milk proteins, proteins involved in vesicle trafficking, membrane fusion, and intracellular transport could indirectly impact the secretion of milk proteins. These processes are crucial for the proper packaging and release of proteins from cells, including the MECs responsible for milk production.

Phosphodiesterases play a crucial role in intracellular signaling by hydrolyzing cyclic nucleotides, particularly cyclic guanosine monophosphate (cGMP) and cyclic adenosine monophosphate (cAMP) [42]. These cyclic nucleotides are involved in signaling pathways that regulate various cellular processes. Although there may not be a direct link between phosphodiesterase and milk protein synthesis, alterations in cyclic nucleotide levels regulated by phosphodiesterases could potentially influence cellular processes and signaling pathways, indirectly impacting milk protein synthesis. Protein Arginine N-Methyltransferase belongs to a family of enzymes involved in the methylation of arginine residues in proteins. They play various roles in cellular processes, including gene expression regulation, signal transduction, and RNA processing [43].

All these upregulated and downregulated proteins were associated with the modified casein composition in GMECs. Doubtlessly, changes in the levels of α-CN or β-CN would alter the properties of milk and the produced cheese, influencing their processing. Furthermore, increasing evidence indicates that β-asomorphin-7 derived from A1 β-casein contributes to milk intolerance syndrome. Our findings provide interesting information for the fields of milk processing and nutrition mechanisms.

The GO annotations found two DEPs to be upregulated, with these associated with cellular process and response to stimulus, and nine to be downregulated, related to RNA metabolic process, cellular metabolic process, macromolecule metabolic process, and nucleobase-containing compound metabolic process (Figure 6).

Milk protein synthesis is a complex process that occurs in the mammary glands and is generally associated with hormonal signals, nutritional factors, and the specific needs of the developing offspring [32,44]. Therefore, to better understand the role of IRS1 in goat mammary glands and milk protein synthesis, further functional studies of the proteins influenced by IRS1 silencing in GMECs are needed.

5. Conclusions

Our findings confirm that the *IRS1* gene influences the casein content of milk in goats and the milk protein synthesis pathways in GMECs. Modifying the expression of *IRS1* could increase the amount of κ-casein and β-casein but decrease the content of α-casein. This study is the first to successfully use a microproteomic approach to analyze the proteins of GMECs with low amount requirements. By identifying the proteins that were differentially expressed in response to *IRS1* silencing, it was possible to gain a new insight into the goat milk protein synthesis network and related signal pathways. Some DEPs were found to indirectly influence milk protein synthesis based on their GO annotation and their KEGG and KOG analysis. These findings may have positive implications for future studies on the milk synthesis system in goats.

Supplementary Materials: The following supporting information can be downloaded at https://www.mdpi.com/article/10.3390/foods13040619/s1: Table S1: Pathway annotation analysis of identified proteins in GMECs; Table S2: Identified DEPs in *IRS1*-silenced GMECs; Table S3: GO enriched DEPs; Table S4: Subcellular localization of DEPs.

Author Contributions: Conceptualization, L.C. and E.B.; methodology, L.C., E.B. and H.T.; writing—original draft preparation, L.C.; writing—review and editing, L.C. and E.B.; supervision, E.B. and H.T.; funding acquisition, L.C. and E.B. All authors have read and agreed to the published version of the manuscript.

Funding: This research was funded by the Horizon 2020 Marie Skłodowska-Curie Action PASIFIC 847639 and the National Nature Science Foundation of China 32101908.

Institutional Review Board Statement: Not applicable.

Informed Consent Statement: Not applicable.

Data Availability Statement: The original contributions presented in the study are included in the article/supplementary material, further inquiries can be directed to the corresponding authors.

Acknowledgments: We offer our gratitude to Jernej Ogorevc, University of Ljubljana, Slovenia, for the GMEC donation and to Hossein Khodadadi for technical support.

Conflicts of Interest: The authors declare no conflicts of interest.

References

1. Prosser, C.G. Compositional and functional characteristics of goat milk and relevance as a base for infant formula. *J. Food Sci.* **2021**, *86*, 257–265. [CrossRef]
2. Dermit, Z.F.; Mikulec, N.; Ljoljic, D.B.; Antunac, N. Nutritional and therapeutic properties of goat's milk. *Mljekarstvo* **2014**, *64*, 280–286. [CrossRef]

3. Osorio, J.S.; Lohakare, J.; Bionaz, M. Biosynthesis of milk fat, protein, and lactose: Roles of transcriptional and posttranscriptional regulation. *Physiol. Genom.* **2016**, *48*, 231–256. [CrossRef]
4. Chen, L.; Bagnicka, E.; Chen, H.; Shu, G. Health potential of fermented goat dairy products: Composition comparison with fermented cow milk, probiotics selection, health benefits and mechanisms. *Food Funct.* **2023**, *14*, 3423–3436. [CrossRef]
5. Ramakrishnan, M.; Eaton, K.; Sermet, O.M.; Savaiano, D.A. Milk containing A2 β-casein only, as a single meal, causes fewer symptoms of lactose intolerance than milk containing A1 and A2 β-caseins in subjects with lactose maldigestion and intolerance: A randomized, double-blind, crossover trial. *Nutrients* **2020**, *12*, 3855. [CrossRef]
6. Pal, S.; Woodford, K.; Kukuljan, S.; Ho, S. Milk intolerance, beta-casein and lactose. *Nutrients* **2015**, *7*, 7285–7297. [CrossRef]
7. ALKaisy, Q.H.; Al-Saadi, J.S.; Al-Rikabi, A.K.J.; Altemimi, A.B.; Hesarinejad, M.A.; Abedelmaksoud, T.G. Exploring the health benefits and functional properties of goat milk proteins. *Food Sci. Nutr.* **2023**, *11*, 5641–5656. [CrossRef]
8. Carrascosa, J.M.; Ramos, P.; Molero, J.C.; Herrera, E. Changes in the Kinase Activity of the Insulin Receptor Account for an Increased Insulin Sensitivity of Mammary Gland in Late Pregnancy. *Endocrinology* **1998**, *139*, 520–526. [CrossRef]
9. Luyimbazi, D.; Akcakanat, A.; McAuliffe, P.F.; Zhang, L.; Singh, G.; Gonzalez-Angulo, A.M.; Chen, H.Q.; Ko, K.A.; Zheng, Y.H.; Hung, M.C.; et al. Rapamycin regulates stearoyl CoA desaturase 1 expression in breast cancer. *Mol. Cancer Ther.* **2010**, *9*, 2770–2784. [CrossRef]
10. Li, Y.; Soos, T.J.; Li, X.; Wu, J.; Degennaro, M.; Sun, X.; Littman, D.R.; Birnbaum, M.J.; Polakiewicz, R.D. Protein kinase C Theta inhibits insulin signaling by phosphorylating IRS1 at Ser(1101). *J. Biol. Chem.* **2004**, *279*, 45304–45307. [CrossRef]
11. Zhu, C.; Zhu, J.; Duan, Q.; Jiang, Y.; Yin, H.; He, Y.L.; Li, F.; An, X.P. Exploration of the lactation function of protein phosphorylation sites in goat mammary tissues by phosphoproteome analysis. *BMC Genom.* **2021**, *22*, 703. [CrossRef]
12. Zhao, Q.; Li, K.; Jiang, K.; Yuan, Z.; Xiao, M.; Wei, G.; Zheng, W.; Wang, X.; Huang, A. Proteomic approach-based comparison of metabolic pathways and functional activities of whey proteins derived from Guishan and Saanen goat milk. *J. Dairy Sci.* **2023**, *106*, 2247–2260. [CrossRef]
13. Wang, C.; Zhao, R.; Zhao, Z.; Liu, N.; Cheng, J.; Guo, M. Proteomic characterization and comparison of milk fat globule membrane proteins of Saanen goat milk from 3 habitats in China using SWATH-MS technique. *J. Dairy Sci.* **2023**, *106*, 2289–2302. [CrossRef]
14. Song, B.; Lu, J.; Hou, Y.; Wu, T.; Tao, X.; Liu, D.; Wang, Y.; Regenstein, J.M.; Liu, X.; Zhou, P. Proteomic Comparisons of Caprine Milk Whole Cream Buttermilk Whey and Cheese Whey Cream Buttermilk. *J. Agric. Food Chem.* **2024**, *72*, 933–945. [CrossRef]
15. Alexovič, M.; Sabo, J.; Longuespée, R. Microproteomic sample preparation. *Proteomics* **2021**, *21*, e2000318. [CrossRef]
16. Chan, C.; Peng, J.; Rajesh, V.; Scott, E.Y.; Sklavounos, A.A.; Faiz, M.; Wheeler, A.R. Digital Microfluidics for Microproteomic Analysis of Minute Mammalian Tissue Samples Enabled by a Photocleavable Surfactant. *J. Proteome Res.* **2023**, *22*, 3242–3253. [CrossRef] [PubMed]
17. Ogorevc, J.; Dovč, P. Expression of estrogen receptor 1 and progesterone receptor in primary goat mammary epithelial cells. *Anim Sci. J.* **2016**, *87*, 1464–1471. [CrossRef] [PubMed]
18. Cox, J.; Mann, M. MaxQuant enables high peptide identification rates, individualized p.p.b.-range mass accuracies and proteome-wide protein quantification. *Nat. Biotechnol.* **2008**, *26*, 1367–1372. [CrossRef]
19. Jin, D.; Liu, H.; Bu, L.; Ke, Q.; Li, Z.; Han, W.; Zhu, S.; Liu, C. Comparative Analysis of Whey Proteins in Human Milk Using a Data-Independent Acquisition Proteomics Approach during the Lactation Period. *J. Agric. Food Chem.* **2021**, *69*, 4319–4330. [CrossRef] [PubMed]
20. Wang, J.; Tan, J.; Qi, Q.; Yang, L.; Wang, Y.; Zhang, C.; Hu, L.; Fang, X. miR-487b-3p Suppresses the Proliferation and Differentiation of Myoblasts by Targeting IRS1 in Skeletal Muscle Myogenesis. *Int. J. Biol. Sci.* **2018**, *14*, 760–774. [CrossRef]
21. Lyu, M.; Wang, X.; Meng, X.; Qian, H.; Li, Q.; Ma, B.; Zhang, Z.; Xu, K. chi-miR-487b-3p Inhibits Goat Myoblast Proliferation and Differentiation by Targeting IRS1 through the IRS1/PI3K/Akt Signaling Pathway. *Int. J. Mol. Sci.* **2022**, *23*, 115. [CrossRef]
22. Ma, X.M.; Blenis, J. Molecular mechanisms of mTOR-mediated translational control. *Nat. Rev. Mol. Cell Biol.* **2009**, *10*, 307–318. [CrossRef]
23. Yoon, M.S. The Role of Mammalian Target of Rapamycin (mTOR) in Insulin Signaling. *Nutrients* **2017**, *9*, 1176. [CrossRef]
24. Shaw, L.M. The insulin receptor substrate (IRS) proteins: At the intersection of metabolism and cancer. *Cell Cycle* **2011**, *10*, 1750–1756. [CrossRef]
25. Giovannone, B.; Scaldaferri, M.L.; Federici, M.; Porzio, O.; Lauro, D.; Fusco, A.; Sbraccia, P.; Borboni, P.; Lauro, R.; Sesti, G. Insulin receptor substrate (IRS) transduction system: Distinct and overlapping signaling potential. *Diabetes Metab. Res. Rev.* **2000**, *16*, 434–441. [CrossRef] [PubMed]
26. Su, J.; Tang, L.; Luo, Y.; Xu, J.; Ouyang, S. Research progress on drugs for diabetes based on insulin receptor/insulin receptor substrate. *Biochem. Pharmacol.* **2023**, *217*, 115830. [CrossRef] [PubMed]
27. Yang, Y.; Chan, J.Y.; Temiz, N.A.; Yee, D. Insulin Receptor Substrate Suppression by the Tyrphostin NT157 Inhibits Responses to Insulin-Like Growth Factor-I and Insulin in Breast Cancer Cells. *Horm. Cancer* **2018**, *9*, 371–382. [CrossRef] [PubMed]
28. Laustsen, P.G.; Michael, M.D.; Crute, B.E.; Cohen, S.E.; Ueki, K.; Kulkarni, R.N.; Keller, S.R.; Lienhard, G.E.; Kahn, C.R. Lipoatrophic diabetes in Irs1(−/−)/Irs3(−/−) double knockout mice. *Genes Dev.* **2002**, *16*, 3213–3222. [CrossRef] [PubMed]
29. Gibson, S.L.; Ma, Z.; Shaw, L.M. Divergent roles for IRS-1 and IRS-2 in breast cancer metastasis. *Cell Cycle* **2007**, *6*, 631–637. [CrossRef]
30. Byron, S.A.; Horwitz, K.B.; Richer, J.K.; Lange, C.A.; Zhang, X.; Yee, D. Insulin receptor substrates mediate distinct biological responses to insulin-like growth factor receptor activation in breast cancer cells. *Br. J. Cancer* **2006**, *95*, 1220–1228. [CrossRef]

31. Hue-Beauvais, C.; Faulconnier, Y.; Charlier, M.; Leroux, C. Nutritional Regulation of Mammary Gland Development and Milk Synthesis in Animal Models and Dairy Species. *Genes* **2021**, *12*, 523. [CrossRef] [PubMed]
32. Burgoyne, R.D.; Duncan, J.S. Secretion of Milk Proteins. *J. Mammary Gland Biol. Neoplasia* **1998**, *3*, 275–286. [CrossRef] [PubMed]
33. Dai, W.; White, R.; Liu, J.; Liu, H. Organelles coordinate milk production and secretion during lactation: Insights into mammary pathologies. *Prog. Lipid Res.* **2022**, *86*, 101159. [CrossRef]
34. Shaw, R.J. LKB1 and AMP-activated protein kinase control of mTOR signalling and growth. *Acta Physiol.* **2009**, *196*, 65–80. [CrossRef] [PubMed]
35. Han, B.; Zhang, L.; Luo, B.; Ni, Y.; Bansal, N.; Zhou, P. Comparison of milk fat globule membrane and whey proteome between Dromedary and Bactrian camel. *Food Chem.* **2022**, *367*, 130658. [CrossRef] [PubMed]
36. Huang, X.; Liu, G.; Guo, J.; Su, Z. The PI3K/AKT pathway in obesity and type 2 diabetes. *Int. J. Biol. Sci.* **2018**, *14*, 1483–1496. [CrossRef]
37. Lee, J.; Pilch, P.F. The insulin receptor: Structure, function, and signaling. *Am. J. Physiol.* **1994**, *266*, 319–334. [CrossRef]
38. Shi, Y.; Ma, Z.; Cheng, Q.; Wu, Y.; Parris, A.B.; Kong, L.; Yang, X. FGFR1 overexpression renders breast cancer cells resistant to metformin through activation of IRS1/ERK signaling. *Biochim. Biophys. Acta Mol. Cell Res.* **2021**, *1868*, 118877. [CrossRef]
39. St-Gelais, D.; Haché, S. Effect of β-casein concentration in cheese milk on rennet coagulation properties, cheese composition and cheese ripening. *Food Res. Int.* **2005**, *38*, 523–531. [CrossRef]
40. Ballabio, C.; Chessa, S.; Rignanese, D.; Gigliotti, C.; Pagnacco, G.; Terracciano, L.; Fiocchi, A.; Restani, P.; Caroli, A.M. Goat milk allergenicity as a function of $αs_1$-casein genetic polymorphism. *J. Dairy Sci.* **2011**, *94*, 998–1004. [CrossRef]
41. Teng, F.Y.; Wang, Y.; Tang, B.L. The syntaxins. *Genome Biol.* **2001**, *2*, 3012. [CrossRef]
42. Levy, I.; Horvath, A.; Azevedo, M.; de Alexandre, R.B.; Stratakis, C.A. Phosphodiesterase function and endocrine cells: Links to human disease and roles in tumor development and treatment. *Curr. Opin. Pharmacol.* **2011**, *11*, 689–697. [CrossRef] [PubMed]
43. Kim, J.H.; Yoo, B.C.; Yang, W.S.; Kim, E.; Hong, S.; Cho, J.Y. The Role of Protein Arginine Methyltransferases in Inflammatory Responses. *Mediat. Inflamm.* **2016**, *2016*, 4028353. [CrossRef] [PubMed]
44. Rezaei, R.; Wu, Z.L.; Hou, Y.Q.; Bazer, F.W.; Wu, G.Y. Amino acids and mammary gland development: Nutritional implications for milk production and neonatal growth. *J. Anim. Sci. Biotechnol.* **2016**, *7*, 20. [CrossRef] [PubMed]

Disclaimer/Publisher's Note: The statements, opinions and data contained in all publications are solely those of the individual author(s) and contributor(s) and not of MDPI and/or the editor(s). MDPI and/or the editor(s) disclaim responsibility for any injury to people or property resulting from any ideas, methods, instructions or products referred to in the content.

Article

Polymorphisms of *CCSER1* Gene and Their Correlation with Milk Quality Traits in Gannan Yak (*Bos grunniens*)

Guowu Yang [1,2,3], Juanxiang Zhang [1,2], Xiaoyong Ma [1,2], Rong Ma [1,2], Jinwei Shen [1,2], Modian Liu [1,2], Daoning Yu [1,2,3], Fen Feng [1,2], Chun Huang [1,2], Xiaoming Ma [1,2], Yongfu La [1,2], Xian Guo [1,2], Ping Yan [1,2] and Chunnian Liang [1,2,*]

[1] Key Laboratory of Animal Genetics and Breeding on Tibetan Plateau, Ministry of Agriculture and Rural Affairs, Lanzhou 730050, China; xueshengyangguowu@163.com (G.Y.); juanxiangzhang@163.com (J.Z.); mxy15609445561@163.com (X.M.); marong202017@163.com (R.M.); shenjw9090@163.com (J.S.); liumodian@126.com (M.L.); ydn9907@163.com (D.Y.); feng990111@163.com (F.F.); johnchun825@163.com (C.H.); maxiaoming@caas.cn (X.M.); layongfu@caas.cn (Y.L.); guoxian@caas.cn (X.G.); pingyanlz@163.com (P.Y.)
[2] Key Laboratory of Yak Breeding Engineering of Gansu Province, Lanzhou Institute of Husbandry and Pharmaceutical Sciences, Chinese Academy of Agricultural Sciences, Lanzhou 730050, China
[3] College of Life Sciences and Engineering, Northwest Minzu University, Lanzhou 730106, China
* Correspondence: liangchunnian@caas.cn

Abstract: Coiled-coil serine-rich protein 1 (*CCSER 1*) gene is a regulatory protein gene. This gene has been reported to be associated with various economic traits in large mammals in recent years. The aim of this study was to investigate the association between *CCSER1* gene single nucleotide polymorphisms (SNPs) and Gannan yaks and to identify potential molecular marker loci for breeding milk quality in Gannan yaks. We genotyped 172 Gannan yaks using Illumina Yak cGPS 7K liquid microarrays and analyzed the correlation between the three SNPs loci of the *CCSER1* gene and the milk qualities of Gannan yaks, including milk fat, protein and casein. It was found that mutations at the g.183,843A>G, g.222,717C>G and g.388,723G>T loci all affected the fat, protein, casein and lactose traits of Gannan yak milk to varying extents, and that the milk quality of individuals with mutant phenotypes was significantly improved. Among them, the milk fat content of AG heterozygous genotype population at g.183,843A>G locus was significantly higher than that of AA and GG genotype populations ($p < 0.05$); the casein and protein content of mutant GG and CG genotype populations at g.222,717C>G locus was significantly higher than that of wild-type CC genotype population ($p < 0.05$); and the g.388,723G>T locus of the casein and protein contents of the mutant TT genotype population were significantly higher ($p < 0.05$) than those of the wild-type GG genotype population. These results provide potential molecular marker sites for Gannan yak breeding.

Keywords: *CCSER1* gene; milk quality; SNPs; Gannan yak

1. Introduction

Yaks (*Bos grunniens*) are mainly distributed in high altitude areas of 2500~6000 m above sea level, and can fully adapt to extremely harsh environments such as those with low temperature, high altitude, and strong ultraviolet radiation [1]. These animals rely entirely on natural grasslands for food and nutrition and do not require supplemental feed [2]. Therefore, yak milk is an exceptionally non-polluting source of green food with great potential for further development and utilization [3]. In the past, Tibetans mainly drank yak milk, known as "liquid gold". Yak milk and its dairy products provide most of the energy, vitamins and nutrients that Tibetans need [4]. Compared with ordinary milk, yak milk contains higher levels of dry matter, milk fat, milk protein and other nutrients. Among them, yak casein is not only a source of antihypertensive peptides, but also a typical dietary protein, which can be used for various high-value-added functional diets [5]. Also, casein is the foremost source of essential amino acids [6,7]. Yak milk is a valuable source of nutrition and therefore ideal for the production of a wide range of dairy products [8]. In recent

years, there has also been a significant increase in interest in yak milk, and more and more research has been conducted on yak milk, such as studies on antioxidant properties [9], lactation mechanisms [10] and yak milk product development [11].

The quantity and quality of milk produced by yaks is regulated by a number of factors, such as seasonal variations, altitude and age. Milk composition is likewise a complex trait influenced by a number of factors, including genetic factors and environmental conditions (such as altitude, temperature, stage of lactation, season, herd and diet) [12,13]. It was found that the regions affecting milk quality traits such as milk protein composition were concentrated on chromosomes 1, 6, 11, 13, 14 and 18 [14]. Various genes have been reported in the literature to be associated with milk production traits, such as *DGAT1*, *HSF1*, *MGST1*, *GHR*, *ABCG2*, *ADCK5* and *CSN1S1* [14–16]. The diacylglycerol acyltransferase 1 (*DGAT1*) gene was shown to synergistically affect the milk quality of Egyptian Zaraibi goats through single nucleotide polymorphism (SNP) in the gene [15]. Teng et al. conducted a genome-wide association study of milk production traits in Holstein cows by using medium density microarray data and found that among the new genes that have not yet been reported, coiled-coil serine-rich protein 1 (*CCSER1*) also showed good potential as a candidate gene for milk quality [16]. The *CCSER1* gene, also named *FAM190A* (family with sequence similarity 190, member A), is a regulatory protein gene [17]. As a regulatory or structural component of normal mitosis, when *CCSER1* gene expression is altered, it causes chromosomal instability. In addition, the yak *CCSER1* gene, with a length of 1,597,616 bp and the presence of 10 transcripts, is one of the important functional genes located on chromosome 6 [18]. Studies have found that the *CCSER1* gene is associated with a variety of economic traits, such as growth traits [18,19], feed efficiency [20] and milk quality. Notably, a sequence analysis revealed that the CCSER1 protein has 23 serines in the first 69 amino acids, but no other identified structural domains or known functions [21]. Notably, the *CCSER1* gene has also been studied in cancer therapy. The results of Kang et al. demonstrated that transcript variants of the *CCSER1* gene can serve as clinical therapeutic targets for cancer patients and that the oncogenic properties of the *CCSER1* gene involve in-frame deletions [22].

Low yak milk yield limits the industrial production of yak milk [23]. Therefore, improving milk yield and milk quality are among the important breeding objectives in the yak industry [10]. It has been found that the bioactive components in yak milk have various functional properties, such as antioxidant, anticancer, antibacterial and blood pressure lowering abilities [24]. Compared to the milk fat of other animals, yaks, because of their specific high-altitude, low-oxygen, etc., grazing environments, make it possible for their milk fat to contain certain unique fatty acids [25]. These fatty acids may be associated with potential health advantages, such as anti-cancer and anti-diabetic effects, as well as positive effects on organs such as the brain, heart and eyes [26,27]. Hydrolysates of yak milk casein were found to have the ability to scavenge free radicals such as superoxide and hydrogen peroxide, as well as to inhibit the secretion of inflammatory actives such as pro-inflammatory cytokines and tumor necrosis factor-alpha [28]. Thus, hydrolysates of yak milk casein may be used for the prevention of oxidative-stress- and inflammation-related diseases. It has also been shown that yak casein differs from cow casein in several functional aspects due to its larger particle size, different amino acid sequence and richer mineral concentration, which results in greater conformational stability [29,30]. Therefore, the interest in yak milk quality has increased significantly. Within this is the improvement of milk quality by means of single nucleotide polymorphisms (SNPs), which belong to the third generation of genetic markers [12]. SNPs are changes in the DNA sequence that are prevalent when a base in a gene is altered. This variation mainly includes four types of single gene transition, transversion, deletion and insertion, but transition and transversion are predominant [31]. SNPs have been used to identify genes related to milk production traits in Holstein cows [32], but in dairy cows they have been used mainly for the identification of loci related to milk yield, with fewer studies on specific milk quality. However, SNPs have hardly been studied in yaks for milk yield or milk quality. The Gannan yak is an ancient and primitive livestock breed on the Tibetan Plateau, a unique local genetic resource in Gansu Province, China, and has strong resistance to adversity through long-term natural selection and artificial breeding. It is able to adapt to the ecological conditions of high altitude, strong radiation, a large temperature difference between day and night, a short growing period of pasture grass, extreme cold and little oxygen. In this study, Gannan yak milk was used as the research object. It aimed to investigate the novel SNPs in the *CCSER1* gene and their relationship with the quality of Gannan yak milk.

2. Materials and Methods

2.1. Ethics Approval

All the animal experiments were approved by the Lanzhou Institute of Husbandry and Pharmaceutical Sciences of the Chinese Academy of Agricultural Sciences (CAAS) with the grant number 1610322020018.

2.2. Animal and Milk Composition Analysis

Gannan yak milk samples were collected in July 2023 in Xiahe County, Gannan Tibetan Autonomous Prefecture, Gansu Province (34.99° E, 102.92° N, altitude 3000~3800 m). The Gannan yaks used for sampling were all healthy, disease-free, in the same body condition, had similar milk production, were not artificially supplemented with concentrate and roughage and grazed on natural summer pastures in Gannan, grazing in the same native pasture. A total of 172 yak milk samples were collected, and the parity of lactating yaks was 2–3 times. The collected yak milk was used for milk composition analysis. The analysis included the determination of fat, protein, lactose, casein, non-fat milk solids (SNF), acidity and total solids (TS). A MilkoScanTM FT120 milk composition analyzer (Danish FUCHS Analytical Instruments Ltd., Hellerup, Denmark) was used for determination.

2.3. Biological Material Sampling and DNA Extraction

Ear tissue samples from 172 Gannan yaks were collected and stored in liquid nitrogen and brought back to the laboratory. They were preserved at -80 °C until DNA was extracted. Tissue sample DNA was extracted using the magnetic bead method using Magnetic Animal Tissue Genomic DNA Kit (DP341, Tiangen Biochemical Technology Co., Ltd., Beijing, China) [33]. The approximate operation was as follows: take 50 mg of ear tissue, cut it into as small pieces as possible, and add 200 μL of tissue digest GHA and 20 μL of Proteinase K for tissue grinding. Then, 300 uL of lysate GHB was added and incubated at 75 °C for 15 min, after which 350 uL of isopropanol was added. Then, 30 uL of Magnetic Bead Suspension G was added and the liquid was aspirated using the function of magnetic bead adsorption. After that, 700 uL of buffer GDA and 700 uL of rinse solution PWD were added sequentially, and magnetic bead adsorption was performed sequentially to absorb the liquid. The sample was then allowed to dry at room temperature for 10–15 min. Then, 200 uL of elution buffer TB was added and incubated at 56 °C for 10 min. The sample was placed on a magnetic rack for 2 min, and after waiting for complete adsorption of the magnetic beads, the DNA solution was carefully transferred to a new centrifuge tube and stored in appropriate conditions. The concentration of DNA samples was detected by a Qubit fluorescence quantitative instrument. The integrity of the DNA samples was detected with 1% agarose gel electrophoresis.

2.4. Genotyping

A total of 172 Gannan yaks were genotyped using an Illumina Yak cGPS 7K (Illumina, Huazhi Biotechnology Co., Ltd., Changsha, China) liquid chip. Genotyping was performed using *cGPS* (Genotyping by Pinpoint Sequencing of liquid captured targets). *cGPS* is based on the optimized thermodynamic stability algorithm model to design specific probes for target interval sequences. The synthesized specific probes are used to capture and enrich multiple different target sequences located in different genomic locations through liquid-phase hybridization, and then the library construction and second-generation sequencing are carried out on the target intervals captured and enriched, so as to obtain the genotypes of all SNP/InDel marker sites in the target interval. Fastp was used to control the quality of the raw reads data. Low-quality reads were filtered, and when the base with a mass value of $Q \leq 20$ in reads accounted for more than 50% of the total base, the reads pair was removed. Reads with too many N bases were filtered, and reads containing more than 5 Ns were removed, and reads with a length less than 100 were filtered. The genomic location of the SNP was derived from the assembly of the yak reference genome Bosgru v3.0 [34] (GCA_005887515.1).

2.5. SNPs Validation

The genotyping results of the Illumina Yak cGPS 7K microarray were validated by amplifying sequences at the g.183,843A>G, g.222,717C>G and g.388,723G>T loci. PCR amplification primers were designed using Primer Premier 5.0 software (Premier Biosoft International, San Francisco, CA, USA) based on the yak *CCSER1* gene published by Ensemble (accession number: ENSBGRG00000023090). All three loci are located in the yak reference genome Bosgru v3.0 on chromosome 6 of the Yak reference genome (accession number GCA_005887515.1). The g.183,843A>G, g.222,717C>G and g.388,723G>T loci in the Gannan yak were amplified using the primer sequences listed in Table 1. The PCR amplification system was 40 µL: 20 µL of 2× Accurate Taq Master Mix (dye plus), 1 µL of DNA template (100 ng/µL), 1 µL each of the upstream and downstream primers (10 µmol/L), and 17 µL of enzyme-free sterile water. PCR amplification program: pre-denaturation at 94 °C for 30 s; denaturation at 98 °C for 10 s; annealing for 30 s (the annealing temperatures of the three pairs of primers were 57 °C, 50 °C and 53 °C, respectively); extension at 72 °C for 1 min, 35 cycles; extension at 72 °C for 2 min; and storage at 4 °C for cooling. The amplification products were detected with electrophoresis under 1% agarose gel at the end of amplification. Samples that showed specific amplification products of the expected length were sent to Qinke Zexi Biotechnology Co., Ltd. (Xi'an, China) for Sanger sequencing. Sequencing results were analyzed using MEGA11 software (version 11.0).

Table 1. SNPs amplification primer sequence of *CCSER1* gene.

SNPs	Primer Sequence (5′–3′)	Product Size
g.183,843A>G	F: TAACAGAACGGGCAGGTAGC R: AAATCAGCATACCTTTGGCAGG	633 bp
g.222,717C>G	F: AATAAATGATGTCGCCAATA R: CTGCGTAGAATACAAAAGAAT	317 bp
g.388,723G>T	F: AGCACCTTCTTCTTACTCAT R: ATTGTTCTGCTGCTGGGATT	404 bp

2.6. Statistical Analysis

Homozygotes (HO) were calculated online using GDICALL (http://www.msrcall.com/gdicall.aspx), (Last accessed on 5 October 2020). The heterozygosity (HE), number of effective alleles (NE), polymorphism (PIC), genotype and allele frequency of the two loci were calculated, and the *p*-values of the chi-square test and Hardy–Weinberg test were calculated.

We used one-way analysis of variance (ANOVA) in IBM SPSS Statistics 25 (IBM, Armonk, NY, USA) to explore the relationship between *CCSER1* gene polymorphisms and yak milk production traits. In order to analyze the influencing factors of yak milk production traits, we used a general linear model, which was appropriately simplified according to the current situation. The simplified model used Equation (1), where Y_i is the phenotypic value of milk quality traits, μ is the population mean of milk fat traits, SNP_i is the fixed effect of the genotypic category of the locus, and e is the random error effect. Differences between means were tested using Duncan's multiple comparisons test and results were expressed as mean ± standard deviation. Differences were considered significant at $p < 0.05$.

$$Y_i = \mu + SNP_i + e \qquad (1)$$

3. Results

3.1. Genotyping Results for CCSER1 and Genetic Parameter Analysis of the Loci in Gannan Yak

The genotype frequency, allele frequency and polymorphism information content of the three SNPs loci of *CCSER1* gene are shown in Table 2. The results showed that the three SNPs loci showed three genotypes in the Gannan yak population. The chip typing results were verified and the results are shown in Figure A1, indicating that the chip typing results are correct. Figure 1 shows the distribution of three genotypes of SNPs in the main milk production traits of the Gannan yak. It can be seen from the figure that the distribution of SNPs is relatively uniform. Among the genotype frequencies of *CCSER1* g.183,843A>G, g.222,717C>G and g.388,723G>T, the genotype frequencies of AG, CG and GT were the highest, and were 0.500, 0.414 and 0.525, respectively. This shows that among the three SNPs sites, heterozygous types are dominant. In g.183,843A>G, the gene frequency of G is 0.571, indicating that the mutant allele accounts for the majority at this site. In g.222,717C>G and g.388,723G>T, the gene frequencies are higher in C and G, respectively, indicating that the unmutated alleles account for the majority in these two loci, but the difference is small. The PICs of g.183,843A>G, g.222,717C>G and g.388,723G>T are 0.370, 0.374 and 0.375, respectively. The polymorphic information contents are between 0.25 and 0.5, and all are moderately polymorphic. Except for the g.222,717C>G site, the other SNPs were consistent with the Hardy–Weinberg equilibrium ($p > 0.05$).

Table 2. Variation information and diversity parameters of *CCSER1* loci.

SNPs	Position	Genotypic Frequencies			Allelic Frequencies		He	Ne	PIC	*p* Value
g.183,843A>G	Exon	AA	AG	GG	A	G	0.490	1.960	0.370	0.793
		0.179	0.500	0.321	0.429	0.571				
g.222,717C>G	Intron	CC	CG	GG	C	G	0.498	1.994	0.374	0.030
		0.321	0.414	0.265	0.528	0.472				
g.388,723G>T	Intron	GG	GT	TT	G	T	0.499	1.996	0.375	0.513
		0.259	0.525	0.216	0.522	0.478				

Note: He: heterozygosity; Ne: effective number of alleles; PIC: polymorphism. PIC < 0.25, low polymorphism; 0.25 < PIC < 0.5, moderate polymorphism; PIC > 0.5, high polymorphism; *p* > 0.05 suggests that the population gene is in the Hardy–Weinberg balance and the sample comes from the same Mendel population.

3.2. Association Analysis between SNPs Genotypes and Milk Traits in Gannan Yak

The correlation between individual loci and yak milk composition was analyzed based on SNPs genotyping data. The correlation analysis of the g.183,843A>G, g.222,717C>G and g.388,723G>T loci of the *CCSER1* gene in Gannan yaks with milk traits is shown in Table 3. As shown in Table 3, the g.183,843A>G locus was significantly associated with milk fat and lactose traits ($p < 0.05$). The AG heterozygous genotype had significantly higher milk fat content than the AA and GG genotypes ($p < 0.05$). The lactose content of GG pure genotypes was significantly higher ($p < 0.05$) than AG genotypes, but the difference from AA genotypes was not significant ($p > 0.05$). The g.222,717C>G locus was significantly associated with casein, protein, SNF and acidity traits ($p < 0.05$). The casein and protein contents of GG and CG genotypes were not significantly different ($p > 0.05$), but both were significantly higher than the CC genotype ($p < 0.05$). The differences in SNF content and acidity between the mutant genotypes GG and CG were not significant ($p > 0.05$), but the SNF content and acidity of the CG genotype was significantly higher than that of the CC genotype ($p < 0.05$). The g.388,723G>T locus was significantly associated with casein and protein traits ($p < 0.05$). The casein and protein contents were significantly higher in the TT genotype than in the GG genotype ($p < 0.05$), but the difference between the TT and GT genotypes was not significant ($p > 0.05$). This showed that the *CCSER1* gene was mainly associated with the traits of casein, protein, fat and lactose in Gannan yak milk, and both the pure and heterozygous types of the mutant were significantly higher than that of the wild type. This indicated that the individual yak milk quality was higher in the mutant type.

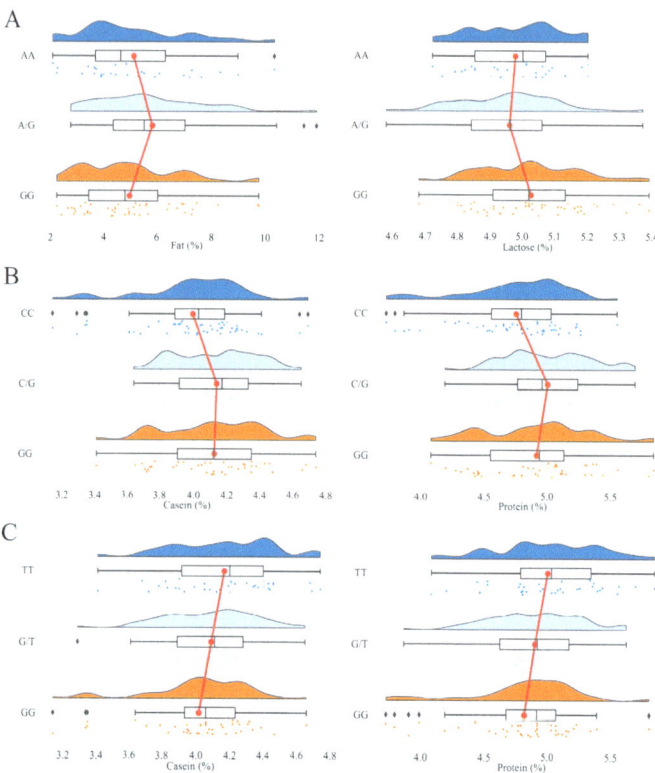

Figure 1. Distribution of all the three genotypes of SNPs (the main significant milk production traits of Gannan yak). (**A**) g.183,843A>G, (**B**) g.222,717C>G, (**C**) g.388,723G>T. The red line represents the mean value of the milk quality trait corresponding to the three genotypes. The three different colored points represent the distribution of data.

Table 3. Correlation analysis between *CCSER1* g.183,843A>G, g.222,717C>G, g.388,723G>T and milk traits in Gannan yak.

Genotype	Casein/%	Protein/%	Fat/%	SNF/%	Lactose/%	Acidity/°T	TS/%
			SNPs g.183,843A>G				
AA	4.10 ± 0.25	4.92 ± 0.34	5.12 ± 2.09 [b]	11.27 ± 0.38	4.97 ± 0.14 [ab]	12.44 ± 1.15	16.57 ± 2.58
AG	4.11 ± 0.28	4.93 ± 0.38	5.82 ± 2.06 [a]	11.28 ± 0.46	4.96 ± 0.16 [b]	12.53 ± 1.25	16.90 ± 2.43
GG	4.05 ± 0.33	4.83 ± 0.45	4.97 ± 1.84 [b]	11.23 ± 0.55	5.03 ± 0.16 [a]	12.15 ± 1.41	16.49 ± 2.85
p-Value	*p* = 0.453	*p* = 0.351	*p* = 0.042	*p* = 0.826	*p* = 0.047	*p* = 0.268	*p* = 0.652
Genotype	Casein/%	Protein/%	Fat/%	SNF/%	Lactose/%	Acidity/°T	TS/%
			SNPs g.222,717C>G				
CC	3.99 ± 0.32 [b]	4.75 ± 0.41 [b]	5.12 ± 1.97	11.13 ± 0.49 [b]	5.02 ± 0.15	12.01 ± 1.18 [b]	16.77 ± 2.86
CG	4.14 ± 0.24 [a]	5.00 ± 0.34 [a]	5.26 ± 2.14	11.37 ± 0.38 [a]	4.96 ± 0.16	12.71 ± 1.14 [a]	16.49 ± 2.09
GG	4.12 ± 0.30 [a]	4.92 ± 0.42 [a]	5.98 ± 2.53	11.25 ± 0.54 [ab]	4.98 ± 0.16	12.33 ± 1.50 [ab]	16.99 ± 2.95
p-Value	*p* = 0.020	*p* = 0.002	*p* = 0.069	*p* = 0.024	*p* = 0.083	*p* = 0.012	*p* = 0.607
Genotype	Casein/%	Protein/%	Fat/%	SNF/%	Lactose/%	Acidity/°T	TS/%
			SNPs g.388,723G>T				
GG	4.02 ± 0.32 [b]	4.82 ± 0.43 [b]	5.37 ± 1.97	11.25 ± 0.47	5.03 ± 0.17	12.20 ± 1.25	16.31 ± 2.09
GT	4.09 ± 0.27 [ab]	4.90 ± 0.36 [ab]	5.51 ± 2.09	11.24 ± 0.47	4.98 ± 0.15	12.39 ± 1.26	16.78 ± 2.68
TT	4.17 ± 0.31 [a]	5.00 ± 0.42 [a]	5.28 ± 2.03	11.32 ± 0.49	4.94 ± 0.14	12.61 ± 1.40	17.00 ± 2.90
p-Value	*p* = 0.038	*p* = 0.043	*p* = 0.852	*p* = 0.761	*p* = 0.582	*p* = 0.375	*p* = 0.406

Note: In the same group of data, different lowercase letters showed significant differences ($p < 0.05$). Data are presented as the mean ± standard deviation.

4. Discussion

Milk fat is a very high-quality lipid in milk, which is better absorbed by the body because of its high fat-soluble fiber content [8]. It has been found that the digestibility of milk fat in the human gastrointestinal tract is more than 98% [35]. The high protein level in milk can provide essential amino acids for the growth of newborns, and some special proteins can also improve immunity and promote the utilization of trace elements [36]. Moreover, yak milk is rich in calcium, protein, peptides, amino acids, iron, phosphorus, vitamins and other nutrients and functional substances, and yak milk has a higher proportion of iron, zinc, manganese, selenium, lactoferrin, immunoglobulin, etc., compared with ordinary cow's milk [9]. In terms of safety and hypoallergenicity, yak milk is closer to breast milk than ordinary cow's milk and goat's milk, and has the effect of lowering blood sugar, promoting growth and development, providing anti-inflammation properties and strengthening immunity [37,38]. Therefore, yak milk has higher nutritional characteristics, with fat and protein content being important indicators of quality in yak milk. Similarly, the intake of protein and fat plays a very important role in the growth and development of mammals. The results of this experiment showed that the g.183,843A>G locus mainly affected the milk fat and lactose of Gannan yaks, in which the AG heterozygous group had significantly higher milk fat content than the two pure groups. However, g.222,717C>G and g.388,723G>T mainly improved the milk protein of the Gannan yak, and both mutant heterozygous populations had significantly higher protein content than the wild-type population. The experimental results indicated that the mutation of SNPs significantly improved the milk quality of the Gannan yak.

SNPs are an important basis for the study of genetic variation in livestock and poultry. SNPs have the characteristics of high reliability, wide distribution and ease of analysis [39]. They play an important role in analyzing the genetic basis of important economic traits of livestock and poultry. In this study, we found that when the base of the *CCSER1* g.183,843A>G locus was mutated from A to G, the mutant genotypes had a significant effect on fat and lactose in Gannan yaks on analyzing the correlation between the loci of the *CCSER1* gene and milk traits. When the base of the *CCSER1* g.222,717C>G locus was mutated from C to G, both heterozygous and pure genotypes of the mutation had a significant improvement on casein, protein and SNF in Gannan yaks. The mutation of the *CCSER1* g.388,723G>T locus was basically the same as that of the *CCSER1* g.222,717C>G locus, and the mutant genotype TT also had a significant improvement on casein and protein and SNF in Gannan yaks, with significant improvement in casein and protein. Overall, the mutations of all three SNPs positively affected the milk quality traits of Gannan yaks to different degrees. Clancey et al. detected SNPs and gene sets associated with 305 standard milk yield in 781 primiparous Holstein cows, and identified new candidate genes that cause milk production variation [32]. Jiang et al. analyzed the SNPs of 480 yaks using a commercial high-density (600 K) yak SNP chip, and identified 12 and 4 SNPs that may be related to the weight of male and female yaks, respectively [40]. Nevertheless, the study of yak SNPs is still in its infancy, and a large amount of experimental data are needed to elucidate the potential relationship between SNPs and economic traits.

It was found that *CCSER1* is localized to the γ-microtubulin ring complex in early mitosis, as well as to an intermediate in late cytokinesis [21]. Therefore, it has been concluded that the *CCSER1* gene is an important regulator or structural molecule that plays an important regulatory role in normal cellular mitosis [16,21], which may alter chromosome stability and somatic cell division. Based on bovine gene expression profiles, it is expressed in 85 tissues or cell types in cattle, while its expression is relatively high in the mammary gland [41]. The results of the present study also confirmed that the *CCSER1* gene had a significant effect on milk quality traits in the yak population. This may provide new insights to improve yak milk quality traits through selection strategies.

Here, we identified three SNPs loci, g.183,843A>G, g.222,717C>G and g.388,723G>T, in the *CCSER1* gene. Among them, the g.183,843A>G locus is located in the coding region of the *CCSER1* gene, which has an impact on gene expression and function, and thus may have a more direct effect on milk quality. In addition, we found that the g.222,717C>G and g.388,723G>T loci are located in the intronic region of the *CCSER1* gene, which does not affect their association with yak milk quality. Although introns are sequences with no coding function in genes, they have important roles in the regulation of gene expression [42], and the important function of SNPs in introns in altering the transcription level of genes has been elucidated [43]. They cause shear abnormalities by altering the structure of the shear site, which may lead to changes in protein structure and function [44]. In addition, several SNPs sites in the intron region of the erb-b2 receptor tyrosine kinase 2 (*ERBB2*) gene were reported to be significantly associated with milk protein content in Chinese Holstein cows [45]. Five SNPs (g.29029T>C, g.29050A>G, g.29245C>T, g.29305C>T and g.29347T>C) in the intron region of sorbin and the SH3 domain-containing 1 (*SORBS1*) gene were reported to be significantly associated with milk fat traits in cattleyak [46]. All of the above studies had results similar to this study. Therefore, the role of mutations in the g.222,717C>G and g.388,723G>T loci of the *CCSER1* gene in Gannan yaks on the genetic characterization of this gene still needs to be further confirmed. The polymorphisms of *CCSER1* g.183,843A>G, g.222,717C>G and g.388,723G>T were 0.370, 0.374 and 0.375, respectively, and the polymorphic information content of the three SNPs was moderate. Among them, the p-value of the *CCSER1* g.222,717C>G locus was greater than 0.05, indicating that the population at this locus was genetically unbalanced and deviated from Hardy–Weinberg. This may be due to a variety of factors such as sampling bias, natural selection or human intervention. In our study, we used liquid-phase microarray genotyping, which is a technique for precise positional sequencing typing based on the liquid-phase capture of target sequences and is specifically designed for genotyping large samples at specific SNP loci. Therefore, genotyping is unlikely to be the cause of this phenomenon.

Consistent with the present study, the functional role of the *CCSER1* gene for milk production traits was also identified as a positional candidate gene for lactation persistence in Holstein cattle. Thus, the results of the present study and previously published studies suggest that *CCSER1* is a promising candidate gene with strong genetic effects on milk production traits in Gannon yaks. In addition, single nucleotide polymorphism-based and correlation analyses also suggest that the novel SNPs in the *CCSER1* gene may be used as potential genetic markers for genetic improvement in yak breeding programs.

5. Conclusions

In this study, we explored the *CCSER1* gene polymorphisms, identified three SNPs loci and analyzed the relationship between these polymorphisms and the milk quality of Gannan yaks. The results showed that all three SNPs of the *CCSER1* gene were moderately polymorphic. Correlation analysis revealed that the mutant genotype of the *CCSER1* g.183,843A>G locus in Gannan yaks resulted in a significant ($p < 0.05$) increase in milk fat content, whereas the mutant genotypes of both *CCSER1* g.222,717C>G and g.388,723G>T loci resulted in a significant increase ($p < 0.05$) in the casein and protein content of Gannan yak milk. Therefore, mutations in the *CCSER1* g.183,843A>G, g.222,717C>G and g.388,723G>T loci all significantly improved the quality traits of Gannan yak milk. The identification of these SNPs opens the way for further research and applications in the selective breeding of Gannan yaks. The results of this study can help to develop and optimize the milk quality of Gannan yaks.

Author Contributions: Conceptualization, G.Y. and C.L.; methodology, G.Y.; software, G.Y. and X.M. (Xiaoming Ma); validation, G.Y., P.Y. and C.L.; formal analysis, G.Y.; investigation, G.Y. and C.L.; resources, J.Z., X.M. (Xiaoyong Ma), R.M., J.S., M.L., D.Y., C.H. and F.F.; data curation, G.Y., X.M. (Xiaoming Ma), Y.L. and C.L.; writing—original draft preparation, G.Y.; writing—review and editing, X.G., P.Y. and C.L.; visualization, C.L.; supervision, C.L.; project administration, C.L.; funding acquisition, C.L. All authors have read and agreed to the published version of the manuscript.

Funding: This research was funded by the National Key Research Program (2021YFD1600200), the Gansu Provincial Basic Research Innovation Group Project, the Hezuo City Yak Germplasm Improvement and Quality Improvement Project (20JR5RA580), the Modern Beef Cattle and Yak Industry Technology System (CARS-37), the Major Science and Technology Projects in Gansu Province (21ZD10NA001, GZGG-2021-1) and the Innovation Project of Chinese Academy of Agricultural Sciences (25-LIHPS-01).

Institutional Review Board Statement: All the animal experiments were approved by the Lanzhou Institute of Husbandry and Pharmaceutical Sciences of the Chinese Academy of Agricultural Sciences (CAAS) with the grant number 1610322020018.

Data Availability Statement: The data presented in this study are available on request from the corresponding author.

Conflicts of Interest: The authors declare no conflict of interest.

Appendix A

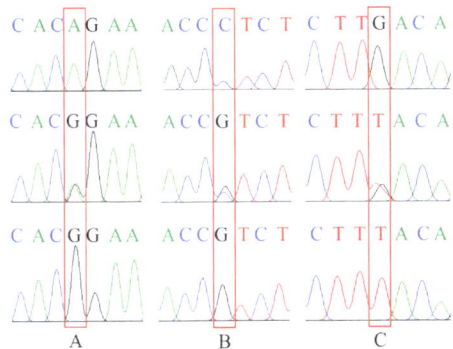

Figure A1. Peak plots of three genotypes at the SNPs locus of the *CCSER1* gene. (**A**) g.183,843A>G. (**B**) g.222,717C>G. (**C**) g.388,723G>T.

References

1. Zhang, Z.; Chu, M.; Bao, Q.; Bao, P.; Guo, X.; Liang, C.; Yan, P. Two different copy number variations of the SOX5 and SOX8 genes in yak and their association with growth traits. *Animals* **2022**, *12*, 1587. [CrossRef] [PubMed]
2. Hongqiang, L.; Xiaoling, Y.; Defu, T.; Bin, X.; Weihong, L.; Zhaohui, C.; Yongqing, B.; Renqing, D.; Yaqin, G.; Peng, W.; et al. Exploring the link between microbial community structure and flavour compounds of traditional fermented yak milk in Gannan region. *Food Chem.* **2023**, *435*, 137553.
3. Kang, Y.; Guo, S.; Wang, X.; Cao, M.; Pei, J.; Li, R.; Bao, P.; Wang, J.; Lamao, J.; Gongbao, D.; et al. Whole-Genome Resequencing Highlights the Unique Characteristics of Kecai Yaks. *Animals* **2022**, *12*, 2682. [CrossRef]
4. Hu, L.; Lizhuang, H.; Xuliang, C.; Guo, Y.; Abraham Allan, D.; Ling, X.; Shujie, L.; Jianwei, Z. Effects of supplementary concentrate and/or rumen-protected lysine plus methionine on productive performance, milk composition, rumen fermentation, and bacterial population in Grazing, Lactating Yaks. *Anim. Feed. Sci. Technol.* **2023**, *297*, 115591.
5. Jiang, J.; Chen, S.; Ren, F.; Luo, Z.; Zeng, S. Yak Milk Casein as a Functional Ingredient: Preparation and Identification of Angiotensin-I-Converting Enzyme Inhibitory Peptides. *J. Dairy Res.* **2007**, *74*, 18–25. [CrossRef]
6. Pan, S.; Chen, G.; Wu, R.; Cao, X.; Liang, Z. Non-sterile Submerged Fermentation of Fibrinolytic Enzyme by Marine Bacillus subtilis Harboring Antibacterial Activity With Starvation Strategy. *Front. Microbiol.* **2019**, *10*, 1025. [CrossRef] [PubMed]
7. Zhang, J.; Wolf, B. Physico-Chemical Properties of Sugar Beet Pectin-Sodium Caseinate Conjugates via Different Interaction Mechanisms. *Foods* **2019**, *8*, 192. [CrossRef] [PubMed]
8. Yang, G.; Zhang, J.; Dai, R.; Ma, X.; Huang, C.; Ren, W.; Ma, X.; Lu, J.; Zhao, X.; Renqing, J.; et al. Comparative Study on Nutritional Characteristics and Volatile Flavor Substances of Yak Milk in Different Regions of Gannan. *Foods* **2023**, *12*, 2172. [CrossRef]
9. Yang, F.; He, X.; Chen, T.; Liu, J.; Luo, Z.; Sun, S.; Qin, D.; Huang, W.; Tang, Y.; Liu, C.; et al. Peptides Isolated from Yak Milk Residue Exert Antioxidant Effects through Nrf2 Signal Pathway. *Oxid. Med. Cell. Longev.* **2021**, *2021*, 9426314. [CrossRef]
10. Wu, X.; Zhou, X.; Xiong, L.; Pei, J.; Yao, X.; Liang, C.; Bao, P.; Chu, M.; Guo, X.; Yan, P. Transcriptome Analysis Reveals the Potential Role of Long Non-coding RNAs in Mammary Gland of Yak During Lactation and Dry Period. *Front. Cell. Dev. Biol.* **2020**, *8*, 579708. [CrossRef]

11. Li, H.; Xi, B.; Yang, X.; Wang, H.; He, X.; Li, W.; Gao, Y. Evaluation of change in quality indices and volatile flavor components in raw milk during refrigerated storage. *LWT—Food Sci. Technol.* **2022**, *165*, 113674. [CrossRef]
12. Ghulam Mohyuddin, S.; Liang, Y.; Ni, W.; Adam Idriss Arbab, A.; Zhang, H.; Li, M.; Yang, Z.; Karrow, N.A.; Mao, Y. Polymorphisms of the IL-17A Gene Influence Milk Production Traits and Somatic Cell Score in Chinese Holstein Cows. *Bioengineering* **2022**, *9*, 448. [CrossRef] [PubMed]
13. Verma, A.; Meitei, N.S.; Gajbhiye, P.U.; Raftery, M.J.; Ambatipudi, K. Comparative Analysis of Milk Triglycerides Profile between Jaffarabadi Buffalo and Holstein Friesian Cow. *Metabolites* **2020**, *10*, 507. [CrossRef] [PubMed]
14. Zhou, C.; Li, C.; Cai, W.; Liu, S.; Yin, H.; Shi, S.; Zhang, Q.; Zhang, S. Genome-Wide Association Study for Milk Protein Composition Traits in a Chinese Holstein Population Using a Single-Step Approach. *Front. Genet.* **2019**, *10*, 72. [CrossRef] [PubMed]
15. Mohamed, S.A.; Teleb, D.F.; Saad El-Deen, H.K.; Eid, J.I.; El-Ghor, A.A. Association of new SNPs at DGAT1 gene with milk quality in Egyptian Zaraibi goat breed. *Anim. Biotechnol.* **2022**, *34*, 2499–2504. [CrossRef] [PubMed]
16. Teng, J.; Wang, D.; Zhao, C.; Zhang, X.; Chen, Z.; Liu, J.; Sun, D.; Tang, H.; Wang, W.; Li, J.; et al. Longitudinal genome-wide association studies of milk production traits in Holstein cattle using whole-genome sequence data imputed from medium-density chip data. *J. Dairy Sci.* **2023**, *106*, 2535–2550. [CrossRef] [PubMed]
17. Scrimieri, F.; Calhoun, E.S.; Patel, K.; Gupta, R.; Huso, D.L.; Hruban, R.H.; Kern, S.E. FAM190A rearrangements provide a multitude of individualized tumor signatures and neo-antigens in cancer. *Oncotarget* **2011**, *2*, 69–75. [CrossRef]
18. Xu, Z.; Wang, X.; Song, X.; An, Q.; Wang, D.; Zhang, Z.; Ding, X.; Yao, Z.; Wang, E.; Liu, X.; et al. Association between the copy number variation of CCSER1 gene and growth traits in Chinese Capra hircus (goat) populations. *Anim. Biotechnol.* **2023**, *34*, 1377–1383. [CrossRef]
19. Smith, J.L.; Wilson, M.L.; Nilson, S.M.; Rowan, T.N.; Schnabel, R.D.; Decker, J.E.; Seabury, C.M. Genome-wide association and genotype by environment interactions for growth traits in U.S. Red Angus cattle. *BMC Genom.* **2022**, *23*, 517. [CrossRef]
20. Abo-Ismail, M.K.; Lansink, N.; Akanno, E.; Karisa, B.K.; Crowley, J.J.; Moore, S.S.; Bork, E.; Stothard, P.; Basarab, J.A.; Plastow, G.S. Development and validation of a small SNP panel for feed efficiency in beef cattle. *J. Anim. Sci.* **2018**, *96*, 375–397. [CrossRef]
21. Patel, K.; Scrimieri, F.; Ghosh, S.; Zhong, J.; Kim, M.S.; Ren, Y.R.; Morgan, R.A.; Iacobuzio-Donahue, C.A.; Pandey, A.; Kern, S.E. FAM190A deficiency creates a cell division defect. *Am. J. Pathol.* **2013**, *183*, 296–303. [CrossRef] [PubMed]
22. Kang, S.U.; Park, J.T. Functional evaluation of alternative splicing in the FAM190A gene. *Genes. Genom.* **2018**, *41*, 193–199. [CrossRef] [PubMed]
23. Li, H.; Ma, Y.; Li, Q.; Wang, J.; Cheng, J.; Xue, J.; Shi, J. The chemical composition and nitrogen distribution of Chinese yak (Maiwa) milk. *Int. J. Mol. Sci.* **2011**, *12*, 4885–4895. [CrossRef] [PubMed]
24. Wang, D.; Zhou, Y.; Zheng, X.; Guo, J.; Duan, H.; Zhou, S.; Yan, W. Yak Milk: Nutritional Value, Functional Activity, and Current Applications. *Foods* **2023**, *12*, 2090. [CrossRef] [PubMed]
25. Liu, H.N.; Ren, F.Z.; Jiang, L.; Ma, Z.L.; Qiao, H.J.; Zeng, S.S.; Gan, B.Z.; Guo, H.Y. Fatty acid profile of yak milk from the Qinghai-Tibetan Plateau in different seasons and for different parities. *J. Dairy Sci.* **2011**, *94*, 1724–1731. [CrossRef]
26. Din, J.N.; Newby, D.E.; Flapan, A.D. Omega 3 fatty acids and cardiovascular disease--fishing for a natural treatment. *BMJ* **2004**, *328*, 30–35. [CrossRef]
27. Moatsou, G.; Sakkas, L. Sheep milk components: Focus on nutritional advantages and biofunctional potential. *Small Rumin. Res.* **2019**, *180*, 86–99. [CrossRef]
28. Mao, X.-Y.; Cheng, X.; Wang, X.; Wu, S.-J. Free-radical-scavenging and anti-inflammatory effect of yak milk casein before and after enzymatic hydrolysis. *Food Chem.* **2011**, *126*, 484–490. [CrossRef]
29. Wang, P.; Liu, H.; Wen, P.; Zhang, H.; Guo, H.; Ren, F. The composition, size and hydration of yak casein micelles. *Int. Dairy J.* **2013**, *31*, 107–110. [CrossRef]
30. Yang, M.; Shi, Y.; Liang, Q. Effect of microbial transglutaminase crosslinking on the functional properties of yak caseins: A comparison with cow caseins. *Dairy Sci. Technol.* **2015**, *96*, 39–51. [CrossRef]
31. Hong, J.; Li, X.; Jiang, M.; Hong, R. Co-expression Mechanism Analysis of Different Tachyplesin I-Resistant Strains in Pseudomonas aeruginosa Based on Transcriptome Sequencing. *Front. Microbiol.* **2022**, *13*, 871290. [CrossRef] [PubMed]
32. Clancey, E.; Kiser, J.N.; Moraes, J.G.N.; Dalton, J.C.; Spencer, T.E.; Neibergs, H.L. Genome-wide association analysis and gene set enrichment analysis with SNP data identify genes associated with 305-day milk yield in Holstein dairy cows. *Anim. Genet.* **2019**, *50*, 254–258. [CrossRef]
33. Gu, X.; Huang, S.; Zhu, Z.; Ma, Y.; Yang, X.; Yao, L.; Guo, X.; Zhang, M.; Liu, W.; Qin, L.; et al. Genome-wide association of single nucleotide polymorphism loci and candidate genes for frogeye leaf spot (Cercospora sojina) resistance in soybean. *BMC Plant. Biol.* **2021**, *21*, 588. [CrossRef] [PubMed]
34. Yang, Y.; Han, L.; Yu, Q.; Gao, Y.; Song, R.; Zhao, S. Phosphoproteomic analysis of longissimus lumborum of different altitude yaks. *Meat Sci.* **2020**, *162*, 108019. [CrossRef] [PubMed]
35. Mohan, M.S.; O'Callaghan, T.F.; Kelly, P.; Hogan, S.A. Milk fat: Opportunities, challenges and innovation. *Crit. Rev. Food Sci. Nutr.* **2020**, *61*, 2411–2433. [CrossRef] [PubMed]
36. Molinari, C.E.; Casadio, Y.S.; Hartmann, B.T.; Livk, A.; Bringans, S.; Arthur, P.G.; Hartmann, P.E. Proteome mapping of human skim milk proteins in term and preterm milk. *J. Proteome Res.* **2012**, *11*, 1696–1714. [CrossRef]

37. Ji, X.; Li, X.; Ma, Y.; Li, D. Differences in proteomic profiles of milk fat globule membrane in yak and cow milk. *Food Chem.* **2016**, *221*, 1822–1827. [CrossRef]
38. Tarun Pal, S.; Shalini, A.; Mihir, S. Yak milk and milk products: Functional, bioactive constituents and therapeutic potential. *Int. Dairy J.* **2023**, *142*, 105637.
39. Wang, X.; Yang, Q.; Zhang, S.; Zhang, X.; Pan, C.; Chen, H.; Zhu, H.; Lan, X. Genetic Effects of Single Nucleotide Polymorphisms in the Goat GDF9 Gene on Prolificacy: True or False Positive? *Animals* **2019**, *9*, 886. [CrossRef]
40. Jiang, H.; Chai, Z.-X.; Cao, H.-W.; Zhang, C.-F.; Zhu, Y.; Zhang, Q.; Xin, J.-W. Genome-wide identification of SNPs associated with body weight in yak. *BMC Genom.* **2022**, *23*, 833. [CrossRef]
41. Fang, L.; Cai, W.; Liu, S.; Canela-Xandri, O.; Gao, Y.; Jiang, J.; Rawlik, K.; Li, B.; Schroeder, S.G.; Rosen, B.D.; et al. Comprehensive analyses of 723 transcriptomes enhance genetic and biological interpretations for complex traits in cattle. *Genome Res.* **2020**, *30*, 790–801. [CrossRef] [PubMed]
42. Casas, E.; White, S.N.; Riley, D.G.; Smith, T.P.; Brenneman, R.A.; Olson, T.A.; Johnson, D.D.; Coleman, S.W.; Bennett, G.L.; Chase, C.C. Assessment of single nucleotide polymorphisms in genes residing on chromosomes 14 and 29 for association with carcass composition traits in Bos indicus cattle. *J. Anim. Sci.* **2005**, *83*, 13–19. [CrossRef] [PubMed]
43. Tokuhiro, S.; Yamada, R.; Chang, X.; Suzuki, A.; Kochi, Y.; Sawada, T.; Suzuki, M.; Nagasaki, M.; Ohtsuki, M.; Ono, M.; et al. An intronic SNP in a RUNX1 binding site of SLC22A4, encoding an organic cation transporter, is associated with rheumatoid arthritis. *Nat. Genet.* **2003**, *35*, 341–348. [CrossRef] [PubMed]
44. Soemedi, R.; Cygan, K.J.; Rhine, C.L.; Wang, J.; Bulacan, C.; Yang, J.; Bayrak-Toydemir, P.; McDonald, J.; Fairbrother, W.G. Pathogenic variants that alter protein code often disrupt splicing. *Nat. Genet.* **2017**, *49*, 848–855. [CrossRef]
45. Li, C.; Wang, M.; Cai, W.; Liu, S.; Zhou, C.; Yin, H.; Sun, D.; Zhang, S. Genetic Analyses Confirm SNPs in HSPA8 and ERBB2 are Associated with Milk Protein Concentration in Chinese Holstein Cattle. *Genes* **2019**, *10*, 104. [CrossRef]
46. Yang, L.; Min, X.; Zhu, Y.; Hu, Y.; Yang, M.; Yu, H.; Li, J.; Xiong, X. Polymorphisms of SORBS1 Gene and Their Correlation with Milk Fat Traits of Cattleyak. *Animals* **2021**, *11*, 3461. [CrossRef]

Disclaimer/Publisher's Note: The statements, opinions and data contained in all publications are solely those of the individual author(s) and contributor(s) and not of MDPI and/or the editor(s). MDPI and/or the editor(s) disclaim responsibility for any injury to people or property resulting from any ideas, methods, instructions or products referred to in the content.

Article

Association between Single Nucleotide Polymorphisms of *PRKD1* and *KCNQ3* Gene and Milk Quality Traits in Gannan Yak (*Bos grunniens*)

Xiaoyong Ma [1,2], Guowu Yang [1,2], Juanxiang Zhang [1,2], Rong Ma [1,2], Jinwei Shen [1,2], Fen Feng [1,2], Daoning Yu [1,2], Chun Huang [1,2], Xiaoming Ma [1,2], Yongfu La [1,2], Xiaoyun Wu [1,2], Xian Guo [1,2], Min Chu [1,2], Ping Yan [1,2,3] and Chunnian Liang [1,2,*]

[1] Key Laboratory of Animal Genetics and Breeding on Tibetan Plateau, Ministry of Agriculture and Rural Affairs, Lanzhou 730050, China; abdullah33@163.com (X.M.); xueshengyangguowu@163.com (G.Y.); juanxiangzhang@163.com (J.Z.); marong202017@163.com (R.M.); shenjw9090@163.com (J.S.); feng990111@163.com (F.F.); ydn9907@163.com (D.Y.); johnchun825@163.com (C.H.); maxiaoming@caas.cn (X.M.); layongfu@caas.cn (Y.L.); wuxiaoyun@caas.cn (X.W.); guoxian@caas.cn (X.G.); chumin@caas.cn (M.C.); pingyanlz@163.com (P.Y.)

[2] Key Laboratory of Yak Breeding of Gansu Province, Lanzhou Institute of Husbandry and Pharmaceutical Sciences, Chinese Academy of Agricultural Sciences, Lanzhou 730050, China

[3] Institute of Western Agriculture, The Chinese Academy of Agricultural Sciences, Changji 831100, China

* Correspondence: liangchunnian@caas.cn

Citation: Ma, X.; Yang, G.; Zhang, J.; Ma, R.; Shen, J.; Feng, F.; Yu, D.; Huang, C.; Ma, X.; La, Y.; et al. Association between Single Nucleotide Polymorphisms of *PRKD1* and *KCNQ3* Gene and Milk Quality Traits in Gannan Yak (*Bos grunniens*). *Foods* **2024**, *13*, 718. https://doi.org/10.3390/foods13050781

Academic Editors: Paolo Formaggioni and Piero Franceschi

Received: 12 January 2024
Revised: 28 February 2024
Accepted: 28 February 2024
Published: 2 March 2024

Copyright: © 2024 by the authors. Licensee MDPI, Basel, Switzerland. This article is an open access article distributed under the terms and conditions of the Creative Commons Attribution (CC BY) license (https://creativecommons.org/licenses/by/4.0/).

Abstract: *Protein kinase D1 (PRKD1)* functions primarily in normal mammary cells, and the potassium voltage-gated channel subfamily Q member 3 (*KCNQ3*) gene plays an important role in controlling membrane potential and neuronal excitability, it has been found that this particular gene is linked to the percentage of milk fat in dairy cows. The purpose of this study was to investigate the relationship between nucleotide polymorphisms (SNPs) of *PRKD1* and *KCNQ3* genes and the milk quality of Gannan yak and to find molecular marker sites that may be used for milk quality breeding of Gannan yak. Three new SNPs were detected in the *PRKD1* (g.283,619T>C, g.283,659C>A) and *KCNQ3* gene (g.133,741T>C) of 172 Gannan lactating female yaks by Illumina yak cGPS 7K liquid-phase microarray technology. Milk composition was analyzed using a MilkoScan^{TM} milk composition analyzer. We found that the mutations of these three loci significantly improved the lactose, milk fat, casein, protein, non-fat milk solid (SNF) content and acidity of Gannan yaks. The lactose content of the TC heterozygous genotype population at g.283,619T>C locus was significantly higher than that of the TT wild-type population ($p < 0.05$); the milk fat content of the CA heterozygous genotype population at g.283,659C>A locus was significantly higher than that of the CC wild-type and AA mutant populations ($p < 0.05$); the casein, protein and acidity of the CC mutant and TC heterozygous groups at the g.133,741T>C locus were significantly higher than those of the wild type ($p < 0.05$), and the SNF of the TC heterozygous group was significantly higher than that of the mutant group ($p < 0.05$). The results showed that *PRKD1* and *KCNQ3* genes could be used as candidate genes affecting the milk traits of Gannan yak.

Keywords: milk quality; *PRKD1* gene; *KCNQ3* gene; single nucleotide polymorphism; Gannan Yak

1. Introduction

Yak is an important species in the Qinghai–Tibet Plateau (QTP). The means of production such as meat, milk and fur produced by yak are closely related to the life and economic income of herdsmen in the Qinghai–Tibet Plateau area [1]. Among them, yak milk is a nutritious and widely consumed food, known as "liquid gold" [2]. Compared with cow milk, yak milk is rich in all kinds of nutrients (except phosphorus) [3], cholesterol and sphingomyelin content is also higher [4]. In the Qinghai–Tibet region, people cannot often eat vegetables under plateau climate conditions; interestingly, people there do not have related nutritional deficiencies. It has been found that yak milk not only supplements most of the energy consumed by people's daily activities, but also provides vitamins and minerals needed by the human body [5,6]. Yak milk contains more protein, casein, and fat than European dairy cows (*Bos taurus*) [7]. In China, tens of thousands of tons of yak milk and its dairy products are produced every year [8]. Gannan yaks are grazed in natural pastures all year round without using additives and antibiotic drugs [9]. Therefore, Yak milk has great potential for development and utilization, and is an ideal raw material for producing various dairy products such as butter, ghee, cheese, yogurt, etc. Among them, ghee is widely used in the religious activities of Tibetan Buddhism, and the amount of use is huge. At present, lactating female yaks grazing in natural pastures still have the disadvantage of low milk yield, with an average milk yield of about 3.18 Kg/d [10]. A lot of research is needed to improve milk yield and milk quality [11].

The *PRKD1* gene encodes protein kinase D1 (*PKD1*) [12]. The protein kinase D (*PKD*) family of serine/threonine protein kinases has three members: *PKD1-3* plays a role in diacylglycerol and related signaling pathways. *PKDs* are involved in cell proliferation, migration, differentiation, angiogenesis and immune response [13]. *PKD1* prevents epithelial-mesenchymal transition (EMT) and maintains epithelial phenotype [14]. In addition, *PKD1* also inhibits directional cell migration by blocking the actin recombination process at the forefront of migrating cells [15,16]. These results indicate that PKD1 is a key protein that inhibits the invasive phenotype in breast cancer, and the expression regulation ability of *PKD1* is down-regulated in breast invasive ductal carcinoma [17], mammary gland health is the key to ensure milk yield and milk quality of dairy cows [18]. At present, a large number of *PRKD1* gene related studies mainly focus on breast cancer. In the yak, *PRKD1* gene is located on chromosome 17, and the correlation between *PRKD1* gene polymorphism and milk quality in yak has not been mentioned. Previously, we found that three SNPS in the *CCSER1* gene were associated with milk quality (milk fat, protein, and casein) in Gannan yaks [19]. Regions reported to affect milk quality traits in dairy cows are concentrated on chromosomes 1, 6, 11, 13, 14, and 18 [20]. The *KCNQ3* gene is located on chromosome 18 of yak. According to the results of Zhou et al. [20], we speculated that the gene may be a candidate gene affecting the milk quality traits of yak. In addition, Kolbehdari et al. [21] found a SNP (rs41580517) related to milk fat content in the intron of the potassium voltage-gated channel subfamily *KQT3-KCNQ3* gene located in bovine BTA14. *KCNQ* channels in the central nervous system neurons, the main subunits are *KCNQ2, 3* and *5* [22], these genes are regulated by various signaling pathways, and they play a crucial role in controlling membrane potential and neuronal excitability [23,24]. These channels produce muscarinic-inhibited potassium currents that control excitability, known as M currents [25]. The information RNA of *KCNQ3* gene exists in the brain, including cerebral cortex, cerebellum, basal ganglia and hippocampus. At present, the research on *KCNQ3* gene mainly focuses on diseases such as epilepsy, neuropathic pain and anxiety disorders [26].

Yak milk has a high concentration of nutrients and has a good therapeutic effect on some diseases [27]. Studies have shown that the milk of multiparous yaks contains more unsaturated fatty acids than that of primiparous yaks [28]. In addition, yak milk also contains other important nutritional values [29,30]. Yak milk contains several important trace elements, such as zinc, iron, copper, manganese and selenium. Zinc plays an important role in immune function, wound healing, cell growth and division [31]. Iron is an important element to produce hemoglobin, and hemoglobin is the carrier of oxygen in animals, which is crucial in the hypoxic plateau environment. It is reported that yak milk can prolong the survival time of mice under normal hypoxic conditions and improve their red blood cells and hemoglobin levels; thus, yak milk may indirectly improve the oxygen-carrying function of plateau people [32,33]. As an important antioxidant, selenium helps to protect cells from damage [32]. In the Qinghai–Tibet Plateau region, people can withstand the environment of strong ultraviolet light. On the one hand, they have adapted to this environment over a long period of evolution. On the other hand, it may be that the selenium contained in yak milk enhances the skin's ability to resist ultraviolet light [34]. In recent years, there has been increasing interest in developing the potential of yak milk [35]. However, due to geographical location and folk culture, yak milk has not been fully developed and utilized. Herders are still mainly manual milking, and the work efficiency is low, in addition, the disadvantage of low milk yield of yak has not been improved. At present, there are few studies on the relationship between milk yield and milk quality in yaks, and there is no study on the relationship between *PRKD1* and *KCNQ3* gene polymorphisms and milk quality in yaks. The purpose of this study was to investigate the genetic polymorphisms of *PRKD1* and *KCNQ3* genes and their relationship with milk quality in Gannan yak, and to provide a scientific basis for improving milk quality and marker-assisted selection (MAS) in yak breeds (population).

2. Materials and Methods

2.1. Ethics Approval

The Lanzhou Institute of Husbandry approved all the animal experiments as well as the Pharmaceutical Sciences of the Chinese Academy of Agricultural Sciences (CAAS) with the grant number: No. 1610322020018.

2.2. Experimental Animals and Sample Analysis

This experiment was conducted in July 2023. The experimental animals were selected from the same natural pasture in Gannan Tibetan Autonomous Prefecture, with an average altitude of 3000 m. Milk samples were collected from 172 lactating yaks with 2 or 3 fetuses. The milk's composition was analyzed using a MilkoScan™ milk composition analyzer (Danish FUCHS Analytical Instruments Co., Ltd., Hellerup, Denmark). The measured indicators included milk fat, lactoprotein, lactose, casein, non-fat milk solid (SNF) content, acidity and total solid (TS) content.

2.3. DNA Samples

The ear tissues of 172 Gannan yaks were collected using U-shaped ear forceps. The collected tissues were placed in a 1.5 mL frozen tube and stored in liquid nitrogen. After being brought back to the laboratory, the animal tissue genomic DNA was extracted using the magnetic bead method (DP341, Tiangen Biochemical Technology Co., Ltd., Beijing, China) [36]. For specific extraction steps, please refer to the kit instructions used in this assay. After extraction, the concentration of DNA samples was detected by quantum bit fluorescence quantitative analyzer. The integrity of DNA samples was checked by electrophoresis on a 1% agarose gel.

2.4. Genotyping

Genotyping was performed on 172 Gannan yaks using the Illumina Huazhi Biotechnology Co., Ltd. (Changsha, China). cGPS 7K liquid chip. The genotyping was carried out by precisely sequencing liquid capture targets. The cGPS applies an optimized thermodynamic stability algorithm model to create a unique probe for a specific interval sequence. The synthesized probes are utilized to capture and enrich multiple target sequences located at different genomic locations via liquid-phase hybridization. The target intervals were captured and enriched, followed by library construction and next-generation sequencing to obtain the genotypes of all SNP/InDel marker loci in the target interval. After identifying the low-quality reads, we screened and removed them. If the ratio of bases with a quality value (Q) of 20 or lower in the reads was over 50% of the total bases, then we deleted those reads. Additionally, we filtered out reads with too many N bases. Only reads with less than or equal to 5 N bases were kept, and we removed reads that were less than 100 bases long. The genomic location of this SNP was determined using the assembly of the yak reference genome Bosgru v3.0 [37] (GCA_005887515.1).

2.5. Statistical Analysis

I used the online software GDICALL (http://www.msrcall.com/GDICALL.aspx, accessed on 7 October 2023) to calculate the homozygosity (HO) of our sample. We use Cervus 3.0 (https://www.softpedia.com/get/Science-CAD/Cervus.shtml accessed on 7 October 2023) to calculate two loci heterozygosity (HE), the effective number of alleles (NE) and polymorphism (PIC), genotype and allele frequency. I also calculated p values for the chi-square test and the Hardy–Weinberg test.

We used one-way analysis of variance (ANOVA) of IBM SPSS Statistics 25 (IBM, Armonk, NY, USA) to explore the relationship between *PRKD1* and *KCNQ3* gene polymorphisms and milk production traits in yaks. To analyze the factors that influence yak milk production traits, a general linear model was used and simplified based on the actual situation. Equation (1) shows the simplified model that was used. Here, Yi represents the phenotypic value of milk traits, μ is the population mean of milk fat traits, $SNPi$ is the fixed effect of genotype category at this locus, and e is the random error effect. Duncan's multiple comparison test was carried out to determine the difference between the mean values. The results were expressed as mean ± standard deviation. $p < 0.05$ was considered significant.

$$Yi = \mu + SNPi + e. \tag{1}$$

3. Results

3.1. Conducted Genotyping Analysis and Examined Genetic Parameters of PRKD1 and KCNQ3 in Gannan Yak

The genotyping analysis of Gannan yak showed that there were T>C substitution (g.283,619T>C) and C>A substitution (g.283,659C>A) in the *PRKD1* gene on chromosome 17, and T>C substitution (g.133,741T>C) in the *KCNQ3* gene on chromosome 18, which were located at 43,090,728, 43,090,768 and 11,881,240, respectively. As presented in Table 1, the 3 SNP loci showed 3 genotypes in the Gannan yak population. The genotypes of three SNP sites, *PRKD1* gene g.283,619T>C, g.283,659C>A, and *KCNQ3* gene g.133,741T>C, were analyzed. The most frequent genotypes observed were TC, CA, and TC; these genotypes had frequencies of 0.512, 0.494, and 0.475, respectively, indicating that the heterozygous type was dominant in all three SNP sites. In the genetic sequence g.283,619T>C, the frequency of the T allele was 0.349, while the frequency of the C allele was 0.651. This suggests that the mutant allele is predominant at this locus. In the genetic sequence g.283,659C>A, the frequency of the unmutated C allele was 0.605, implying that there are a large number of unmutated alleles. In the genetic sequence g.133,741T>C, C allele frequency was higher than T, indicating more mutant alleles. According to the genetic diversity classification of PIC (PIC < 0.25, low polymorphism; 0.25 < PIC < 0.5, moderate polymorphism; PIC > 0.5, high polymorphism) analysis found that the PIC values of g.283,619 T>C, g.283,659 C>A and g.133,741 T>C were 0.351, 0.346, 0.362, respectively, which were moderate polymorphisms. The genotype frequencies of these three loci in the Gannan yak population conformed to the Hardy–Weinberg equilibrium ($p > 0.05$).

Table 1. Information regarding the variation and diversity parameters of the *PRKD1* and *KCNQ3* loci.

SNPs	Genotypic Frequencies			Allelic Frequencies		He	Ne	PIC	x^2	p Value
g.283,619T>C	TT (15)	TC (83)	CC (64)	T (113)	C (211)	0.454	1.833	0.351	2.649	0.104
	0.093	0.512	0.395	0.349	0.651					
g.283,659C>A	CC (58)	CA (80)	AA (24)	C (196)	A (128)	0.478	1.916	0.364	0.178	0.673
	0.358	0.494	0.148	0.605	0.395					
g.133,741T>C	TT (24)	TC (77)	CC (61)	T (125)	C (199)	0.474	1.901	0.362	0.001	0.970
	0.148	0.475	0.377	0.386	0.614					

Note: He: heterozygosity; Ne: effective number of alleles; PIC: polymorphism. PIC < 0.25, low polymorphism; 0.25 < PIC < 0.5, moderate polymorphism; PIC > 0.5, high polymorphism; $p > 0.05$ suggests that the population gene is in the Hardy–Weinberg balance and the sample comes from the same Mendel population.

3.2. Association Analysis of Milk Traits and Genotypes of SNPs in Gannan Yak

The association analysis of single nucleotide polymorphisms (SNPs) genotypes of *PRKD1* and *KCNQ3* genes with yak milk composition in Gannan yak was shown in Table 2. The results showed that the g.283,619T>C locus was significantly correlated with the lactose traits of Gannan yak milk, and the lactose content of TC type yak population was significantly higher than that of TT type and CC type ($p < 0.05$), indicating that the mutation of *PRKD1* gene improved the milk traits of Gannan yak; there were no significant differences in casein, protein, milk fat, non-fat milk solids (SNF), acidity and total milk solids (TS) among the three genotypes of yak milk ($p > 0.05$). The g.283,659 C > A locus was significantly correlated with milk fat content; the milk fat content of CA heterozygous genotype animal population was significantly higher than that of CC and AA genotypes ($p < 0.05$). The correlation analysis between g.133,741T>C locus and milk composition showed that the casein, protein and acidity of CC and TC groups were significantly higher than those of TT group ($p < 0.05$). The SNF of heterozygote TC group was significantly higher than that of CC group ($p < 0.05$). The results showed that the site mutations of *PRKD1* (g.283,619T>C, g.283,659C>A) gene and *KCNQ3* (g.133,741T>C) gene increased the casein, protein, milk fat and lactose traits of Gannan yak milk, and improved the milk quality of Gannan yak.

Table 2. Association analysis between *PRKD1* gene g.283,619T>C, g.283,659C>A loci, *KCNQ3* gene g.133,741T>C loci and milk traits of Gannan yak.

Genotype	Casein/%	Protein/%	Fat/%	SNF/%	Lactose/%	Acidity/°T	TS/%
			SNPs g.283,619T>C				
TT	4.14 ± 0.22	4.92 ± 0.33	6.77 ± 1.80	11.13 ± 0.55	4.90 ± 0.11 b	12.46 ± 1.02	17.96 ± 2.05
TC	4.09 ± 0.32	4.89 ± 0.43	7.10 ± 2.25	11.28 ± 0.49	5.00 ± 0.16 a	12.28 ± 1.39	16.77 ± 2.62
CC	4.08 ± 0.027	4.90 ± 0.37	6.80 ± 2.34	11.26 ± 0.47	4.98 ± 0.15 ab	12.39 ± 1.29	16.33 ± 2.59
p Value	*p* = 0.777	*p* = 0.958	*p* = 0.538	*p* = 0.494	*p* = 0.024	*p* = 0.516	*p* = 0.082
			SNPs g.283,659C>A				
CC	4.12 ± 0.25	4.89 ± 0.35	4.93 ± 0.95 b	11.21 ± 0.50	4.98 ± 0.15	12.23 ± 1.34	16.61 ± 1.75
CA	4.0 ± 0.33	4.89 ± 0.45	5.61 ± 1.64 a	11.27 ± 0.48	4.99 ± 0.17	12.41 ± 1.31	16.43 ± 2.58
AA	4.10 ± 0.25	4.96 ± 0.33	4.84 ± 1.38 b	11.37 ± 0.37	4.97 ± 0.16	12.70 ± 1.10	15.80 ± 1.99
p Value	*p* = 0.595	*p* = 0.685	*p* = 0.035	*p* = 0.378	*p* = 0.816	*p* = 0.875	*p* = 0.345
			SNPs g.133,741T>C				
TT	3.95 ± 0.32 b	4.61 ± 0.43 b	5.50 ± 3.00	11.02 ± 0.50 ab	5.04 ± 0.12	11.47 ± 1.30 b	16.41 ± 2.90
TC	4.14 ± 0.25 a	5.00 ± 0.34 a	5.30 ± 2.62	11.38 ± 0.43 a	5.05 ± 0.11	12.68 ± 1.14 a	16.54 ± 2.55
CC	4.12 ± 0.27 a	4.89 ± 0.39 a	5.92 ± 2.54	11.22 ± 0.48 b	4.98 ± 0.14	12.39 ± 1.32 a	17.03 ± 2.52
p Value	*p* = 0.011	*p* = 0.000	*p* = 0.396	*p* = 0.003	*p* = 0.006	*p* = 0.000	*p* = 0.453

Note: In the same column of data, the difference between different lowercase letters was statistically significant ($p < 0.05$). The data were expressed as mean ± standard deviation.

4. Discussion

Yak milk is a naturally concentrated milk that is softer and sweeter than regular milk [38]. Yak milk protein contains 80% casein, which is much higher than human milk [39]. These casein proteins consist of a group of phosphoproteins that form a gel structure during the coagulation process, effectively improving the texture and nutritional value of dairy products such as cheese and yogurt, which is the main reason why yak milk is whiter in color than cow milk [32]. In this study, the mutation of g.133,741T>C locus had a significant effect on the content of casein and protein in Gannan yak, the content of casein and protein in CC mutant and TC heterozygous yak populations was significantly higher than that in TT wild type, which indicated that the SNPs mutation of *KCNQ3* gene had a significant effect on the improvement of yak milk quality. Studies have shown that the protein of yak milk is divided into four caseins ($α_{S1}$-CN, $α_{S2}$-CN, β-CN, k-CN) and five major whey proteins (α-lactalbumin, β-lactoglobulin, lactoferrin, serum albumin and immunoglobulin) [40,41]. As a dietary protein, casein is a source of antihypertensive peptides, which is positive for conditioning the body of patients with cardiovascular disease [42]. It has been reported that whey protein can be rapidly digested in the digestive tract and can quickly supplement amino acids for the body [43]. Previously, researchers used iTRAQ-labelling proteomics to find 183 proteins in the whey of yak milk and colostrum, of which 86 proteins were significantly different between colostrum and milk [44]. Bioactive peptides found in yak milk play crucial roles in the metabolic and overall health of humans. They perform various physiological functions and can enhance the resistance of neonates and adults against diseases and pathogens [45]. We found that the mutations of g.283,619T>C and g.283659C>A had no significant effect on casein and protein of Gannan yak. It is worth noting that the mutation of g.283,659C>A site had an effect on milk fat, and the milk fat content of CA heterozygous type was significantly higher than that of CC wild type and AA mutant type. Yak milk has a high milk fat content and large fat globules, so it can produce high-quality ghee products [4]. Yak milk is rich in monounsaturated fatty acids (MUFAs, 20–25%) and polyunsaturated fatty acids (PUFAs, 3–6%), which play an important role in reducing the risk of cardiovascular disease and improving lipid metabolism [46]. Since yak graze in natural pastures all year round, this means yak milk may contain unique fatty acids [47]. Our results showed that the TC heterozygote at the g.133,741T>C locus had significantly higher non-fat milk solid than the CC mutant, and the acidity of the TC and CC genotypes was significantly higher than that of the TT wild type.

Advances in molecular biotechnology have enabled the detection of the contribution of some genes to genetic variation in important economic quantitative traits [21]. Georges et al. first performed a genome-wide scan of dairy cows to determine the genomic region of the genes responsible for phenotypic variation in production traits in dairy cows [48,49]. It can be seen from the quantitative trait loci (QTL) that there are many loci affecting the quantitative traits of dairy cows. At present, single nucleotide polymorphism (SNP) has been used for QTL detection and mapping of complex traits in many species [50]. In our study, when analyzing the correlation between *PRKD1* and *KCNQ3* gene loci and milk traits, it was found that when the *PRKD1* g.283,619 T>C locus was mutated from T to C, the heterozygote of the mutation site had a significant effect on the lactose of Gannan yak, and the heterozygote was significantly higher than the wild type. The heterozygote genotype composed of *PRKD1* g.283659 C>A mutation site had a positive effect on milk fat. There was no significant difference in casein, lactoprotein, SNF, acidity and TS between the *PRKD1* mutant population and the wild type. When the base of *KCNQ3* g.133,741T>C mutation site changed from T to C, it had a positive effect on the content of casein, lactoprotein, SNF and acidity of Gannan yak. In a word, the SNPs mutation sites we found have different effects on the milk quality of Gannan yak.

As a serin/threonine kinase expressed in normal breast cells, the *PRKD1* gene has been proved to be a key protein to inhibit the invasive phenotype in breast cancer [17]. In our study, we also confirmed the significant effect of *PRKD1* gene on milk quality traits of yak. In addition, *KCNQ3* has been identified as a candidate gene that determines the association of SNP markers (rs41580517) with milk fat content [21]. This study also found that the *KCNQ3* g.133,741T>C locus was associated with several milk components in Gannan yak. Therefore, our findings provide candidate genes for improving milk quality traits of Gannan yak, and also lay a foundation for subsequent research on improving milk yield of yak through screening. It is worth noting that the three loci g.283,619T>C, g.283659C>A and g.133,741T>C were located on the intron, but this does not exclude its effect on yak milk quality. The intronic regions of genes cannot be encoded and expressed, but they also play an important role in the regulation of gene expression [51]. It has been reported that the intron SNPs of *SLC22A4* gene affect the transcription efficiency of the gene [52], which strongly indicates the important role of introns in the gene. In previous studies, five SNPs in the intron region of *SORBS1* gene were significantly associated with milk fat traits [53]. This is similar to the results of our study. In addition, the polymorphism information content of the three loci g.283,619T>C, g.283659C>A and g.133,741T>C were 0.351, 0.364 and 0.362, respectively, and the p values of the three loci were 0.104, 0.673 and 0.970, respectively, all of which were greater than 0.05, indicating that each locus was in the Hardy–Weinberg equilibrium. Therefore, PRKD1 and KCNQ3 genes can be used as DNA molecular markers to improve milk production traits in yaks using marker-assisted selection.

5. Conclusions

This study is the first to investigate the association between *PRKD1* and *KCNQ3* gene polymorphisms and dairy traits in Gannan yaks. The results of association analysis showed that the polymorphisms of *PRKD1* gene and *KCNQ3* gene were significantly related to the milk quality of yaks. Therefore, genotyping *PRKD1* and *KCNQ3* genes is helpful to improve the milk quality of Gannan yaks.

Author Contributions: Conceptualization, X.M. (Xiaoyong Ma) and C.L.; Data curation, X.M. (Xiaoyong Ma), X.G. and P.Y.; Formal analysis, C.H., X.M. (Xiaoming Ma), Y.L. and M.C.; Funding acquisition, C.L.; Investigation, F.F., D.Y. and X.W.; Methodology, X.M. (Xiaoyong Ma); Project administration, C.L.; Resources, C.L.; Software, X.M. (Xiaoyong Ma), G.Y. and R.M.; Supervision, C.L.; Validation, X.M. (Xiaoyong Ma), J.Z. and J.S.; Visualization, C.L.; Writing—original draft, X.M. (Xiaoyong Ma); Writing—review and editing, X.M. (Xiaoyong Ma). All authors have read and agreed to the published version of the manuscript.

Funding: This research was funded by the National Key Research Program (2021YFD1600200), the Gansu Provincial Basic Research Innovation Group Project (20JR5RA580), the Modern Beef Cattle and Yak Industry Technology System (CARS-37), the Major Science and Technology Projects in Gansu Province (21ZD10NA001, GZGG-2021-1) and the Innovation Project of Chinese Academy of Ag-ricultural Sciences (25-LIHPS-01).

Institutional Review Board Statement: All the animal experiments were approved by the Lanzhou Institute of Husbandry and Pharmaceutical Sciences of the Chinese Academy of Agricultural Sciences (CAAS) with the grant number 1610322020018.

Informed Consent Statement: Not applicable.

Data Availability Statement: The original contributions presented in the study are included in the article; further inquiries can be directed to the corresponding author.

Conflicts of Interest: The authors declare no conflicts of interest.

References

1. Qiu, Q.; Wang, L.; Wang, K.; Yang, Y.; Ma, T.; Wang, Z.; Zhang, X.; Ni, Z.; Hou, F.; Long, R.; et al. Yak whole-genome resequencing reveals domestication signatures and prehistoric population expansions. *Nat. Commun.* **2015**, *6*, 10283. [CrossRef]
2. Long, R.J. Milk performance of Yak cows under traditional feeding and management on small Tibetan farms. *Acta Prataculturae Sin.* **1994**, *3*, 71–76.
3. Li, H.; Ma, Y.; Li, Q.; Wang, J.; Cheng, J.; Xue, J.; Shi, J. The Chemical Composition and Nitrogen Distribution of Chinese Yak (Maiwa) Milk. *Int. J. Mol. Sci.* **2011**, *12*, 4885–4895. [CrossRef]
4. Luo, J.; Huang, Z.; Liu, H.; Zhang, Y.; Ren, F. Yak milk fat globules from the Qinghai-Tibetan Plateau: Membrane lipid composition and morphological properties. *Food Chem.* **2018**, *245*, 731–737. [CrossRef]
5. Liu, H.; Hao, L.; Cao, X.; Yang, G.; Degen, A.A.; Xiao, L.; Liu, S.; Zhou, J. Effects of supplementary concentrate and/or rumen-protected lysine plus methionine on productive performance, milk composition, rumen fermentation, and bacterial population in Grazing, Lactating Yaks. *Anim. Feed Sci. Technol.* **2023**, *297*, 115591. [CrossRef]
6. Cui, G.X.; Yuan, F.; Degen, A.A.; Liu, S.M.; Zhou, J.W.; Shang, Z.H.; Ding, L.M.; Mi, J.D.; Wei, X.H.; Long, R.J. Composition of the milk of yaks raised at different altitudes on the Qinghai–Tibetan Plateau. *Int. Dairy J.* **2016**, *59*, 29–35. [CrossRef]
7. Nikkhah, A. Science of Camel and Yak Milks: Human Nutrition and Health Perspectives. *Food Nutr. Sci.* **2011**, *2*, 667–673. [CrossRef]
8. Luo, J.; Liu, L.; Liu, T.; Shen, Q.; Liu, C.; Zhou, H.; Ren, F. Simulated in vitro infant gastrointestinal digestion of yak milk fat globules: A comparison with cow milk fat globules. *Food Chem.* **2020**, *314*, 126160. [CrossRef] [PubMed]
9. Li, H.; Yang, X.; Tang, D.; Xi, B.; Li, W.; Chen, Z.; Bao, Y.; Dingkao, R.; Gao, Y.; Wang, P.; et al. Exploring the link between microbial community structure and flavour compounds of traditional fermented yak milk in Gannan region. *Food Chem.* **2024**, *435*, 137553. [CrossRef] [PubMed]

10. Long, R.; Kang, M. Milking performance of China yak (*Bos grunniens*): A preliminary report. *Afr. J. Agric. Res.* **2007**, *2*, 52–57. Available online: http://www.academicjournals.org/AJAR (accessed on 25 February 2024).
11. Wu, X.; Zhou, X.; Xiong, L.; Pei, J.; Yao, X.; Liang, C.; Bao, P.; Chu, M.; Guo, X.; Yan, P. Transcriptome Analysis Reveals the Potential Role of Long Non-coding RNAs in Mammary Gland of Yak During Lactation and Dry Period. *Front. Cell Dev. Biol.* **2020**, *8*, 579708. [CrossRef] [PubMed]
12. Steinberg, S.F. Regulation of protein kinase D1 activity. *Mol. Pharmacol.* **2012**, *81*, 284–291. [CrossRef] [PubMed]
13. Zhang, X.; Connelly, J.; Chao, Y.; Wang, Q.J. Multifaceted Functions of Protein Kinase D in Pathological Processes and Human Diseases. *Biomolecules* **2021**, *11*, 483. [CrossRef] [PubMed]
14. Bastea, L.I.; Döppler, H.; Balogun, B.; Storz, P. Protein kinase D1 maintains the epithelial phenotype by inducing a DNA-bound, inactive SNAI1 transcriptional repressor complex. *PLoS ONE* **2012**, *7*, e30459. [CrossRef]
15. Eiseler, T.; Döppler, H.; Yan, I.K.; Kitatani, K.; Mizuno, K.; Storz, P. Protein kinase D1 regulates cofilin-mediated F-actin reorganization and cell motility through slingshot. *Nat. Cell Biol.* **2009**, *11*, 545–556. [CrossRef]
16. Eiseler, T.; Hausser, A.; De Kimpe, L.; Van Lint, J.; Pfizenmaier, K. Protein kinase D controls actin polymerization and cell motility through phosphorylation of cortactin. *J. Biol. Chem.* **2010**, *285*, 18672–18683. [CrossRef]
17. Eiseler, T.; Döppler, H.; Yan, I.K.; Goodison, S.; Storz, P. Protein kinase D1 regulates matrix metalloproteinase expression and inhibits breast cancer cell invasion. *Breast Cancer Res.* **2009**, *11*, R13. [CrossRef]
18. Coulona, J.B.; Gasquib, P.; Barnouin, J.; Ollier, A.; Pradel, P. Effect of mastitis and related-germ on milk yield and composition during naturally-occurring udder infections in dairy cows. *Anim. Res.* **2002**, *51*, 383–393. [CrossRef]
19. Yang, G.; Zhang, J.; Ma, X.; Ma, R.; Shen, J.; Liu, M.; Yu, D.; Feng, F.; Huang, C.; Ma, X.; et al. Polymorphisms of *CCSER1* Gene and Their Correlation with Milk Quality Traits in Gannan Yak (*Bos grunniens*). *Foods* **2023**, *12*, 4318. [CrossRef]
20. Zhou, C.; Li, C.; Cai, W.; Liu, S.; Yin, H.; Shi, S.; Zhang, Q.; Zhang, S. Genome-Wide Association Study for Milk Protein Composition Traits in a Chinese Holstein Population Using a Single-Step Approach. *Front. Genet.* **2019**, *10*, 72. [CrossRef]
21. Kolbehdari, D.; Wang, Z.; Grant, J.R.; Murdoch, B.; Prasad, A.; Xiu, Z.; Marques, E.; Stothard, P.; Moore, S.S. A whole genome scan to map QTL for milk production traits and somatic cell score in Canadian Holstein bulls. *J. Anim. Breed. Genet.* **2009**, *126*, 216–227. [CrossRef]
22. De la Rosa, V.; Guzmán-Hernández, M.L.; Carrillo, E. Triclosan is a KCNQ3 potassium channel activator. *Pflügers Arch.—Eur. J. Physiol.* **2022**, *474*, 721–732. [CrossRef]
23. Robbins, J. KCNQ potassium channels: Physiology, pathophysiology, and pharmacology. *Pharmacol. Ther.* **2001**, *90*, 1–19. [CrossRef]
24. Hernandez, C.C.; Zaika, O.; Tolstykh, G.P.; Shapiro, M.S. Regulation of neural KCNQ channels: Signalling pathways, structural motifs and functional implications. *J. Physiol.* **2008**, *586*, 1811–1821. [CrossRef] [PubMed]
25. Vigil, F.A.; Bozdemir, E.; Bugay, V.; Chun, S.H.; Hobbs, M.; Sanchez, I.; Hastings, S.D.; Veraza, R.J.; Holstein, D.M.; Sprague, S.M.; et al. Prevention of brain damage after traumatic brain injury by pharmacological enhancement of KCNQ (Kv7, "M-type") K$^+$ currents in neurons. *J. Cereb. Blood Flow Metab.* **2020**, *40*, 1256–1273. [CrossRef]
26. Rían, W.M.; Geoffrey, W.A. Gabapentin Is a Potent Activator of KCNQ3 and KCNQ5 Potassium Channels. *Mol. Pharmacol.* **2018**, *94*, 1155. [CrossRef]
27. Park, Y.W.; Haenlein, G.F.W.; Wendorff, W.L. Overview of Milk of Non-Bovine Mammals. In *Handbook of Milk of Non-Bovine Mammals*, 2nd ed.; John Wiley & Sons, Ltd.: Hoboken, NJ, USA, 2017; pp. 1–9. [CrossRef]
28. Peng, Y.S.; Brown, M.A.; Wu, J.P.; Wei, L.X.; Wu, J.L.; Sanbei, D.Z. Fatty Acid Profile in Milk Fat from Qinghai Plateau Yak at Different Altitudes and Parities. *Prof. Anim. Sci.* **2008**, *24*, 479–487. [CrossRef]
29. Wang, T.T.; Guo, Z.W.; Liu, Z.P.; Feng, Q.Y.; Wang, X.L.; Tian, Q.; Ren, F.Z.; Mao, X.Y. The aggregation behavior and interactions of yak milk protein under thermal treatment. *J. Dairy Sci.* **2016**, *99*, 6137–6143. [CrossRef]
30. DePeters, E.J.; Ferguson, J.D. Nonprotein Nitrogen and Protein Distribution in the Milk of Cows. *J. Dairy Sci.* **1992**, *75*, 3192–3209. [CrossRef]
31. Cerbulis, J.; Farrell, H.M., Jr. Composition of milks of dairy cattle. I. Protein, lactose, and fat contents and distribution of protein fraction. *J. Dairy Sci.* **1975**, *58*, 817–827. [CrossRef] [PubMed]
32. Kalwar, Q.; Ma, X.; Xi, B.; Korejo, R.A.; Bhuptani, D.K.; Chu, M.; Yan, P. Yak milk and its health benefits: A comprehensive review. *Front. Vet. Sci.* **2023**, *10*, 1213039. [CrossRef]
33. Zhang, W.; Wu, S.; Cao, J.; Li, H.; Li, Y.; He, J.; Zhang, L. A preliminary study on anti-hypoxia activity of yak milk powder in vivo. *Dairy Sci. Technol.* **2014**, *94*, 633–639. [CrossRef]
34. Guo, X.; Long, R.; Kreuzer, M.; Ding, L.; Shang, Z.; Zhang, Y.; Yang, Y.; Cui, G. Importance of functional ingredients in yak milk-derived food on health of Tibetan nomads living under high-altitude stress: A review. *Crit. Rev. Food Sci. Nutr.* **2014**, *54*, 292–302. [CrossRef] [PubMed]
35. Singh, T.P.; Arora, S.; Sarkar, M. Yak milk and milk products: Functional, bioactive constituents and therapeutic potential. *Int. Dairy J.* **2023**, *142*, 105637. [CrossRef]
36. Gu, X.; Huang, S.; Zhu, Z.; Ma, Y.; Yang, X.; Yao, L.; Gao, X.; Zhang, M.; Liu, W.; Qiu, L.; et al. Genome-wide association of single nucleotide polymorphism loci and candidate genes for frogeye leaf spot (*Cercospora sojina*) resistance in soybean. *BMC Plant Biol.* **2021**, *21*, 588. [CrossRef] [PubMed]

37. Yang, Y.; Han, L.; Yu, Q.; Gao, Y.; Song, R.; Zhao, S. Phosphoproteomic analysis of longissimus lumborum of different altitude yaks. *Meat Sci.* **2020**, *162*, 108019. [CrossRef] [PubMed]
38. Faccia, M.; D'Alessandro, A.G.; Summer, A.; Hailu, Y. Milk Products from Minor Dairy Species: A Review. *Animals* **2020**, *10*, 1260. [CrossRef] [PubMed]
39. Malacarne, M.; Martuzzi, F.; Summer, A.; Mariani, P. Protein and fat composition of mare's milk: Some nutritional remarks with reference to human and cow's milk. *Int. Dairy J.* **2002**, *12*, 869–877. [CrossRef]
40. Li, H.; Ma, Y.; Dong, A.; Wang, J.; Li, Q.; He, S.; Maubois, J.-L. Protein composition of yak milk. *Dairy Sci. Technol.* **2010**, *90*, 111–117. [CrossRef]
41. Mao, Y.J.; Zhong, G.H.; Zheng, Y.C.; Pen, X.; Yang, Z.P.; Wang, Y.; Jiang, M.F. Genetic Polymorphism of Milk Protein and Their Relationships with Milking Traits in Chinese Yak. *Asian-Australas. J. Anim. Sci.* **2004**, *17*, 1479–1483. [CrossRef]
42. Din, J.N.; Newby, D.E.; Flapan, A.D. Omega 3 fatty acids and cardiovascular disease—Fishing for a natural treatment. *BMJ (Clin. Res. Ed.)* **2004**, *328*, 30–35. [CrossRef]
43. Wang, L.; Ma, Y.; Li, H.; Yang, F.; Cheng, J. Identification and characterization of yak α-lactalbumin and β-lactoglobulin. *J. Dairy Sci.* **2021**, *104*, 2520–2528. [CrossRef]
44. Yang, Y.; Zhao, X.; Yu, S.; Cao, S. Quantitative proteomic analysis of whey proteins in the colostrum and mature milk of yak (*Bos grunniens*). *J. Sci. Food Agric.* **2015**, *95*, 592–597. [CrossRef] [PubMed]
45. Park, Y.W.; Nam, M.S. Bioactive Peptides in Milk and Dairy Products: A Review. *Korean J. Sci. Food Agric.* **2015**, *35*, 831–840. [CrossRef] [PubMed]
46. Ogunka-Nnoka, C.; Igwe, F.; Okorosaye-Orubite, K. Comparative Study of the Physicochemical Properties and Fatty Acid Composition of Some Indigenous Spices in Nigeria. *Int. J. Biochem. Res. Rev.* **2015**, *5*, 171–177. [CrossRef] [PubMed]
47. Simopoulos, A.P. Omega-3 fatty acids in health and disease and in growth and development. *Am. J. Clin. Nutr.* **1991**, *54*, 438–463. [CrossRef] [PubMed]
48. Georges, M.; Nielsen, D.; Mackinnon, M.; Mishra, A.; Okimoto, R.; Pasquino, A.T.; Sargeant, L.S.; Sorensen, A.; Steele, M.R.; Zhao, X.; et al. Mapping quantitative trait loci controlling milk production in dairy cattle by exploiting progeny testing. *Genetics* **1995**, *139*, 907–920. [CrossRef] [PubMed]
49. Smaragdov, M.; Prinzenberg, E.-M.; Zwierzchowski, L. QTL mapping in cattle: Theoretical and empirical approach. *Anim. Sci. Pap. Rep.* **2006**, *24*, 69–110. Available online: https://api.semanticscholar.org/CorpusID:83094926 (accessed on 25 February 2024).
50. Daw, E.W.; Heath, S.C.; Lu, Y. Single-nucleotide polymorphism versus microsatellite markers in a combined linkage and segregation analysis of a quantitative trait. *BMC Genet.* **2005**, *6*, S32. [CrossRef] [PubMed]
51. Casas, E.; White, S.N.; Riley, D.G.; Smith, T.P.; Brenneman, R.A.; Olson, T.A.; Johnson, D.D.; Coleman, S.W.; Bennett, G.L.; Chase, C.C., Jr. Assessment of single nucleotide polymorphisms in genes residing on chromosomes 14 and 29 for association with carcass composition traits in *Bos indicus* cattle. *J. Anim. Sci.* **2005**, *83*, 13–19. [CrossRef]
52. Tokuhiro, S.; Yamada, R.; Chang, X.; Suzuki, A.; Kochi, Y.; Sawada, T.; Suzuki, M.; Nagasaki, M.; Ohtsuki, M.; Ono, M.; et al. An intronic SNP in a RUNX1 binding site of SLC22A4, encoding an organic cation transporter, is associated with rheumatoid arthritis. *Nat. Genet.* **2003**, *35*, 341–348. [CrossRef] [PubMed]
53. Yang, L.; Min, X.; Zhu, Y.; Hu, Y.; Yang, M.; Yu, H.; Li, J.; Xiong, X. Polymorphisms of *SORBS1* Gene and Their Correlation with Milk Fat Traits of Cattleyak. *Animals* **2021**, *11*, 3461. [CrossRef] [PubMed]

Disclaimer/Publisher's Note: The statements, opinions and data contained in all publications are solely those of the individual author(s) and contributor(s) and not of MDPI and/or the editor(s). MDPI and/or the editor(s) disclaim responsibility for any injury to people or property resulting from any ideas, methods, instructions or products referred to in the content.

Article

Distribution of Calcium, Phosphorus and Magnesium in Yak (*Bos grunniens*) Milk from the Qinghai Plateau in China

Piero Franceschi [1], Wancheng Sun [2,*], Massimo Malacarne [1,*], Yihao Luo [2], Paolo Formaggioni [1], Francesca Martuzzi [3] and Andrea Summer [1]

[1] Department of Veterinary Science, University of Parma, Via del Taglio 10, 43126 Parma, Italy
[2] Animal Science Department, College of Animal Husbandry and Veterinary Medicine, Qinghai University, Nig Da Road 251, Xining 810016, China
[3] Department of Food and Drug, University of Parma, Parco Area delle Scienze, 27/A, 43124 Parma, Italy
* Correspondence: sun.wancheng0108@aliyun.com (W.S.); massimo.malacarne@unipr.it (M.M.)

Abstract: This research was aimed to assess the distribution of calcium, phosphorus and magnesium within the casein micelles of yak milk. To this aim, nine bulk yak milk samples (Y-milk), collected in three yak farms located in the Chinese province of Qinghai, were compared to nine bulk cow milk samples used as a reference. A quite similar content of colloidal calcium (0.80 vs. 0.77 mmol/g of casein; $p > 0.05$), a higher content of magnesium (0.05 vs. 0.04 mmol/g of casein; $p \leq 0.01$) and a lower content of colloidal phosphorus (0.48 vs. 0.56 mmol/g of casein; $p \leq 0.01$) between yak and cow casein micelles were found. Moreover, the yak casein micelles showed a lower value of prosthetic phosphorus (0.20 vs. 0.26 mmol/g of casein; $p \leq 0.05$) compared to the cow micelles. The lower values of colloidal and prosthetic phosphorus in yak casein micelles suggest that the yak casein is less phosphorylated than the cow one.

Keywords: yak milk; protein fractions; mineral content; salt equilibria; casein micelles mineralisation

1. Introduction

The heat stability of milk and its ability to transfer high quantities of Ca and P in a highly assimilable chemical form by the human organism depend on casein micelles structure. Moreover, the casein micelles are the substrate of the coagulation of milk (acid or enzymatic). This process is an essential step in milk cheesemaking and allows the efficient release of biological active components of milk during digestion. Casein micelles are organised roughly in spherical particles constituted by the four caseins and by an amorphous mineral cluster defined as colloidal calcium phosphate. In bovine milk, micellar minerals represent about 6% of the dry matter of casein micelles. Although there are several models of casein micelles structure, there is a general agreement about the stabilising effect played by k-casein on the surface of the micelles and by the nanoclusters of calcium phosphate in the internal zones of the micelles. A quantitative model of the bovine milk casein micelle, characterized by ion equilibrium and calcium phosphate sequestration by individual caseins, was recently proposed by Bijl et al. [1].

The degree of mineralisation (or mineralisation level) of micelles can be defined as the concentration of Ca, Mg and P within the casein micelle, in the form of calcium phosphate nanoclusters or via ionic bonds with amino acid residues [1,2]. The degree of mineralisation of casein micelles influences the processing and nutritional properties of milk and of the products derived from it. In particular, a high level of mineralisation of casein improves the rennet coagulation ability of milk but seems to decrease the degradation of casein during in vitro gastric digestion [3,4].

Studies on the degree of mineralisation of casein micelles mainly concerned bovine milk casein, although milk from other species was also examined. For example, in the last years, some studies investigated the degree of mineralisation of casein in milk from

sheep [5,6], goat [6,7], buffalo [7–9], camel [10], donkey [11–13] and wild ruminants [14] such as red deer (*Cervus elaphus*), fallow deer (*Dama dama*), roe deer (*Capreulus capreulus*).

Yaks are extensively raised in the plateau of the western Tibetan region of China at altitudes ranging approximately from 2000 to 5000 m above the sea level [15,16]. Generally, yaks are raised mainly for their milk, their meat and their wool, which are a vital part of the local economy in the Tibetan region of China [16]. In particular, yak milk is a food product of great value for the population of the Qinghai plateau, where yaks are the only raised animals producing milk [15]. Indeed, yak milk, besides being a beverage, is used to produce many products. The main product is butter, but this milk is used for a variety of other products as well, such as yogurt and fermented beverages, hard and soft cheeses and other traditional products [17].

Nowadays, approximately 25% of yak milk is processed at industrial level [18] and, for its relevance in the nutrition of the people living in the Tibetan plateau, during the last years, several studies were carried out aimed to characterise the milk yield, chemical composition, cheesemaking aptitude and nutritional properties.

Yak average daily milk yield ranges from about 0.8 to 3.2 kg/d, according to the breed and rearing zone. Furthermore, also the length of lactation is variable, depending on the same factors, from 100 to 180 days, and this leads to an average milk production from 150 to 500 kg/lactation [16].

Compared to cow milk, yak milk has a higher concentration of milk constituents such as fat, protein and casein [15,19–21], a larger casein micelles size, better rennet coagulation properties and a higher cheese yield [22,23]. Moreover, it was observed that yak casein is less soluble than the cow one [24] and that these proteins differ from each other in composition and hydration [25].

However, to date, only a scarce number of studies were carried out to analyse yak milk mineral composition. Yak milk has a higher content of Ca, P and Mg than cow milk [26–28]. Most of Ca and P are in the colloidal phase, whereas 3/4 of Mg is in the soluble phase [27]. Nowadays, no one has yet investigated casein micelles mineralisation in yak milk in its native state.

For this reason, the knowledge of the characteristics of casein micelles and their mineral content can be useful to exploit the dairy potential and address the transformation technology of yak milk for its valorisation. Thus, the characterisation of milk from yak was carried out in this study, focusing on its mineral content and the distribution of Ca, P and Mg between its soluble and micellar phases, as well as their concentration within casein micelles.

2. Materials and Methods

2.1. Experimental Design and Sampling Procedure

Nine bulk yak milk samples were collected in three yak farms located in three different zones of the Chinese province of Qinghai. All animals belonged to Plateau yak breed. One herd was from the province of Guoluo (3719 m a.s.l.), one herd from the province of Hainan (2835 m a.s.l.), and one herd from the province of Wulan (2960 m a.s.l.). Samples were collected monthly from June to August in each herd. For the purpose of comparison, nine bulk milk samples of Italian Friesian cows were collected and analysed with the same method. These latter were taken from three farms located in the north of Italy raising only Italian Friesian cows. As for the yak milk samples, the Italian Friesian milk samples were collected monthly from June to August in each herd.

The milk samples were representative of the herd bulk milk and were collected at the end of the morning milking. After sampling, they were frozen, transported to the laboratories and kept at $-20\ °C$ until the analysis.

2.2. Analytical Methods

For each milk sample, total N (TN), non-casein N (SN) and non-protein N (NPN) were determined by the Kjeldahl method [29–31] in milk, milk acid whey at pH 4.6 and trichloroacetic acid filtered whey (TCA 120 g/L; Carlo Erba Reagents, I-20010, Milan, Italy), respectively. The Kjeldahl method was performed using a DK6 Digestor and an UDK126A Distiller (VELP Scientifica, Usmate, Italy).

Moreover, for each milk sample, also by the Kjeldahl method, not-coagulable N (NCN) on the rennet whey was determined. The rennet whey was obtained from milk, according to Franceschi et al. [32], adding 2 mL of diluted 1:100 rennet (Christian Hansen, DK-7172 Hørsholm, Denmark) into 100 mL of milk previously thermostated at 35 °C. After coagulation of the milk, the whey outcome by curd syneresis was filtered on Whatman 1 paper filter (Merck Millipore Corporation, D-64293, Darmstadt, Germany).

From these data, crude protein (TN × 6.38/1000), whey protein (SN × 6.38/1000), casein ((TN-SN) × 6.38/1000), casein number ((TN-SN) × 100/TN) and NPN × 6.38 (NPN × 6.38/1000) were calculated as described by Franceschi et al. [33], and paracasein was calculated ((TN-NCN) × 6.38/1000) according to Franceschi et al. [34].

The lactose and fat contents were determined by the mid-infrared method [35] using a MilkoScan FT 6000 (Foss Electric, DK-3400, Hillerød, Denmark). Moreover, somatic cells and total bacterial count were assessed using the fluoro-opto-electronic method [36] with Fossomatic (Foss Electric, DK-3400, Hillerød, Denmark) and the flow cytometry method [37] with BactoScan FC (Foss Electric, Hillerød, Denmark), respectively, and dry matter (DM) was obtained after oven-drying 20 g of milk at 102 °C [38].

Furthermore, each milk sample was skimmed and subsequently submitted to ultrafiltration process in Amicon 8200 ultrafiltration cells (Merck Millipore Corporation, Darmstadt, Germany). The ultrafiltration process was performed as described by Petrera et al. [39] with a Millipore membrane with 30 kDa cut-off in a N_2 flow at 75 psi (polyethersulfone ultrafiltration membranes, Merck Millipore Corporation, Darmstadt, Germany).

The ash content was obtained by muffle calcination at 530 °C of 20 g of milk and of 10 g of ultrafiltered whey [40].

The ashes were solubilised in hydrochloric acid to obtain a hydrochloric ash solution [41] and, from this, by a colorimetric method [42], total P and soluble P were assessed in the hydrochloric ash solution of milk and in the hydrochloric ash solution of ultrafiltered whey, respectively. Moreover, also by the colorimetric method of Allen [42], the content of total acid-soluble P was assessed in trichloroacetic (TCA) acid-filtered whey digested at 240 °C by a DK6 digestion unit (VELP Scientifica, Usmate, Italy) for 1 h in perchloric acid 65% (Carlo Erba Reagents, I-20010, Milan, Italy). The colorimetric method of Allen, in brief, was performed by adding to 10 mL of hydrochloric ash solution, diluted 40 times, 2 mL of perchloric acid 65%, 2 mL of a solution with 20 g/L of 2,4-diaminophenol dihydrochloride, 200 g/L of sodium metabisulfite and 1 mL of ammonium molybdate (83 g/L) solution (all reagents came from Carlo Erba Reagents, I-20010, Milan, Italy). After 25 min, 1 mL of this solution was read by a Helios spectrophotometer (Thermo Fisher Scientific, Waltham, MA 02451, USA) at 750 nm. For the determination of P 5, standard solutions from 25 mg/100 g to 400 mg/100 g were used (KH_2PO_4, Carlo Erba Reagents, I-20010, Milan, Italy).

Moreover, starting from the hydrochloric ash solution of the milk and of the ultrafiltered whey, opportunely diluted ten thousand times, the total content of Ca and Mg and their content in the solution were determined [41] by atomic absorption spectrometry using a Perkin-Elmer 1100 B instrument (Perkin-Elmer, Waltham, MA 02451, USA). For both Ca and Mg determination, calibration curves were obtained using 5 standard solutions, ($CaCl_2 \cdot 6H_2O$ and $MgCl_2 \cdot 6H_2O$ of Carlo Erba Reagents, I-20010, Milan, Italy) ranging from 0.5 to 8 ppm for Ca determination and ranging from 0.05 to 0.8 ppm for Mg determination.

From these data, the colloidal fractions, namely, the minerals inside the casein micelles, of Ca and Mg were calculated as the difference between their total and soluble contents. Differently from Ca and Mg, colloidal P, within casein micelles, is present in two different chemical forms, i.e., as part of the phosphorylated residues of caseins (casein P) and as a constituent of colloidal inorganic P. These fractions were calculated according to Malacarne et al. [12] as follows:

$$\text{Colloidal P} = \text{TP} - \text{SP} \tag{1}$$

$$\text{Casein P} = \text{TP} - \text{TASP} \tag{2}$$

$$\text{Colloidal inorganic P} = \text{TASP} - \text{SP} \tag{3}$$

where TP = total phosphorus; SP = soluble phosphorus; TASP = total acid-soluble phosphorus.

Furthermore, colloidal P content was corrected for the quota of P in phospholipids according to Bonaga and Mascolo [43].

Then, the ratios of the mineral soluble contents with respect to their total amounts were calculated and, according to Malacarne et al. [12], as well as the ratios between colloidal minerals and casein, expressed in millimoles per gram of casein, were calculated.

Finally, the pH was measured by a Crison potentiometer (Crison Instruments, E-08328, Barcelona, Spain), and the density at 15 °C by means of a Quevenne lactometer.

2.3. Statistical Analysis

The data collected were tested by analysis of variance, using the general linear model procedure of SPSS (IBM SPSS Statistics 26, Armonk, NY 10504-1722, USA), according to the following hierarchical model:

$$Y_{ijkl} = \mu + S_i + T_j + H_{ik} + \varepsilon_{ijkl} \tag{4}$$

where Y_{ijkl} = dependent variable; μ = overall mean; S_i = effect of the species (i = 1, 2); T_j = effect of the trial (j = 1, ..., 3); H_{ik} = effect of herd nested within species (1, ..., 5); ε_{ijkl} = residual error.

The significance of the differences between yak milk and cow milk was tested by the Bonferroni post-hoc test.

3. Results

In Table 1, the chemical composition, physicochemical properties and counts of somatic cells and total bacteria of yak bulk milk (Y-milk) and cow bulk milk (C-milk) are shown.

Table 1. Least-square means of chemical composition, physico-chemical properties and counts of somatic cells and total bacteria for yak (*Bos grunniens*, Y-milk) and cow (*Bos taurus*, C-milk) milk.

Parameters	Unit of Measure	Y-Milk n[1] = 9	C-Milk n[1] = 9	SE[2]	p[3]
Dry matter	g/100 g	18.57	12.25	0.98	***
Ash	g/100 g	0.78	0.71	0.03	*
Lactose	g/100 g	4.79	4.95	0.04	*
Fat	g/100 g	6.17	3.46	0.34	**
Crude protein	g/100 g	4.57	3.16	0.10	***
Whey protein	g/100 g	1.05	0.71	0.02	***
Casein	g/100 g	3.53	2.45	0.07	***
Casein number	%	77.12	77.58	0.17	NS
NPN × 6.38	g/100 g	0.26	0.17	0.01	***
True protein	g/100 g	4.31	2.99	0.10	***
True whey protein	g/100 g	0.79	0.54	0.02	***
Paracasein	g/100 g	3.05	2.87	0.08	***
Density	Kg/L	1.034	1.032	0.01	NS
pH-value	value	6.69	6.70	0.02	NS
Somatic cells count	10³ cells/mL	101.69	311.88	29.14	**
Total bacterial count	10³ FCU/mL	890.99	82.11	285.25	**

[1] Number of samples; [2] Standard error; [3] p-value: NS $p > 0.05$; * $p \leq 0.05$; ** $p \leq 0.01$; *** $p \leq 0.001$.

The average values of dry matter, crude protein, whey protein, casein, NPN × 6.38 true protein, true whey protein and paracasein were different in Y-milk and C-milk, with $p \leq 0.001$. Furthermore, Y-milk and C-milk showed different average values of fat, somatic cells count and total bacterial count, with $p \leq 0.01$, and of lactose and ash, with $p \leq 0.05$.

The content of dry matter and its main constituents (lactose, fat, crude protein and ash), as well as of crude protein and its fractions (whey protein, casein, paracasein and NPN × 6.38) and the total bacterial count were higher in Y-milk than in C-milk. On the contrary, the somatic cell count was found to be lower in Y-milk than in C-milk. The casein number and the values of density and pH were not different between the two types of milk.

In Table 2, the contents of Ca, P and Mg and their distribution between the colloidal and the soluble phases of Y-milk and C-milk are reported.

Table 2. Least-square means of Ca, P and Mg contents and distribution between the colloidal and the soluble phases of yak (*Bos grunniens*, Y-milk) and cow (*Bos taurus*, C-milk) milk.

Parameters	Unit of Measure	Y-Milk n[1] = 9	C-Milk n[1] = 9	SE[2]	p[3]
Ash of ultrafiltered whey	g/100 g	0.56	0.54	0.02	NS
Total Ca	mg/100 g	160.74	113.65	7.25	**
Colloidal Ca	mg/100 g	112.62	76.01	6.21	**
Soluble Ca	mg/100 g	48.12	37.64	1.33	***
Total P	mg/100 g	113.88	87.70	3.91	**
Colloidal P	mg/100 g	51.96	42.39	1.68	**
Colloidal inorganic P	mg/100 g	30.27	22.34	1.15	**
Casein P	mg/100 g	21.70	20.05	1.70	NS
Soluble P	mg/100 g	58.40	43.20	3.06	**
Total Mg	mg/100 g	12.52	9.49	0.51	**
Colloidal Mg	mg/100 g	4.40	2.28	0.20	***
Soluble Mg	mg/100 g	8.12	7.22	0.33	*
Soluble Ca to total Ca ratio	g/100 g of total Ca	30.24	33.13	0.87	*
Soluble P to total P ratio	g/100 g of total P	50.70	49.24	1.03	NS
Soluble Mg to total Mg ratio	g/100 g of total Mg	64.88	76.06	0.52	***

[1] Number of samples; [2] Standard error; [3] p-value: NS $p > 0.05$; * $p \leq 0.05$; ** $p \leq 0.01$; *** $p \leq 0.001$.

Among the considered minerals, both in Y-milk and in C-milk, Ca was the most abundant mineral, P being the second, and Mg the third.

The total contents of Ca, P and Mg and their fractions as well, except for casein P that was not significantly different, were higher in Y-milk than in C-milk.

On the other hand, the percentage ratios of soluble Ca and soluble Mg with respect to their total contents were lower in Y-milk than in C-milk, whereas no significant differences were observed for the percentage ratio of soluble P with respect to total P.

In Table 3, the concentrations of Ca, P and Mg within the casein micelles of Y-milk and C-milk, expressed as mmol per gram of casein, are reported.

The contents of colloidal Ca and colloidal Mg and the colloidal-Ca-to-colloidal-P ratio showed significant differences between Y-milk and C-milk ($p \leq 0.01$), while casein P contents showed a significant difference between Y-milk and C-milk, with $p \leq 0.05$.

The casein micelle of Y-milk had lower content of P and higher content of Mg, than C-milk. Finally, no differences between the two types of micelles were observed for Ca.

Table 3. Least-square means of Ca, P and Mg concentrations within the casein micelles of yak (*Bos grunniens*, Y-milk) and cow (*Bos taurus*, C-milk) milk.

Parameters	Unit of Measure	Y-Milk n[1] = 9	C-Milk n[1] = 9	SE[2]	p[3]
Casein	g 100 g^{-1}	3.53	2.45	0.07	***
Colloidal Ca	mmol/g of casein	0.80	0.77	0.05	NS
Colloidal P	mmol/g of casein	0.48	0.56	0.02	**
Colloidal inorganic P	mmol/g of casein	0.28	0.29	0.01	NS
Casein P	mmol/g of casein	0.20	0.26	0.02	*
Colloidal Mg	mmol/g of casein	0.05	0.04	0.01	**
Colloidal Ca to colloidal P	value	1.67	1.39	0.05	**
Colloidal Ca to colloidal inorganic P	value	2.95	2.66	0.24	NS
Soluble Ca to soluble P	value	0.67	0.68	0.02	NS

[1] Number of samples; [2] Standard error; [3] p-value: NS $p > 0.05$; * $p \leq 0.05$; ** $p \leq 0.01$; *** $p \leq 0.001$.

4. Discussion

4.1. Yak Milk Chemical Composition and Physico-Chemical Properties

In general, the results of the main studies concerning the chemical composition of yak milk were reported in several comparative reviews [44–48]. The concentration of the main constituents of Y-milk (ash, lactose, fat and crude protein) were within the ranges reported by previous studies [15,16,49]. Differently from C-milk, in which the principal constituent of dry matter was lactose, in Y-milk the main constituent of dry matter was fat (33.22 g 100 g^{-1} of dry matter). In addition, the values of the protein fractions were comparable with those reported in previous studies [18,20,49]. The higher casein content of Y-milk, compared to C-milk, resulted in a higher paracasein content, which is the rennet-coagulable fraction of casein.

Nevertheless, since the cheesemaking process consists in the formation of a three-dimensional network of paracasein, in which fat globules and part of the whey are entrapped [50], milk the casein and paracasein contents are very important traits, for their repercussions on the yield of both soft cheeses [51,52] and hard cheeses [53,54]. Indeed, the cheesemaking yield is directly proportional to the milk casein and paracasein contents for both soft and hard cheese production, as reported by many authors [23,33,51–54]

From this point of view, the yak milk high contents of casein, paracasein and fat can result in a high cheese-yielding ability, as demonstrated by Zhang et al. [23]. The casein number of Y-milk in the present research was higher than that reported by Li et al. [15], who found an average value of 74.63% from 104 individual milk samples collected from Maiwa breed yaks. This difference may depend on genetic differences between Maiwa and Plateau yak breeds.

The Y-milk somatic cells average content was slightly higher than 100,000 cells/mL of milk. Currently, there is not a clear threshold limit to assess intra-mammary infections (IMI) in yak. If we consider the threshold limit for somatic cells commonly accepted for individual cows reared in an intensive system (200,000 cells/mL), the value in yak bulk milk observed seemed to indicate a low prevalence of IMI in the yak herds sampled here. Moreover, it is important to highlight that a high somatic cells content in cow milk has negative effects on its rennet coagulation aptitude [55–57] and, consequentially on the cheese-making efficiency [50] and milk cheese-yielding ability [33,58].

In contrast, the total bacterial count in Y-milk was very high when compared to that in C-milk one. This was probably due to differences between the raising systems of the yaks and cows involved in this research. Indeed, C-milk was collected from a free-stall herd, the more widespread housing system in Italy, and milking was mechanised and carried out in a milking parlor [59]. With this system, the collected milk is transported through pipes to a refrigeration tank where it is cooled. The limited contact between the milk and the environment and the cooling of the milk keeps the total bacterial count low [60], and refrigeration contributes to reducing the bacterial growth [61]. On the opposite, yaks were raised in high mountain pastures and were not reared in a stall, the milking procedure was manually performed, and the milk was not immediately cooled after milking, conditions that can all promote bacterial growth.

Clearly, improving the hygiene of the milking practices and cooling the yak milk at the farm could be the best method to reduce somatic cell count and bacterial growth and to limit the proteolytic and lipolytic activity of enzymes that can alter the milk quality [62].

Finally, the average pH value was consistent with that reported by Zhang et al. [63] in a study on the factors influencing the rennet-induced coagulation properties of yak milk.

4.2. Yak Milk Mineral Content, Salt Equilibrium and Casein Micelles Mineralisation

Cui et al. [28], in a study carried out on milk from the same yak breed considered here, found mineral contents quite higher than those observed in this study, using a different method to assess mineral concentration (inductively coupled plasma atomic emission spectroscopy), i.e., 227, 170 and 14.5 mg/100 g of milk for Ca, P and Mg, respectively. The contents of total Ca and P were slightly higher, and that of Mg slightly lower than those showed by Li et al. [15], reporting for Maiwa Y-milk average values of 1545.45, 922.04 and 154.10 mg/kg of milk for Ca, P and Mg, respectively.

Moreover, the percentages of soluble Ca and P were lower and higher, respectively, than those reported by Wang et al. [27] for Y-milk produced by yak raised in the Qinghai-Tibetan Plateau.

The higher concentrations of Ca, P and Mg in Y-milk compared to C-milk depended mainly on the high amount of casein in the former milk with respect to the latter one. Indeed, Ca, P and Mg contribute to the casein micelle structure, and thus, the milk casein content positively affects the colloidal contents of Ca, P and Mg [64].

In general, the data om micelle mineralisation in C-milk are in agreement with the results of previous studies carried out on individual and bulk milk samples [2,3]. Actually, colloidal P is composed of two different fractions: inorganic P, which represents P in inorganic calcium phosphate within the casein micelles, and casein P, corresponding to phosphorus in the phosphorylated amino acid residues of caseins [3].

Differences were observed for inorganic P, while Y-milk showed a lower content of casein P than C-milk. Thus, the micelles of the two types of milk had the same concentration of inorganic salts, and Y-milk had a lower number of phosphorylated amino acid residues. This observation was confirmed by the values of the colloidal-Ca-to-colloidal-P ratio and of the colloidal-Ca-to-colloidal-inorganic-P ratio. Indeed, the first value was higher in Y-milk than in C-milk, while the second did not show a significant difference when comparing the milk of the two species.

5. Conclusions

Yak milk appeared to be characterised by a high content of casein and, therefore, of minerals that contribute to the casein micelle structure. In particular, the milk of yak seems to be extremely rich in Ca, P and Mg, especially in their colloidal forms, which should positively influence the bioavailability of Ca and P during the digestion process.

Finally, in yak milk, the lower casein P content per casein unit suggests that the yak caseins are less phosphorylated than the cow ones and have a lower number of phosphorylated amino acid residues within the casein micelles. This feature may affect the casein micelle structure, with repercussion on the processability and digestion of yak milk casein when compared to the cow one.

However, it should be considered that this study was conducted on bulk milk samples, using milk from only one breed (Plateau yak breeds) of yak.

Therefore, it could be important, in the future, to expand this research, examining milk from other yak breeds.

Author Contributions: Conceptualisation, P.F. (Piero Franceschi), W.S., M.M. and A.S.; methodology, P.F. (Piero Franceschi); F.M., Y.L. and A.S.; software, P.F. (Piero Franceschi), M.M. and P.F. (Paolo Formaggioni); formal analysis, P.F. (Piero Franceschi), W.S., Y.L., M.M., F.M. and A.S.; investigation, P.F. (Paolo Formaggioni), P.F. (Piero Franceschi), M.M. and A.S.; resources, A.S. and M.M.; data curation, P.F. (Piero Franceschi), F.M., P.F. (Paolo Formaggioni), W.S. and Y.L.; writing—original draft preparation, P.F. (Piero Franceschi), P.F. (Paolo Formaggioni), M.M., F.M., W.S., Y.L. and A.S.; writing—review and editing, P.F. (Piero Franceschi), P.F. (Paolo Formaggioni), M.M., W.S., Y.L., F.M. and A.S.; visualisation, P.F. (Piero Franceschi), M.M., F.M. and A.S.; supervision, A.S., F.M. and M.M.; project administration, P.F. (Piero Franceschi), M.M. and A.S. All authors have read and agreed to the published version of the manuscript.

Funding: This research received a grant, funding number 22167020, from the National Natural Science Foundation of China.

Data Availability Statement: The data presented in this study are available on request from the corresponding author.

Conflicts of Interest: The authors declare that they have no conflict of interest for this research article.

References

1. Bijl, E.; Huppertz, T.; van Valenberg, H.; Holt, C. A quantitative model of the bovine casein micelle: Ion equilibria and calcium phosphate sequestration by individual caseins in bovine milk. *Eur. Biophys. J.* **2019**, *48*, 45–59. [CrossRef] [PubMed]
2. Huppertz, T.; Heck, J.; Bijl, E.; Poulsen, N.A.; Larsen, L.B. Variation in casein distribution and mineralisation in the milk from Holstein-Friesian cows. *Int. Dairy J.* **2021**, *108*, 105064. [CrossRef]
3. Malacarne, M.; Franceschi, P.; Formaggioni, P.; Sandri, S.; Mariani, P.; Summer, A. Influence of micellar calcium and phosphorus on rennet coagulation properties of cows milk. *J. Dairy Res.* **2014**, *81*, 129–136. [CrossRef] [PubMed]
4. Huppertz, T.; Lambers, T.T. Influence of micellar calcium phosphate on in vitro gastric coagulation and digestion of milk proteins in infant formula model systems. *Int. Dairy J.* **2020**, *107*, 104717. [CrossRef]
5. Polychroniadou, A.; Vafopoulou, A. Salt distribution between the colloidal and soluble phases of ewes' milk. *J. Dairy Res.* **1986**, *53*, 353–358. [CrossRef]
6. Park, Y.W.; Juárez, M.; Ramos, M.; Haenlein, G.F.W. Physico-chemical characteristics of goat and sheep milk. *Small Rumin. Res.* **2007**, *68*, 88–113. [CrossRef]
7. Singh, M.; Sharma, R.; Ranvir, S.; Gandhi, K.; Mann, B. Profiling and distribution of minerals content in cow, buffalo and goat milk. *Indian J. Dairy Sci.* **2019**, *72*, 480–488. [CrossRef]
8. Sabarwal, P.K.; Oomen, S.; Ganguli, N.C. Inorganic constituents of casein micelles from cow and buffalo milk. *J. Food Sci. Technol.* **1972**, *9*, 144–146.
9. Hewdey, M.M.; Nassib, T.A.; El Din, M.M.Z.; El Sokkary, E.S. Studies on buffalo casein. I. Composition of casein micelles. *Egypt. J. Dairy Sci.* **1989**, *17*, 87–92.
10. Farah, Z.; Ruegg, M.W. The size distribution of casein micelles in camel milk. *Food Struct.* **1989**, *8*, 211–216.
11. Li, C.P.; Peng, J.C.; Zhu, G.H.; Zhao, H.; Li, C. Micellar calcium phosphate cross-linkage in donkey casein micelles. *Milchwissenschaft* **2010**, *65*, 274–276.
12. Malacarne, M.; Criscione, A.; Franceschi, P.; Tumino, S.; Bordonaro, S.; Di Frangia, F.; Marletta, D.; Summer, A. Distribution of Ca, P and Mg and casein micelle mineralisation in donkey milk from the second to ninth month of lactation. *Int. Dairy J.* **2017**, *66*, 1–5. [CrossRef]
13. Fantuz, F.; Ferraro, S.; Todini, L.; Cimarelli, L.; Fatica, A.; Marcantoni, F.; Salimei, E. Distribution of calcium, phosphorus, sulphur, magnesium, potassium, and sodium in major fractions of donkey milk. *J. Dairy Sci.* **2020**, *103*, 8741–8749. [CrossRef]
14. Malacarne, M.; Franceschi, P.; Formaggioni, P.; Pisani, G.M.; Petrera, F.; Abeni, F.; Soffiantini, C.S.; Summer, A. Mineral content and distribution in milk from red deer (*Cervus elaphus*) fallow deer (*Dama dama*) and roe deer (*Capreolus capreolus*). *Small Rumin. Res.* **2015**, *13*, 208–215. [CrossRef]
15. Li, H.; Ma, Y.; Li, Q.; Wang, J.; Cheng, J.; Xue, J.; Shi, J. The Chemical composition and nitrogen distribution of Chinese Yak (Maiwa) milk. *Int. J. Mol. Sci.* **2011**, *12*, 4885–4895. [CrossRef]
16. Dong, S.; Long, R.; Kang, M. Milking performance of China yak (*Bos grunniens*): A preliminary report. *Afr. J. Agric. Res.* **2007**, *2*, 52–57.
17. Yang, M.; Zhang, W.; Wen, P.; Zhang, Y.; Liang, Q. Heat stability of yak micellar casein as affected by heat treatment temperature and duration. *Dairy Sci. Technol.* **2014**, *94*, 469–481. [CrossRef]
18. Wang, L.; Ma, Y.; Cui, J.; Oyeyinka, S.A.; Cheng, J.; He, S. Yak milk whey protein denaturation and casein micelle disaggregation/aggregation at different pH and temperature. *Int. Dairy J.* **2017**, *71*, 131–135. [CrossRef]
19. Li, Q.; Luo, X.L.; Xu, J.T.; Li, Z.Q. Measurement of milk performance in domestic yak in Guoluo Prefecture. *Chin. Qinghai J. Anim. Vet. Sci.* **2000**, *30*, 10–11.

20. Li, H.M.; Ma, Y.; Dong, A.J.; Wang, J.Q.; Li, Q.M.; He, S.H.; Maubois, J.L. Protein composition of yak milk. *Dairy Sci. Technol.* **2010**, *90*, 111–117. [CrossRef]
21. Sun, W.; Formaggioni, P.; Franceschi, P.; Luo, Y.; Sandri, S.; Malacarne, M.; Mariani, P.; Kong, Q.; Summer, A. Physico-chemical properties, gross composition and nitrogen fractions of Chinese Qinghai plateau yak (*Bos grunniens*) reared in two different altimetric zones. *Milchwissenschaft* **2012**, *67*, 389–391.
22. Zhang, Y.; Li, Y.; Wang, P.; Tian, Y.; Liang, Q.; Ren, F. Rennet-induced coagulation properties of yak casein micelles: A comparison with cow casein micelles. *Food Res. Intern.* **2017**, *102*, 25–31. [CrossRef]
23. Zhang, Y.; Yang, M.; Dongyan, C.; Yijiang, H.; Xiao, Z.; Yuanhua, Z.; Hong, Z.; Zhennai, Y. Composition, coagulation characteristics, and cheese making capacity of yak milk. *J. Dairy Sci.* **2020**, *103*, 1276–1288. [CrossRef]
24. Yang, M.; Zhang, G.D.; Yang, J.T.; Sun, D.; Wen, P.C.; Zhang, W.B. Effect of pH on dissociation of casein micelles in yak skim milk. *J. Dairy Sci.* **2018**, *101*, 2998–3007. [CrossRef]
25. Zhang, Y.; Ren, F.; Wang, P.; Liang, Q.; Peng, Y.; Song, L.; Wen, P. The influence of yak casein micelle size on rennet-induced coagulation properties. *J. Sci. Food Agric.* **2021**, *101*, 327–333. [CrossRef]
26. Sun, W.; Ghidini, S.; Luo, Y.; Zanardi, E.; Ma, H.; Ianieri, A. Macro and micro element profile of yak (*Bos grunniens*) milk from Quilian Qinghai plateau. *Ital. J. Anim. Sci.* **2012**, *11*, 180–183. [CrossRef]
27. Wang, P.; Liu, H.; Wen, P.; Zhang, H.; Guo, H.; Rena, F. The composition, size and hydration of yak casein micelles. *Int. Dairy J.* **2013**, *31*, 107–110. [CrossRef]
28. Cui, G.X.; Yuan, F.; Degen, A.A.; Liu, S.M.; Zhou, J.W.; Shang, Z.H.; Ding, L.M.; Mi, J.D.; Wei, X.H.; Long, R.J. Composition of the milk of yaks raised at different altitudes on the Qinghai-Tibetan Plateau. *Int. Dairy J.* **2016**, *59*, 29–35. [CrossRef]
29. Association of Official Analytical Chemists (AOAC). Nitrogen (Total) in Milk, Method No. 991.20. In *Official Methods of Analysis of AOAC International*, 18th ed.; Horowitz, W., Ed.; AOAC International: Gaithersburg, MD, USA, 2005; pp. 10–12.
30. Association of Official Analytical Chemists (AOAC). Non-Casein Nitrogen Content of Milk, Method No. 998.05. In *Official Methods of Analysis of AOAC International*, 18th ed.; Horowitz, W., Ed.; AOAC International: Gaithersburg, MD, USA, 2005; pp. 50–51.
31. Association of Official Analytical Chemists (AOAC). Nonprotein Nitrogen in Whole Milk, Method No. 991.21. In *Official Methods of Analysis of AOAC International*, 18th ed.; Horowitz, W., Ed.; AOAC International: Gaithersburg, MD, USA, 2005; pp. 12–13.
32. Franceschi, P.; Brasca, M.; Malacarne, M.; Formaggioni, P.; Faccia, M.; Natrella, G.; Summer, A. Effects of the cooling temperature at the farm on milk maturation and cheesemaking process in the manufacture of Parmigiano Reggiano PDO Cheese. *Animals* **2021**, *11*, 2835. [CrossRef]
33. Franceschi, P.; Faccia, M.; Malacarne, M.; Formaggioni, P.; Summer, A. Quantification of cheese yield reduction in manufacturing Parmigiano Reggiano from milk with non-compliant somatic cells count. *Foods* **2020**, *9*, 212. [CrossRef]
34. Franceschi, P.; Sandri, S.; Pecorari, M.; Vecchia, P.; Sinisi, F.; Mariani, P. Effects of milk storage temperature at the herd on cheesemaking losses in the manufacture of Parmigiano-Reggiano cheese. *Vet. Res. Commun.* **2008**, *32* (Suppl. 1), S339–S341. [CrossRef]
35. IDF 141/ISO9622; Milk and Liquid Milk Products, Guidelines for the Application of Mid-Infrared Spectrometry. International Dairy Federation: Brussels, Belgium, 2013.
36. IDF 148-2/ISO13366-2; Milk, Enumeration of Somatic Cells, Part. 2: Guidance on the Operation of Fluoro-Opto-Electronic Counters. International Dairy Federation Standard: Brussels, Belgium, 2006.
37. IDF 196/ISO21187; Milk, Quantitative Determination of Bacteriological Quality. Guidance for Establishing and Verifying a Conversion Relationship between Routine Method Results and Anchor Method Results. International Dairy Federation Standard: Brussels, Belgium, 2004.
38. ISO 6731:2010; Milk, Cream and Evaporated Milk—Determination of Total Solids Content. International Organisation for Standardisation: Geneva, Switzerland, 2010.
39. Petrera, F.; Catillo, G.; Napolitano, F.; Malacarne, M.; Franceschi, P.; Summer, A.; Abeni, F. New insights into the quality characteristics of milk from Modenese breed compared with Italian Friesian. *Ital. J. Anim. Sci.* **2016**, *15*, 559–567. [CrossRef]
40. IDF 27; Determination of Ash Content of Processed Cheese Products. International Dairy Federation Standard: Brussels, Belgium, 1964.
41. Malacarne, M.; Criscione, A.; Franceschi, P.; Bordonaro, S.; Formaggioni, P.; Marletta, D.; Summer, A. New insights into chemical and mineral composition of donkey milk throughout nine months of lactation. *Animals* **2019**, *9*, 1161. [CrossRef]
42. Allen, R.J.L. The estimation of phosphorous. *Biochem. J.* **1940**, *34*, 856–858. [CrossRef]
43. Bonaga, G.; Mascolo, V. I lipidi polari del latte: I fosfolipidi del latte nella zona di produzione del Parmigiano-Reggiano. *Sci. Tec. Latt.-Casearia* **1977**, *28*, 257–265.
44. Barlowska, J.; Szwajowska, M.; Litwinczuk, Z.; Krol, J. Nutritional value and technological suitability of milk from various animal species used for dairy production. *Compr. Rev. Food Sci. Food Saf.* **2011**, *10*, 291–302. [CrossRef]
45. Nikkhah, A. Science of camel and yak milks: Human nutrition and health perspectives. *Food Nutr. Sci.* **2011**, *2*, 667–673. [CrossRef]
46. Medhammar, E.; Wijesinha-Bettoni, R.; Stadlmayr, B.; Nilsson, E.; Charrondiere, U.R.; Burlingame, B. Composition of milk from minor dairy animals and buffalo breeds: A biodiversity perspective. *J. Sci. Food Agric.* **2012**, *92*, 445–474. [CrossRef]

47. Claeys, W.L.; Verraes, C.; Cardoen, S.; De Block, J.; Huyghebaert, A.; Raes, K.; Dewettinck, K.; Herman, L. Consumption of raw or heated milk from different species: An evaluation of the nutritional and potential health benefits. *Food Control* **2014**, *42*, 188–201. [CrossRef]
48. Faccia, M.; D'Alessandro, A.G.; Summer, A.; Hailu, Y. Milk products from minor dairy species: A review. *Animals* **2020**, *10*, 1260. [CrossRef]
49. Singh, T.P.; Deshwal, G.K.; Bam, J.; Paul, V. A Comparative appraisal of traditional "ghee" derived from the three genotypes (Arunachali yak, yak-cow hybrid, and cow) reared under semi-intensive conditions. *Eur. J. Lipid Sci. Technol.* **2022**, *124*, 2100101. [CrossRef]
50. Franceschi, P.; Malacarne, M.; Formaggioni, P.; Cipolat-Gotet, C.; Stocco, G.; Summer, A. Effect of season and factory on cheese-making efficiency in Parmigiano Reggiano manufacture. *Foods* **2019**, *8*, 315. [CrossRef]
51. Klei, L.; Yun, J.; Sapru, A.; Lynch, J.; Barbano, D.M.; Sears, P.; Galton, D. Effects of milk somatic cell count on cottage cheese yield and quality. *J. Dairy Sci.* **1998**, *81*, 1205–1213. [CrossRef]
52. Franceschi, P.; Malacarne, M.; Faccia, M.; Rossoni, A.; Santus, E.; Formaggioni, P.; Summer, A. New insights of cheese yield capacity between Italian Brown and Italian Friesian milks in the production of high moisture mozzarella. *Food Technol. Biotechnol.* **2020**, *58*, 91–97. [CrossRef]
53. Formaggioni, P.; Summer, A.; Malacarne, M.; Franceschi, P.; Mucchetti, G. Italian and Italian-style hard cooked cheeses: Predictive formulas for Parmigiano-Reggiano 24 h cheese yield. *Int. Dairy J.* **2015**, *51*, 52–58. [CrossRef]
54. Franceschi, P.; Malacarne, M.; Formaggioni, P.; Faccia, M.; Summer, A. Quantification of the effect of the cattle breed on milk cheese yield: Comparison between Italian Brown Swiss and Italian Friesian. *Animals* **2020**, *10*, 1331. [CrossRef] [PubMed]
55. Dang, A.K.; Suman, K.; Charan, S.; Sehgal, J.P. Milk differential cell counts and compositional changes in cows during different physiological stages. *Milchwissenschaft* **2008**, *63*, 239–242.
56. Franceschi, P.; Summer, A.; Sandri, S.; Formaggioni, P.; Malacarne, M.; Mariani, P. Effects of the full cream milk somatic cell content on the characteristics of vat milk in the manufacture of Parmigiano-Reggiano cheese. *Vet. Res. Commun.* **2009**, *3* (Suppl. 1), 281–283. [CrossRef]
57. Le Maréchal, C.; Thiéry, R.; Vautor, E.; Le Loir, Y. Mastitis impact on technological properties of milk and quality of milk products—A review. *Dairy Sci. Technol.* **2011**, *91*, 247–282. [CrossRef]
58. Summer, A.; Franceschi, P.; Formaggioni, P.; Malacarne, M. Influence of milk somatic cell content on Parmigiano-Reggiano cheese yield. *J. Dairy Res.* **2015**, *82*, 222–227. [CrossRef]
59. Summer, A.; Franceschi, P.; Formaggioni, P.; Malacarne, M. Characteristics of raw milk produced by free-stall or tie-stall cattle herds in the Parmigiano-Reggiano cheese production area. *Dairy Sci. Technol.* **2014**, *94*, 581–590. [CrossRef]
60. Franceschi, P.; Malacarne, M.; Formaggioni, P.; Righi, F.; Summer, A. Chemical composition, hygiene characteristics, and coagulation aptitude of milk for Parmigiano Reggiano cheese from farms yielding different milk levels. *Rev. Brasil. Zootec.* **2020**, *49*, e20180113. [CrossRef]
61. Malacarne, M.; Summer, A.; Franceschi, P.; Formaggioni, P.; Pecorari, M.; Panari, G.; Vecchia, P.; Sandri, S.; Fossa, E.; Scotti, C.; et al. Effects of storage conditions on physico-chemical characteristics, salt equilibria, processing properties and microbial development of raw milk. *Int. Dairy J.* **2013**, *29*, 36–41. [CrossRef]
62. Franceschi, P.; Malacarne, M.; Formaggioni, P.; Faccia, M.; Summer, A. Effects of milk storage temperature at the farm on the characteristics of Parmigiano Reggiano cheese: Chemical composition and proteolysis. *Animals* **2021**, *11*, 879. [CrossRef]
63. Zhang, Y.; Li, Y.; Wang, P.; Liang, Q.; Zhang, Y.; Ren, F. The factors influencing rennet-induced coagulation properties of yak milk: The importance of micellar calcium during gelation. *LTW-Food Sci. Technol.* **2019**, *111*, 500–505. [CrossRef]
64. Bijl, E.; van Valenberg, J.F.; Huppertz, T.; van Hooijdonk, A.C.M. Protein, casein, and micellar salts in milk: Current content and historical perspectives. *J. Dairy Sci.* **2013**, *96*, 5455–5464. [CrossRef]

Disclaimer/Publisher's Note: The statements, opinions and data contained in all publications are solely those of the individual author(s) and contributor(s) and not of MDPI and/or the editor(s). MDPI and/or the editor(s) disclaim responsibility for any injury to people or property resulting from any ideas, methods, instructions or products referred to in the content.

Systematic Review

Milk Quality Conceptualization: A Systematic Review of Consumers', Farmers', and Processing Experts' Views

Greta Castellini [1,2,3], Serena Barello [4,*] and Albino Claudio Bosio [1,5]

[1] EngageMinds HUB—Consumer, Food & Health Engagement Research Center, Università Cattolica del Sacro Cuore, 20123 Milan, Italy; greta.castellini@unicatt.it (G.C.); claudio.bosio@unicatt.it (A.C.B.)
[2] Faculty of Agriculture, Food and Environmental Sciences, Università Cattolica del Sacro Cuore, Via Bissolati, 74, 26100 Cremona, Italy
[3] Department of Psychology, Università Cattolica del Sacro Cuore, Largo Agostino Gemelli, 1, 20123 Milan, Italy
[4] Department of Brain and Behavioural Sciences, University of Pavia, Via Agostino Bassi, 21, 27100 Pavia, Italy
[5] IRCAF (Invernizzi Reference Center on Agri-Food), Campus Santa Monica, Via Bissolati, 74, 26100 Cremona, Italy
* Correspondence: serena.barello@unipv.it; Tel.: +39-348-9334-458

Abstract: Milk consumption has traditionally been recognized as a fundamental element of global dietary patterns due to its perceived nutritional advantages. Nonetheless, a substantial decrease in milk consumption has been identified within diverse populations in recent times. Specifically, consumers' expectations and representations of milk quality have undergone notable transformations, contributing to the observed reduction in consumption. The objective of this systematic review was to conduct a comprehensive examination and categorization of the conceptual attributes associated with milk quality, considering the representations of citizen-consumers, farmers, and processing experts. This review was conducted following the Preferred Reporting Items for Systematic reviews and Meta-Analyses (PRISMA) guidelines. The titles and abstracts of 409 articles were screened, and 20 full-text articles were assessed for eligibility. The results demonstrate the existence of a dual articulation in the conceptual definition of milk quality. Farmers and processing experts exhibited a relatively similar representation of milk quality, focusing on technical indicators. In contrast, citizen-consumers held more simplistic and subjective concepts that are challenging to quantify. This study emphasized the critical need for establishing a platform for communication and knowledge exchange to foster shared representations and expectations regarding milk quality.

Keywords: milk quality; representations; citizen-consumer psychology; farmer; concept mapping; processing expert

1. Introduction

Milk consumption has long been regarded as a fundamental element of global dietary patterns due to its perceived nutritional advantages. However, a notable decline in milk consumption has been observed across various populations in recent times [1]. More specifically, there has been a 2% decline in milk consumption in the EU between 2013 and 2018, and this decrease is expected to continue [2]. The reduction in milk purchases is particularly relevant in Italy, where its consumption has been decreasing in a progressive way, from 56.4 L per capita in 2009 to 50.2 L in 2014 (6%) [3]. Research indicates that this decline is influenced by several significant factors, with particular emphasis on the profound shifts in consumers' perceptions of milk quality. These altered expectations and representations of milk quality have contributed to the observed consumption decrease [4]. Firstly, there has been a growing emphasis on health and nutrition as key dimensions of milk quality. Consumers are increasingly focused on nutritional content, the absence of harmful additives, and the overall health and environmental impact of the foods they consume. For instance, recent studies have demonstrated how health and animal welfare concerns can impact the hedonic and emotional response to milk and subsequently affect consumption [5]. Additionally, sustainability and ethical considerations have taken center stage as crucial aspects of food quality. Consumers now prioritize environmentally friendly production methods, fair trade practices, and animal welfare in their definitions of quality. These aspects seem to be particularly important for those countries that have limited natural resources and are densely populated [6]. Asian countries such as China and India are increasing the attention paid to food sustainability as they perceive the risk of not having enough resources to meet the needs of the entire population [6,7]. Sensory attributes and taste, though still crucial, are now being sought after for more diverse and authentic taste experiences, often linked to cultural preferences and personal enjoyment. A study conducted in Latin America (Mexico and Chile), Europe (Italy, Spain, Greece, and Denmark), and Asia (Bangladesh) showed that, in European and Asian countries, sheep and goat dairy products are not consumed because consumers dislike them, while in Mexico a higher percentage of people do not consume these dairy products because they are unfamiliar with them [8]. Moreover, convenience, affordability, and transparency in the food supply chain are emerging as significant factors shaping consumer perceptions of food quality, leading to profound shifts in how milk quality is defined.

From a legislative point of view, the rules introduced to protect the quality of milk are many and vary from country to country [9]. European Union regulations encompass a series of legislative measures that comprehensively cover various aspects of the dairy sector. The production of dairy products adheres to general hygiene prerequisites outlined in several European regulations: Regulation (EC) No 178/2002 [10], Regulation (EC) No 852/2004 [11], and Regulation (EC) No 853/2004 [12,13]. Processed milk must meet stringent hygiene criteria, including limits on microorganisms, somatic cell counts, the absence of veterinary drug residues, and not surpassing acceptable levels of specific contaminants. Moreover, compliance with public health standards is imperative. For instance, non-EU nations must possess an approved monitoring scheme for "residues". Items introduced into the EU market must adhere to food law requisites, notably Regulation (EC) No 178/2002 [10]. The legislation also incorporates specifications for product labeling. Variations exist in standards and labels for milk fat and spread products across different global regions [14]. Several authors have assessed the implications of the new EU Regulation No. 1169/2011 [15,16]. Within the EU, Regulation No 931/2011 [10] pertaining to the traceability of animal products, Regulation No 1169/2011 [17] addressing consumer information provision, and Regulation No 1308/2013 [18] governing the organization of agricultural markets collectively serve as the principal legislative frameworks overseeing milk labeling.

However, current marketing strategies reveal a gap in adopting a comprehensive approach that considers the perspectives of both dairy experts and citizen-consumers regarding milk quality. This fragmentation in milk quality definitions has resulted in the formulation of marketing and communication strategies that have proven to be ineffective and unsuccessful, ultimately negatively impacting milk consumption [19]. Built upon these premises, it is imperative for the dairy industry to grasp and explore the societal perspective regarding milk quality as underscored by the Food and Agriculture Organization of the United Nations (FAO) [20]. Specifically, it is of utmost importance to investigate the novel representations and quality attributes of citizen-consumers pertaining to milk and ascertain whether these are aligned with those of experts such as farmers and processing experts. This comprehension plays a pivotal role in the development of products and the formulation of marketing strategies that cater to the ever-evolving needs and demands of consumers [2,21]. Notably, for citizen-consumers, it is crucial that certain characteristics of milk are visible and comprehensible in order to minimize uncertainty and prevent dissatisfaction.

However, the scope of research that focuses on the concept of milk quality beyond the existing technological and hygienic definitions remains limited [22]. While current knowledge about milk quality is valuable, it does not encompass all possible ways of representing and conceptualizing its meaning.

To bridge this knowledge gap, the objective of this systematic review was to undertake a comprehensive examination and categorization of the conceptual attributes associated with milk quality, considering the viewpoints of citizen-consumers, farmers, and processing experts.

The specific objectives of this review are as follows: (a) to identify the primary attributes that define milk quality, taking into account the perspectives and distinct representations of citizen-consumers, farmers, and dairy processing experts (advisors and processors); (b) to examine the differences and similarities in the representation of milk quality among these key stakeholders in the dairy industry; (c) to categorize these attributes of milk quality conceptualization utilizing an ecological framework to provide a comprehensive description and analysis.

2. Materials and Methods

This systematic review was conducted and reported following the Preferred Reporting Items for Systematic reviews and Meta-Analyses (PRISMA) guidelines [23].

2.1. Search Strategy

A comprehensive search strategy was formulated to identify relevant peer-reviewed publications pertaining to the determinants influencing the perception of milk quality among farmers, citizen-consumers, and processing experts (advisors and processors). In the context of the milk supply chain, farmers are individuals or entities primarily engaged in dairy farming. They manage farms where dairy animals, such as cows, goats, or sheep, are raised for the purpose of producing milk. Processors are entities responsible for collecting, pasteurizing, processing, and packaging milk. They play a vital role in ensuring that raw milk is transformed into a safe, shelf-stable, and consumer-ready product through processes that involve heat treatment, separation, and other techniques. [24]. The strategy employed a combination of keywords extracted from titles and abstracts. The search terms were grouped into three categories: (I) the concept of milk quality, which was searched as a single term while excluding closely related concepts to ensure conceptual clarity; (II) specific domains of interest such as perception, attitude, and expectation; and (III) the target subjects of interest, namely, farmers, citizen-consumers, and processing experts (identified as processors and advisors). The following search string was developed: (milk quality) and (acceptance*) OR (opinion*) OR (perception*) OR (attitud*) OR (evaluation) OR (valuation) OR (adopt*) OR (defin*) OR (expectation*) OR (determinant*) OR (criteri*) OR (factor*) OR (representation*) OR (attribute*) and (consumer*) OR (citizen*) OR (shopper*) OR (user*) OR (public) OR (buyer) OR (farmer∗) OR (processor∗) OR (stakeholder∗) OR (supply chain∗) OR (producer).

This search strategy was adapted to the thesaurus characteristics of each considered database (i.e., SCOPUS, PSYCINFO, WEB OF SCIENCE, and PUBMED) and launched in December 2022. Literature search was limited to peer-reviewed studies published in English or Italian. No time restriction was applied, so as to be as inclusive as possible. Reference lists of eligible studies and review articles were scanned to identify any missed articles.

2.2. Study Selection and Data Extraction

A three-step screening process was implemented to identify suitable studies for inclusion in this review, as described by [25]. In cases where there was disagreement between the two reviewers, all three researchers discussed the articles until a consensus was reached.

For all selected studies, the authors extracted information included study author(s), year of publication, countries where the study was carried out, sample characteristics (including sample size, age, and percentage of females involved), and study design. Moreover, the type of milk investigated and the type of participants (farmers, citizen-consumers, or processing experts) involved in the studies were extracted. In addition, attribute categories of milk quality were mapped for citizen-consumers, farmers, and processing experts (advisors and processors). Since the selected studies considered different attributes to define the concept of milk quality, they have been reviewed, selected, and grouped into macro-categories.

The data were extracted systematically using a standardized data extraction form as described by [25]. The extracted data were summarized in tables and a narrative synthesis was developed using a textual approach to synthesis the findings [26].

Procedure of Grouping Variables

The included studies reported several attributes (namely, "micro-categories") to describe the concept of milk quality. These micro-categories were then grouped into broader macro-categories to allow for an effective synthesis of the results (Figure 1).

In particular, a qualitative content analysis procedure, widely implemented to analyze textual data [27], was adapted to reduce the number of categories. More specifically, conventional content analysis [27,28], also described as inductive category development [29], was applied because this procedure allows categories and their names to flow from the data instead of using preconceived categories [27]. The procedure for developing the categories of the extracted attributes is presented in Figure 1 and was carried out by three researchers independently (GC, SB and CB).

In order to handle the large amount of data, all the micro-categories that impact the concept of milk quality were transcribed into Excel. After that, the micro-categories were carefully re-read and those that referred to the same key concept were grouped under the same macro-category (e.g., all variables that mentioned worker hygiene, animal hygiene, or farm hygiene were grouped under the same macro-category), identifying labels that were consistent with the micro-categories grouped (e.g., hygiene quality).

Finally, the macro-categories were further validated (formative check of reliability) by the three researchers (GC, SB, and CB), checking the level of agreement among the categories created by the researchers independently and discussing cases of doubt and overlapping labels.

The validated macro-categories were used to compare differences and similarities among the different actors (citizen-consumers, farmers, and processing experts) with respect to the concept of milk quality.

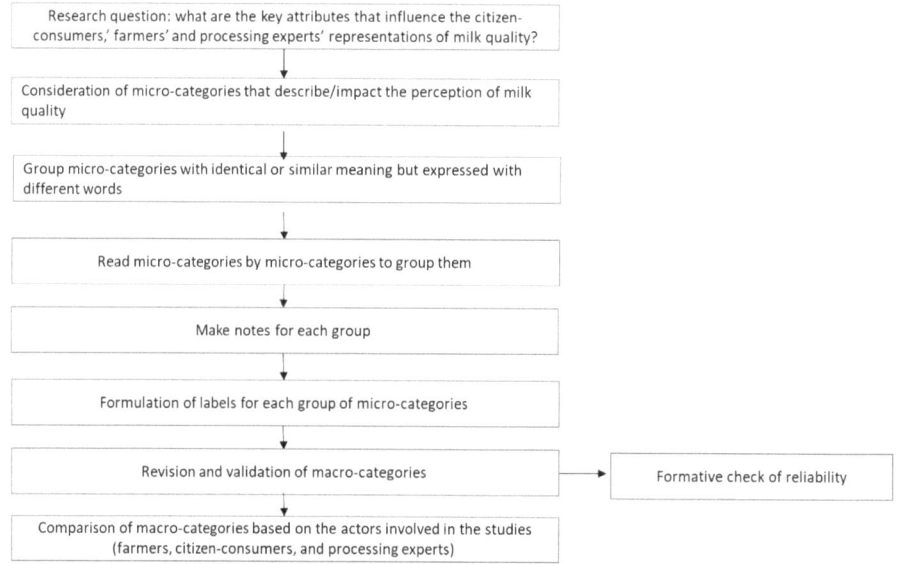

Figure 1. Procedure of inductive category development, adapted from Mayring [30] and Schilling [13].

2.3. Data Analysis

The macro-categories have been integrated and organized according to the framework of Story et al. [31] which is based on the socio-ecological framework of Bronfenbrenner and Capruso [32] and Bronfenbrenner [33]. This framework presents four systems within which people act, and these systems can be paramount in influencing the formation of one's own opinions about social phenomena: (I) The Individual system, identified as the place where people generate opinions based on their experiences with the phenomenon (e.g., attributes related to milk-related sensory aspects); (II) The Microsystem, which is the context where opinions are structured and formed through comparisons with others (e.g., attributes related to the concept of trust towards milk producers); (III) The Mesosystem, which is the context where one's own opinions are shaped by considering the tangible features of a phenomenon (e.g., milk's nutritional value on the label, packaging features) or context (e.g., milk-related hygiene conditions, technological systems); and finally, (IV) the Macrosystem, which relates to the context of social norms (e.g., attributes concerning the legislative or policy systems related to milk). Subsequently, a comprehensive diagram was created to offer a visual depiction of the outcomes. This diagram encompassed the macro-categories linked to the concept of milk quality, which were subsequently classified and distinguished in alignment with the four systems of the socio-ecological framework. Moreover, the diagram portrayed the percentage distribution of micro-categories within each macro-category and system. Furthermore, the diagram facilitated a comparative analysis of the macro-categories, accentuating the distinctions and similarities among citizen-consumers, farmers, and processing experts (advisors and processors).

3. Results

3.1. Search Results

A total of 729 records were retrieved. A first screening round was conducted, eliminating 320 duplicate records. A further round of screening was applied to the title and abstracts on the remaining 409 records. After applying the eligibility criteria, 49 records were judged as potentially relevant. Another screening phase was applied to the remaining full-text articles to exclude articles not in line with the study's objectives. Finally, according to the pre-defined eligibility criteria, 20 studies were identified as coherent with the inclusion/exclusion criteria as they focused on the attributes of the milk quality concept. Figure 2 describes the selection and screening process.

3.2. Studies' Overview

Table 1 provides an overview of the studies included in this analysis. The publications spanned from 2000 to 2022, with an increase in the number of studies observed in the recent years (2021–2022), as shown in Figure 3. Geographically, most studies were conducted in the Americas, including South America ($n = 4$, 20%) and the North America ($n = 2$, 10%), followed by Europe ($n = 5$, 25%), Africa ($n = 5$, 25%), and Asia ($n = 4$, 20%), as illustrated in Figure 4. Quantitative research designs were predominantly utilized in most studies ($n = 11$, 55%), as indicated in Figure 5. The sample sizes across the studies varied from $n = 40$ to $n = 1646$, as detailed in Table 1. The focus of nearly all studies was on cow's milk quality ($n = 20$, 95%), with a significant involvement of farmers ($n = 13$, 65%). When considering the study design and participants involved, recent research conducted in 2021–2022 primarily employed qualitative methods (5 out of 7, 71%) and focused on the perspective of citizen-consumers (4 out of 7, 57%), while earlier studies conducted from 2000 to 2020 predominantly employed quantitative designs (9 out of 13, 69%) and mainly focused on the viewpoint of farmers (8 out of 13, 62%).

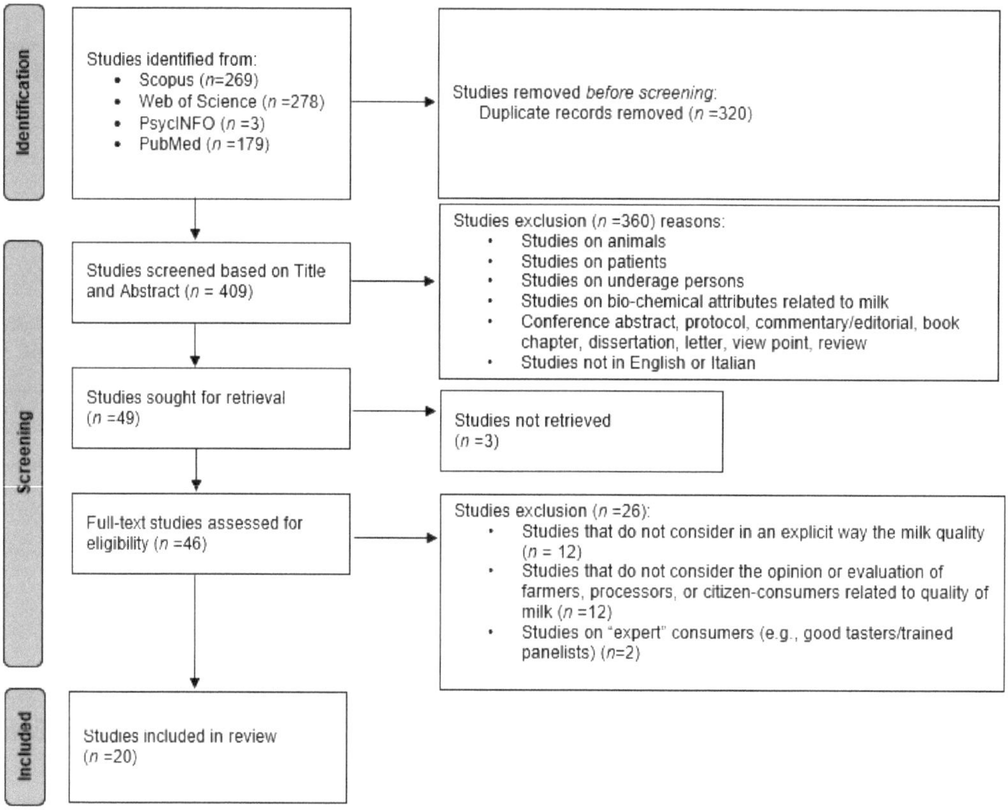

Figure 2. PRISMA flow diagram of study selection.

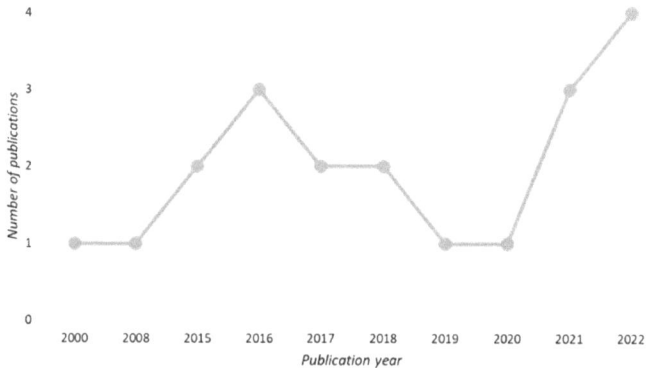

Figure 3. Time distribution of papers on milk quality.

Table 1. General features of included studies.

Study	Country	Study Design	Sample Size	Age Range (Mean Years, SD)	Gender (% Female)	Type of Milk	Point of View
[34]	USA	Qualitative research	93	18–25 (20, NR)	41%	Cow milk	Lay citizens Farmers Intermediaries Retailers Traditional processors Consumers
[35]	India	Mixed methods (qualitative and quantitative)	120	NR	NR	Cow milk	Key informants from government regulatory bodies Private and non-profit sectors
[36]	Mexico	Quantitative research	40	NR (52.65 ± 12.15)	NR	Cow milk	Farmers
[37]	Tanzania	Mixed methods (qualitative and quantitative)	208	NR	NR	Cow milk	Farmers Intermediaries Vendors Consumers Government officials Private sector donors
[38]	France	Participative approach (focus group meetings/Delphi)	44	N.R.	N.R.	Cow milk	Processor
[39]	Kenya	Mixed methods (qualitative and quantitative)	723	Most of the respondents were aged between 30–60 years	50%	Cow milk	Farmers
[40]	Italy	Quantitative research	1216	Most of the respondents were aged <65 years	68%	Cow milk	Lay citizens
[41]	India	Quantitative research	300	19–76 (40, NR)	3%	Cow milk	Farmers
[42]	Brazil	Qualitative research	557	>18 years old	35%	Cow milk	Dairy farmers Agricultural advisors Lay citizens
[43]	USA	Quantitative research	217	NR	NR	Cow milk	Farmers
[44]	Zimbabwe	Quantitative research	344	Most of the respondents were aged >30 years	NR	Cow milk	Farmers
[45]	Brazil	Mixed methods (quantitative and qualitative)	336	>18 years old, most of the respondents were aged 25–34	54%	Cow milk	Lay citizens

Table 1. *Cont.*

Study	Country	Study Design	Sample Size	Age Range (Mean Years, SD)	Gender (% Female)	Type of Milk	Point of View
[46]	Colombia	Quantitative research	46	NR	NR	Cow milk	Farmers
[47]	Germany	Quantitative research	1646	>18 years old, the majority of the respondents were aged >60 (32%)	18%	Cow milk	Lay citizens
[48]	Ethiopia	Quantitative research	160	NR (42.14;-4.50)	32%	Cow milk	Lay citizens
[49]	Ireland	Participative approach (focus group meetings/Delphi)	112	NR	NR	Cow milk	Farmers Stakeholders
[50]	Tanzania	Quantitative research	105	>18 years old, most of the respondents were aged <45 (83%)	46%	Cow milk	Dairy farmers Milk vendors Milk retailers
[51]	Indonesia	Quantitative research	33	NR	NR	Goat milk	Farmers Lay citizens
[52]	Denmark and Netherlands	Qualitative research (focus group approach)	25	30–60 (NR; NR)	NR	Cow milk	Farmers Advisors
[53]	Indonesia	Quantitative research	1225	NR	NR	Cow milk	Lay citizens

Note: NR = Note Reported, SD = standard deviation.

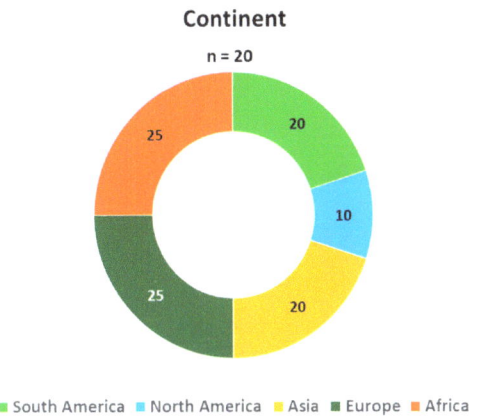

Figure 4. Geographical distribution of papers.

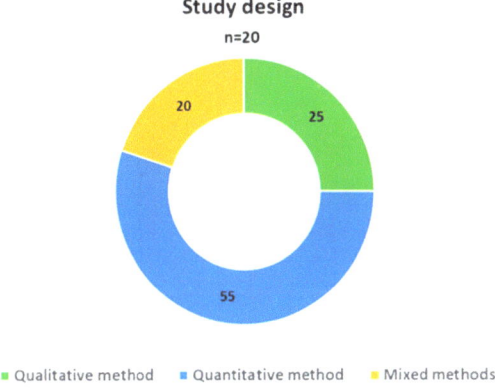

Figure 5. Study design of papers.

3.3. Attributes and Macro-Categories Related to the Concept of Milk Quality

A total of 70 attributes (micro-categories) of milk quality were identified (see Supplementary Table S1). By employing the procedure of inductive category development adapted from Mayring [30], these 70 micro-categories were grouped into 12 macro-categories (Figure 6). Specifically, the results showed that the concept of milk quality is related to the following: (I) policy quality (i.e., transparency of the regulations ruling milk production and processing); (II) relation quality with expert (i.e., trust that people have in the producers and distributors of milk); (III) sensory quality (i.e., perceived organoleptic properties of milk); (IV) packaging quality (i.e., clarity and comprehensiveness of information on the milk pack); (V) nutritional quality/healthiness (i.e., nutritional value/perceived healthiness of milk); (VI) animal welfare quality (i.e., animal welfare protection); (VII) animal safety quality (i.e., animal health protection); (VIII) transport quality (i.e., speed and safety of product transportation/distribution); (IX) company quality (i.e., company reputation); (X) workers' knowledge and attitudes quality (i.e., knowledge and experience of the producing company's workers); (XI) hygiene quality (i.e., product hygiene protection); and (XII) technological quality (i.e., level of technological advancement of the producing company).

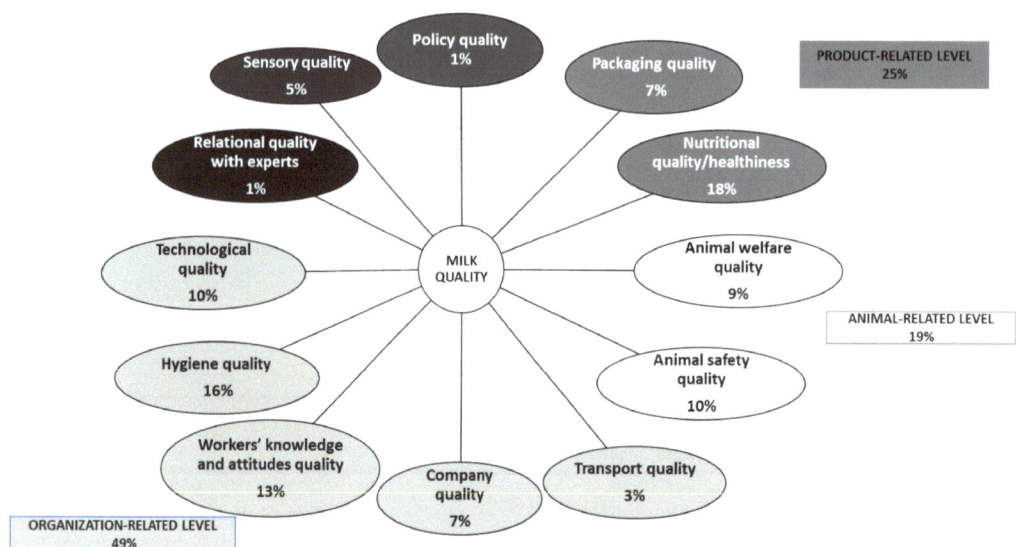

Figure 6. Macro-categories related to the concept of milk quality.

Some of these macro-categories are linked together by overarching dimensions. In particular, the macro-categories "packaging quality" and "nutritional quality/healthiness" refer to milk quality attributes related to the product; the macro-categories "animal welfare quality" and "animal safety quality" concern attributes related to animals; and "technological quality", "hygiene quality", "workers' knowledge and attitudes quality", "company quality", "transport quality" are attributes related to the organizational context in which milk is produced or processed. Most of the micro-categories that connote the concept of milk quality are attributes related to the organization level (49%) and the product level (25%). In particular, the nutritional quality/healthiness (18%), hygiene quality (16%), and workers' knowledge and attitudes quality (13%) are the most salient attributes in defining the concept of milk quality (Figure 6). Moreover, the results of this study showed that sustainability and in particular welfare and health of animals are becoming paramount aspects in defining quality in milk. Indeed, 19 % of the micro-categories analyzed considered this issue.

3.4. Classification of Micro- and Macro-Categories about the Concept of Milk Quality According to Bronfenbrenner's Socio-Ecological Framework

In accordance with Bronfenbrenner's socio-ecological framework (1979), a significant proportion of macro-categories and their corresponding micro-categories associated with the concept of milk quality are situated within the Mesosystem (93%) (Figure 7). These findings highlight the predominant influence of beliefs and perceptions concerning the physical environment where milk is produced and processed on the understanding of milk quality. Conversely, less emphasis is placed on individual factors such as personal inclinations or taste preferences (Individual system; 5%), social norms encompassing trust and social influence (Microsystem; 1%), and cultural norms and agricultural policies (Macrosystem; 1%).

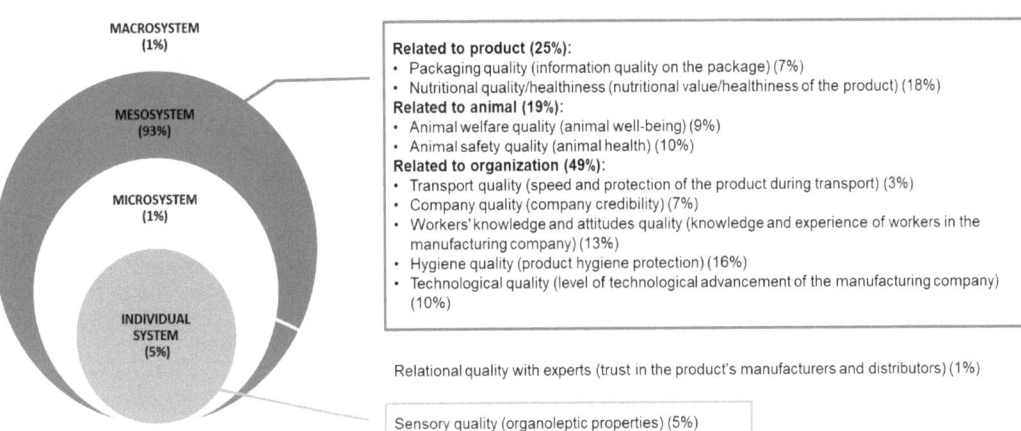

Figure 7. Classification of micro- and macro-categories regarding the concept of milk quality.

3.5. Milk Quality through the Lens of Citizen-Consumers, Farmers, and Processing Experts (Advisors and Processors)

In this section, we describe the semantic attributes associated with the representations of the three main targets examined in this study. Figure 8 provides a detailed analysis of the main overlaps and thematic content concerning the conceptualizations of milk quality among these three representations.

		TARGET	
MACRO-CATEGORIES		FARMER	STAKEHOLDER	CITIZEN-CONSUMER
Policy quality	Focus on regulatory transparency		
Packaging quality			Focus on the quality of the information on the packaging
Nutritional quality/healthiness	Focus on nutritional quality		Focus on naturalness ("without adding")
Animal welfare quality	Focus on healthy growth of the animal		Focus on animal welfare (absence of suffering)
Animal safety quality	Focus on animal health		
Transport quality	Focus on speed of transport and protection of the product during transport		
Company quality	Focus on company credibility		
Workers' knowledge and attitudes quality	Focus on the knowledge and experience of the manufacturing company's workers		
Hygiene quality	Focus on product hygiene protection		
Technological quality	Focus on the level of technological advancement of the manufacturing company		
Relational quality with experts			Focus on trust in the producers and vendors of the product
Sensory quality			Focus on sensory quality

Figure 8. Farmer, processing experts, and citizen-consumer thematic content related to milk quality.

Regarding the overlapping thematic content, our literature analysis reveals that several attributes of milk quality are relevant across different actors. For citizen-consumers, farmers, and processing experts (advisors and processors), milk quality is linked to transparency in regulations regarding milk quality requirements, production, and distribution processes. Additionally, all actors highlight the importance of an approach to milk quality and safety that ensures integrity from farm to glass. Furthermore, the results suggest a need to enhance farmers' knowledge and attitudes and implement hygienic control in the milk production process to meet the required milk quality and food safety standards. Moreover, the conceptualization of milk quality appears to be influenced by the level of technological advancement of the production company. The more a company adopts innovation to ensure a high-quality chain from farm to glass, the more the milk is perceived as a quality product. Finally, citizen-consumers, farmers, and processing experts (advisors and processors) converge in defining milk quality as a product that guarantees certified animal health protection and exhibits high nutritional quality. However, the content of these attributes/themes related to milk quality varies among actors (Figure 8). Dairy experts (farmers and processing experts) assert that milk can be considered a quality product if animal welfare is upheld, including proper disease identification, milk culturing for pathogen detection, appropriate treatment options, and effective management techniques to reduce mastitis incidence. On the other hand, citizen-consumers contend that milk is of good quality when animals have not suffered and continue to live according to their natural behaviors (e.g., grazing, eating grass). Furthermore, while experts (farmers and processing experts) associate high nutritional value with milk quality based on its energy content, protein source, and calcium content, citizen-consumers perceive milk quality as determined by the absence of added ingredients and the naturalness of the product.

This duality in thematic and content perspectives characterizing the representations of milk quality by dairy experts and citizen-consumers, particularly regarding nutritional aspects and animal welfare, highlights how the former prioritize technical aspects such as animal diseases and somatic cell counts, whereas the latter hold simpler and more naïve concepts (e.g., absence of animal suffering or "free-from" products) in their representation of quality milk. Additionally, certain conceptual attributes of the milk quality definition appear to be target-specific. For example, only citizen-consumers identify clear and transparent labels related to nutritional properties, trust in dairy experts, and organoleptic qualities (e.g., appearance, taste, smell) as attributes of milk quality. Conversely, farmers and processing experts (advisors and processors) share similar perspectives on the definition of milk quality, emphasizing two attributes: the speed and protection of milk during transportation from the farm to the industry and the credibility of the production company.

4. Discussion

The decrease in milk consumption can be attributed to several multifaceted factors, among which the evolving notion of milk quality among citizen-consumers plays a pivotal role in contributing to this decline [4]. To address this concern, it is essential to understand how citizen-consumers perceive milk quality and ascertain whether their perception aligns with that of experts, including processing experts and farmers. As a result, we undertook a systematic review with the objective of identifying the crucial attributes that shape the concept of milk quality across three key stakeholder groups: farmers, processing experts (advisors and processors), and citizen-consumers.

The findings reveal that, while milk quality is a relatively new research area, there has been a notable surge in studies conducted on this topic in recent years (2021–2022). Additionally, recent studies have predominantly adopted qualitative methodologies, focusing on the perspective of citizen-consumers, in contrast to earlier research trends. This shift can be attributed to evolving consumer demands, which have significantly reshaped the broader notion of food quality [54]. Currently, food quality is not solely linked to functional parameters like nutritional value, appearance, and taste; it is also deeply intertwined with the ethical, identity, and emotional values of citizen-consumers [55,56]. Furthermore, in terms of the geographical distribution of the studies, the results indicate a heightened interest in the subject of milk quality in Africa and Asia. This observation could be attributed to the necessity of these regions to enhance and promote high-protein foods as a strategy to address malnutrition rates [57], where milk emerges as a potential key solution [58].

The findings show the presence of 12 main attributes (macro-categories) that characterize the concept of milk quality. Many of these attributes pertain to the organizational and product levels. Notably, nutritional quality/healthiness, hygiene quality, and workers' knowledge and attitudes emerge as the most prominent attributes in defining the concept of milk quality. These findings are consistent with previous research indicating that milk quality is primarily associated with its nutritional and hygienic aspects [59] and the skills of workers, which significantly impact the economic efficiency of dairy farms [60]. In particular, hygiene standards are defined and regulated differently depending on the country of reference. As for the nations belonging to the European Union, appendix III, section IX, chapter I of Regulation (EC) No 853/2004 of the European Parliament [11] and of the Council of 29 April 2004 [12] describe the acceptable numbers of bacterial and somatic cells in milk to define it as safe and therefore saleable. However, in some countries outside the European Union, food safety legislation is poor, resulting in scarce hygienic practices in the treatment of milk. In the South African territories, for example, the lack of regulation with respect to hygienic standards in the treatment and sale of milk is considered the main reason for losses, resulting in reduced income for the farmers and for the smallholder dairies [61]. Also, in Ethiopia, there is no hygiene standard followed by producers during milk production. Hygiene conditions vary depending on the production system. In most cases, under small-scale farming conditions, the common hygiene measures adopted during milk production, especially during milking, are limited to allowing the calf to suckle for a few minutes and/or washing the udder before milking [62]. However, the aspect of sustainability, particularly animal welfare and health, is increasingly recognized as a crucial component in defining quality milk. In line with this, several studies have highlighted that controlling cow mastitis and somatic cell count (SCC) is a significant concern for farmers in maintaining milk quality [63]. Furthermore, there are some studies that claim farmers are very attentive to the animals' diets, as they are aware that it impacts the features of milk [64,65]. For instance, it has been demonstrated that pasture feeding positively influences the nutritional profile of milk, enhancing its health benefits [66], which is highly valued by consumers. Additionally, the animals' diets affect the organoleptic qualities of milk [67]. For example, the ratio between maize silage and lucerne silage can impact the milk's color, creaminess, and density, indirectly influencing quality assessment [68]. Moreover, the environmental and welfare conditions to which cows are exposed can influence the organoleptic and nutritional characteristics of milk [69]. Specifically, subjecting cows to significant stress due to poor welfare conditions results in decreased milk production with lower levels of fat and protein, thus rendering the milk less nutritious and of inferior quality [70]. Finally, paying attention to the well-being and health of animals is not only important for producing quality milk, but also for achieving a positive economic return. In fact, dirty and poorly maintained environments can increase the likelihood of animals getting sick and requiring antibiotic treatments, resulting in additional and often prohibitive costs for the farmers [71]. Considering citizen-consumers, recent research indicates that they associate milk quality with factors such as free cow grazing, natural feed, and the absence of medical treatment for cows [72,73]. Moreover, technological development and automation of breeding and milking processes are relevant in defining milk quality, as shown by past studies [74]. However, it is interesting to note that traceability technologies are not mentioned. This aspect points out that, although such technologies have been implemented to increase milk controls in order to ensure a quality product [75], these, in the imaginations of the targets considered, are not linked to the attributes of milk quality. Supporting these findings, some studies showed that perceptions of and interest in traceability change across countries [76]. Although in most cases traceability is strongly perceived as synonymous with genuine and safe product, those who do not trust certifying bodies, technology, and have little knowledge do not consider them as part of the safety- and quality-assurance strategy in the food industry [77].

Regarding the classification of micro- and macro-categories within Bronfenbrenner's socio-ecological framework, it is evident that attributes related to taste preference received minimal mention from the study participants. These findings appear to contradict previous research, which commonly associates food quality with personal evaluations based on taste and liking. However, the attributes utilized by individuals to describe quality are dynamic and subject to change based on their interests, concerns, or needs [4]. Several studies [4,21,78] have observed a recent shift wherein extrinsic quality attributes, which pertain to characteristics associated with a product but are not physically inherent to it [79], have gained increasing importance in defining food quality, alongside intrinsic attributes, which are related to the physical composition of the product itself and cannot be altered without changing its nature, such as aroma, taste, and color [79,80]. Of particular significance are the extrinsic quality attributes known as "Search Qualities," which individuals can determine before purchasing a food product through direct examination (e.g., nutritional value or packaging size and features), and "Credence Qualities" [81], which require additional information for evaluation and cannot be directly experienced from the product itself (e.g., environmental impact). These extrinsic qualities have gained importance as sustainability and company practices have become priorities in food choices [82]. For example, many quality food characteristics have been associated with farming practices and the entire distribution chain, including the processes from farm to fork and how crops and livestock are managed [4,83,84]. Therefore, the extrinsic quality aspects related to product features and the physical environment in which food is processed and produced are the most utilized attributes in defining food product quality, as affirmed by the present study. In summary, it can be concluded that the perceived quality of milk is primarily shaped by extrinsic attributes associated with the production and processing of milk, while intrinsic attributes tied to individual sensory perceptions appear to have less prominence in the representation of milk quality. Lastly, the attributes employed to define the concept of milk quality vary among the study participants. Farmers and processing experts appear to share a relatively similar perception of milk quality, marked by technical indicators and a strong emphasis on knowledge and expertise. In contrast, citizen-consumers hold a representation of quality milk rooted in simplified and less sophisticated concepts (such as the absence of animal suffering or "free-from" products). For example, low-fat milk, milk without additives, and milk derived from animals not treated with antibiotics are among the aspects that consumers pay attention to [41,85]. These aspects can be challenging to quantify and are primarily tied to their individual perceptions, which may not always be based on concrete evidence. Previous research has highlighted citizen-consumers' concerns regarding farming practices that they believe impact the emotional well-being of animals, the treatment of animals, and the idea of naturalness [86]. Furthermore, even when various stakeholders share a common conceptual category for defining milk quality (such as "animal welfare"), they diverge in the interpretations assigned to it (like "physical health of the animal" versus "well-being and safeguarding of the animal's quality of life"). This could suggest only an apparent alignment of perspectives, but it reveals a profound fragmentation of the semantic framework within which the representations of quality milk are generated by the different social actors involved in the milk production and consumption arena. From our standpoint, this study offers valuable insights for future research in the field. Primarily, it underscores the necessity to delve deeper into the fundamental attributes that shape the concept of milk quality through both qualitative and quantitative investigations. This endeavor will contribute to a more comprehensive grasp of the representations held by various stakeholders, encompassing both experts and non-experts within the dairy industry, especially given the notable disparities revealed in this study. Moreover, it is imperative for scholarly experts in the dairy domain to adopt a more holistic research approach when addressing these matters. Moving away from a self-referential perspective, an interdisciplinary approach should be embraced to scrutinize the concept of milk quality. This approach should encompass an ecological perspective that integrates a variety of disciplines, aiming to present a more cohesive portrayal of milk quality. Furthermore, this perspective should be reassessed and harmonized through a bottom-up strategy in conjunction with the viewpoint of citizen-consumers. To facilitate this, initiatives that facilitate dialogue and collaboration between citizens and industry experts, such as participatory and citizen science methods, should be encouraged. These initiatives will serve to educate and involve citizen-consumers in conversations about milk quality, ultimately fostering the development of a shared comprehension and addressing the dual fragmentation present between the interpretations of milk quality among dairy experts and citizen-consumers. Moreover, in order to have a more complete view related to the "milk quality" topic, it might be interesting to conduct new research involving other supply chain actors not included in this study, such as sellers. Lastly, it is also crucial to investigate spontaneous discourse and social communication related to milk quality. This analysis will enable a deeper insight into the ongoing conversations surrounding this subject and yield valuable concepts and perspectives for further exploration in this field.

5. Conclusions

This systematic review marks the inaugural scientific effort dedicated to exploring the psychosocial discourse surrounding milk quality as documented in the literature, yielding promising outcomes. Specifically, the study delved into the core attributes associated with the concept of milk quality across three key stakeholders: citizen-consumers, farmers, and processing experts (including advisors and processors). The findings unveil that the definition of milk quality revolves around 12 major conceptual categories, which can be organized within the framework of Bronfenbrenner's ecological theory into four distinct systems. Notably, the representation of milk quality exhibits significant variation among the three targeted groups, particularly between expert figures in the dairy system (processing experts and farmers) and citizen-consumers. The study outcomes contribute to establishing a more methodical comprehension of the representations connected to the concept of milk quality, as perceived by all social actors involved in its production and consumption. Moreover, the findings underscore the necessity of fostering transdisciplinary and cross-sectoral links among perspectives stemming from diverse paradigms. Furthermore, the results underscore the importance of instigating a collaborative process to construct a shared social representation on this topic that effectively merges societal impact with a solid scientific foundation. To bridge the gap in perception and align milk quality representation, several educational strategies can be employed. For example, the experts (farmers and processors) can be encouraged to provide consumers with insights into their farming/production practices. This could involve hosting farm tours, workshops, or online videos that showcase the daily routines, animal welfare standards, and quality control measures undertaken on the farm and in the company. Moreover, organizing workshops for both experts and consumers can serve as a platform for knowledge exchange. Experts can gain insights into consumer preferences and concerns, while consumers can learn about the complexities of milk production. This two-way dialogue can bridge understanding and highlight the efforts that farmers and processors put into ensuring quality. Finally, introducing educational programs in schools that highlight the journey of milk from farm to table can cultivate informed consumer choices from a young age. Engaging activities, like farm visits or virtual tours, can make the learning experience more interactive and memorable. By implementing these educational strategies, farmers, processors, and consumers can collaborate to build a shared perception of milk quality. These efforts will not only foster transparency and trust but also contribute to the sustainability of the dairy industry by ensuring that products meet the expectations of both experts and consumers.

Supplementary Materials: The following supporting information can be downloaded at: https://www.mdpi.com/article/10.3390/foods12173215/s1, Table S1. Micro and Macro categories about milk quality representation.

Author Contributions: This paper derives from a collaboration of the authors. G.C.: conceptualization, methodology, data curation, formal analysis, writing—original draft. S.B.: conceptualization, methodology, data curation, formal analysis, writing—original draft. A.C.B.: writing—review and editing, supervision. All authors have read and agreed to the published version of the manuscript.

Funding: This work was supported by IRCAF (Centro di Riferimento Agro-Alimentare Romeo e Enrica Invernizzi) special project "Milk Quality: a multifaceted approach".

Data Availability Statement: Data are fully available upon request to the corresponding author.

Acknowledgments: We want to thank IRCAF (Centro di Riferimento Agro-Alimentare Romeo e Enrica Invernizzi) that supported this special project "Milk Quality: a multifaceted approach".

Conflicts of Interest: The authors declare no conflict of interest. The funders had no role in the design of the study; in the collection, analyses, or interpretation of data; in the writing of the manuscript; or in the decision to publish the results.

References

1. Castellini, G.; Graffigna, G. Evolution of Milk Consumption and Its Psychological Determinants: A Mini-Review. *Front. Nutr.* **2022**, *9*, 845154. [CrossRef] [PubMed]
2. Bentivoglio, D.; Finco, A.; Bucci, G.; Staffolani, G. Is There a Promising Market for the A2 Milk? Analysis of Italian Consumer Preferences. *Sustainability* **2020**, *12*, 6763. [CrossRef]
3. Zingone, F.; Bucci, C.; Iovino, P.; Ciacci, C. Consumption of milk and dairy products: Facts and figures. *Nutrition* **2017**, *33*, 322–325. [CrossRef] [PubMed]
4. Petrescu, D.C.; Vermeir, I.; Petrescu-Mag, R.M. Consumer Understanding of Food Quality, Healthiness, and Environmental Impact: A Cross-National Perspective. *Int. J. Environ. Res. Public Health* **2019**, *17*, 169. [CrossRef]
5. Jiang, R.; Sharma, C.; Bryant, R.; Mohan, M.S.; Al-Marashdeh, O.; Harrison, R.; Torrico, D.D. Animal welfare information affects consumers' hedonic and emotional responses towards milk. *Food Res. Int.* **2021**, *141*, 110006. [CrossRef] [PubMed]
6. Gao, Z.; Li, C.; Bai, J.; Fu, J. Chinese consumer quality perception and preference of sustainable milk. *China Econ. Rev.* **2020**, *59*, 100939. [CrossRef]
7. Kumar, A.; Joshi, P.K.; Kumar, P.; Parappurathu, S. Trends in the consumption of milk and milk products in India: Implications for self-sufficiency in milk production. *Food Secur.* **2014**, *6*, 719–726. [CrossRef]
8. Vargas-Bello-Pérez, E.; Tajonar, K.; Foggi, G.; Mele, M.; Simitzis, P.; Mavrommatis, A.; Tsiplakou, E.; Habib, M.R.; Gonzalez-Ronquillo, M.; Toro-Mujica, P. Consumer attitudes toward dairy products from sheep and goats: A cross-continental perspective. *J. Dairy Sci.* **2022**, *105*, 8718–8733. [CrossRef]
9. Čapla, J.; Zajác, P.; Ševcová, K.; Čurlej, J.; Fikselová, M. Overview of the milk and dairy products legislation in the European Union. *Legestic* **2023**, *1*, 1–16. [CrossRef]
10. European Parliament Regulation (EC), No. 178/2002 of the European Parliament and of the Council. *Off. J. Eur. Communities* **2002**, *31*, 1–24.
11. EC Regulation (EC) No 852/2004, Hygiene of foodstuffs. *Off. J. Eur. Union* **2004**, *139*, 1–54.
12. EC Commission Regulation (EC) No 853/2004 of 29 April 2004 laying down specific hygiene rules for food of animal origin. *Off. J. Eur. Union* **2004**, *139*, 55–205.
13. Vinet, L.; Zhedanov, A. A 'missing' family of classical orthogonal polynomials. *J. Phys. A Math. Theor.* **2011**, *44*, 085201. [CrossRef]
14. Lee, C.-L.; Liao, H.-L.; Lee, W.-C.; Hsu, C.-K.; Hsueh, F.-C.; Pan, J.-Q.; Chu, C.-H.; Wei, C.-T.; Chen, M.-J. Standards and labeling of milk fat and spread products in different countries. *J. Food Drug Anal.* **2018**, *26*, 469–480. [CrossRef]
15. Fransvea, A.; Celano, G.; Pagliarone, C.N.; Disanto, C.; Balzaretti, C.; Celano, G.V.; Bonerba, E. Food labelling: A brief analysis of European Regulation 1169/2011. *Ital. J. Food Saf.* **2014**, *3*. [CrossRef]
16. Henderikx, F. Labelling of food: A challenge for many. *Vet. Glas.* **2017**, *71*, 16–23. [CrossRef]
17. European Parliament REGULATION (EU) No 1169/2011 OF THE EUROPEAN PARLIAMENT AND OF THE COUNCIL of 25 October 2011. *Off. J. Eur. Union* **2011**, *304*, 18–63.
18. European Parliament and Council Regulation (EU) No 1308/2013: Establishing a common organization of the markets in agricultural products. *Off. J. Eur. Union* **2013**, *347*, 671–854.
19. Lombardi, G.V.; Berni, R.; Rocchi, B. Environmental friendly food. Choice experiment to assess consumer's attitude toward "climate neutral" milk: The role of communication. *J. Clean. Prod.* **2017**, *142*, 257–262. [CrossRef]
20. McLaren, S.; Berardy, A.; Henderson, A.; Holden, N.; Huppertz, T.; Jolliet, O.; De Camillis, C.; Renouf, M.; Rugani, B.; Saarinen, M.; et al. *Integration of Environment and Nutrition in Life Cycle Assessment of Food Items: Opportunities and Challenges*; FAO: Rome, Italy, 2021; ISBN 978-92-5-135532-9.
21. Verdú Jover, A.J.; Lloréns Montes, F.J.; Fuentes Fuentes, M.d.M. Measuring perceptions of quality in food products: The case of red wine. *Food Qual. Prefer.* **2004**, *15*, 453–469. [CrossRef]
22. Andersen, H. The issue 'Raw milk quality' from the point of view of a major dairy industry. *J. Anim. Feed Sci.* **2007**, *16*, 240–254. [CrossRef]
23. Moher, D.; Liberati, A.; Tetzlaff, J.; Altman, D.G.; Altman, D.; Antes, G.; Atkins, D.; Barbour, V.; Barrowman, N.; Berlin, J.A.; et al. Preferred reporting items for systematic reviews and meta-analyses: The PRISMA statement. *PLoS Med.* **2009**, *151*, 264–269. [CrossRef]

24. Moreira-Dantas, I.R.; Martínez-Zarzoso, I.; Torres-Munguía, J.A. Sustainable Food Chains to Achieve SDG-12 in Europe: Perspectives from Multi-stakeholders Initiatives. In *SDGs in the European Region*; Springer: Cham, Switzerland, 2022; pp. 1–26.
25. Atkinson, L.Z.; Cipriani, A. How to carry out a literature search for a systematic review: A practical guide. *BJPsych Adv.* **2018**, *24*, 74–82. [CrossRef]
26. Popay, J.; Roberts, H.; Sowden, A.; Petticrew, M.; Arai, L.; Rodgers, M.; Britten, N.; Roen, K.; Duffy, S.; Arai, L.; et al. Developing guidance on the conduct of narrative synthesis in systematic reviews. *A Prod. ESRC Methods Program Lanc. Inst. Health Res.* **2005**, *1*, 92.
27. Hsieh, H.-F.; Shannon, S.E. Three Approaches to Qualitative Content Analysis. *Qual. Health Res.* **2005**, *15*, 1277–1288. [CrossRef]
28. Schilling, J. On the Pragmatics of Qualitative Assessment. *Eur. J. Psychol. Assess.* **2006**, *22*, 28–37. [CrossRef]
29. Mayring, P. Qualitative Content Analysis. Available online: http://www.qualitative-research.net/index.php/fqs/index (accessed on 10 March 2023).
30. Mayring, P.A.E. Qualitative content analysis. In *International Encyclopedia of Education(Fourth Edition)*; Elsevier: Amsterdam, The Netherlands, 2023; pp. 314–322. ISBN 9780128186299.
31. Story, M.; Kaphingst, K.M.; Robinson-O'Brien, R.; Glanz, K. Creating healthy food and eating environments: Policy and environmental approaches. *Annu. Rev. Public Health* **2008**, *29*, 253–272. [CrossRef]
32. Bronfenbrenner, U.; Capurso, M. *Rendere Umani Gli Esseri Umani. Bioecologia Dello Sviluppo*; Erickson: Trento, Italy, 2010.
33. Bronfenbrenner, U. *The Ecology of Human Development*; Harvard University Press: Cambridge, MA, USA, 1979.
34. Connors, P.L.; Schuelke, W.C. Date labels and college student perceptions of milk drinkability. *J. Food Compos. Anal.* **2022**, *105*, 104249. [CrossRef]
35. Nicolini, G.; Guarin, A.; Deka, R.P.; Vorley, B.; Alonso, S.; Blackmore, E.; Grace, D. Milk quality and safety in the informal sector in Assam, India: Governance, perceptions, and practices. *Cogent Food Agric.* **2022**, *8*, 2137897. [CrossRef]
36. Romo-Bacco, C.E.; Parga-Montoya, N.; Valdivia-Flores, A.G.; Carranza-Trinidad, R.G.; Montoya Landeros, M.D.C.; Llamas-Martínez, A.A.; Aguilar Romero, M.M. Perspectivas sobre la continuidad, calidad de leche y entorno en unidades de producción de leche en el estado de Aguascalientes, México. *Rev. Mex. Ciencias Pecu.* **2022**, *13*, 357–374. [CrossRef]
37. Blackmore, E.; Guarin, A.; Kinyua, C.; Vorley, W.; Grace, D.; Alonso, S. The governance of quality and safety in Tanzania's informal milk markets. *Front. Sustain. Food Syst.* **2022**, *6*, 971961. [CrossRef]
38. Rey-Cadilhac, L.; Botreau, R.; Ferlay, A.; Hulin, S.; Hurtaud, C.; Lardy, R.; Martin, B.; Laurent, C. Co-construction of a method for evaluating the intrinsic quality of bovine milk in relation to its fate. *Animal* **2021**, *15*, 100264. [CrossRef]
39. Nyokabi, S.; Luning, P.A.; de Boer, I.J.M.; Korir, L.; Muunda, E.; Bebe, B.O.; Lindahl, J.; Bett, B.; Oosting, S.J. Milk quality and hygiene: Knowledge, attitudes and practices of smallholder dairy farmers in central Kenya. *Food Control* **2021**, *130*, 108303. [CrossRef]
40. Merlino, V.M.; Massaglia, S.; Borra, D.; Mimosi, A.; Cornale, P. Which Factors Drive Consumer Decisions during Milk Purchase? New Individuals' Profiles Considering Fresh Pasteurized and UHT Treated Milk. *Foods* **2021**, *11*, 77. [CrossRef] [PubMed]
41. Singh, H.; Singh, J.; Verma, H.; Kansal, S. Milk quality and safety issues inside the farm gate of dairy farmers of Punjab (India). *Indian J. Dairy Sci.* **2020**, *73*, 614–624. [CrossRef]
42. Cardoso, C.S.; von Keyserlingk, M.G.; Hötzel, M.J. Views of dairy farmers, agricultural advisors, and lay citizens on the ideal dairy farm. *J. Dairy Sci.* **2019**, *102*, 1811–1821. [CrossRef] [PubMed]
43. Tse, C.; Barkema, H.W.; DeVries, T.J.; Rushen, J.; Pajor, E.A. Impact of automatic milking systems on dairy cattle producers' reports of milking labour management, milk production and milk quality. *Animal* **2018**, *12*, 2649–2656. [CrossRef]
44. Paraffin, A.S.; Zindove, T.J.; Chimonyo, M. Perceptions of Factors Affecting Milk Quality and Safety among Large- and Small-Scale Dairy Farmers in Zimbabwe. *J. Food Qual.* **2018**, *2018*, 5345874. [CrossRef]
45. Cardoso, C.; von Keyserlingk, M.; Hötzel, M. Brazilian Citizens: Expectations Regarding Dairy Cattle Welfare and Awareness of Contentious Practices. *Animals* **2017**, *7*, 89. [CrossRef] [PubMed]
46. Múnera-Bedoya, O.D.; Cassoli, L.D.; Machado, P.F.; Cerón-Muñoz, M.F. Influence of attitudes and behavior of milkers on the hygienic and sanitary quality of milk. *PLoS ONE* **2017**, *12*, e0184640. [CrossRef]
47. Pieper, L.; Doherr, M.G.; Heuwieser, W. Consumers' attitudes about milk quality and fertilization methods in dairy cows in Germany. *J. Dairy Sci.* **2016**, *99*, 3162–3170. [CrossRef] [PubMed]
48. Bekele, A.D.; Beuving, J.; Ruben, R. Food choices in Ethiopia: Does nutritional information matter? *Int. J. Consum. Stud.* **2016**, *40*, 625–634. [CrossRef]
49. Henchion, M.; McCarthy, M.; Resconi, V.C.; Berry, D.P.; McParland, S. Stakeholder involvement in establishing a milk quality sub-index in dairy cow breeding goals: A Delphi approach. *Animal* **2016**, *10*, 878–891. [CrossRef] [PubMed]
50. Ngasala, J.u.B.; Nonga, H.E.; Mtambo, M.M.A. Assessment of raw milk quality and stakeholders' awareness on milk-borne health risks in Arusha City and Meru District, Tanzania. *Trop. Anim. Health Prod.* **2015**, *47*, 927–932. [CrossRef]
51. Cyrilla, L.; Purwanto, B.P.; Atabany, A.; Astuti, D.A.; Sukmawati, A. Improving Milk Quality for Dairy Goat Farm Development. *Media Peternak.* **2015**, *38*, 204–211. [CrossRef]
52. Oudshoorn, F.W.; Renes, R.J.; De Boer, I.J.M. Systems In Organic Dairy Production. *J. Agric. Environ. Ethics* **2008**, *21*, 205–228. [CrossRef]
53. Viaene, J.; Verbeke, W.; Sufiati, M. Quality Guidance Toward Milk in West Java, Indonesia. *J. Int. Food Agribus. Mark.* **2000**, *11*, 73–88. [CrossRef]

54. Yan, M.R.; Hsieh, S.; Ricacho, N. Innovative Food Packaging, Food Quality and Safety, and Consumer Perspectives. *Processes* **2022**, *10*, 747. [CrossRef]
55. Gil-Giménez, D.; Rolo-González, G.; Suárez, E.; Muinos, G. The Influence of Environmental Self-Identity on the Relationship between Consumer Identities and Frugal Behavior. *Sustainability* **2021**, *13*, 9664. [CrossRef]
56. Costa, I.; Gill, P.R.; Morda, R.; Ali, L. "More than a diet": A qualitative investigation of young vegan Women's relationship to food. *Appetite* **2019**, *143*, 104418. [CrossRef]
57. Kalu, R.E.; Etim, K.D. Factors associated with malnutrition among underfive children in developing countries: A review. *Glob. J. Pure Appl. Sci.* **2018**, *24*, 69. [CrossRef]
58. Headey, D. Can dairy help solve the malnutrition crisis in developing countries? An economic analysis. *Anim. Front.* **2023**, *13*, 7–16. [CrossRef] [PubMed]
59. Schukken, Y.H.; Wilson, D.J.; Welcome, F.; Garrison-Tikofsky, L.; Gonzalez, R.N. Monitoring udder health and milk quality using somatic cell counts. *Vet. Res.* **2003**, *34*, 579–596. [CrossRef] [PubMed]
60. Edgar, S. *Milk and Dairy Product Technology*; Axel, M., Ed.; Routledge: Abingdon, UK, 2017; ISBN 9780203747162.
61. Gran, H.M.; Mutukumira, A.N.; Wetlesen, A.; Narvhus, J.A. Smallholder dairy processing in Zimbabwe: Hygienic practices during milking and the microbiological quality of the milk at the farm and on delivery. *Food Control* **2002**, *13*, 41–47. [CrossRef]
62. Bereda, A.; Yesuf Kurtu, M.; Yilma, Z. Handling, Processing and Utilization of Milk and Milk Products in Ethiopia: A Review. *World J. Dairy Food Sci.* **2014**, *9*, 105–112.
63. Hamann, J. *Mastitis in Dairy Production*; Hogeveen, H., Ed.; Wageningen Academic Publishers: Wageningen, The Netherlands, 2005; ISBN 978-90-76998-70-1.
64. Duncan, A.J.; Teufel, N.; Mekonnen, K.; Singh, V.K.; Bitew, A.; Gebremedhin, B. Dairy intensification in developing countries: Effects of market quality on farm-level feeding and breeding practices. *Animal* **2013**, *7*, 2054–2062. [CrossRef]
65. Duplessis, M.; Pellerin, D.; Robichaud, R.; Fadul-Pacheco, L.; Girard, C.L. Impact of diet management and composition on vitamin B12 concentration in milk of Holstein cows. *Animal* **2019**, *13*, 2101–2109. [CrossRef]
66. Alothman, M.; Hogan, S.A.; Hennessy, D.; Dillon, P.; Kilcawley, K.N.; O'Donovan, M.; Tobin, J.; Fenelon, M.A.; O'Callaghan, T.F. The "Grass-Fed" Milk Story: Understanding the Impact of Pasture Feeding on the Composition and Quality of Bovine Milk. *Foods* **2019**, *8*, 350. [CrossRef]
67. Manzocchi, E.; Martin, B.; Bord, C.; Verdier-Metz, I.; Bouchon, M.; De Marchi, M.; Constant, I.; Giller, K.; Kreuzer, M.; Berard, J.; et al. Feeding cows with hay, silage, or fresh herbage on pasture or indoors affects sensory properties and chemical composition of milk and cheese. *J. Dairy Sci.* **2021**, *104*, 5285–5302. [CrossRef]
68. Larsen, M.K.; Kidmose, U.; Kristensen, T.; Beaumont, P.; Mortensen, G. Chemical composition and sensory quality of bovine milk as affected by type of forage and proportion of concentrate in the feed ration. *J. Sci. Food Agric.* **2013**, *93*, 93–99. [CrossRef]
69. Sinclair, K.D.; Garnsworthy, P.C.; Mann, G.E.; Sinclair, L.A. Reducing dietary protein in dairy cow diets: Implications for nitrogen utilization, milk production, welfare and fertility. *Animal* **2014**, *8*, 262–274. [CrossRef] [PubMed]
70. Kawonga, B.S.; Chagunda, M.G.G.; Gondwe, T.N.; Gondwe, S.R.; Banda, J.W. Characterisation of smallholder dairy production systems using animal welfare and milk quality. *Trop. Anim. Health Prod.* **2012**, *44*, 1429–1435. [CrossRef]
71. Hogeveen, H.; Huijps, K.; Lam, T. Economic aspects of mastitis: New developments. *N. Z. Vet. J.* **2011**, *59*, 16–23. [CrossRef] [PubMed]
72. de Graaf, S.; Van Loo, E.J.; Bijttebier, J.; Vanhonacker, F.; Lauwers, L.; Tuyttens, F.A.M.; Verbeke, W. Determinants of consumer intention to purchase animal-friendly milk. *J. Dairy Sci.* **2016**, *99*, 8304–8313. [CrossRef] [PubMed]
73. Crump, A.; Jenkins, K.; Bethell, E.J.; Ferris, C.P.; Arnott, G. Pasture Access Affects Behavioral Indicators of Wellbeing in Dairy Cows. *Animals* **2019**, *9*, 902. [CrossRef]
74. Klungel, G.H.; Slaghuis, B.A.; Hogeveen, H. The Effect of the Introduction of Automatic Milking Systems on Milk Quality. *J. Dairy Sci.* **2000**, *83*, 1998–2003. [CrossRef]
75. Rocchetti, G.; O'Callaghan, T.F. Application of metabolomics to assess milk quality and traceability. *Curr. Opin. Food Sci.* **2021**, *40*, 168–178. [CrossRef]
76. Hansstein, F.V. Consumer Knowledge and Attitudes towards Food Traceability: A Comparison between the European Union, China and North America Francesca. In Proceedings of the 2014 International Conference on Food Security and Nutrition, Shanghai, China, 29–30 March 2014.
77. Corallo, A.; Latino, M.E.; Menegoli, M.; Striani, F. The awareness assessment of the Italian agri-food industry regarding food traceability systems. *Trends Food Sci. Technol.* **2020**, *101*, 28–37. [CrossRef]
78. Caswell, J.A.; Noelke, C.M.; Mojduszka, E.M. Unifying Two Frameworks for Analyzing Quality and Quality Assurance for Food Products. In *Global Food Trade and Consumer Demand for Quality*; Springer: Boston, MA, USA, 2002; pp. 43–61.
79. Olson, J.C. Price as an informational cue: Effects in product evaluation. In *Consumer and Industrial Buying Behaviour*; Woodside, A.G., Sheth, J.N., Bennet, P.D., Eds.; North Holland Publishing Company: New York, NY, USA, 1977; pp. 267–286.
80. Olson, J.C.; Jacoby, J. Cue utilization in the quality perception process. In Proceedings of the Third Annual Conference of the Association for Consumer Research, Chicago, IL, USA, 3–5 November 1972; Venkatesan, M., Ed.; Association for Consumer Research: Lowa City, IA, USA, 1972; pp. 167–179.
81. Darby, M.R.; Karni, E. Free Competition and the Optimal Amount of Fraud. *J. Law Econ.* **1973**, *16*, 67–88. [CrossRef]

82. Saba, A.; Sinesio, F.; Moneta, E.; Dinnella, C.; Laureati, M.; Torri, L.; Peparaio, M.; Saggia Civitelli, E.; Endrizzi, I.; Gasperi, F.; et al. Measuring consumers attitudes towards health and taste and their association with food-related life-styles and preferences. *Food Qual. Prefer.* **2019**, *73*, 25–37. [CrossRef]
83. McInerney, J. The production of food: From quantity to quality. *Proc. Nutr. Soc.* **2002**, *61*, 273–279. [CrossRef] [PubMed]
84. Yu, H.; Gibson, K.E.; Wright, K.G.; Neal, J.A.; Sirsat, S.A. Food safety and food quality perceptions of farmers' market consumers in the United States. *Food Control* **2017**, *79*, 266–271. [CrossRef]
85. Vargas-Bello-Pérez, E.; Faber, I.; Osorio, J.S.; Stergiadis, S. Consumer knowledge and perceptions of milk fat in Denmark, the United Kingdom, and the United States. *J. Dairy Sci.* **2020**, *103*, 4151–4163. [CrossRef] [PubMed]
86. Henchion, M.M.; Regan, Á.; Beecher, M.; MackenWalsh, Á. Developing 'Smart' Dairy Farming Responsive to Farmers and Consumer-Citizens: A Review. *Animals* **2022**, *12*, 360. [CrossRef]

Disclaimer/Publisher's Note: The statements, opinions and data contained in all publications are solely those of the individual author(s) and contributor(s) and not of MDPI and/or the editor(s). MDPI and/or the editor(s) disclaim responsibility for any injury to people or property resulting from any ideas, methods, instructions or products referred to in the content.

Review

Fermented Mare Milk and Its Microorganisms for Human Consumption and Health

Francesca Martuzzi [1], Piero Franceschi [2,*] and Paolo Formaggioni [2]

[1] Department of Food and Drug Science, University of Parma, Via delle Scienze, 43124 Parma, Italy; francesca.martuzzi@unipr.it
[2] Department of Veterinary Science, University of Parma, Via del Taglio 10, 43126 Parma, Italy; paolo.formaggioni@unipr.it
* Correspondence: piero.franceschi@unipr.it; Tel.: +39-0521032615

Abstract: Mare milk is consumed by approximatively 30 million people in the world. In countries in Asia and East Europe, mare milk is mainly consumed as source of fermented products, called koumiss, airag or chigee, alcoholic beverages obtained by means of a culture of bacteria and lactose-fermenting yeasts. Recent research concerning mare milk and its derivatives deals mainly with their potential employment for human health. Studies about the isolation and characterization of *Lactobacillus* spp. and yeasts from koumiss have been aimed at assessing the potential functional properties of these micro-organisms and to find their employment for the industrial processing of mare milk. The aim of this literature review is to summarize recent research about microorganisms in fermented mare milk products and their potential functional properties.

Keywords: mare milk; koumiss; airag; lactic acid bacteria; yeasts

1. Introduction

Mare milk is a fundamental aliment for the people of the Central Asia steppes. It is consumed also in Europe, in particular in Hungary and in the Netherlands. It was estimated that nearly 30 million people consume it regularly [1]. Besides the use of this milk as a source of valuable nutrients, since ancient times it has been considered as a sort of medication for its health-promoting characteristics in the regions of the former USSR and Western Asia. It has been extolled for its many healing properties in some papers, but clinical studies effectively proving its positive effects are scarce [2–4].

In the recent years of the 20th century, in Europe, studies about equine milk have mostly dealt with protein compounds—the identification and characterization of caseins and whey proteins (see reviews by Martuzzi and Doreau [5] and by Uniacke-Lowe et al. [6])—and to a lesser extent other milk components [7], with some interest for its possible use as a substitute for bovine milk for children with intolerance or allergy [8–10]. In particular, whereas most studies in the world have dealt with horse's milk, in Italy interest is more focused on donkey's milk, traditionally used in the past for orphaned or abandoned children, when formulas were not yet available [11–13]. An extensive review about equid milk (horse and donkey) composition and tolerability in human nutrition was published by Salimei and Fantuz [14].

Whereas milk from most domestic ruminants is widely consumed as cheese after processing, cheesemaking from equine milk is not possible, mainly due to its scarce casein content. No curd is formed on the addition of rennet; under acidic conditions only a weak coagulum appears, manufactured especially in the Netherlands for the production of yoghurt-type products with the addition of fruit extract [6,15].

Equine milk is rich in its lactose content, and since most populations in Asia present lactose malabsorption [16,17], in this part of the world mare milk undergoes fermentation before consumption, and the resulting products are called koumiss [18] or qymyz (Eastern

Europe, former Soviet Union Republics) [19,20], airag (Mongolia) [21,22] or chigee (Inner Mongolia and Xinjiang, China) [23,24], which are alcoholic beverages obtained by means of a mixed culture of bacteria and yeasts. During fermentation, the lactose is converted into lactic acid, ethanol and carbon dioxide, and the milk becomes an accessible nutriment for lactose-intolerant people. In addition, acidifying fermentation is the oldest method of milk conservation [25]. Organic residue analysis, using $\delta^{13}C$ and deuterium isotope (δD) values of fatty acids, has revealed processing of mare milk products in ancient potsherds dating back to about 3500 B.C.E., found in Northern Kazakhstan [26]. Kazakhstan is still at present the nation with the largest koumiss production in the world [27,28].

Nevertheless, until recently, little was known about the numerous studies carried out into koumiss in the former USSR, where this product has been consumed for centuries, due to the difficult accessibility of these papers, written in the Russian language. This lack of knowledge was overcome by a review by Kondybayev et al. [19] which surveyed many studies of soviet authors.

The reader can find a lot of data about the composition of and variability in mare milk in the aforementioned reviews. Recent research about mare milk deals mainly with its derivatives, in particular fermented milk products and the potential probiotic properties of the microorganisms living within [29].

The aim of this review is to present recent advancements in fermented mare milk research, with a particular focus on its potential functional properties and effects on human health.

The following search strategy for the review was adopted: to conduct the literature search and include relevant references, we scanned the databases Medline, PubMed, ScienceDirect, and CNKI for Chinese papers. For the search in the online databases, we used the following keywords: (koumiss), (kumiss), (airag), and (mare milk). In addition, since mare milk is one of the topics of our research group, we had collected most of the considered papers during the preceding years and therefore they were present in our collection of studies already. Regarding the recent literature, at the first screening process, we scanned titles and abstracts of the yielded articles or book chapters. Afterwards, if the publication did not show any signs of incoherence, we screened the full text and searched the sections "similar articles" or "related documents" according to the diverse databases. Three papers were excluded as they studied several traditional fermented foods obtained by the milk of diverse species (such as yak, cow, and camel) together, and it was not possible to select data regarding the specific product obtained by mare milk.

2. Microorganisms in Fermented Mare Milk Products

The most abundant microorganisms naturally present in milk can be classified in order of their possible roles (i) microorganisms involved in dairy fermentation (e.g., *Lactococcus*, *Lactobacillus*, *Streptococcus*, *Propionibacterium* and fungal populations); (ii) involved in spoilage (e.g., *Pseudomonas*, *Clostridium*, *Bacillus*, and other spore-forming bacteria); (iii) involved in food disease (e.g., *Listeria*, *Salmonella*, *Escherichia coli*, *Campylobacter* and mycotoxin-producing fungi); and (iv) involved in promoting health (e.g., lactobacilli and bifidobacteria) [30]. Human milk from healthy women contains up to 10^9 microbes L^{-1} [31]. These organisms come mainly from the nipple and surrounding skin [31]. Ward et al. studied the complexity of the bacterial community in human milk, finding over 360 prokaryotic genera, mainly belonging to the phyla of Proteobacteria (65%) and Firmicutes (34%), and the *genera* of *Pseudomonas* (61.1%), *Staphylococcus* (33.4%) and *Streptococcus* (0.5%) [32]. Microbial colonization during the first few weeks of life in the gastrointestinal tracts of humans and farm animals is remarkably similar. Bifidobacteria are the predominant lactic acid bacteria (LAB) in infants, whereas lactobacilli are the primary LAB in the tract of new-born foals [33].

The microbial composition of raw mare milk, like that in other similarly less common milk types, has not been studied in depth: nevertheless, it is known to be widely variable, depending on many factors, such as the breed, season, and region [34]. However, many studies regarding the microbiological composition of fermented mare milk are available.

Koumiss, airag and chigee are the main products obtained through mare milk fermentation. Mare milk fermentation is due to lactic acid bacteria (LAB) and yeast interaction. LAB are mainly involved in milk acidification, whereas the yeasts partly modify its titratable acidity due to the production of acidic compounds, such as acetic acid. On the other hand, yeasts produce ethanol that is very important for determining the properties and increasing the stability of fermented milk [21,35].

The interaction between LAB and yeast can be complex and has to be studied in more depth; however, different theories have been suggested. Positive relationships between the two types of microorganisms can occur because lactic acid bacteria are responsible for the lowering of the pH due to the secretion of organic acids, allowing the yeast population to become competitive in the immediate environment, followed by yeast fermentation [36]. Sudun et al. observed a positive correlation regarding glucose and galactose produced by the lactase of LAB, and consumed by yeast for their growth [21].

In addition, the interaction between the two types of microorganisms is reflected in the product. Regarding its safety, the combination of the acidic condition, saturated with carbon dioxide and alcohol, is inhibitory to many spoilage bacteria [36]. Regarding quality, yeast proteolytic and lypolitic metabolism can stimulate LAB growth and play an important role in aroma development [37].

Koumiss is a lactic acid–alcoholic beverage produced traditionally by the fermentation of mare's milk by indigenous organisms or by a starter culture. First, lactic acid is formed, and then alcoholic fermentation of the residual sugar content occurs. Koumiss generally contains about 2% alcohol, 0.5–1.5% lactic acid, 2–4% sugar and 2% fat (Kerr and McHale [38]; quoted by Danova et al. [39]). The alcohol has been known to reach a level of 3.5% and, depending on its content, in Russia there is a differentiation between weak koumiss with 0.7–1% ethanol, normal koumiss with 1–1.75% and strong koumiss with 1.75–2.5% [40].

According to Danova et al., another distinction exists: depending on the lactic acid content, three types of koumiss can be distinguished: 'strong', 'moderate' and 'light' koumiss. Lactic acid bacteria (*Lactobacillus delbrueckii* subsp. *bulgaricus* and *Lactobacillus rhamnosus*) acidifying the milk to pH 3.6–3.3 and converting about 80–90% of lactose into lactic acid produce 'strong' koumiss. In 'moderate' koumiss, other *Lactobacillus* (*Lb.*) bacteria (*Lb. acidophilus, Lb. plantarum, Lb. casei*, and *Lb. fermentum*) with lesser acidification properties and a conversion ratio of about 50% lower the pH to 4.5–3.9 at the end of the process. 'Light' koumiss, produced by *Streptococcus thermophilus* and *Str. cremoris*, is slightly acidic (pH 4.5–5.0) [39]. According to several authors, "moderate" koumiss presents a sweet–sour taste and a yeasty odour and is the most appreciated [39,41]. It was demonstrated that acetaldehyde is the most important substance determining aroma in koumiss and the suitable range of its content is $78.25-257.07~\mu L~L^{-1}$ [42,43].

Airag, which is called also tsege in Inner Mongolia, contains *Lb. helveticus, Lactobacillus fermentum*, and *Saccharomyces cerevisiae* [44]. The traditional technique of production in a cowhide vessel of this product in Mongolia was included in 2019 in the Representative List of the Intangible Cultural Heritage of Humanity.

In an attempt to identify which properties are related to the most appreciated sensory characteristics of airag, it was found that only its electrical conductivity has a statistically significant relationship with its taste score: higher electrical conductivity values are associated with lower taste scores; even though the mean pH values were not different, calcium and phosphorus concentrations were lower in the airag samples rated with high scores, taking in account 51 different airag samples exhibited in a competition in Mongolia [22].

Many mesophilic LAB and yeasts have been detected in koumiss. According to tradition, fresh milk is inoculated with a small quantity of already-fermented milk as natural starter, but many other different substances could be used as well [19]. The use of raw milk and natural undefined starter cultures causes a strong variability in the microbial composition. Many authors have observed a wide variability in species and strains [27,45]. It has been evidenced that milk derivatives from each family have their particular microbiota [46].

A few studies have investigated the relationship between koumiss microbiota and the production of volatile flavour compounds or organic acids, and their effects on taste. Despite the wide variability of microbes and yeasts involved, a "core" microbiota was identified, represented by four bacterial *genera* (*Lactobacillus*, *Acetobacter*, *Lactococcus*, and *Pseudomonas*), and two yeast *genera* (*Kazachstania* and *Candida*), and it was observed that notwithstanding the differences depending on the region and production techniques, the basic volatile flavours in traditional koumiss are similar [45,47].

Since this product is administered in hospitals, in the former USSR guidelines were started in 1969 to regulate its production according to a standard: in this case, a pure culture should be used as starter (with *L. bulgaricus* and *S. lactis*, which have antibacterial properties). Regardless, according to soviet authors, afterwards old koumiss was commonly used as starter [19].

Like in other fermented dairy products, the genus *Lactobacillus* (*Lb.*) plays an important role in affecting the aroma, texture, and acidity of koumiss. Recently, much interest has been focused on the usage and safety of these strains, as the properties of probiotics are more known and appreciated. For this reason, special attention has been paid to the accurate identification and characterization of a potential probiotic microorganism to use as the selected starter. Regarding probiotics selection, it is necessary to assess properties affecting specific health benefits, such as the modulation of the immune system, survival and persistence in the host, and proven safety and stability. Regarding LAB identification, the comparison of molecular sequences, mainly 16S rRNA-encoding genes, is commonly used even if it is not always effective to identify genetically close species. It was stated that if the 16S rRNA gene sequence identity shared by two microorganism is lower than 97%, at the genomic level they are considered to belong to different species. If the shared identity values are higher than 97% or the sequences are identical, the organisms appear closely related and total DNA–DNA hybridization data or more discriminative analyses are needed for species identification [48].

Regarding LAB characterization, several techniques can be used, from the more traditional phenotypic approach like whole-cell protein and cell wall composition analysis, and other morphological, physiological, and biochemical analyses [49] to more recent molecular and genomic characterization [50]. Methods for characterizing probiotics were recommended in the advisory report of the Working Group "8651 Probiotics" of the Belgian Superior Health Council (SHC) by Huys et al. [51].

2.1. Lactic Acid Bacteria

Studies regarding the isolation, identification, and characterization of lactic acid bacteria (LAB) in fermented mare milk have been carried out mainly by Asian research groups in the regions where this production has been a tradition for centuries, namely, Mongolia and Inner Mongolia and the autonomic Region of China.

These kinds of studies began with a notable frequency in the early years of the 2000s, and are still actively going on, probably with the intent to find the most suitable combination of LAB for the industrial production of horse milk derivatives.

The main LAB genera and species characterized in recent years are summarized in Table 1. Different methodological approaches were used, also in consideration of the advent of new increasingly powerful methods.

Table 1. Main LAB species isolated in different studies

			Main LAB Species	Identification Method	Product	Sampling Location	References
1	2						
10	258		Enterococcus faecium, Lactobacillus casei, Lactobacillus paracasei, Lactobacillus plantarum, Leuconostoc mesenteroides subsp. dextranicum, Streptococcus thermophilus	Growth condition, gas production, ammonia production, sugar fermentation, hippurate hydrolysis	chigee	Inner Mongolia	Burentegusi et al., 2002 [52]
14	117		Lactobacillus pentosus, Lactobacillus plantarum, Lactococcus lactis subsp. cremoris	Chemotyping, sugar fermentation, 16S rDNA sequencing	chigee	Inner Mongolia (Silinguole, Wulanchabu)	An et al., 2004 [23]
5	30		Lactobacillus acidophilus, Lactobacillus casei, Lactobacillus plantarum	Not mentioned	koumiss	Mongolia	Menghe et al., 2004 [53]
21	80		Lactobacillus acidophilus, Lactobacillus casei, Lactobacillus coryniformis, Lactobacillus curvatus, Lactobacillus fermentum, Lactobacillus kefiranofaciens, Lactobacillus paracasei, Lactobacillus plantarum, Weissella kandleri, Weissella paramesenteroides	Not mentioned	koumiss	Inner Mongolia	Menghe et al., 2004 [53]
2	7		Lactobacillus buchneri, Lactobacillus plantarum, Lactobacillus salivarius	Growth condition, gas production, ammonia production, sugar fermentation, hippurate hydrolysis	lyophilized koumiss	Mongolia	Danova et al., 2005 [39]
3	66		Lactobacillus acetotolerans, Lactobacillus casei, Lactobacillus homohiochii, Lactobacillus kefiranofaciens, Lactobacillus plantarum, Lactococcus lactis subsp. cremoris, Lactococcus lactis spp. lactis, Lactococcus raffinolactis, Leuconostoc mesenteroides subsp. cremoris	Growth condition, gas production, sugar fermentation	hurunge	Inner Mongolia	Shungquan et al., 2006 [54]
3			Lactobacillus farciminis, Lactobacillus helveticus, Lactobacillus kefiri, Lactobacillus paracasei, Lactobacillus plantarum	sugar fermentation, 16S rDNA sequencing	airag	Mongolia	Uchida et al., 2007 [46]
	12		Lactobacillus casei, Lactobacillus fermentum, Lactobacillus helveticus, Lactobacillus plantarum	Physiological tests and 16S rDNA sequencing	koumiss	China	Wang et al., 2008 [49]
22	183		Enterococcus faecium, Lactobacillus casei, Lactobacillus curvatus, Lactobacillus farciminis, Lactobacillus helveticus, Lactobacillus hilgardii, Lactobacillus kefiranofaciens, Lactobacillus kefiri, Lactobacillus parafarraginis, Lactobacillus plantarum, Lactococcus lactis subsp. lactis, Lactococcus spp, Leuconostoc mesenteroides, Leuconostoc pseudomesenteroides, Streptococcus thermophilus	RAPD-PCR multiplex and 16S rDNA sequencing	airag	Mongolia	Watanabe et al., 2008 [55]
16	48		Lactobacillus casei, Lactobacillus coryniformis, Lactobacillus curvatus, Lactobacillus fermentum, Lactobacillus helveticus, Lactobacillus kefiranofaciens, Lactobacillus paracasei, Lactobacillus plantarum, Weissella kandleri	Growth condition, sugar fermentation, acid resistance, bile tolerance, 16S rDNA sequencing	koumiss	Inner Mongolia	Wu et al., 2009 [56]
5	26		Lactobacillus casei, Lactobacillus helveticus, Lactobacillus plantarum	Growth condition, gas production, ammonia production, sugar fermentation, lactic acid isomers, 16S rRNA sequencing	airag	Mongolia	Sun et al., 2010 [57]
7	7		Lactobacillus delbrueckii subsp. Lactis, Lactobacillus fermentum, Lactobacillus helveticus, Lactobacillus kefiri, Lactobacillus pentosus, Lactobacillus plantarum, Lactobacillus sakei, Lactococcus lactis subsp. lactis, Leuconostoc citreum, Leuconostoc garlicum, Weissella confusa, Weissella viridescens	16S rDNA sequencing, acid resistance, bile tolerance, adhesion to Caco2 cells, RAPD-PCR	airag	Mongolia	Takeda et al., 2011 [58]

Table 1. Cont.

1	2	Main LAB Species	Identification Method	Product	Sampling Location	References
	18	Lactobacillus helveticus, Lactobacillus kefiranofaciens, Lactobacillus kefiri, Lactobacillus delbrueckii, Enterococcus faecium, Enterococcus durans	Growth condition, sugar fermentation, acid resistance, bile tolerance 16S rDNA sequencing BLAST search program	airag	Mongolia	Choi, 2016 [59]
	11	Lactococcus lactis, Lactobacillus buchneri, Enterococcus italicus, Lactobacillus homohiochii, Lactobacillus hilgardii, Lactobacillus helveticus, Leuconostoc mesenteroides, Streptococcus paurauberis	16S rDNA sequencing RAPD-PCR	koumiss	Inner Mongolia	Guo et al., 2019 [60]
	109	Lacticaseibacillus paracasei, Limosilactobacillus fermentum, Lacticaseibacillus casei, Lentilactobacillus diolivorans, Lactobacillus helveticus, Schleiferilactobacillus harbinensis, Leuconostoc mesenteroides, Lactobacillus kefiranofaciens, Lentilactobacillus parabuchneri, Staphylococcus epidermidis [3]	16S rRNA gene sequencing, PacBio SMRT sequencing technology	koumiss	Kyrgyzstan	Sun et al., 2022 [61]
15	26	Lactobacillus helveticus, Lactobacillus delbrueckii, Lactobacillus kefiranofaciens, Lentilactobacillus diolivorans, Lentilactobacillus kefiri [3]	16S rRNA gene sequencing, PacBio SMRT sequencing technology	koumiss	Uzbekistan	Sun et al., 2022 [61]
	14	Lacticaseibacillus paracasei, Lacticaseibacillus casei, Lacticaseibacillus rhamnosus, Lactobacillus delbrueckii, Lactobacillus delbrueckii subsp. bulgaricus, Enterococcus faecalis, Streptococcus thermophilus [3]	16S rRNA gene sequencing	koumiss	Kazakhstan (Almaty and Zhambyl regions)	Oleinikova et al., 2024 [62]

[1] Number of samples; [2] Number of isolates; [3] In the studies after 2022, a new nomenclature was adopted for LAB. See http://lactobacillus.ualberta.ca/, (accessed on 19 December 2023).

Most of the knowledge of LAB in fermented mare milk has been gained through culture-dependent methods, and the subsequent isolation and identification of these microorganisms. The number of isolated and identified strains is strongly variable in the manuscripts, depending on the aim of each piece of research, varying from 2 for koumiss [63] to 258 for chigee [52]. In Table 1, the results of 18 studies carried out with several methods are shown. Briefly summarizing the species identified with these approaches, it is possible to conclude that 45 different species have been identified, but the most common were *Lb. plantarum*, identified in 12 of the 18 studies; *Lb. casei*, identified in 10 studies; *Lb. helveticus*, found in 9 studies; *Lb. kefiranofaciens*, found in 7 studies; and *Lb. paracasei*, found in 6 studies (Table 1). Less frequent were the species *Lb. fermentum*, *Lb. kefiri*, and *Leuconostoc mesenteroides*, found in 5 studies; *Lb. diolivorans*, found in 4 studies; *Lc. lactis* subsp. *lactis*, *Enterococcus faecium* and *S. thermophilus*, found in 3 studies; and *Lactococcus lactis* subsp. *cremoris*, *Lb. acidophilus*, *Lb. coryniformis*, *Lb. curvatus*, *Lb. farciminis*, *Lb. pentosus*, *Weissella kandleri* and *Lb. delbrueckii*, identified in two studies (Table 1). The other species were found only one time (Table 1).

Recently, considerable efforts have been made to develop more rapid, culture-independent methods. One study was carried out with this approach and in particular by means of denaturing gradient gel electrophoresis (DGGE) [64]. The authors concluded that the biodiversity of ten samples of collected koumiss, made by nomadic families in one region of China, was high. In particular, the dominant species identified by DGGE were *Lb. acidophilus*, *Lb. helveticus*, *Lb. fermentum*, and *Lb. kefiranofaciens*. Less frequent were *Enterococcus faecalis*, *Lactococcus lactis*, *Lb. paracasei*, *Lb. kitasatonis*, and *Lb. kefiri*. *Leuconostoc mesenteroides*, *Streptococcus thermophilus*, *Lb. buchneri*, and *Lb. jensenii* were occasionally found [64].

Comparing all the data, it is possible to highlight that the only species found with high frequency using the two approaches was *Lb. helveticus*, a thermophilic, homofermentative, proteolytic species traditionally used both for the manufacture of Swiss-type cheeses and long-ripened Italian cheeses or in the production of fermented beverages in north Europe [65]. However, due to its high proteolytic activity, *Lb. helveticus* is very effective in the production of bioactive peptides such as angiotensin-converting enzyme (ACE) inhibitory peptides [66]. The strain heterogeneity of this species isolated from koumiss was evidenced by an intra-species genotypic and phenotypic characterization [67].

The other three dominant species for DGGE have also been isolated and identified by culture-dependent research, but surprisingly, the species isolated in 12 of the 15 culture-dependent studies, *Lb. plantarum*, has not been recognized in the samples of koumiss analysed with DGGE. Even if this aspect suggests some considerations about the advantages and disadvantages of both approaches, other culture-independent research should be conducted to discuss this comparison.

More recently, new next-generation sequencing technologies have been applied to explore genes implicated in microbial metabolism: shotgun metagenomic analyses provide taxonomic and functional data about complex microbial communities, with culture-independent methods [68]. In particular, a study investigated bacterial function during several phases of koumiss fermentation by metagenomics. It was observed that the microbial composition of koumiss changes mostly in the first 36 h of fermentation, and the predominant species is *Lb. helveticus* [68].

The predominance of this species was observed in another study, which analysed the metagenomes of 23 koumiss samples collected in several regions of China: sequences representing 216 different species were found and *Lb. helveticus* comprised 78.9% of the total sequences in koumiss, followed by *Lactobacillus kefiranofaciens* (6.0%) and *Lactococcus lactis* (4.2%) [69].

Analysing the bacterial metagenomes of koumiss from Mongolia and Inner Mongolia by the single-cell genomics technique allowed the identification of rare bacterial species never detected in koumiss before, such as *Lb. otakiensis* and *Streptococcus macedonicus*, both present in other fermented foods or dairy products [70].

Since new genetic methods provide more precise metrics to classify bacteria, the genus *Lactobacillus* was recently re-evaluated and new genus and species names were recently adopted [71]. Therefore, in the more recent studies, LAB names are in accordance with the new taxonomy rules (see note for Table 1) [61,62].

Technological properties and probiotic aptitudes have been considered for strains isolated from koumiss and airag. In particular, probiotic aptitude, such as bile tolerance and other preliminary tests, has been evaluated for strains of *Lb. casei*, *Lb. helveticus* and *Lb. plantarum* [56] and for different strains of *Lb. acidophilus* [72]. Different technological properties have been investigated in several species. The fermentation properties of four *Lb. casei* strains were studied by Xu et al. [73], and in *Lb. rhamnosus* and *Lb. paracasei* by Zuo et al. [74]. The latter authors were also able to verify the performances of those strains in cheese and yogurt manufacturing [74]. The effects of distinctive proteolytic activity on casein degradation have been studied for six *Lb. helveticus* strains isolated from home-made airag samples [75].

In airag collected in Mongolia, Watanabe et al. isolated two novel microorganisms belonging to the genus *Bifidobacterium*. Their phenotypic and genotypic characteristics demonstrated that the strains can be considered a single *Bifidobacterium* species never observed before, and its proposed name was *Bifidobacterium mongoliense* sp. nov. [76].

2.2. Yeasts

The yeasts in koumiss are the main aspect responsible for the presence of ethanol and carbon dioxide. The amount of ethanol in koumiss is slightly higher than in kefir because the amount of lactose in mare milk is higher than in cow milk [9]. However, not only are lactose-fermenting yeast species, belonging to the genus *Kluyveromices* (K.), mainly *K. marxianus* and *K. fragilis*, and *Candida* (C.) *kefir*, common in koumiss, but also non-lactose-fermenting species, such as *Saccharomyces unisporus* (S.), are found [77]. A study which explored the correlation between microflora and volatile flavour substances demonstrated a correlation with the *genus Candida* and ethanol, considered the most important alcohol flavour compound [47].

In the manufacture of koumiss, a considerable amount of free amino acids are produced by yeasts, ranging around 470–490 mg kg^{-1} [40,43].

Few studies are available about the yeast composition of fermented mare milk. To the authors' knowledge, six studies have been conducted on koumiss [24,37,47,60,77,78], one on chigee [79], one on airag [46] and one on hurunge which is a the starter culture for fermented traditional dairy products such as chigee [54]. To briefly summarise, 11 genera and 24 species were isolated but only non-lactose-fermenting species *S. cerevisiae* was isolated in all the studies, and lactose-fermenting *K. marxianus* was found in seven of the nine studies. Lactose-fermenting species *C. krusei* and non-lactose-fermenting yeasts such as *Pichia* (P.) *membranaefaciens*, *C. kefyr*, *C. valida*, *Dekkera anomala*, *Kazachstania unispora*, and *Issatchenkia orientalis* were found in two studies, and the other species, *C. buinensis*, *C. pararugosa*, *Geotrichum* sp., *K. wickerham*, *P. cactophila*, *P. deserticola*, *P. fermentans*, *P. manshurica*, *P. membranaefaciens*, *S. dairensis*, *S. servazzii*, *S. unisporus*, *Trichosporum asaii*, *Penicillium carneum*, *Clavispora lusitaniae* and *Torulaspora delbrueckii*, were found in only one study (Table 2).

Interestingly, a polyphasic approach was used in a complete and complex study, using a culture-independent and also culture-dependent method, to study yeasts present in koumiss sampled from three representative regions of China, Mongolia, Xin Jiang and Qing Hai [37]. Using 96 samples, 655 isolates and also DGGE, the authors were able to show how the yeast community in koumiss is complex and rich in different species [37].

Table 2. Main yeast species isolated from fermented mare milk in different studies.

[1]	[2]	Main Yeast Species	Identification Method	Product	Sampling Location	References
94	417	Candida buinensis, Kluyveromyces marxianus, Saccharomyces cerevisiae, Saccharomyces unisporus	morphological and physiological tests	koumiss	Kazakhstan	Montanari et al., 1996 [77]
3	nd	Kulyveromyces wickerham, Issatchenkia orientalis, Saccharomyces cerevisiae, Saccharomyces dairensis	sugar fermentation, 26S rDNA sequencing	airag	Mongolia	Uchida et al., 2007 [46]
96	655	Candida pararugosa, Dekkera anomala, Geotrichum sp., Issatchenkia orientalis, Kazachstania unispora, Kluyveromyces marxianus, Pichia deserticola, Pichia fermentans, Pichia manshurica, Pichia membranaefaciens, Saccharomyces cerevisiae, Torulaspora delbrueckii	5.8S-ITS rDNA and 26S rDNA	koumiss	Mongolia, Xin Jiang, Qing Hai	Mu et al., 2012 [37]
5	108	Candida kefyr, Candida krusei, Candida valida, Kluyveromyces marxianus, Pichia cactophila, Saccharomyces cerevisiae, Saccharomyces seroazzii	API ID and physiological tests	chigee	Mongolia	Sudun et al., 2010 [79]
28	87	Kluyveromyces marxianus, Pichia membranaefaciens, Saccharomyces cerevisiae, Saccharomyces unisporus	biochemical tests, 26S rDNA sequencing	koumiss	Xin Jiang	Ni et al., 2007 [24]
3	30	Candida kefyr, Candida krusei, Candida valida, Kluyveromyces marxianus, Saccharomyces cerevisiae	API ID and physiological tests	hurunge	Inner Mongolia	Shungquan et al., 2006 [54]
11		Kluyveromyces marxianus, Kazachstania unispora, Dekkera anomala, Saccharomyces cerevisiae, Trichosporum asaii, Penicillium carneum, Pichia membranaefaciens, Clavispora lusitaniae	ITS rDNA, 16S rDNA sequencing	koumiss	Inner Mongolia	Guo et al., 2019 [60]

[1] Number of samples; [2] number of isolates.

Another research article showed that the prevalent yeast found at high altitudes in Kazakhstan was *S. unisporus*, which was different from lower zones, where lactose-fermenting yeasts are more common, mostly belonging to the *Kluyveromyces genus* [77]. The authors hypothesized that a high altitude could affect the LAB population composition during the preparation of koumiss, with a selection of LAB not metabolizing galactose, causing therefore an enrichment of this sugar and the prevalence of *C. buinensis*, which ferments galactose but not lactose, and of the *Saccharomyces genus* compared with *Kluyveromyces* [77].

In traditional koumiss from Inner Mongolia, Guo et al. identified 57 fungal species, also, among them the genera *Penicillium*, *Cladosporium* and *Aspergillus* were detected in all samples. These are filamentous fungi considered the cause of spoilage in dairy products. Therefore, the production process of traditional koumiss requires sanitation measures and a better control of environmental hygiene [60].

In conclusion, it could be observed that in comparison with fermented dairy products from other species, in mare milk the higher lactose content (6–7% compared to 4–5% of cow, yak, goat and camel's milk) usually determines the prevalence of lactose-fermenting yeast strains, but this trait could be modified by the altitude effect, which, when operating in a selection of LAB, can increase the galactose content, causing therefore the major presence of galactose-fermenting yeast strains.

3. Potential Functional Properties of Microorganisms in Fermented Mare Milk for Human Consumption and Health

Probiotics are defined as "living microorganisms, which upon ingestion in certain numbers, exert health effects beyond inherent basic nutrition" [80]. Probiotic cultures in nutritional supplements, pharmaceuticals and functional foods are mainly constituted by LAB belonging to the genera *Lactobacillus* and *Bifidobacterium*. The proteolytic systems of LAB are exploited for the production of bioactive peptides in fermented dairy products, because milk proteins need to be hydrolysed into peptides to exert some effect [81].

It is believed, but not yet definitively proven, that probiotics compete with undesirable microorganisms and can inhibit their growth in the intestine. Therefore, probiotics must survive through the stomach and upper intestine, tolerating the acidic and protease-rich environment and the action of bile salts, and need to be numerous enough to exert their effect in the colon [82]. According to several authors, a concentration of at least 10^6 c.f.u.g^{-1} of viable and active microorganisms is necessary in the product throughout its specified shelf life [83].

Many recent studies regard koumiss as a source of microorganisms with potential probiotic activities, and trials have been carried out to provide data about these properties [84].

In Figure 1, a synthesis of the main potential functional properties of microorganisms in fermented mare milk is shown.

A Chinese research group has been particularly active in this research field: during the years 2004–2009, 240 *Lactobacillus* strains were isolated from koumiss and investigated. Among these strains, a novel strain, called *Lb. casei* Zhang, was screened out and its potential probiotic properties were investigated in many trials, reported initially in national and successively in international journals. According to the combined analyses of a phylogenetic dendrogram and partial sequences of 16S rDNA, this strain was classified as *Lb. casei* subsp. *casei* [1,63,73]. The complete genome sequence of *Lb. casei* Zhang was successively investigated by a whole-genome shotgun strategy, and comparative analyses with other *Lb* strains evidenced a richer enzyme abundance, which could explain some of its favourable properties in terms of its use of different sugar sources and behaviour in the host environment [85]. Moreover, proteins expressed by *Lb. casei* Zhang in the exponential and stationary phases were identified and characterized, mainly as stress response proteins, in a proteomics study, evidencing their role in its adaptation to the environment [56]. Therefore, the survival capacity of this strain was initially tested: yoghurt samples fermented with *Lb. casei* Zhang showed a similar viable count (1.0×10^9 c.f.u. mL^{-1}) as other samples inoculated with selected commercial probiotics after 28 d of refrigerated storage [83]. Similar results were obtained by Zhou et al. [1].

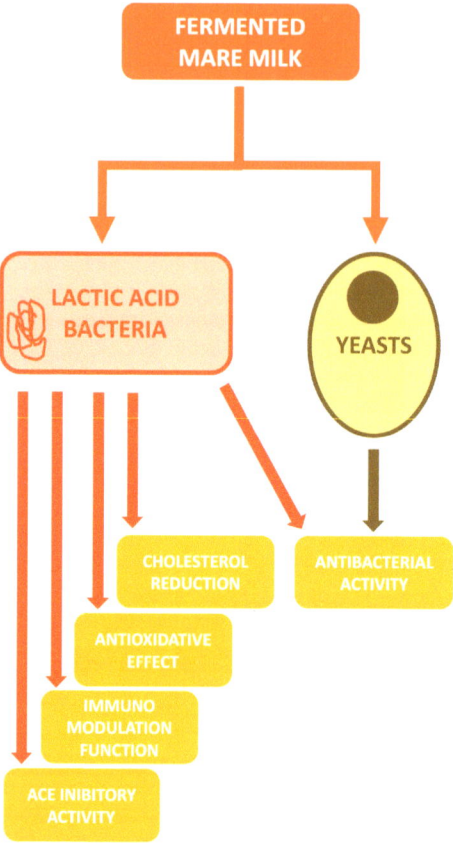

Figure 1. Main potential functional properties of microorganisms in fermented mare milk for human consumption and health.

In the following years, the properties of this strain were investigated with in vivo experiments on rats and on humans; some of the results of these studies, showing positive effects on several health problems, are listed by He et al. in the introduction of their paper, concerning the effects of the long-term administration of *Lb. casei* Zhang on human gut microbiota (see next section) [86].

Particular emphasis about the potential properties of the microorganisms from koumiss in the prevention of chronic diseases is shown in a recent review by Xue et al. [84].

The following sections consider the results of studies aimed at assessing several properties of LAB strains from koumiss as potential probiotics.

3.1. Survival of LAB through the Human Digestive Tract and Cholesterol Reduction Effect

The survival capacity of *Lactobacillus* strains isolated from mare milk products through the human digestive tract is considered in several studies. In particular, survival under acidic conditions, similar to gastric juice, and resistance to bile salts have been investigated. Bile salts have antibacterial action but do not damage resident microflora. The physiological concentration of human bile ranges from 0.3 to 0.5% (wt vol^{-1}) [87]. It was recently observed that some *Lactobacillus* strains excrete bile salt hydrolase, an enzyme that catalyses the hydrolysis of conjugated bile salts. This action could reduce serum cholesterol levels, together with various other mechanisms [88]. According to Guo et al. [83], in vitro tests demonstrated *Lb. casei* Zhang's tolerance to simulated gastric and intestinal juices and in the presence of 0.3% bile salts. The molecular mechanisms involved in the adaptation of the bacterial cells under bile salt stress were investigated considering the growth and protein expression patterns of *Lb. casei* Zhang with and without bile salts. It was observed that twenty-six proteins were differentially expressed using two-dimensional gel electrophoresis. These proteins were identified by peptide mass fingerprinting [87].

In the empty human stomach, pH is lower than 3.0 [89]. The tolerance to a low pH of *Lb. casei* strains from home-made koumiss was investigated considering the expression of H^+-ATPase, which is supposed to play an important role in the maintenance of physiological cytoplasmic pH, by means of a mechanism controlling the H^+ concentration through the cell membrane. The survival of the *Lb. casei* Zhang strain in artificial digestion was observed and the expression of H^+-ATPase, detected by the reverse transcription–polymerase chain reaction method, increased accordingly to the pH lowering, in accordance with what has been observed regarding other microorganisms. Therefore, it seems that the acid tolerance of *Lb. casei* could have some relationship with the H^+-ATPase gene [89].

It is assumed that an LAB strain could tolerate gastric acid if, after exposure to pH 3.0 with 0.04% pepsin for 3 h, it is detected at over 7 log c.f.u. mL^{-1}. *Lb. plantarum* 05AM23 isolated from Mongolian airag showed good bile acid tolerance (97% viable colonies), viability in low pH (8.1 log c.f.u. mL^{-1}) and a high capacity of adhesion on Caco-2 cells (72.0×10^3 c.f.u. mL^{-1}) [58].

There is an increasing interest in the research of natural compounds in food with effective abilities to decrease serum cholesterol concentrations, especially in countries where coronary heart disease is the principal cause of death, such as China. It has been known for many years that several *Lactobacillus* strains exert this property by means of different mechanisms, mostly studied by in vitro experiments [90,91]. According to some studies, which were published in Chinese national journals, several *Lb* strains have been observed in diverse acidic conditions (pH 3.0 and pH 4.0) and bile salt concentrations (e.g., in media containing 0.6%, 0.4% and 0.3% bile salts) in vitro: some of them showed sufficient tolerance and the cholesterol removed from the growth media was around 50% [72,92]. These properties were confirmed for the strain *Lb. helveticus* MG2-1, which, moreover, showed a good adhesiveness to Caco-2 cells, an important property to be assessed as a potential probiotic [93].

Evaluation of the cholesterol-reducing effects of LAB has been performed in vivo as well by several studies; e.g., the acid and bile tolerance and cholesterol reduction activity of *Lb. fermentum* SM-7, isolated from home-made koumiss in Xinjiang, China, were observed in vitro and in vivo in artificially induced hyperlipidaemial ICR mice. The cholesterol reduction rate in vitro observed in this study was 66.8%, and a significant decrease in serum total cholesterol was also observed in vivo, in the groups treated with high and low doses of *Lb. fermentum* SM-7 [90].

Nevertheless, observing the behaviour of three different strains, screened out from 68 *Lb* strains from koumiss produced in the same area mentioned above, it was concluded that the results in vitro were not consistent with what was observed in vivo in mice, and a precise explanation of the mechanisms of the hypocholesterolemic action of LAB is seemingly quite difficult, also because different strains show different cholesterol-lowering actions [91].

Regarding this issue, Zhong et al. [94], carrying out one of the aforementioned numerous studies conducted in China into *Lb. casei* Zhang, investigated in rats how the expression of several genes involved in fatty acid metabolic processes was differently affected by the administration of *Lb. casei* Zhang, therefore providing some information about the molecular mechanism of the cholesterol-lowering effect exerted by this bacterium [94].

Another quite recent study demonstrated, by metagenomic and metatranscriptomic profiling, that in the human gut, the gene expression of *Lb. casei* Zhang was very different from what was observed in vitro. Individual variabilities in the intestinal microbiomes among the healthy volunteers were higher than those induced by the probiotic ingestion. Therefore, interactions between probiotics and resident microbiota are a further step in the study of the effects of these promising strains [95]. Moreover, gut microbiota vary along the different life phases, and the differences in the represented phyla in the elderly gut, in comparison with those of younger adults, are well known.

In relation to this issue, He et al. studied the effects of the long-term administration of this strain on the gut microbiota of healthy adults, divided into two groups according to age. Modulating and stabilizing effects of *Lb. casei* Zhang on gut microbiota were observed in both groups, but with age-dependent differences. According to some indices, the long-time consumption of *Lb. casei* Zhang changed the microbiota composition in the older adult group, making the microbial community more similar to that of younger people [86].

Another study [96] considered thirteen patients, diagnosed with severe hyperlipidaemia, who consumed 750 g of koumiss every day for 60 days. The composition of the koumiss microbiota and the effects of koumiss consumption on patients' faecal sample microbiota were investigated by the Pacific Biosciences single-molecule, real-time sequencing technology (SMRT), a state-of-the-art tool which permits the profiling of microbiota. This technique produces long-read sequences, allowing high-resolution taxonomic identification by sequencing the full-length 16S rRNA gene. After 60 days of koumiss consumption, HDL cholesterol values increased significantly, and hyperlipidaemia-associated symptom scores decreased significantly. These results were attributed to the bacterial population of koumiss and its metabolites. According to several indices, participants' faecal samples reflected a significant increase in the abundance and diversity of gut microbiota. This was the first study which observed the relationship between koumiss consumption and gut microbiota at the species level in hyperlipidaemic patients [96].

3.2. Antioxidative Effect

The action of binding toxins is among the favourable properties attributed to fermented mare milk and its microorganisms, operating like natural chelators to remove the pollutants from the body. This property and the antioxidative effect of koumiss and its LAB strains have been investigated in a rat model: the strain *Lb. acidophilus* MG2-1, heat-killed or living, extracted from koumiss, was orally administered for 28 days and an antioxidative effect was observed in serum and in rat liver tissue homogenate. In rats fed living *Lb. acidophilus* MG2-1, the activity of superoxide dismutase (SOD) and of glutathione peroxidase (GSH-PX) significantly increased in liver tissue and in serum, while the content of malondialdehyde in serum and liver tissue homogenate decreased [97].

The oxidative stress effects and damage induced in the rats, which received 25 ppm mercury ($HgCL_2$) in drinking water for 6 weeks, was alleviated in the rat group fed fermented mare's milk prepared with starter cultures of *Streptococcus thermophilus*, *Lb. acidophilus* and *Bifidobacterium bifidum* and supplemented with fibre (6% extract of Dandelium root, *Taraxacum officinalis*). Nevertheless, the mechanism of action was not explained, and the authors declared that it was still under investigation [98].

3.3. Immuno-Modulation Function

Some studies have been performed to assess a potential immuno-modulating function of probiotic species isolated from koumiss [99]. Among them, five studies considered the effects of live or heat-killed probiotic *Lb. casei* Zhang administration on healthy or liver-injured rats or mice.

For example, varying doses of *Lb. casei* Zhang, isolated from koumiss, were orally administered to healthy BALB/c mice and positive effects on several parameters of immune response were observed, such as the increased production of interferon-γ (IFN-γ) and decreasing levels of tumour necrosis factor-α (TNF-α). Moreover, interleukin-2 (IL-2) and IL-2 receptor gene transcription increased, and the production of secretory Immunoglobulin A (sIgA) was enhanced. It was concluded that the dose-dependent administration of living

Lb. casei Zhang influences immune responses in mice and could be taken into account for the probiotic's use for humans as well [100,101].

The research group of Wang et al. also observed several protective actions of *Lb. casei* Zhang in rats with induced liver injury, and demonstrated a positive effect of *Lb. casei* Zhang upon pro-inflammatory cytokine and hepatic inflammation in rats with acute liver failure [102–104].

3.4. ACE Inhibitory Activity

Angiotensin I-converting enzyme (ACE; dipeptidyl carboxypeptidase, EC3.4.15.1) is a multifunctional enzyme which plays a role in the conversion of angiotensin-I to angiotensin-II and in the degradation of bradykinin, causing increased peripheral blood pressure [105]. It has been demonstrated that the proteolytic systems of several LAB produce bioactive peptides exerting an inhibitory activity on the ACE, with a hypotensive effect in the rat model and in clinical studies. In particular, *Lb. helveticus* cell-wall proteases can activate antihypertensive tripeptides from the hydrolysis of casein. This capacity is strain-dependent and the effect can be exerted by milk fermented by these strains [106,107].

In koumiss collected in the Xilingole region in Inner Mongolia, four ACE-inhibitory peptides, called P_I, P_K, P_M and P_P, were identified and analysed: P_I was derived from mare milk β-casein (f213–241), and the peptide P_K was derived from *Actinobacillus succinogenes*, which had not been previously found in koumiss, whereas the other two peptides did not correspond to any known milk peptides or proteins. The in vitro studies demonstrated the thermal stability of the four peptides and a high ACE-inhibitory activity, maintained under various pH and ACE treatments [18].

Another antihypertensive property of some LAB strains is the production of gamma-aminobutyric acid (GABA), a neurotransmitter which is effective in lowering blood pressure. Sixteen *Lactobacillus* strains isolated from koumiss collected in Xinjiang, China, demonstrated high ACE inhibitory activity, and among them two strains produced GABA: the *Lb.*–ND01 strain, which showed 99% homology to *Lb. helveticus* according to the sequence of 16S rDNA, possesses both properties, and after further studies regarding its resistance to acidic conditions and its production of free amino nitrogen, it was considered interesting for its possible employment in probiotic dairy products [106].

3.5. Antibacterial Activity of LAB and Yeasts

The growth of pathogenic bacteria could be inhibited by particular toxins, produced by LAB or yeasts present in several fermented foods.

Many LAB produce a high diversity of bacteriocins, proteinaceous or peptidic toxins categorized in several ways, which kill or inhibit the growth of similar bacterial strains. Antimicrobial activity was observed in as many as 53 LAB strains isolated from airag. Class II bacteriocins of LAB are considered to have a large potential for food preservation, due to their anti-*Listeria* activity and physicochemical properties [108,109]. Bacteriocins produced by LAB isolated from fermented mare's milk were characterized and tested for their antibacterial effect against yeasts and spoilage bacteria. It was concluded that bacteriocins A5-11A and B, isolated from *Enterococcus durans* issued from airag, could be used in food preservation due to their anti-pathogenic-bacteria activity, whereas yeasts were not inhibited [110]. Another interesting bacteriocin was identified in airag from the strain *Leuc. mesenteroides* 406, which showed a narrow antimicrobial spectrum but is effective against *Listeria monocytogenes* and *Clostridium botulinum* [108].

Recently, it was demonstrated that a bacteriocin produced by *Lb. plantarum* NMD-17, isolated from koumiss from Inner Mongolia, has a broad spectrum against Gram-positive and Gram-negative bacteria [111], and a bacteriocin produced by *Lb. rhamnosus* 1.0320, also extracted from koumiss, besides its wide-spectrum action, is particularly active against *E. coli* UB1005 [112].

Whereas the antimicrobial properties of LAB are widely documented, less is known about similar properties in yeasts extracted from koumiss. Particular yeast strains produce toxins called mycocins: compounds extracted by *Saccharomyces cerevisiae* from koumiss exerted antibacterial activity on pathogenic *Escherichia coli* in in vitro tests [113], and crude extracts of mycocins from *K. marxianus* showed efficacy against *Escherichia coli* in mice in vivo as well [114].

4. Conclusions

Research about mare's milk and derivatives for human consumption is very active especially in Asian countries. The identification and characterization of LAB and yeasts have been performed in many studies: a wide variability of species and strains have been observed. Even if fermented mare's milk is a very ancient product, a lot of research is still needed: whereas the yeast flora are quite well known, a wide range of scientific activity remains to be carried out to clarify the interdependency and cooperation of its various microbial components, because in fermented mare milk products really complex population structures of yeasts, white moulds, lactic acid bacteria and acetic acid bacteria are present and interacting.

Studies are actively going on to assess the properties of different strains for their optimal employment in the production, at an industrial scale, of koumiss, in particular.

Trials in vitro and in vivo are confirming some of the beneficial effects and healing properties traditionally attributed to mare milk derivatives: some probiotic properties of LAB in fermented milk have been demonstrated, but more necessary experimental steps must be performed to clearly assess their efficacy.

Author Contributions: Conceptualisation, F.M. and P.F. (Piero Franceschi); methodology, F.M., P.F. (Piero Franceschi) and P.F. (Paolo Formaggioni); software, P.F. (Piero Franceschi), F.M. and P.F. (Paolo Formaggioni); formal analysis, F.M. and P.F. (Piero Franceschi); investigation, F.M., P.F. (Piero Franceschi) and P.F. (Paolo Formaggioni); resources, F.M., P.F. (Paolo Formaggioni) and P.F. (Piero Franceschi); data curation, F.M. and P.F. (Piero Franceschi); writing—original draft preparation, F.M., P.F. (Piero Franceschi) and P.F. (Paolo Formaggioni); writing—review and editing, F.M., P.F. (Piero Franceschi) and P.F. (Paolo Formaggioni); visualisation, F.M. and P.F. (Paolo Formaggioni); supervision, F.M. and P.F. (Piero Franceschi); project administration, F.M., P.F. (Piero Franceschi) and P.F. (Paolo Formaggioni). All authors have read and agreed to the published version of the manuscript.

Funding: This research received no external funding.

Institutional Review Board Statement: Not applicable.

Informed Consent Statement: Not applicable.

Data Availability Statement: Data are contained within the article.

Acknowledgments: The authors wish to thank Monica Gatti for her advice and valuable references.

Conflicts of Interest: The authors declare that there are no conflicts of interest in this research article.

References

1. Zhou, Q.; Wang, J.; Guo, Z.; Yan, L.; Zhang, Q.; Chen, W.; Liu, X.M.; Zhang, H. Fermentation characteristics and transit tolerance of *Lactobacillus casei* Zhang in reconstituted mare milk during storage. *Int. J. Dairy Technol.* **2009**, *62*, 249–254. [CrossRef]
2. Foekel, C.; Schubert, R.; Kaatz, M.; Schmidt, I.; Bauer, A.; Hipler, U.C.; Vogelsang, H.; Rabe, K.; Jahreis, G. Dietetic effects of oral intervention with mare's milk on the Severity Scoring of Atopic Dermatitis on faecal microbiota and on immunological parameters in patients with atopic dermatitis. *Int. J. Food Sci. Nutr.* **2009**, *60*, 41–52. [CrossRef]
3. Schubert, R.; Kahle, C.; Kauf, E.; Hofmann, J.; Hobert, I.; Gruhn, B.; Häfer, R.; Vogelsang, H.; Jahreis, G. Dietetic efficacy of mare's milk for patients with chronic inflammatory bowel diseases—Clinical study. *Ernährung* **2009**, *33*, 314–321.
4. Kushugulova, A.; Kozhakhemtov, S.; Sattybayeva, R.; Nurgozhina, A.; Ziyat, A.; Yadav, H.; Marotta, F. Mare's milk as a prospective functional product. *Funct. Foods Health Dis.* **2018**, *8*, 537–543. [CrossRef]
5. Martuzzi, F.; Doreau, M. Mare milk composition: Recent findings about protein fractions and mineral content. In *Nutrition and Feeding of the Broodmare*; Miraglia, N., Martin-Rosset, W., Eds.; EAAP Scientific Series; Wageningen Pers: Wageningen, The Netherlands, 2006; Volume 120, pp. 65–76.
6. Uniacke-Lowe, T.; Huppertz, T.; Fox, P.F. Equine milk proteins: Chemistry, structure and nutritional significance. *Int. Dairy J.* **2010**, *20*, 609–629. [CrossRef]

7. Doreau, M.; Martuzzi, F. Fat content and composition of mare's milk. In *Nutrition and Feeding of the Broodmare*; Miraglia, N., Martin-Rosset, W., Eds.; EAAP Scientific Series; Wageningen Pers: Wageningen, The Netherlands, 2006; Volume 120, pp. 77–87.
8. Curadi, M.C.; Giampietro, P.G.; Lucenti, P.; Orlandi, M. Use of mare milk in pediatric allergology. In Proceedings of the XIV Congress of Associazione Scientifica di Produzione Animale (ASPA), Firenze, Italy, 12–15 June 2001; pp. 647–649.
9. Malacarne, M.; Martuzzi, F.; Summer, A.; Mariani, P. Nitrogen fractions and fat composition of mare milk: Some nutritional remarks with reference to woman and cow milk. *Int. Dairy J.* **2002**, *12*, 869–877. [CrossRef]
10. Tidona, F.; Sekse, C.; Criscione, A.; Jacobsen, M.; Bordonaro, S.; Marletta, D.; Vegarud, G.E. Antimicrobial effect of donkeys' milk digested in vitro with human gastrointestinal enzymes. *Int. Dairy J.* **2011**, *21*, 158–165. [CrossRef]
11. Muraro, M.A.; Giampietro, P.G.; Galli, E. Soy formulas and nonbovine milk. *Ann. Allergy Asthma Immunol.* **2002**, *89* (Suppl. S1), 97–101. [CrossRef]
12. Monti, G.; Bertino, E.; Muratore, M.C.; Coscia, A.; Cresi, F.; Silvestro, L.; Fabris, C.; Fortunato, D.; Giuffrida, M.G.; Conti, A. Efficacy of donkey's milk in treating highly problematic cow's milk allergic children: An in vivo and in vitro study. *Pediatr. Allergy Immunol.* **2007**, *18*, 258–264. [CrossRef] [PubMed]
13. La Torre, G.L.; Saitta, M.; Potortì, A.G.; Di Bella, G.; Dugo, G. High performance liquid chromatography coupled with atmospheric pressure ionization mass spectrometry for sensitive determination of bioactive amines in donkey milk. *J. Chromatogr. A* **2010**, *1217*, 5215–5224. [CrossRef] [PubMed]
14. Salimei, E.; Fantuz, F. Equid milk for human consumption. *Int. Dairy J.* **2012**, *24*, 130–142. [CrossRef]
15. Bornaz, S.; Guizani, N.; Sammari, J.; Allouch, W.; Sahli, A.; Attia, H. Physicochemical properties of fermented Arabian mares' milk. *Int. Dairy J.* **2010**, *20*, 500–505. [CrossRef]
16. Wang, Y.G.; Yan, Y.S.; Xu, J.J.; Du, R.F.; Flatz, S.D.; Kühnau, W.; Flatz, G. Prevalence of primary adult lactose malabsorption in three populations of northern China. *Hum. Genet.* **1984**, *67*, 103–106.
17. Itan, Y.; Powell, M.A.; Beaumont, M.A.; Burger, J.; Thomas, M.G. The origins of lactase persistence in Europe. *PLoS Comput. Biol.* **2009**, *5*, e1000491. [CrossRef]
18. Chen, Y.; Wang, Z.; Chen, X.; Liu, Y.; Zhang, H.; Sun, T. Identification of angiotensin I-converting enzyme inhibitory peptides from koumiss, a traditional fermented mare's milk. *J. Dairy Sci.* **2010**, *93*, 884–892. [CrossRef]
19. Kondybayev, A.; Loiseau, G.; Achir, N.; Mestres, C.; Konuspayeva, G. Fermented mare milk product (Qymyz, Koumiss). *Int. Dairy J.* **2021**, *119*, 105065. [CrossRef]
20. Konuspayeva, G.; Baubekova, A.; Akhmetsadykova, S.; Faye, B. Traditional dairy fermented products in Central Asia. *Int. Dairy J.* **2023**, *137*, 105514. [CrossRef]
21. Sudun, W.; Arakawa, K.; Miyamoto, M.; Miyamoto, T. Interaction between lactic acid bacteria and yeasts in airag, an alcoholic fermented milk. *Anim. Sci. J.* **2013**, *84*, 66–74. [CrossRef] [PubMed]
22. Tsuchiya, R.; Kawai, T.; Bat-Oyun, T.; Shinoda, M.; Morinaga, Y. Electrical conductivity, pH, minerals, and sensory evaluation of airag (Fermented Mare's Milk). *Foods* **2020**, *9*, 333. [CrossRef] [PubMed]
23. An, Y.; Adachi, Y.; Ogawa, Y. Classification of lactic acid bacteria isolated from chigee and mare milk collected in Inner Mongolia. *Anim. Sci. J.* **2004**, *75*, 245–252. [CrossRef]
24. Ni, H.J.; Bao, H.H.; Sun, T.S.; Chen, X.; Zhang, H.P. Identification and biodiversity of yeasts isolated from koumiss in Xinjiang of China. *Wei Sheng Wu Xue Bao (Acta Microbiol. Sin.)* **2007**, *47*, 578–582.
25. Alais, C. *Scienza del Latte*, 4th ed.; Tecniche Nuove: Milano, Italy, 2000; p. 717.
26. Outram, A.K.; Stear, N.A.; Bendrey, R.; Olsen, S.; Kasparov, A.; Zaibert, V.; Thorpe, N.; Evershed, R.P. The earliest horse harnessing and milking. *Science* **2009**, *323*, 1332–1335. [CrossRef] [PubMed]
27. Dugan, F.M. Dregs of our forgotten ancestors. Fermentative microorganism in the prehistory of Europe, the steppes and Indo-Iranian Asia and their contemporary use in traditional and probiotic beverages. *Fungi* **2009**, *24*, 16–39.
28. Siddiqui, S.A.; Salman, S.H.M.; Redha, A.A.; Zannou, O.; Chabi, I.B.; Oussou, K.F.; Bhowmik, S.; Nirmal, N.P.; Maqsood, S. Physicochemical and nutritional properties of different non-bovine milk and dairy products: A review. *Int. Dairy J.* **2023**, *148*, 105790. [CrossRef]
29. Martuzzi, F.; Vaccari Simonini, F. Advances on equine milk and derivatives for human consumption. In Proceedings of the 61st Annual Meeting of the European Association of Animal Production, Heraclion, Greece, 23–27 August 2010; Wageningen Academic Publisher: Wageningen, The Netherlands, 2010; p. 40.
30. Quigley, L.; O'Sullivan, O.; Stanton, C.; Beresford, T.P.; Ross, R.P.; Fitzgerald, G.F.; Cotter, P.D. The complex microbiota of raw milk. *FEMS Microbiol. Rev.* **2013**, *37*, 664–698. [CrossRef] [PubMed]
31. Mackie, R.I.; Abdelghani, S.; Gaskins, H.R. Developmental microbial ecology of the neonatal gastrointestinal tract. *Am. J. Clin. Nutr.* **1999**, *69*, 1035S–1045S. [CrossRef] [PubMed]
32. Ward, T.L.; Hosid, S.; Ioshikhes, I.; Altosaar, I. Human milk metagenome: A functional capacity analysis. *BMC Microbiol.* **2013**, *13*, 116. [CrossRef] [PubMed]
33. Lawrence, L.A.; Lawrence, T.J. Development of the equine gastrointestinal tract. In *Advances in Equine Nutrition IV*; Pagan, J.D., Ed.; Nottingham University Press—Kentucky Equine Research: Versailles, Kentucky, USA, 2008; pp. 173–184.
34. Kondybayev, A.; Konuspayeva, G.; Strub, C.; Loiseau, G.; Mestres, C.; Grabulos, J.; Manzano, M.; Akhmetsadykova, S.; Achir, N. Growth and metabolism of *Lacticaseibacillus casei* and *Lactobacillus kefiri* isolated from qymyz, a traditional fermented central Asian beverage. *Fermentation* **2022**, *8*, 367. [CrossRef]

35. Guo, L.; Xu, W.; Li, C.; Ya, M.; Guo, Y.; Qian, J.; Zhu, J. Production technology, nutritional, and microbiological investigation of traditionally fermented mare milk (Chigee) from Xilin Gol in China. *Food Sci. Nutr.* **2020**, *1*, 257–264. [CrossRef]
36. Viljoen, B. Yeast ecological interactions. Yeast-yeast, yeast-bacteria, yeast-fungi interactions and yeasts as biocontrol agents. In *Yeasts in Food and Beverages*; Querol, A., Fleet, G., Eds.; Springer: Berlin, Germany, 2006; pp. 83–110.
37. Mu, Z.; Yang, X.; Yuan, H. Detection and identification of wild yeast in Koumiss. *Food Microbiol.* **2012**, *31*, 301–308. [CrossRef]
38. Kerr, T.J.; McHale, B.B. *Applications in General Microbiology: A Laboratory Manual*, 6th ed.; Hunter Textbooks: Winston-Salem, NC, USA, 2001; p. 398.
39. Danova, S.; Petrov, K.; Pavlov, P.; Petrova, P. Isolation and characterization of *Lactobacillus* strains involved in koumiss fermentation. *Int. J. Dairy Technol.* **2005**, *58*, 100–105. [CrossRef]
40. Seiler, H. A review: Yeasts in kefir and koumiss. *Milchwissenschaft* **2003**, *58*, 392–396.
41. Lozovich, S. Medical Uses of Whole and Fermented mare Milk in Russia. *Cult. Dairy Prod. J.* **1995**, *30*, 18–21.
42. Zhang, L. Studies on orthogonal test of blend fermenting of three phases of intermediate culture of mare's milk and flavour evaluation of koumiss. *China Dairy Ind.* **1990**, *18*, 246–251.
43. Di Cagno, R.; Tamborrino, A.; Gallo, G.; Leone, C.; de Angelis, M.; Faccia, M.; Amirante, P.; Gobbetti, M. Uses of mare's milk in manufacture of fermented milks. *Int. Dairy J.* **2004**, *14*, 767–775. [CrossRef]
44. Akuzawa, R.; Miura, T.; Surono, I.S. Fermented Milks | Asian Fermented Milks. In *Encyclopedia of Dairy Sciences*, 2nd ed.; Fuquay, J.W., Fox, P.F., McSweeney, P.L.H., Eds.; Academic Press: San Diego, CA, USA, 2011; Volume 2, pp. 507–511.
45. Tang, H.; Ma, H.; Hou, Q.; Li, W.; Xu, H.; Liu, W.; Sun, Z.; Haobisi, H.; Menghe, B. Profiling of koumiss microbiota and organic acids and their effects on koumiss taste. *BMC Microbiol.* **2020**, *20*, 85. [CrossRef] [PubMed]
46. Uchida, K.; Hirata, M.; Motoshima, H.; Urashima, T.; Arai, I. Microbiota of 'airag', 'tarag' and other kinds of fermented dairy products from nomad in Mongolia. *Anim. Sci. J.* **2007**, *78*, 650–658. [CrossRef]
47. Meng, Y.; Chen, X.; Sun, Z.; Li, Y.; Chen, D.; Fang, S.; Chen, J. Exploring core microbiota responsible for the production of volatile flavor compounds during the traditional fermentation of Koumiss. *LWT Food Sci. Technol.* **2021**, *135*, 110049. [CrossRef]
48. Felis, G.E.; Dellaglio, F. Taxonomy of Lactobacilli and Bifidobacteria. *Curr. Issues Intest. Microbiol.* **2007**, *8*, 44–61.
49. Wang, J.; Chen, X.; Liu, W.; Yang, M.; Airidengcaicike; Zhang, H. Identification of *Lactobacillus* from koumiss by conventional and molecular methods. *Eur. Food Res. Technol.* **2008**, *227*, 1555–1561. [CrossRef]
50. de Vos, W. Systems solutions by lactic acid bacteria: From paradigms to practice. *Microb. Cell Factories* **2011**, *10* (Suppl. S1), S2. [CrossRef]
51. Huys, G.; Botteldoorn, N.; Delvigne, F.; De Vuyst, L.; Heyndrickx, M.; Pot, B.; Dubois, J.J.; Daube, G. Microbial characterization of probiotics—Advisory report of the Working Group "8651 Probiotics" of the Belgian Superior Health Council (SHC). *Mol. Nutr. Food Res.* **2013**, *57*, 1479–1504. [CrossRef]
52. Burentegusi; Miyamoto, T.; Nakamura, S.; Nozaka, Y.; Aoishi, A. Identification of Lactic Acid bacteria isolated from fermented mare's milk "chigee" in Inner Mongolia China. *Nihon Chikusan Gakkaiho* **2002**, *73*, 441–448. [CrossRef]
53. Menghe, B.; Wu, R.; Wang, L.; Yang, X.; Xu, J.; Dong, Y.; Sun, Z.; Zang, H. Isolation and identification of *Lactobacillus* from koumiss collected in Inner Mongolia and People's Republic of Mongolia. *China Dairy Ind.* **2004**, *32*, 6–11.
54. Shuangquan; Burentegusi; Yu, B.; Miyamoto, T. Microflora in traditional starter cultures for fermented milk hurunge from Inner Mongolia China. *Anim. Sci. J.* **2006**, *77*, 235–241. [CrossRef]
55. Watanabe, K.; Fujimoto, J.; Sasamoto, M.; Dugersuren, J.; Tumursuh, T.; Demberel, S. Diversity of lactic acid bacteria and yeasts in Airag and Tarag traditionally fermented milk products of Mongolia. *World J. Microbiol. Biotechnol.* **2008**, *24*, 1313–1325. [CrossRef]
56. Wu, R.; Wang, L.; Wang, J.; Li, H.; Menghe, B.; Wu, J.; Guo, M.; Zhang, H. Isolation and preliminary selection of lactobacilli from koumiss in Inner Mongolia. *J. Basic Microbiol.* **2009**, *49*, 318–326. [CrossRef] [PubMed]
57. Sun, Z.H.; Liu, W.J.; Zhang, J.C.; Yu, J.; Gao, W.; Jiri, M.; Menghe, B.; Sun, T.S.; Zhang, H.P. Identification and characterization of the dominant lactic acid bacteria isolated from traditional fermented milk in Mongolia. *Folia Microbiol.* **2010**, *55*, 270–276. [CrossRef] [PubMed]
58. Takeda, S.; Yamasaki, K.; Takeshita, M.; Kikuchi, Y.; Tsend-Ayush, C.; Dashnyam, B.; Ahhmed, A.M.; Kawahara, S.; Muguruma, M. The investigation of probiotic potential of lactic acid bacteria isolated from traditional Mongolian dairy products. *Anim. Sci. J.* **2011**, *82*, 571–579. [CrossRef]
59. Choi, S. Characterization of airag collected in Ulaanbaatar, Mongolia with emphasis on isolated lactic acid bacteria. *J. Anim. Sci. Technol.* **2016**, *58*, 10. [CrossRef] [PubMed]
60. Guo, L.; Ya, M.; Guo, Y.S.; Li, C.D.; Sun, J.P.; Zhu, J.J.; Qian, J.P. Study of bacterial and fungal community structures in traditional koumiss from Inner Mongolia. *Int. Dairy Sci.* **2019**, *102*, 1972–1984. [CrossRef]
61. Sun, Y.; Yang, J.; Yuan, J.; Cong, L.; Dang, N.; Sun, T.; Liu, W. Evaluation of lactic acid bacterial communities in spontaneously-fermented dairy products from Tajikistan, Kyrgyzstan and Uzbekistan using culture-dependent and culture-independent methods. *Int. Dairy J.* **2022**, *130*, 105281. [CrossRef]
62. Oleinikova, Y.; Alybayeva, A.; Daugaliyeva, S.; Alimzhanova, M.; Ashimuly, K.; Yermekbay, Z.; Khadzhibayeva, I.; Saubenova, M. Development of an antagonistic active beverage based on a starter including *Acetobacter* and assessment of its volatile profile. *Int. Dairy J.* **2024**, *148*, 105789. [CrossRef]
63. Wu, R.; Zhang, H.; Menghe, B. 16S rDNA sequence and cluster analysis of *Lb. casei* Zhang and ZL12-1 isolated from koumiss. *China Dairy Ind.* **2005**, *33*, 4–9.

64. Hao, Y.; Zhao, L.; Zhang, H.; Zhai, Z.; Huang, Y.; Liu, X.; Zhang, L. Identification of the bacterial biodiversity in koumiss by denaturing gradient gel electrophoresis and species-specific polymerase chain reaction. *J. Dairy Sci.* **2010**, *93*, 1926–1933. [CrossRef]
65. Slattery, L.; O'Callaghan, J.; Fitzgerald, G.F.; Beresford, T.; Ross, R.P. Invited review: Lactobacillus helveticus—A thermophilic dairy starter related to gut bacteria. *J. Dairy Sci.* **2010**, *93*, 4435–4454. [CrossRef]
66. Foster, L.M.; Tompkins, T.; Dahl, W. A comprehensive post-market review of studies on a probiotic product containing *Lactobacillus helveticus* R0052 and *Lactobacillus rhamnosus* R0011. *Benef. Microbes* **2011**, *2*, 319–334. [CrossRef] [PubMed]
67. Sedláček, I.; Yansanjav, A.; Nováková, D.; Švec, P. Ribotyping of *Lactobacillus helveticus* from the Koumiss. In Proceedings of the Indonesian Society for Microbiology and Indonesian Society for Lactic Acid Bacteria, Sanur, Bali, Indonesia, 22 October 2015; p. 151.
68. Wu, Y.; Li, Y.; Gesudu, Q.; Zhang, J.; Sun, Z.; Halatu, H.; Menghe, B.; Liu, W. Bacterial composition and function during fermentation of Mongolia koumiss. *Food Sci. Nutr.* **2021**, *9*, 4146–4155. [CrossRef] [PubMed]
69. You, L.; Yang, C.; Jin, H.; Kwok, L.Y.; Sun, Z.; Zhang, H. Metagenomic features of traditional fermented milk products. *Lwt* **2022**, *155*, 112945. [CrossRef]
70. Yao, G.; Yu, J.; Hou, Q.; Hiu, W.; Liu, W.; Kwok, L.; Menghe, B.; Sun, T.; Zhang, H.; Zhang, W. A perspective study of koumiss microbiome by metagenomics analysis based on single-cell amplification technique. *Front. Microbiol.* **2017**, *8*, 165. [CrossRef] [PubMed]
71. Oberg, T.S.; McMahon, D.J.; Culumber, M.D.; McAuliffe, O.; Oberg, C.J. Invited review: Review of taxonomic changes in dairy-related lactobacilli. *J. Dairy Sci.* **2022**, *105*, 2750–2770. [CrossRef]
72. Wang, L.; Xu, J.; Yun, Y.; Wu, R.; Menghe, B.; Zhang, H. Assessment of potential probiotic properties of *Lactobacillus* isolated from traditionally home-made koumiss in Mongolia. *China Dairy Ind.* **2005**, *33*, 4–10.
73. Xu, J.; Yun, Y.; Zhang, W.; Shao, Y.; Menghe, B.; Zhang, H. Fermentation properties of 4 strains of *Lactobacillus casei* isolated from traditionally home-made koumiss in Inner Mongolia of China. *China Dairy Ind.* **2006**, *34*, 23–27.
74. Zuo, F.L.; Feng, X.J.; Chen, L.L.; Chen, S.W. Identification and partial characterization of lactic acid bacteria isolated from traditional dairy products produced by herders in the western Tianshan Mountains of China. *Lett. Appl. Microbiol.* **2014**, *59*, 549–556. [CrossRef] [PubMed]
75. Myamoto, M.; Ueno, H.M.; Watanabe, M.; Tatsuna, Y.; Seto, Y.; Miyamoto, T.; Nakajima, H. Distinctive proteolytic activity of cell envelope proteinase of *Lactobacillus helveticus* isolated from airag, a traditional Mongolian fermented mare's milk. *Int. J. Food Microbiol.* **2015**, *197*, 65–71. [CrossRef]
76. Watanabe, K.; Makino, H.; Sasamoto, M.; Kudo, Y.; Fujimoto, J.; Demberel, S. *Bifidobacterium mongoliense* sp. nov. from airag a traditional fermented mare's milk product from Mongolia. *Int. J. Syst. Evol. Microbiol.* **2009**, *59*, 1535–1540. [CrossRef]
77. Montanari, G.; Zambonelli, C.; Grazia, L.; Kamesheva, G.K.; Shigaeva, M.K. *Saccharomyces unisporus* as the principal alcoholic fermentation microorganism of traditional koumiss. *J. Dairy Res.* **1996**, *63*, 327–331. [CrossRef]
78. Rakhmanova, A.; Wang, T.; Xing, G.; Ma, L.; Hong, Y.; Lu, Y.; Xin, L.; Xin, W.; Zhu, Q.; Lü, X. Isolation and identification of microorganisms in *Kazakhstan koumiss* and their application in preparing cow-milk koumiss. *J. Dairy Sci.* **2020**, *104*, 151–166. [CrossRef] [PubMed]
79. Yu, B.; Miyamoto, T. Isolation and identification of yeasts in chigee, fermented mare's milk, a traditional drink of Inner Mongolia, China. *Milk Sci.* **2010**, *59*, 231–236.
80. Guarner, F.; Schaafsma, G.J. Probiotics. *Int. J. Food Microbiol.* **1998**, *39*, 237–238. [CrossRef]
81. Choi, J.; Sabikhi, L.; Hassan, A.; Anand, S. Bioactive peptides in dairy products. *Int. J. Dairy Technol.* **2012**, *65*, 11–12. [CrossRef]
82. Del Piano, M.; Morelli, L.; Strozzi, G.P.; Allesina, S.; Barba, M.; Deidda, F.; Lorenzini, P.; Ballare, M.; Montino, F.; Orsello, M.; et al. Probiotics, from research to consumer. *Dig. Liver Dis.* **2006**, *38* (Suppl. S2), S248–S255. [CrossRef]
83. Guo, Z.; Wang, J.; Yan, L.; Chen, W.; Zhang, H. In vitro comparison of probiotic properties of *Lactobacillus casei* Zhang a potential new probiotic with selected probiotic strains. *LWT Food Sci. Technol.* **2009**, *42*, 1640–1646. [CrossRef]
84. Xue, W.; Yuan, X.; Ji, Z.; Li, H.; Yao, Y. Nutritional ingredients and prevention of chronic diseases by fermented koumiss: A comprehensive review. *Front. Nutr.* **2023**, *10*, 1270920. [CrossRef]
85. Zhang, W.; Yu, D.; Sun, Z.; Wu, R.; Chen, X.; Chen, W.; Meng, H.; Hu, S.; Zhang, H. Complete genome sequence of *Lactobacillus casei* Zhang, a new probiotic strain isolated from traditional homemade koumiss in Inner Mongolia, China. *J. Bacteriol.* **2010**, *192*, 5268–5269. [CrossRef]
86. He, Q.; Hou, Q.; Wang, Y.; Shen, L.; Sun, Z.; Zhang, H.; Liong, M.T.; Kwok, L.Y. Long-term administration of *Lactobacillus casei* Zhang stabilized gut microbiota of adults and reduced gut microbiota age index of older adults. *J. Funct. Foods* **2020**, *64*, 103682. [CrossRef]
87. Wu, R.; Sun, Z.; Wu, J.; Meng, H.; Zhang, H. Effect of bile salts stress on protein synthesis of *Lactobacillus casei* Zhang revealed by 2-dimensional gel electrophoresis. *J. Dairy Sci.* **2010**, *93*, 3858–3868. [CrossRef]
88. Liong, M.T.; Shah, N.P. Bile salt deconjugation ability, bile salt hydrolase activity and cholesterol co-precipitation ability of lactobacilli strains. *Int. Dairy J.* **2005**, *15*, 391–398. [CrossRef]
89. Chen, X.; Sun, Z.; Meng, H.; Zhang, H. The acid tolerance association with expression of H^+-ATPase in *Lactobacillus casei*. *Int. J. Dairy Technol.* **2009**, *62*, 272–276. [CrossRef]
90. Pan, D.D.; Zeng, X.Q.; Yan, Y.T. Characterisation of *Lactobacillus fermentum* SM-7 isolated from koumiss, a potential probiotic bacterium with cholesterol-lowering effects. *J. Sci. Food Agric.* **2011**, *91*, 512–518. [CrossRef]

91. Wang, J.; Zhang, H.; Chen, X.; Chen, Y.; Menghebilige; Bao, Q. Selection of potential probiotic lactobacilli for cholesterol-lowering properties and their effect on cholesterol metabolism in rats fed a high-lipid diet. *J. Dairy Sci.* **2012**, *95*, 1645–1654. [CrossRef]
92. Zhang, H.; Menghe, B.; Wang, J.; Sung, T.; Xu, J.; Wang, L.; Yun, Y.; Wu, R. Assessment of potential probiotic properties of *casei Zhang* isolated from traditionally home-made koumiss in Mongolia. *China Dairy Ind.* **2006**, *34*, 4–10.
93. Bilighe, M.; Liu, W.; Rina, W.; Wang, L.; Sun, T.; Wang, J.; Li, H.; Zhang, H. Evaluation of potential probiotics properties of the screened Lactobacilli isolated from home-made koumiss in Mongolia. *Ann. Microbiol.* **2009**, *59*, 493–498. [CrossRef]
94. Zhong, Z.; Zhang, W.; Du, R.; Meng, H.; Zhang, H. *Lactobacillus casei* Zhang stimulates lipid metabolism in hypercholesterolemic rats by affecting gene expression in the liver. *Eur. J. Lipid Sci. Technol.* **2012**, *114*, 244–252. [CrossRef]
95. Wang, J.; Zhang, J.; Liu, W.; Zhang, H.; Sun, Z. Metagenomic and metatranscriptomic profiling of *Lactobacillus casei* Zhang in the human gut. *NPJ Biofilms Microbiomes* **2021**, *7*, 55. [CrossRef] [PubMed]
96. Hou, Q.; Li, C.; Liu, Y.; Li, W.; Chen, Y.; Siqinbateer; Bao, Y.; Saqila, W.; Zhang, H.; Menghe, B.; et al. Koumiss consumption modulates gut microbiota, increases plasma high density cholesterol, decreases immunoglobulin G and albumin. *J. Funct. Foods* **2019**, *52*, 469–478. [CrossRef]
97. Menghe, B.; Zhou, Y.; Zhang, H.; Chen, Y.; Wang, L.; Xu, J.; Zhou, D. Antioxidative effect of Lactobacillus MG2-1 in koumiss. *China Dairy Ind.* **2005**, *33*, 21–24.
98. Abdel-Salam, A.M.; Al-Dekheil, A.; Babkr, A.; Farahna, M.; Mousa, H.M. High fiber probiotic fermented mare's milk reduces the toxic effects of mercury in rats. *N. Am. J. Med. Sci.* **2010**, *2*, 569–575. [CrossRef] [PubMed]
99. Rong, J.; Zheng, H.; Liu, M.; Hu, X.; Wang, T.; Zhang, X.; Jin, F.; Wang, L. Probiotic and anti-inflammatory attributes of an isolate *Lactobacillus helveticus* NS8 from Mongolian fermented koumiss. *BMC Microbiol.* **2015**, *15*, 196. [CrossRef] [PubMed]
100. Ya, T.; Su, Y.; Zhang, H. Effect of *Lb. casei* Zhang on koumiss on the production of cytokines in sera of mouse. *Food Sci.* **2006**, *27*, 488–491.
101. Ya, T.; Zhang, Q.; Chu, F.; Merritt, J.; Menhe, B.; Sun, T.; Du, R.; Zhang, H. Immunological evaluation of *Lactobacillus casei* Zhang, a newly isolated strain from koumiss in Inner Mongolia, China. *BMC Immunol.* **2008**, *9*, 68. [CrossRef]
102. Wang, Y.; Li, Y.; Xie, J.; Zhang, Y.; Wang, J.; Sun, X.; Zhang, H. Protective effects of probiotic *Lactobacillus casei* Zhang against endotoxin- and D-galactosamine-induced liver injury in rats via anti-oxidative and anti-inflammatory capacities. *Int. Immunopharmacol.* **2013**, *15*, 30–37. [CrossRef]
103. Wang, Y.; Xie, J.; Wang, N.; Li, Y.; Sun, X.; Zhang, Y.; Zhang, H. *Lactobacillus casei* Zhang modulate cytokine and Toll-like receptor expression and beneficially regulate poly I:C-induced immune responses in RAW264.7 macrophages. *Microbiol. Immunol.* **2013**, *57*, 54–62. [CrossRef] [PubMed]
104. Wang, Y.; Xie, J.; Li, Y.; Dong, S.; Liu, H.; Chen, J.; Wang, Y.; Zhao, S.; Zhang, Y.; Zhang, H. Probiotic *Lactobacillus casei* Zhang reduces pro-inflammatory cytokine production and hepatic inflammation in a rat model of acute liver failure. *Eur. J. Nutr.* **2016**, *55*, 821–831. [CrossRef] [PubMed]
105. Özer, B.H.; Kirmaci, H.A. Functional milks and dairy beverages. *Int. J. Dairy Technol.* **2009**, *63*, 1–15.
106. Sun, T.S.; Zhao, S.P.; Wang, H.K.; Cai, C.K.; Chen, Y.F.; Zhang, H.P. ACE-inibitory activity and gamma-aminobutyric acid content of fermented skim milk by *Lactobacillus helveticus* isolated from Xinjiang koumiss in China. *Eur. Food Res. Technol.* **2009**, *228*, 607–612. [CrossRef]
107. Aquilanti, L.; Carbini, A.; Strappati, R.; Santarelli, S.; Silvestri, G.; Garofalo, C.; Clementi, F. Characterisation of *Lactobacillus helveticus* strains producing antihypertensive peptides by RAPD and inverse-PCR of IS elements. *Benef. Microbes* **2010**, *1*, 229–242. [CrossRef] [PubMed]
108. Wulijideligen; Asahina, T.; Hara, K.; Arakawa, K.; Nakano, H.; Miyamoto, T. Production of bacteriocin by *Leuconostoc mesenteroides* 406 isolated from Mongolian fermented mare's milk, airag. *Anim. Sci. J.* **2012**, *83*, 704–711. [CrossRef] [PubMed]
109. Cui, Y.; Zhang, C.; Wang, C.; Lv, X.; Chen, Z.; Ding, Z.; Wang, Y.; Cui, H. A screening of class IIA bacteriocins produced by lactic acid bacteria isolated from fermented mare milk. *J. Food Saf.* **2013**, *33*, 433–439. [CrossRef]
110. Batdorj, B.; Dalgalarrondo, M.; Choiset, Y.; Pedroche, J.; Métro, F.; Prévost, H.; Chobert, J.M.; Haertlé, T. Purification and characterization of two bacteriocins produced by lactic acid bacteria isolated from Mongolian airag. *J. Appl. Microbiol.* **2006**, *101*, 837–848. [CrossRef]
111. Li-Li, M.; Xiang, D.J. Effect of LuxS/AI-2-mediated quorum sensing system on bacteriocin production of *Lactobacillus plantarum* NMD-17. *Folia Microbiol.* **2023**, *68*, 855–866.
112. Xu, C.; Fu, Y.; Liu, F.; Liu, Z.; Ma, J.; Jiang, R.; Song, C.; Jiang, Z.; Hou, J. Purification and antimicrobial mechanism of a novel bacteriocin produced by *Lactobacillus rhamnosus* 1.0320. *LWT Food Sci. Technol.* **2021**, *137*, 110338. [CrossRef]
113. Chen, Y.; Wang, C.; Hou, W.; Wang, X.; Gali, B.; Huasai, S.; Yang, S.; Wu, A.; Zhao, Y.; Wu, Y.; et al. Effects of antibacterial compounds produced by *Saccharomyces cerevisiae* in Koumiss on pathogenic *Escherichia coli* O_8 and its cell surface characteristics. *J. Integr. Agric.* **2017**, *16*, 742–748. [CrossRef]
114. Chen, Y.; Aorigele, C.; Wang, C.; Simujide, H.; Yang, S. Screening and extracting mycocin secreted by yeast isolated from koumiss and their antibacterial effect. *J. Food Nutr. Res.* **2015**, *3*, 52–56. [CrossRef]

Disclaimer/Publisher's Note: The statements, opinions and data contained in all publications are solely those of the individual author(s) and contributor(s) and not of MDPI and/or the editor(s). MDPI and/or the editor(s) disclaim responsibility for any injury to people or property resulting from any ideas, methods, instructions or products referred to in the content.

Review

Production, Composition and Nutritional Properties of Organic Milk: A Critical Review

Kevin Linehan [1,2,3,†], Dhrati V. Patangia [1,2,3,†], Reynolds Paul Ross [2,3] and Catherine Stanton [1,2,4,*]

1. Teagasc Food Research Centre, Moorepark, Fermoy, P61 C996 Cork, Ireland; kevin.linehan@teagasc.ie (K.L.); dhrati.patangia@teagasc.ie (D.V.P.)
2. APC Microbiome Ireland, University College Cork, T12 Y120 Cork, Ireland; p.ross@ucc.ie
3. School of Microbiology, University College Cork, T12 XF62 Cork, Ireland
4. VistaMilk Research Centre, Teagasc Moorepark, Fermoy, P61 C996 Cork, Ireland
* Correspondence: catherine.stanton@teagasc.ie; Tel.: +353-(0)25-42606
† These authors contributed equally to this work.

Abstract: Milk is one of the most valuable products in the food industry with most milk production throughout the world being carried out using conventional management, which includes intensive and traditional systems. The intensive use of fertilizers, antibiotics, pesticides and concerns regarding animal health and the environment have given increasing importance to organic dairy and dairy products in the last two decades. This review aims to compare the production, nutritional, and compositional properties of milk produced by conventional and organic dairy management systems. We also shed light on the health benefits of milk and the worldwide scenario of the organic dairy production system. Most reports suggest milk has beneficial health effects with very few, if any, adverse effects reported. Organic milk is reported to confer additional benefits due to its lower omega-6–omega-3 ratio, which is due to the difference in feeding practices, with organic cows predominantly pasture fed. Despite the testified animal, host, and environmental benefits, organic milk production is difficult in several regions due to the cost-intensive process and geographical conditions. Finally, we offer perspectives for a better future and highlight knowledge gaps in the organic dairy management system.

Keywords: organic; milk; dairy; composition; milk production systems

1. Introduction

Milk is among the most versatile and valuable foods in the food industry. In 2018, global milk production reached 843 billion liters, with an estimated value of USD 307 billion and is projected to grow by 22% by 2027 [1]. Approximately 80% of yearly milk production comes from cows, with the rest from other dairy animals like buffaloes, goats, camels, and sheep, according to the Food and Agriculture Organization [2]. Milk is also an essential component of the human diet, consumed by 80% of the world's population [3]. Milk and dairy products are important sources of macro and micronutrients, including high-quality proteins, fats, calcium, potassium, phosphorus, vitamin D, riboflavin, and vitamin B12 [4,5]. The majority of global milk production is carried out using traditional and intensive systems, collectively referred to as the conventional milk production system [5]. For the purposes of this review, the conventional milk production system, unless otherwise defined, will refer to milk produced from traditional and intensive milk production systems. The intensive use of mechanization, artificial fertilizers, pesticides and antibiotics within the conventional milk production system has raised substantial concerns for the environment, animal welfare, and consumer health [6]. Misuse of these practices can lead to soil, water and air pollution, increased antibiotic resistance spread, loss of biodiversity, and elevated greenhouse gas (GHG) emissions [7]. Moreover, the conventional milk production system, which prioritizes high productivity and profitability, may compromise the nutritional quality of milk and

Citation: Linehan, K.; Patangia, D.V.; Ross, R.P.; Stanton, C. Production, Composition and Nutritional Properties of Organic Milk: A Critical Review. *Foods* **2024**, *13*, 550. https://doi.org/10.3390/foods13040550

Academic Editors: Paolo Formaggioni and Piero Franceschi

Received: 5 January 2024
Revised: 27 January 2024
Accepted: 2 February 2024
Published: 11 February 2024

Copyright: © 2024 by the authors. Licensee MDPI, Basel, Switzerland. This article is an open access article distributed under the terms and conditions of the Creative Commons Attribution (CC BY) license (https://creativecommons.org/licenses/by/4.0/).

dairy products [8]. The intensification and environmental repercussions of conventional agriculture, coupled with heightened consumer awareness of animal welfare and demand for safer and healthier food options, have prompted a re-evaluation of agricultural policy [9]. This shift has given rise to more environmentally and animal-friendly practices, such as organic agriculture [10,11].

The FAO of the United Nations (UN) broadly defines organic agriculture as "a system that relies on ecosystem management rather than external agricultural inputs" [12]. Organic agricultural production is an alternative farming system rooted in the ethos of sustainable production [13]. The objective is to prioritize the health and welfare of animals, ensuring clean and sanitary conditions for their shelter and nourishment, along with effective waste management [14]. Organic production promotes preventive health measures without the constant use of stimulants or antibiotics, allowing animal access to pastures and providing them with a diet consisting entirely of organic ingredients for optimal nutrition and wellbeing [15]. In contrast to conventional agricultural production, the use of artificial fertilizers, pesticides, herbicides, genetically modified organisms (GMOs), and antibiotics is banned or restricted in organic agricultural production [16,17]. According to the International Foundation for Organic Agriculture (IFOAM) in 2021, organic agriculture was practiced in 191 countries, on more than 76 million hectares of agricultural land by at least 3.7 million farmers, and the size of the organic market reached 125 billion euros [18].

The intake of organic milk, whether in its natural state or as part of dairy products such as pasteurized whole milk, yogurt, cheese, curd, cream cheese and butter continues to grow worldwide [19]. Today, milk and dairy are the most in-demand organic products after organic fruits and vegetables in the organic food market [20]. Organic milk and dairy products, once available only in a few specialized shops, are now widely available to meet increasing consumer demand [21]. In recent years, research on organic milk and dairy products has also increased [22]. Several studies have reported compositional differences between organic and conventional milk [23,24]. For instance, organic milk has consistently been reported to contain significantly higher levels of whey proteins, total polyunsaturated fatty acids (PUFA), n-3 PUFA and vitamin E (α-tocopherol) [25,26]. Organic milk production has also been reported to influence the microbial content of milk [27]. Compositional differences have been linked to conditions associated with organic production such as breed, environment, health status, and feeding regime [25]. The health benefits of milk are associated with the various bioactives mentioned and can be direct, such as contributing to nutrient uptake, bone health and bone density development, and immunomodulatory potential with effects reported starting from as early as childhood [28], while other benefits can be indirect through the gut microbiota by exerting probiotic potential. Organic dairy production is free from antibiotics and chemicals, thus helping in the reduction in antibiotic resistance gene generation and spread. Further, the low ratio of omega 6 to omega 3 fatty acids, and the higher PUFA content are associated with health benefits, though some results are debated [29,30].

Despite the known benefits of organic farming practices, the debate over the advantages of organic milk and dairy products compared to their conventional counterparts persists [31]. Organic farming presents challenges for farmers involving changes in animal husbandry, land, and crop management [21]. Furthermore, the switch is cost intensive, resulting in comparatively low yields and higher estimated product prices [32]. In addition, adhering to strict, mandatory, and country-specific regulations for organic farming and food production, makes the transition a demanding process [33]. Consumers also often express skepticism due to the high prices of organic dairy products and the lack of definitive studies showcasing their benefits [34]. However, as sustainability concerns continue to gain global attention, the organic dairy market is expected to grow [21].

In light of this information, this review aims to explore the latest research on the production and composition of milk produced using organic agricultural practices. We compare organic and conventional milk production systems in terms of practices and impact on the quality of milk. Furthermore, we discuss the human health benefits of organic milk and dairy products and the future challenges and prospects of organic dairy management.

2. An Introduction to Organic Milk Production

In this section, we provide an introduction to organic milk production and present the regulatory frameworks and principles that guide this farming practice. For this review, organic milk production, unless otherwise defined, will refer to milk from dairy cattle.

Organic Milk Production Regulations

Organic milk production is permissible exclusively on certified farms, depending on individual countries' regulations or organizational certifications [35]. Despite sharing fundamental principles, the specifics of organic milk production regulations exhibit notable variations globally, primarily regarding the rules governing pasture access, nutrition, use of antibiotics, and conventional to organic status conversion period, as detailed in Table 1. Subsequently, organic milk products produced in one country may not retain their organic status when exported to another country with distinct legal requirements [21]. Therefore, the diversity in organic regulations may contribute to the variability in organic milk composition between countries [36].

Table 1. Country-specific organic dairy farming regulations regarding pasture access, forage feeding, antibiotic usage, and conventional to organic status conversion period. Adapted from [36].

Country	Pasture Access	Nutrition	Antibiotics Use	Organic Conversion Period	Regulation
European Union	Year-round, weather permitting	≥60% of daily dry matter intake must consist of roughage, fresh or dried fodder, or silage.	Permitted under veterinary recommendation. ≥2 day milk withdrawal. ≥3 treatments or ≥1 treatment (if productive lifecycle is <1 y) will cause animal to lose its organic status.	Land conversion period of 24-months. Animals must be under organic management ≥6 months.	Regulation (EU) 2018/848 of the European Parliament and of the Council.
United States	≥120 days annually	≥30% of daily dry matter intake must come from pasture during grazing season.	Prohibited. Usage will cause animal to lose its organic status.	Animals must be under organic management ≥12 months.	Organic foods production act provisions 2023.
Canada	≥120 days annually	≥30% of daily dry matter intake must come from pasture during grazing season. 60% of dry matter intake consists of hay, fresh/dried fodder, or silage.	Permitted under veterinary recommendation. ≥30 day milk withdrawal. ≥2 treatments, 12 month transition period before regaining organic status.	Animals must be under organic management ≥12 months.	Organic Production Systems General Principles and Management Standards 2021.
Japan	≥2 days per week, year-round	≥50% of daily dry matter intake must consist of roughage, fresh or dried fodder, or silage.	Permitted under veterinary recommendation.	Animals must be under organic management ≥6 months.	Japanese Agricultural Standard for Organic Livestock Products, 2018.
New Zealand	≥150 days annually	≥50% of daily dry matter intake must consist of roughage, fresh or dried fodder, or silage.	Prohibited. Usage will cause animal to lose its organic status.	Animals must be under organic management ≥12 months.	AsureQuality Organic Standard For Primary Producers, 2018.
Australia	Year-round, weather permitting	100% of daily dry matter intake must be sourced from organic or bio-dynamic feed.	Permitted under veterinary recommendation. 180 day transition period before regaining organic status.	Animals must be under organic management ≥6 months.	National Standard for Organic and Bio-Dynamic Produce, 2022.
China	Year-round, weather permitting	≥60% of daily dry matter intake must consist of roughage, fresh or dried fodder, or silage.	Permitted under veterinary recommendation.	Animals must be under organic management ≥6 months.	China Organic Standard GB/T 19630-2019.
India	Year-round, weather permitting	≥85% of daily dry matter intake must be sourced from organic feed	Permitted under veterinary recommendation.	Land conversion period of 24 months. Animals must be under organic management ≥6 months.	Agricultural and Processed Food Products Export Development Authority (APEDA) 2018.

3. Milk Production Systems

Traditional and intensive milk production systems are collectively referred to as the conventional milk production system. The conventional system dominates milk production practices worldwide, primarily focusing on high productivity [21]. The intensive system is principally performed in developed countries, while milk production in developing countries is carried out in an extensive (traditional) manner [37,38]. The organic milk production sector is experiencing rapid growth, surpassing the expansion rate of other dairy sectors worldwide [21,39]. A summary of the major distinctions between organic and conventional milk production systems is provided in Table 2.

3.1. Conventional Systems

3.1.1. Traditional System

The traditional system relies on pasture as a low-cost primary feed source [40]. Farming practices are primarily determined by the climate and available resources in a given region. Therefore, the traditional system is primarily employed in temperate climates, such as in Ireland and New Zealand, which leads to a seasonal milk supply. Cows are kept outdoors, grazing on pasture during the warmer months of the year. In the winter months, cows are dried off and housed indoors and are fed a diet of primarily pasture-based silage and hay, which is cut and ensiled from surplus pasture earlier in the year. Their feed is typically administered ad libitum (without specialized equipment and calculation of feed rations). When pasture-based feeds alone fail to meet energy requirements of the animal, concentrate supplements are also provided. The ration is not consistent in this feeding system, making it challenging to achieve a balanced diet and can potentially hinder high milk yields [41]. The traditional system offers cows a more natural environment than the intensive system, allowing the expression of normal behaviors [42]. Pasture-based feeding systems have also been demonstrated to beneficially affect the nutritional quality of milk and dairy products [43]. Milk and dairy products obtained from pasture-based diets have larger proportions of beneficial nutrients for human consumption such as PUFA, conjugated linoleic acid (CLA) and n-3 fatty acids than cows fed concentrate diets [44–48]. While the intensive milk production system is supplanting the traditional system, the latter is expected to dominate for the foreseeable future in developing countries [37,38].

3.1.2. Intensive System

The intensive system is based on the use of a total mixed ration (TMR) diet administered using a feed truck. The intensive system is primarily performed in countries with climates which make pasture difficult to grow, including the United States, China, and large areas of Europe [49,50]. TMR is a mixture of roughage (grass/maize/corn silage) as well as concentrate feeds supplemented with vitamins and minerals [45]. TMR feeding offers greater opportunities to enhance intake rates and meet nutritional needs more effectively [51]. Furthermore, this system protects animals from extreme weather conditions [52]. The number of dairy cow farms employing the intensive milk production system has grown significantly over the last 20 years [49]. Animal welfare concerns continue to grow regarding indoor TMR feeding systems. These include increased incidences of lameness [53], mastitis [54], mortality [55] and aggressive behavior due to reduced space [42]. Indoor TMR feeding systems also restrict the animals' ability to express their natural foraging behavior [42]. The development of the partial mixed ration (PMR) feeding system may alleviate some of these concerns. The PMR system combines indoor TMR feeding with the outdoor grazing of fresh pasture by alternating the feeding approaches. PMR feeding has been shown to increase levels of CLA, α-linolenic acid (ALA), vaccenic acid and PUFA significantly compared to TMR feeding [56], in addition to non-significant differences in milk yield and protein content [56].

3.2. Organic System

Organic milk production is based on maximizing milk production in an environmentally sustainable way, while prioritizing the health and wellbeing of animals. Distinctive variances exist between organic and conventional milk production systems, each presenting its own set of advantages and drawbacks. No single production system can be deemed ideal, as milk production is an ongoing process. The merits of either system hinge on a comprehensive evaluation encompassing longitudinal sustainability, environmental impact, economic factors, and social considerations. There are several fundamental differences between organic and conventional milk production systems [21,23]. In contrast to the conventional systems, the organic system prioritizes the utilization of native cattle breeds [57]. Crops must be fertilized organically, and the use of synthetic and chemical fertilizers, herbicides and pesticides is prohibited, which has been shown to have beneficial effects on soil composition and functionality compared to conventional systems [58,59]. Animals must be provided with organic feed containing ingredients sourced from organic agricultural production, while the inclusion of natural non-agricultural substances is also permitted. For example, vitamins and minerals are sourced from natural substances such as sprouted grains, cod liver oil, and brewer's yeast. In general, a minimum of 60% of the feed must be sourced from the corresponding farm. Additionally, a minimum of 60% of the dry matter in the feed ration must consist of roughage, green fodder, dried fodder, or silage. During the summer season, cows are provided unrestricted access to pasture vegetation, predominantly comprising low grasses (50%), tall grasses (30%), and legumes (10–20%) [23].

On organic farms, the duration of pasture feeding frequently extends beyond 180 days, whereas on traditional farms, it typically does not exceed 140 days [36]. Organic pastures stand out for their rich sward biodiversity, encompassing various species of grasses, legumes, and herbs. This diversity directly contributes to the nutritional value and quality of fodder and milk produced [43,59]. During autumn and winter, cattle are required to be provided with roughage, comprising silage made from combinations of cereals and legumes or haylage. The inclusion of beets or potatoes in the cattle's diet is reserved for the winter season [60]. Similar to traditional farming methods, grazing access has been shown to benefit the welfare and behavior of organic cows compared to conventional systems [61,62]. However, organic farms still show a need for improvement, especially regarding animal health [62]. The main problems faced by organic and conventional systems are analogous, with mastitis and lameness identified as particular areas for improvement [63].The use of GMOs, growth stimulants, and synthetic amino acids is also prohibited in organic agriculture [33], while antibiotics may only be utilized in emergencies for veterinary indication. The rise of antimicrobial resistance (AMR) attributed to the excessive use of antibiotics in food-producing animals has become a significant concern [64], especially concerning the risk of developing newly resistant bacteria that could be transmitted from animals to humans [65]. Encouragingly, organic farming has been demonstrated to markedly decrease the occurrence of AMR in dairy cattle compared to conventional farming, globally [66]. Finally, pasture-based systems have been demonstrated to emit less GHG emissions, such as methane (CH_4) and carbon dioxide (CO_2), than conventional farms [67,68].

Table 2. Management Practices of Organic and Conventional Milk Production Systems. Adapted from [69].

	Milk Production System	
Management Practice	Organic	Conventional
Pasture access	Required	Not required
Nutrition	All feed must be certified organic	Concentrate feed
Antibiotics use	In emergencies, for veterinary indication	Allowed, for veterinary indication

Table 2. *Cont.*

	Milk Production System	
Parasiticide use	In emergencies, for veterinary indication	Allowed, for veterinary indication
Growth Hormone use	Prohibited	Allowed, for veterinary indication
Weed Management	Crop rotation, hand weeding, mulches	Chemical Herbicides
Pest Management	Crop rotation, Companion Planting, trap crops, promotion of beneficial insects and natural predators	Chemical Pesticides
Green House Gas Emissions	Lower per unit of area	Higher per unit of area
Fertilizers	Organic fertilizers only	High dependence on synthetic NPK fertilizers
Genetically Modified Organisms	Prohibited	Allowed
Synthetic food Additives	Prohibited	Allowed
Milk Yields	Lower on average	Higher on average
Shelf Life	Higher on average	Lower on average
Product Price	Higher on average	Lower on average
Soil Impact	Reduced soil loss, increased organic matter, water-holding capacity and microbial diversity	Increased soil loss and erosion, lower water holding capacity, lower carbon storage and microbial diversity
Water Consumption	Lower	Higher
Energy Usage	Low intensity of energy use (higher energy efficiency)	High intensity of non-renewable energy use (agrochemicals, machinery, water pumping etc.)
Impact on Landscape	Larger floral and faunal biodiversity. Diverse agricultural landscapes	Loss of biodiversity in agricultural landscapes, Unified agricultural landscapes (monocultures)

4. Impact of Production Systems on Farm Performance and Raw Milk Composition

The composition and physical characteristics of milk exhibit considerable variability, influenced by factors such as environment, age, breed, nutrition, parity, stage of lactation, and health [70]. Numerous studies have compared the quantity and quality of raw milk produced using organic or conventional milk production systems [23]. The gross chemical and physical composition of raw milk produced using organic and conventional milk production systems is shown in Table 3.

4.1. Milk Yield

The primary determinant of the financial success of dairy cow farms is their level of productivity. Organic dairy production has consistently been reported to have reduced milk yields compared to conventional milk production systems [68]. Organic herds generally attain lower milk yields, ranging from 15% to 28% less compared to the yields of a typical conventional cow [23]. Such stark differences in milk yield are typically traced to lower energy intake, through either less concentrated feeding or lower energy content in forages from organic systems [71]. Furthermore, practices such as adjusting grain feeding levels, selecting breeds to enhance cow milk yield, and employing fossil fuel-based fertilizers to boost forage yields are typically linked to the increased yields in conventional systems [36]. Therefore, lower milk yield, and thus lower profitability of organic milk production, could pose obstacles to the continued growth of the organic dairy industry worldwide [72].

4.2. Udder Health and Somatic Cell Count (SCC)

Somatic cell count (SCC) serves as a crucial diagnostic parameter for assessing the wellbeing of the mammary gland [73]. An SCC surpassing 400,000 cells per milliliter of milk signifies gland inflammation. Inflammation has detrimental consequences on the overall productivity of cows, impacting both the nutritional quality of milk and its suitability for processing [74]. Factors related to management, such as milking hygiene and the cleanliness of cows, play a role in the occurrence of udder infections. These infections can impact both milk yield and composition [75]. Elevated SCC can exhibit a negative correlation with both the yields and percentages of milk protein and fat [76]. Therefore, any conclusions regarding compositional differences between organic and conventionally produced milk should consider udder health as a contributing factor [23]. Conflicting results have been reported regarding increased or decreased SCCs when comparing organic and conventional dairy production systems [23]. Importantly, in most studies which reported significant differences in SCC, the levels were still below 400,000 cells per mL in both conventional and organic milk. At present, the employed farming system appears to have less influence on udder health compared with management factors (e.g., routine teat dipping and seeking veterinary treatment) and animal level variables (e.g., parity, breed) [36,77–79]. Hence, making a generalization about whether organic farmers have a lower tolerance for poor udder health is not feasible due to potential variations in ethical considerations and divergent regulations regarding the use of antibiotics as a treatment option for organic cows among different countries [63]. Therefore, establishing a definitive relationship between SCC and the production system is challenging.

4.3. Microbiological Quality

The total bacterial count (TBC) is the most widely used measure of microbial quality of raw milk and is measured using several methods including the standard plate count (SPC), plate loop count (PLC), Petrifilm (3M) aerobic count, and flow cytometry methodologies (e.g., Bactoscan, Foss Analytica, Hillerød, Denmark l) [80]. While specific values for SPC vary worldwide, high-quality raw milk should always have a low TBC [81]. Similar to milk SCCs, contradictory results have been reported regarding increased or decreased TBCs when comparing raw milk produced using organic and conventional systems [23]. Differing TBCs across studies have been attributed to management factors and animal-level variables [36,77–79].

The microbiome consists of the microbiota and its "theatre of activity," encompassing the collective nucleic acids (including viruses and bacteriophages), structural components, and microbial metabolites associated with the microbiota [82]. The existence of a commensal microbiota on the bovine teat canal and teat skin is widely acknowledged [83,84]. However, the demonstration of a commensal bovine milk microbiome has been disputed by methodological issues, sampling difficulty, and a lack of consistency among studies [85,86]. Previous studies have shown that diet has a direct impact on the gut, rumen, and milk microbiota of bovines [87–89]. To our knowledge, only one study has compared the microbiota of dairy cows from conventional and organic farming [27]. This study demonstrated that the microbiome of the cow's gut and milk was significantly different between agricultural management systems, while no differences were found in the microbial communities of soil and silage [27]. Milk samples from organic farms were significantly associated with the family *Rhodobacteraceae* and elevated levels of *Ruminococcaceae*. Furthermore, there was a notable association of the fungi Dothideomycetes, Tremellomycetes, and Pleosporales with milk samples from organic farms. Fungi within these classes are commonly associated with plant pathogens that thrive on wood debris or decaying leaves. Nevertheless, their presence has been reported in the dairy farm environment [90] and on shelves used for ripening cheese [91].

4.4. Mastitis

Mastitis stands out as the most widespread and economically impactful disease in dairy cattle globally, primarily attributed to diminished milk production, discarded milk, premature culling, and associated treatment expenses [92]. Bovine mastitis is a polymicrobial disease with the principal etiological agents being *Staphylococcus aureus*, *Streptococcus dysgalactiae*, and *Streptococcus uberis* [93]. Although treatment with antibiotics is the last resort for organic farmers, their usage is permitted under the prescription of a veterinarian [78]. Antibiotics are currently the preferred treatment for mastitis control on both organic and conventional farms [94]. While the epidemiology of mastitis on organic farms has not been extensively studied, available reports suggest that organic farms have an elevated prevalence of *Staphylococcus aureus* compared with conventional dairy farms [95–98]. The incidence of clinical mastitis on organic dairy farms has been reported to be lower than on conventional farms [99–101]. Additionally, no differences have been found in the incidence of subclinical mastitis [102] or individual SCC [103] on organic versus conventional farms. Such reports suggest that there may be differences in mastitis epidemiology between conventional and organic dairy farms. Future studies are needed to assess the antimicrobial resistance profiles and ubiquity of antibiotic-resistant bacteria, such as Methicillin-resistant *Staphylococcus aureus* (MRSA) in mastitic milk from conventional and organic dairy farms.

4.5. Volatile Organic Compounds

Milk contains low concentrations of volatile organic compounds (VOCs) which are influenced by several variables, such as environment, breed, and lactation stage [104]. VOCs have consistently been associated with the sensory profiles of milk products, especially odors and flavors [105]. VOCs emanate in milk via metabolic processes of the cow (e.g., rumen gases, blood, etc.) or can be infused into milk through animal feed, which influences the flavor of dairy products [106–108]. There have been conflicting reports regarding variation in the VOC composition of milk produced using the organic and intensive production systems have been reported [36,109–111]. VOCs markers, such as terpenes, warrant further exploration for their potential to authenticate dairy products [26,70,112].

Animal feeding systems (pasture or TMR) have been shown by numerous studies to alter the sensory characteristics of milk and dairy products [113]. Some studies have found little differences in the flavor and texture of milk and dairy products produced using organic and conventional systems [36,109,114]. Studies have also indicated that raw organic milk was creamier and tended to have greater 'hay' and 'grass' flavor notes than conventional milk [115]. Irrespective of production system, a stronger odor of milk, butter, and cheese (more intense 'animal' notes) has repeatedly been reported when cows are pasture-fed vs. fed on conserved forages [46,116].

4.6. Protein

The total protein and casein content of organic and conventionally produced milk is typically reported to not differ significantly [117–119]. Whey proteins, while making up only 20–25% of the total protein, constitute a crucial group of milk proteins (the remaining 75–80% is casein). Albumins, i.e., α-lactalbumin (α-LA), β-lactoglobulin (β-LG), and bovine serum albumin (BSA), make up approximately 75% of whey proteins. Other minor whey proteins include bacteriostatic substances, i.e., immunoglobulins, lactoferrin, lactoperoxidase, and lysozyme, which constitute 1–2% of total milk proteins. These proteins exhibit diverse positive effects on the human body, encompassing antimicrobial (antiviral and antibacterial), anticancer, immunomodulatory, and antioxidant properties. Whey proteins serve as an excellent source of energy, essential amino acids, and peptides [120]. The concentration of whey proteins and albumins in organic and conventionally produced milk is largely similar in studies to date [117,121,122]. Recent studies have indicated that concentrations of lactoferrin and lysozyme are significantly higher in milk on organic farms than on conventional farms [121,123]. Lysozyme is an antimicrobial enzyme that induces cell lysis by hydrolyzing the peptidoglycan layer of both gram-positive and gram-negative bacterial cells. When ingested, lactoferrin induces various beneficial biological effects, such as enhancing iron absorption, modulating the immune system, boosting the antimicrobial activity of lysozyme, and promoting the growth of epithelial cells and fibroblasts [124].

4.7. Vitamins

Vitamin A (retinol) serves as the precursor to a group of compounds known as retinoids, which exhibit the biological activity associated with vitamin A. Vitamin A encompasses a group of analogous fat-soluble vitamins that play ubiquitous roles in the human body, such as enhancing vision, cell differentiation, embryogenesis, reproduction, growth and immune system functioning [125]. In general, foods of animal origin provide preformed vitamin A as retinyl esters while plant-derived foods provide precursors of vitamin A, i.e., carotenoids. Only carotenoids with a β-ionone ring (e.g., β-carotene) can function as precursors of vitamin A [126]. In cow's milk, vitamin A is typically found in the forms of retinol or β-carotene [126]. The concentration of vitamin A and carotenoids in milk is significantly influenced by the carotenoid content of the animal's diet. Milk from animals fed on pasture generally contains higher levels of carotenes compared to milk from animals fed on concentrate feeds [117]. Vitamin E constitutes a group of fat-soluble molecules which primarily act as antioxidants in cell membranes where the primary function is to prevent oxidative damage by trapping reactive oxyradicals [127]. Vitamin E is also essential for body functions in both bovines and humans such as growth, reproduction, immunity prevention, and protection of tissues [128]. β-carotene and Vitamin E concentrations differ significantly in raw milk depending on the diet [25]. The dairy industry is interested in a high content of vitamin E and β-carotene, as they can prevent the spontaneous oxidation of milk and fatty acids [129]. Vitamin D_3 plays a crucial role in the metabolism of calcium and phosphorus, contributing to the proper mineralization of bones and teeth. Additionally, it exhibits immunomodulatory and anti-cancer properties. In the case of animals spending time at pasture, ultraviolet (UV) rays from sunlight induce the synthesis of vitamin D_3 from 7-dehydrosterol present in the skin. Therefore, milk from cows that spend more time outdoors at pasture is expected to be a more valuable source of this vitamin. Numerous studies have reported higher vitamin D_3 levels in milk from cows of organic and traditional production systems compared to intensive systems [117,122,130].

4.8. Carbohydrates

Lactose is the main carbohydrate in milk and is generally reported to not significantly differ in the feeding system [23]. Oligosaccharides are the third most abundant solid component found in milk, after lactose and lipids [131]. These structurally and biologically diverse molecules, despite being resistant to human digestive enzymes, are linked to numerous beneficial functions [132]. Organic and conventional pasture-based farming systems have been demonstrated to not significantly influence oligosaccharide abundance [110]. However, levels of specific oligosaccharides were increased in organic milk irrespective of sampling date or farm set [110], specifically, trisaccharides with three hexose units (3 Hex), trisaccharides with three hexose units and one N-acetylneuraminic acid unit (3 Hex, 1 NeuAc), tetrasaccharides with four hexose units and one N-acetylhexosamine unit (4 Hex, 1 HexNAc), and trisaccharides with three hexose units and two N-acetylhexosamine units (3 Hex, 2 HexNAc) [110].

4.9. Fats

The total fat content of organic and conventionally produced milk is typically reported to not differ significantly. Milk fat consists of over 400 different fatty acids. The predominant fatty acids in milk are saturated fatty acids (SFA), with unsaturated fatty acids, including monounsaturated fatty acids (MUFA) and polyunsaturated fatty acids (PUFA), following. Nevertheless, recent scientific advancements have suggested that trans fatty acids and certain saturated fatty acids in milk may have beneficial effects [113,133,134]. The concentrations of individual fatty acids in milk fat are affected by factors such as cow breed, stage of lactation, genetics, and diet [113,135]. The composition and quantity of fatty acids in milk are primarily dictated by the feeding system [25,136].

Fresh herbs and grasses in the cow's diet contribute a significantly higher quantity of unsaturated fatty acids, whereas maize silage has a greater concentration of linoleic acid [113]. The TMR feeding system markedly diminishes the fat and fatty acid content in milk, attributed to the insufficient dietary fiber and elevated starch levels in the diet [137,138]. Organic milk has consistently been shown to contain a more favorable fatty acid profile than conventional milk [139], containing more PUFAs, including omega-6 and omega-3 [24,139–141] and a lower ratio of omega-6 to omega-3 fatty acids, which is beneficial for human health [139,142–146]. The omega-6 to omega-3 fatty acid ratio in bovine milk essentially characterizes the concentrations of linoleic acid versus α-linolenic acid, as they represent the most abundant omega-6 and omega-3 fatty acids, respectively.

Forage is abundant in α-linolenic acid, while cereals such as barley, maize, oats, and soybean contain higher quantities of linoleic acid [147]. A lower omega-6 to omega-3 fatty acid ratio is therefore suggestive of a forage-based diet [113]. Organic milk has also been shown to contain higher Conjugated Linoleic Acid (CLA) content than conventional milk [148,149]. The consumption of milk and dairy products rich in CLA is linked to beneficial effects on human health, including improved brain function, antiatherogenic effects, and lower levels of blood lipids [150,151]. CLA also demonstrates anti-carcinogenic, immunostimulatory, and weight-reducing properties [151].

4.10. Minerals and Heavy Metals

The mineral content of milk is influenced by a variety of factors including animal diet, genetics, breed, feeding system, and the surrounding environment [152]. The concentration of minerals in milk is primarily contingent on their levels in fodder [153,154]. The mineral content of forage is determined by the mineral content of soil and pasture, which is influenced by fertilizers, the amount of sewage sludge generated, soil type, or the proximity of mining and industrial areas [36]. In conventional farming, soil fertility can be increased by using mineral fertilizers enriched with selected microelements [155]. Cow diets are also supplemented with mineral mixtures to increase the mineral content of milk produced [23]. Both of these methods are restricted in organic farming; therefore, on-farm fodder is the main source of minerals [23]. Green forage from legume plants offers substantial amounts of calcium and magnesium. Cereal grains provide phosphorus, wheat bran serves as a source of magnesium, and green forage contains smaller amounts of sodium [23]. Organic milk has generally been reported to contain a marginally lower mineral content than conventionally produced milk, with the difference primarily attributed to management practices [24,25,156,157]. These practices include selenium supplementation to improve reproductive performance, iodine-containing teat dipping as a disinfectant after milking, and mineral supplementation [24,25,156,157]. Toxic elements, including heavy metals, such as lead, chromium, mercury, and cadmium may also be present in milk and dairy products [158]. Such heavy metals are non-essential elements, have no biological role in mammals, and can cause toxic effects even at very low concentrations [159]. The main source of heavy metals in agricultural systems is fertilizer [160]. Numerous studies have reported significantly higher levels of heavy metals in conventionally produced milk [119,161,162].

Table 3. Concentrations of select macronutrients, micronutrients, and general antimicrobial peptides present in raw milk produced using organic, traditional and intensive systems. Traditional milk refers to milk produced using the traditional milk production system. Intensive milk refers to milk produced using the intensive milk production system.

	Organic System	**Conventional Systems**	
Proteins	Organic Milk	Traditional Milk	Intensive Milk
Total Protein (%)	3.1–3.26	3.1–3.24	3.48

Table 3. Cont.

	Organic System	Conventional Systems	
Casein (%)	2.54	2.52	2.78
Whey protein (%)	0.72–0.84	0.72–0.84	0.70–0.82
β-Lactoglobulin (g/L)	3.32–3.35	3.26–3.58	3.01–3.28
α-Lactalbumin (g/L)	1.07–1.19	1.05–1.21	0.98–1.14
Bovine serum albumin (g/L)	0.43	0.44	0.41–0.49
Lactoferrin (mg/L)	123.8–125.9	109.80–130.62	94.01–121.23
Lysozyme (µg/L)	11.14	9.92–10.71	6.90–12.13
Vitamins	Organic Milk	Traditional Milk	Intensive Milk
Vitamin A (retinol) (mg/L)	0.468–0.800	0.410–0.556	0.347–0.465
β-carotene (mg/L)	0.195–0.580	0.231–0.252	0.175–0.190
Vitamin E (α-tocopherol) (mg/L)	1.358–2.655	1.656–1.953	1.075–1.302
Vitamin D_3 (cholecalciferol) (µg/L)	0.461–0.768	0.610–1.212	0.589–0.700
Carbohydrates	Organic Milk	Traditional Milk	Intensive Milk
Lactose (%)	4.80–5	4.7–5	nd
3 Hex (Trisa) (m/z)	60.82–61.11	51.37–55.86	nd
3 Hex, 1 NeuAc (m/z)	11.83–14.60	9.24–12.42	nd
4 Hex, 1 HexNAc (m/z)	0.87–0.93	0.63–0.69	nd
3 Hex, 2 HexNAc (m/z)	0.31–0.33	0.25	nd
Fat	Organic Milk	Traditional Milk	Intensive Milk
Fat (%)	3.7–4	3.8–4	3.8–4
SFAs (g/100 g)	66.28	59.03–64.74	67.69–71.41
MUFAs (g/100 g)	26.11–34.07	30.33–32.16	21.87–28.15
Oleic acid (c9 C18:1)	20	16.10–22.66	16.16–17.20
Vaccenic acid (t11 C18:1) (g/100 g)	1.22–2.00	1.18–7.00	0.80–2.00
PUFAs (g/100 g)	3.85–5.36	3.69–5.32	1.65–3.77
Eicosapentaenoic acid, EPA (C20:5 n-3) (g/100 g)	0.05	0.08	0.05
Conjugated linoleic acid, CLA (cis9 trans11) (g/100 g)	0.83–1.53	0.54–0.93	0.42–1.19
Linoleic acid, LA (C18:2 n-6) (g/100 g)	0.59–2.08	1.17–2.18	1.4–2.39
α-linolenic acid, ALA (C18:3 n-3) (g/100 g)	0.44–1.05	0.49–1.25	0.39–0.42
γ-linolenic acid, GLA (C18:3 n-6) (g/100 g)	0.11	0.13	0.12
Proportion 18:3n3: 18:3n6	1.35	0.60–2.77	1.26
Minerals and Heavy Metals	Organic Milk	Conventional Milk	
Calcium (mg/L)	971.33–1161	1170–1417.76	
Iron (mg/L)	0.26–0.67	0.26–0.47	
Manganese (mg/L)	0.023–0.047	0.022–0.139	
Copper (mg/L)	0.023–0.084	0.038–0.161	
Iodine (mg/L)	0.013–0.283	0.071–6.540	
Aluminium (mg/L)	0.76	0.63	
Potassium (mg/L)	1509–1896.92	1514–1844.37	
Sodium (mg/L)	366.59	476.35	

Table 3. *Cont.*

	Organic System	Conventional Systems
Magnesium (mg/L)	86.21	113.87–118.50
Zinc (mg/L)	2.86–3.96	2.96–4.39
Selenium (mg/L)	0.002–0.020	0.008–0.040
Cobalt (mg/L)	0.001	0.001
Strontium (mg/L)	0.166	0.202

Ranges are shown where available. Values for the traditional system and intensive system are shown where available. Values obtained from [23,110,142,144,163–165]. Abbreviations: Trisa, Trisaccharides; Hex, glucose or galactose; HexNAc, N-acetylglucosamine or N-acetylgalactosamine; NeuAc, N-acetylneuraminic acid (sialic acid); (m/z), mass divided by charge number; MUFAs, monounsaturated fatty acids; PUFAs, polyunsaturated fatty acids; SFAs, saturated fatty acids; nd, no data.

5. Perceived Health Benefits of Organic and Conventional Milk

Milk and dairy products provide several health benefits beginning from the early stages of life. Recommendations based on guidelines for several countries across the globe include milk and other dairy products as part of daily healthy eating [166]. Research over the past two decades has delineated the associations between milk consumption and health benefits. Some of the highest increases in the numbers of diseases worldwide are seen concerning obesity, type 2 diabetes, and cancer. A recent meta-analysis reported that children consuming higher dairy intake had lower incidence of overweight compared to those having lower dairy intake [167]. However, another meta-analysis failed to show any association in children, though a slight positive association with a protective effect of dairy consumption was reported in adolescence [167]. Other comprehensive short-term studies even report the role of dairy products in facilitating weight loss in an energy-restricted diet, though long-term studies fail to provide convincing results for the same [168,169]. Similarly, other meta-analyses reported only a slight positive effect or no effect of dairy on diabetes [170,171] and mixed or no association with risk of cardiovascular diseases [170,172]. Another meta-analysis reported a positive role of dairy (particularly yogurt) in preventing the risk of type 2 diabetes; however, no association with milk and a negative association with cheese consumption was reported [173,174]. Furthermore, as observed from several meta-analyses, total dairy, full-fat dairy, low-fat dairy, milk, cheese, and yogurt consumption have no association with the risk of coronary heart disease [170,175]; though, controversial results with slightly positive effects of dairy consumption on preventing risk of cardiovascular disease were reported based on prospective cohort studies [176,177]. Variations in results based on different dairy products can be because of their potential impact on the host microbiome, the variations in practices to prepare them (such as fermented vs. not fermented), and the levels of nutrients in different dairy products. Inconsistent results observed can also be attributed to the varying nutrient content of the milk due to the varying laws and practices used by farms around the world.

Similarly, due to the high calcium and magnesium levels in milk, several studies have associated milk intake in early life with a lower risk of osteoporosis and fracture incidence [178,179]. Another meta-analysis reported that high dietary calcium through dairy with or without vitamin D supplementation increases body and lumbar bone mineral content, though this effect was only seen in children in the low baseline group as opposed to the high baseline dairy intake group [180]. Similar results of the possible effects of calcium along with vitamin D supplementation in reducing the risk of osteoporosis and bone fractures are reported in adults [28]. This points towards the need for further studies to understand the optimum levels of dairy and calcium intake to support bone mineral content and density in children. Furthermore, calcium, magnesium, and other nutrients from milk have similar benefits in adults and contribute to bone health and maintain bone structure [181,182].

In population studies, the relationship between dairy consumption and cancer risk has yielded mixed results, with limited and often inconclusive data. The bioactive compounds in dairy could have both positive (linked to calcium, lactoferrin, and fermentation products) and negative (linked to insulin-like growth factor I (IGF-1)) effects on cancer development. The World Cancer Research Fund (WCRF) continually reviews evidence on diet and cancer prevention, and some findings suggest that dairy, particularly milk and calcium, may reduce the risk of colorectal cancer [183]. Similar results were reported by other meta-analyses [180,184]. However, the evidence regarding breast cancer is inconclusive [183], although some studies suggest a potential protective effect of dairy intake, especially yogurt and low-fat dairy [185,186]. While according to the WCRF, 2014 and 2015 reports, and other observational studies, mixed results with limited evidence have been reported for associations between dairy and risk of prostrate and bladder cancer [183,187].

As reviewed earlier in Section 3, though the nutritional composition of conventional and organic milk is very similar, studies have reported differences in the levels of these nutrients in the two milk types. These differences can lead to enhanced health benefits as perceived and claimed by organic milk. It is important to consider these nutrients in the recommended daily reference intake, so as to understand the benefits, if any. Studies regarding the fatty acid composition are fairly consistent due to the direct effect of diet on milk fatty acid composition; however, protein and carbohydrate compositional results vary between studies. For instance, as mentioned above, it is now confirmed by several studies that organic milk contains higher n-3 PUFAs, CLA, and a lower omega-6 to omega-3 ratio than conventional milk [24,188]. The meta-analysis by Średnicka-Tober et al. also reported higher levels of α-tocopherol, β-carotene, lutein, and vitamin E in organic than conventional milk—an imbalance in omega 6–omega 3 ratio is associated with cardiovascular disease risk, cancer, and hypertension and disease pathogenesis [189]. As reviewed by Givens and Lovegrove, the differences in fatty acids between organic and conventional systems in the context of overall diets are important but are minimal, thus further studies with larger sample sizes are needed to underline the association between organic milk and health benefits [24]. Similarly, Średnicka-Tober et al. also report lower levels of iodine and higher levels of iron in organic milk compared to conventional milk [24]. However, milk is not the primary source of iodine or iron for humans and an otherwise balanced diet must be used to maintain levels of these nutrients. If milk is the source of iron for individuals, then those consuming organic milk must consume the appropriate supplements to avoid deficiency. These differences in nutrients are predicted to be observed under the circumstances of the switch to organic dairy and can impact health [30]. Some studies suggest the positive associations of organic dairy consumption with a lower risk of eczema in children [190] and a higher prevalence of hypospadias in the male offspring of mothers consuming conventional over organic dairy products [191].

A key characteristic of organic milk farming is avoiding the use of antibiotics and pesticides, as this can help enhance the efficacy of existing antibiotics in animals and humans. Even though this might not affect the nutritional composition of milk, it must be noted that this will lead to a reduction in the generation of new antibiotic resistance genes (ARGs) and the selection of antibiotic-resistant bacteria, thus lowering the chances of the spread of ARGs to the calf and the environment. Furthermore, the limited or prohibited use of antimicrobials and chemicals can positively impact the microbial quality of organic dairy with lower numbers of antibiotic-resistant bacteria, but safety concerns are prevalent, such as the risk of foodborne illness [192].

6. Global Market for Organic Milk Products

In the past decade, increasing awareness of self-health and the environment has given much importance to the holistic approach to organic food production. Consumption of organic products by consumers relies partly on their behavior with optimistic consumers more inclined towards organic products than pessimistic consumers, with environmental concerns driving higher consumption by pessimistic consumers [193]. Overall, the health benefits, sensory appeal, and quality of organic food products are some of the prominent factors along with environmental concerns for consumers [194]. Owing to these concerns, organic dairy products are no longer confined to first-world countries but form a big market across populations throughout the world. The global organic dairy market is estimated to be worth about $54 billion US dollars by 2030, nearly more than double the $24 billion US dollars in 2021 [195]. Asia is one of the biggest contenders in organic dairy production, followed by North America.

Organic dairy certification varies between countries; however, some standards as laid out by the United States Department of Agriculture (USDA) under the Organic Foods Production Act of 1990 and the National Organic Program (NOP), the European Union Regulation (2018/848) (https://eur-lex.europa.eu/eli/reg/2018/848/oj, accessed on 14 November 2024), Draft Guidelines of Codex/WHO/FAO and the International Federation of Organic Agricultural Movements (IFOAM) are similar and accepted on a large scale. In the East, Southeast, and South Asia, the Asia Regional Organic Standard (AROS) sets the regulation (https://www.fao.org/family-farming/detail/en/c/282204/, accessed on 8 November 2024). However, to be certified as organic, each farm must comply with the regulations governing that area. These regulations can vary widely, such as the use of antimicrobials for the treatment of mastitis in organic dairy farms is strictly prohibited in the US but permitted under veterinary recommendation in the European Union [196].

Organic dairy is the second most consumed category after fresh fruits and vegetables in the US, with retail sales totaling approximately $6 billion in 2020. In the US, an increase in certified organic dairy cows from 2000 to 360, 000 has been observed from 1990 to 2019. The increase in organic fluid milk sales from 1.92% to 5.5% of the total sales was observed from 2009 to 2021; though the increase almost plateaued by 2014 in the US [197]. By 2019, European countries including Austria, Denmark, Germany, and France had the highest numbers of organic dairy cows in the total dairy herd (Eurostat, ING research). In Europe, Germany, France, Denmark and Austria also produced the highest volume of organic milk in 2017 [198]. However, an increase in the number of organic dairy cows from 2012 to 2019 was only about 2%, which is forecasted to be much higher by 2030 [198]. The total increase in organic milk production from 2007 to 2015 also doubled in Europe, just like in the US [199]. By 2019, organic milk production in the EU represented 3.5% of total EU milk, which is also expected to be around 8% by 2031 [200].

India is the largest milk producer in the world, with an increasing demand for organic dairy. The geographical and climatic conditions in some regions, and the disease-resistant native breeds in particular provide additional benefits and are well suited for organic farming. However, cost concerns and limited knowledge of organic farming in small farms leave a large gap and a plethora of opportunities to maximize the capacity of dairy farming in India [201,202].

Despite New Zealand and Australia being leading milk producers and exporters in the world, their contributions to global organic milk were ranked 15th and 20th, respectively, in 2017 [203]. A decreasing trend in milk production has been reported in Australia over the last decade. Though the majority of dairies in Australia are in coastal areas, allowing access to fresh grazed grass, a shift to concentrate feeding has been observed due to cost and climatic conditions [43]. This could be because of the weather conditions restricting the production system.

The Japanese Agricultural Standards (JAS) for Organic Agricultural Products was established in 2000 with further development in 2005, with the addition of the JAS for Organic Livestock Products. Organic farming is faced with several difficulties in Japan due to the nature of the climate and crops prone to pests making the use of pesticides unavoidable to an extent. However, it is strictly regulated by the JAS in Japan and due to the environment-friendly approach involved, it is projected to increase [204].

7. Future Challenges and Perspectives

The organic dairy production system is associated with several perceived health benefits for consumers, and most importantly with animal health and welfare. However, it is crucial to acknowledge that the level of animal welfare can differ greatly within each production system. At present, there is no substantial evidence to support the claims that animal welfare is better in the organic or conventional system [63]. The prohibitive/restricted use of antimicrobials has invoked farmers to use various antibiotic alternatives such as aloe vera and whey-based products for disease treatment [96], and has created opportunities for several antimicrobial alternatives to be used in the dairy industry. However, the transition from conventional to organic farming is expensive and requires changes to animal husbandry, farm practices and land management. Along with these changes, certification and compliance work is very challenging accompanied by higher costs for animal maintenance [205]. Thus, incentives and support from the government are required, especially in developing countries, to provide a boost for farmers to switch to organic farming. This will help meet the challenge of demand and supply of organic dairy, which might rise with the growing demand for organic food products. Further, studies are also needed to evaluate the nutritional benefits of organic milk with the recommended daily dietary intake. Along with this, appropriate education for consumers is necessary, allowing a well-informed decision among consumers.

The benefits of organic dairy are suggested to be associated with the feed type; however, other factors including farm management and breed type are also variables in the composition of organic and conventional milk [110]. Moreover, the claimed differences in the composition of organic and conventional milk are sometimes associated with differences in the abundance of bacteria such as lactic acid bacteria, which is reported to be higher in organic dairy favored by higher concentrations of peptides and long-chain PUFAs [206]. This points towards the need for further studies to understand the additional probiotic, growth-promoting effects, and microbiological safety of organic dairy products.

Author Contributions: Conceptualization, C.S., R.P.R., K.L. and D.V.P. writing—original draft preparation, K.L. and D.V.P. writing—review and editing, K.L., D.V.P., C.S. and R.P.R. visualization, K.L. and D.V.P. supervision, C.S. and R.P.R. project administration, C.S., R.P.R. funding acquisition, C.S. and R.P.R. All authors have read and agreed to the published version of the manuscript.

Funding: This publication has been supported by grants from Science Foundation Ireland to Vistamilk [grant number 16/RC/3835] and APC Microbiome Ireland [grant numbers SFI/12/RC/2273-P2] and by the European Union (ERC, BACtheWINNER, Project No. 101054719). Views and opinions expressed are however those of the author(s) only and do not necessarily reflect those of the European Union or the European Research Council Executive Agency. Neither the European Union nor the granting authority can be held responsible for them.

Institutional Review Board Statement: Not applicable.

Informed Consent Statement: Not applicable.

Data Availability Statement: No new data were created for the production of this manuscript. All of the data here discussed and presented are available in the relative references here cited and listed.

Acknowledgments: SFI for funding from VISTAMILK and APC.

Conflicts of Interest: The authors declare no conflicts of interest.

References

1. OECD-FAO. Agricultural Outlook 2018–2027. Available online: https://www.fao.org/documents/card/en/c/I9166EN (accessed on 1 February 2024).
2. FAO. The Global Dairy Sector: Facts. 2019. Available online: http://www.dairydeclaration.org/Portals/153/Content/Documents/DDOR%20Global%20Dairy%20Facts%202019.pdf (accessed on 1 February 2024).
3. Fusco, V.; Chieffi, D.; Fanelli, F.; Logrieco, A.F.; Cho, G.S.; Kabisch, J.; Böhnlein, C.; Franz, C.M.A.P. Microbial quality and safety of milk and milk products in the 21st century. *Compr. Rev. Food Sci. Food Saf.* **2020**, *19*, 2013–2049. [CrossRef]
4. Cimmino, F.; Catapano, A.; Petrella, L.; Villano, I.; Tudisco, R.; Cavaliere, G. Role of Milk Micronutrients in Human Health. *Front. Biosci. -Landmark* **2023**, *28*, 41. [CrossRef]
5. Lin, T.; Meletharayil, G.; Kapoor, R.; Abbaspourrad, A. Bioactives in bovine milk: Chemistry, technology, and applications. *Nutr. Rev.* **2021**, *79*, 48–69. [CrossRef]

6. Meier, M.S.; Stoessel, F.; Jungbluth, N.; Juraske, R.; Schader, C.; Stolze, M. Environmental impacts of organic and conventional agricultural products–Are the differences captured by life cycle assessment? *J. Environ. Manag.* **2015**, *149*, 193–208. [CrossRef] [PubMed]
7. Zhu, Z.; Chu, F.; Dolgui, A.; Chu, C.; Zhou, W.; Piramuthu, S. Recent advances and opportunities in sustainable food supply chain: A model-oriented review. *Int. J. Prod. Res.* **2018**, *56*, 5700–5722. [CrossRef]
8. Gomiero, T. Food quality assessment in organic vs. conventional agricultural produce: Findings and issues. *Appl. Soil Ecol.* **2018**, *123*, 714–728. [CrossRef]
9. Reganold, J.P.; Wachter, J.M. Organic agriculture in the twenty-first century. *Nat. Plants* **2016**, *2*, 1–8. [CrossRef]
10. Knorr, D. Organic agriculture and foods: Advancing process-product integrations. *Crit. Rev. Food Sci. Nutr.* **2023**, 1–13. [CrossRef] [PubMed]
11. Lambotte, M.; De Cara, S.; Brocas, C.; Bellassen, V. Organic farming offers promising mitigation potential in dairy systems without compromising economic performances. *J. Environ. Manag.* **2023**, *334*, 117405. [CrossRef]
12. Organic Agriculture: FAQ. 2021. Available online: http://www.fao.org/organicag/oa-faq/oa-faq1/en/ (accessed on 1 February 2024).
13. Yu, Y.; He, Y. Information disclosure decisions in an organic food supply chain under competition. *J. Clean. Prod.* **2021**, *292*, 125976. [CrossRef]
14. Kumar, D.; Willer, H.; Ravisankar, N.; Kumar, A. Current Scenario of Organic Farming Worldwide. In *Transforming Organic Agri-Produce into Processed Food Products*; Apple Academic Press: Palm Bay, FL, USA, 2023; pp. 1–24.
15. David, R.A.R.; Silva, M.A.d.; Lopes, C.F.; Ferreira, M.C.d.Q.; Carvalho, F.V.d.B.; Santos, C.A.F.d.; Lima, N.S.; Vieira, T.A.d.S.; Moreira, T.M.d.O.; Moreira, E.d.O. O LEITE ORGÂNICO: ASPECTOS GERAIS E COLABORAÇÃO PARA O DESENVOLVIMENTO SUSTENTÁVEL. *Avanços Ciênc. Tecnol. Aliment.* **2021**, *5*, 267–284.
16. Nechaev, V.; Mikhailushkin, P.; Alieva, A. Trends in demand on the organic food market in the European countries. In Proceedings of the 2018 International Scientific Conference "Investment, Construction, Real Estate: New Technologies and Special-Purpose Development Priorities", Irkutsk, Russia, 26–27 April 2018; p. 07008.
17. Giampieri, F.; Mazzoni, L.; Cianciosi, D.; Alvarez-Suarez, J.M.; Regolo, L.; Sánchez-González, C.; Capocasa, F.; Xiao, J.; Mezzetti, B.; Battino, M. Organic vs conventional plant-based foods: A review. *Food Chem.* **2022**, *383*, 132352. [CrossRef] [PubMed]
18. Willer, H.; Lernoud, J. *The World of Organic Agriculture. STATISTICS and Emerging Trends 2023*; Research Institute of Organic Agriculture FiBL: Frick, Switzerland; IFOAM-Organics International: Bonn, Germany, 2023.
19. Kapsdorferova, Z.; Čereš, M.; ŠVikruhová, P.; ZÁBojnÍKovÁ, V.; Kataniková, R. Challenges and innovative approaches in the agricultural and food industry and changing consumer behaviour in the milk and milk products market: Case of Slovakia. *Agric. Econ./Zemědělská Ekon.* **2023**, *69*. [CrossRef]
20. Kumar, S.R.; Prajapati, S.; Parambil, J.V. Current Status of Organic Processed Food Products in the World. In *Transforming Organic Agri-Produce into Processed Food Products: Post-COVID-19 Challenges and Opportunitie*; CRC Press: Boca Raton, FL, USA, 2023.
21. Grodkowski, G.; Gołębiewski, M.; Slósarz, J.; Grodkowska, K.; Kostusiak, P.; Sakowski, T.; Puppel, K. Organic Milk Production and Dairy Farming Constraints and Prospects under the Laws of the European Union. *Animals* **2023**, *13*, 1457. [CrossRef] [PubMed]
22. Manuelian, C.L.; Penasa, M.; da Costa, L.; Burbi, S.; Righi, F.; De Marchi, M. Organic livestock production: A bibliometric review. *Animals* **2020**, *10*, 618. [CrossRef] [PubMed]
23. Brodziak, A.; Wajs, J.; Zuba-Ciszewska, M.; Król, J.; Stobiecka, M.; Jańczuk, A. Organic versus Conventional Raw Cow Milk as Material for Processing. *Animals* **2021**, *11*, 2760. [CrossRef]
24. Średnicka-Tober, D.; Barański, M.; Seal, C.J.; Sanderson, R.; Benbrook, C.; Steinshamn, H.; Gromadzka-Ostrowska, J.; Rembiałkowska, E.; Skwarło-Sońta, K.; Eyre, M. Higher PUFA and n-3 PUFA, conjugated linoleic acid, α-tocopherol and iron, but lower iodine and selenium concentrations in organic milk: A systematic literature review and meta-and redundancy analyses. *Br. J. Nutr.* **2016**, *115*, 1043–1060. [CrossRef] [PubMed]
25. Manuelian, C.L.; Vigolo, V.; Burbi, S.; Righi, F.; Simoni, M.; De Marchi, M. Detailed comparison between organic and conventional milk from Holstein-Friesian dairy herds in Italy. *J. Dairy Sci.* **2022**, *105*, 5561–5572. [CrossRef]
26. Liu, N.; Parra, H.A.; Pustjens, A.; Hettinga, K.; Mongondry, P.; Van Ruth, S.M. Evaluation of portable near-infrared spectroscopy for organic milk authentication. *Talanta* **2018**, *184*, 128–135. [CrossRef]
27. Gomes, S.I.F.; van Bodegom, P.M.; van Agtmaal, M.; Soudzilovskaia, N.A.; Bestman, M.; Duijm, E.; Speksnijder, A.; van Eekeren, N. Microbiota in dung and milk differ between organic and conventional dairy farms. *Front. Microbiol.* **2020**, *11*, 1746. [CrossRef]
28. Rozenberg, S.; Body, J.-J.; Bruyere, O.; Bergmann, P.; Brandi, M.L.; Cooper, C.; Devogelaer, J.-P.; Gielen, E.; Goemaere, S.; Kaufman, J.-M. Effects of dairy products consumption on health: Benefits and beliefs—A commentary from the Belgian Bone Club and the European Society for Clinical and Economic Aspects of Osteoporosis, Osteoarthritis and Musculoskeletal Diseases. *Calcif. Tissue Int.* **2016**, *98*, 1–17. [CrossRef]
29. Barański, M.; Rempelos, L.; Iversen, P.O.; Leifert, C. Effects of organic food consumption on human health; the jury is still out! *Food Nutr. Res.* **2017**, *61*, 1287333. [CrossRef]
30. Butler, G.; Stergiadis, S. Organic milk: Does it confer health benefits? In *Milk and Dairy Foods*; Elsevier: Amsterdam, The Netherlands, 2020; pp. 121–143.

31. Calabro, G.; Vieri, S. Limits and potential of organic farming towards a more sustainable European agri-food system. *Br. Food J.* **2023**, *126*, 223–236. [CrossRef]
32. Hirsch, S.; Koppenberg, M. Market power and profitability of organic versus conventional dairy farmers in the EU. In Proceedings of the 2023 Annual Meeting, Washington, DC, USA, 23–25 July 2023.
33. Kononets, Y.; Konvalina, P.; Bartos, P.; Smetana, P. The evolution of organic food certification. *Front. Sustain. Food Syst.* **2023**, *7*, 1167017. [CrossRef]
34. Dinçer, M.A.M.; Arslan, Y.; Okutan, S.; Dil, E. An inquiry on organic food confusion in the consumer perception: A qualitative perspective. *Br. Food J.* **2023**, *125*, 1420–1436. [CrossRef]
35. Banerjee, M.; Shanthakumar, S. International and National Policies on Organic Agriculture. In *Transforming Organic Agri-Produce into Processed Food Products: Post-COVID-19 Challenges and Opportunitie*; CRC Press: Boca Raton, FL, USA, 2023.
36. Schwendel, B.H.; Wester, T.J.; Morel, P.C.H.; Tavendale, M.H.; Deadman, C.; Shadbolt, N.M.; Otter, D.E. Invited review: Organic and conventionally produced milk—An evaluation of factors influencing milk composition. *J. Dairy Sci.* **2015**, *98*, 721–746. [CrossRef] [PubMed]
37. Adesogan, A.T.; Dahl, G.E. MILK Symposium Introduction: Dairy production in developing countries. *J. Dairy Sci.* **2020**, *103*, 9677–9680. [CrossRef] [PubMed]
38. Reardon, T.; Lu, L.; Zilberman, D. Links among innovation, food system transformation, and technology adoption, with implications for food policy: Overview of a special issue. *Food Policy* **2019**, *83*, 285–288. [CrossRef]
39. Clay, N.; Garnett, T.; Lorimer, J. Dairy intensification: Drivers, impacts and alternatives. *Ambio* **2020**, *49*, 35–48. [CrossRef] [PubMed]
40. Cele, L.P.; Hennessy, T.; Thorne, F. Evaluating farm and export competitiveness of the Irish dairy industry: Post-quota analysis. *Compet. Rev. Int. Bus. J.* **2022**, *32*, 1–20. [CrossRef]
41. McAuliffe, S.; Gilliland, T.J.; Hennessy, D. Comparison of pasture-based feeding systems and a total mixed ration feeding system on dairy cow milk production. In Proceedings of the 26th General Meeting of the European Grassland Federation, Trondheim, Norway, 4–8 September 2016.
42. Smid, A.-M.C.; Weary, D.M.; Von Keyserlingk, M.A.G. The influence of different types of outdoor access on dairy cattle behavior. *Front. Vet. Sci.* **2020**, *7*, 257. [CrossRef] [PubMed]
43. Joubran, A.M.; Pierce, K.M.; Garvey, N.; Shalloo, L.; O'Callaghan, T.F. Invited review: A 2020 perspective on pasture-based dairy systems and products. *J. Dairy Sci.* **2021**, *104*, 7364–7382. [CrossRef] [PubMed]
44. Magan, J.B.; O'Callaghan, T.F.; Kelly, A.L.; McCarthy, N.A. Compositional and functional properties of milk and dairy products derived from cows fed pasture or concentrate-based diets. *Compr. Rev. Food Sci. Food Saf.* **2021**, *20*, 2769–2800. [CrossRef] [PubMed]
45. O'Callaghan, T.F.; Hennessy, D.; McAuliffe, S.; Kilcawley, K.N.; O'Donovan, M.; Dillon, P.; Ross, R.P.; Stanton, C. Effect of pasture versus indoor feeding systems on raw milk composition and quality over an entire lactation. *J. Dairy Sci.* **2016**, *99*, 9424–9440. [CrossRef]
46. O'Callaghan, T.F.; Faulkner, H.; McAuliffe, S.; O'Sullivan, M.G.; Hennessy, D.; Dillon, P.; Kilcawley, K.N.; Stanton, C.; Ross, R.P. Quality characteristics, chemical composition, and sensory properties of butter from cows on pasture versus indoor feeding systems. *J. Dairy Sci.* **2016**, *99*, 9441–9460. [CrossRef] [PubMed]
47. O'Callaghan, T.F.; Mannion, D.T.; Hennessy, D.; McAuliffe, S.; O'Sullivan, M.G.; Leeuwendaal, N.; Beresford, T.P.; Dillon, P.; Kilcawley, K.N.; Sheehan, J.J. Effect of pasture versus indoor feeding systems on quality characteristics, nutritional composition, and sensory and volatile properties of full-fat Cheddar cheese. *J. Dairy Sci.* **2017**, *100*, 6053–6073. [CrossRef]
48. Timlin, M.; Fitzpatrick, E.; McCarthy, K.; Tobin, J.T.; Murphy, E.G.; Pierce, K.M.; Murphy, J.P.; Hennessy, D.; O'Donovan, M.; Harbourne, N. Impact of varying levels of pasture allowance on the nutritional quality and functionality of milk throughout lactation. *J. Dairy Sci.* **2023**, *106*, 6597–6622. [CrossRef]
49. Schingoethe, D.J. A 100-Year Review: Total mixed ration feeding of dairy cows. *J. Dairy Sci.* **2017**, *100*, 10143–10150. [CrossRef]
50. van den Pol-van Dasselaar, A.; Hennessy, D.; Isselstein, J. Grazing of dairy cows in Europe—An in-depth analysis based on the perception of grassland experts. *Sustainability* **2020**, *12*, 1098. [CrossRef]
51. Charlton, G.L.; Rutter, S.M.; East, M.; Sinclair, L.A. Preference of dairy cows: Indoor cubicle housing with access to a total mixed ration vs. access to pasture. *Appl. Anim. Behav. Sci.* **2011**, *130*, 1–9. [CrossRef]
52. Legrand, A.L.; Von Keyserlingk, M.A.G.; Weary, D.M. Preference and usage of pasture versus free-stall housing by lactating dairy cattle. *J. Dairy Sci.* **2009**, *92*, 3651–3658. [CrossRef] [PubMed]
53. Armbrecht, L.; Lambertz, C.; Albers, D.; Gauly, M. Does access to pasture affect claw condition and health in dairy cows? *Vet. Rec.* **2018**, *182*, 79. [CrossRef] [PubMed]
54. Firth, C.L.; Laubichler, C.; Schleicher, C.; Fuchs, K.; Käsbohrer, A.; Egger-Danner, C.; Köfer, J.; Obritzhauser, W. Relationship between the probability of veterinary-diagnosed bovine mastitis occurring and farm management risk factors on small dairy farms in Austria. *J. Dairy Sci.* **2019**, *102*, 4452–4463. [CrossRef] [PubMed]
55. Mee, J.F.; Boyle, L.A. Assessing whether dairy cow welfare is "better" in pasture-based than in confinement-based management systems. *N. Z. Vet. J.* **2020**, *68*, 168–177. [CrossRef] [PubMed]

56. Morales-Almaráz, E.; Soldado, A.; González, A.; Martínez-Fernández, A.; Domínguez-Vara, I.; de la Roza-Delgado, B.; Vicente, F. Improving the fatty acid profile of dairy cow milk by combining grazing with feeding of total mixed ration. *J. Dairy Res.* **2010**, *77*, 225–230. [CrossRef] [PubMed]
57. Rodríguez-Bermúdez, R.; Miranda, M.; Baudracco, J.; Fouz, R.; Pereira, V.; López-Alonso, M. Breeding for organic dairy farming: What types of cows are needed? *J. Dairy Res.* **2019**, *86*, 3–12. [CrossRef]
58. Santoni, M.; Verdi, L.; Imran Pathan, S.; Napoli, M.; Dalla Marta, A.; Dani, F.R.; Pacini, G.C.; Ceccherini, M.T. Soil microbiome biomass, activity, composition and CO_2 emissions in a long-term organic and conventional farming systems. *Soil Use Manag.* **2023**, *39*, 588–605. [CrossRef]
59. Gamage, A.; Gangahagedara, R.; Gamage, J.; Jayasinghe, N.; Kodikara, N.; Suraweera, P.; Merah, O. Role of organic farming for achieving sustainability in agriculture. *Farming Syst.* **2023**, *1*, 100005. [CrossRef]
60. Orjales, I.; Lopez-Alonso, M.; Miranda, M.; Alaiz-Moretón, H.; Resch, C.; López, S. Dairy cow nutrition in organic farming systems. Comparison with the conventional system. *Animal* **2019**, *13*, 1084–1093. [CrossRef]
61. Health, E.P.o.A.; Animal, W.; Nielsen, S.S.; Alvarez, J.; Bicout, D.J.; Calistri, P.; Canali, E.; Drewe, J.A.; Garin-Bastuji, B.; Gonzales Rojas, J.L.; et al. Welfare of dairy cows. *EFSA J.* **2023**, *21*, e07993.
62. Wagner, K.; Brinkmann, J.; Bergschmidt, A.; Renziehausen, C.; March, S. The effects of farming systems (organic vs. conventional) on dairy cow welfare, based on the Welfare Quality® protocol. *Animal* **2021**, *15*, 100301. [CrossRef]
63. Åkerfeldt, M.P.; Gunnarsson, S.; Bernes, G.; Blanco-Penedo, I. Health and welfare in organic livestock production systems—A systematic mapping of current knowledge. *Org. Agric.* **2021**, *11*, 105–132. [CrossRef]
64. Xiong, W.; Sun, Y.; Zeng, Z. Antimicrobial use and antimicrobial resistance in food animals. *Environ. Sci. Pollut. Res.* **2018**, *25*, 18377–18384. [CrossRef] [PubMed]
65. Foutz, C.A.; Godden, S.M.; Bender, J.B.; Diez-Gonzalez, F.; Akhtar, M.; Vatulin, A. Exposure to antimicrobials through the milk diet or systemic therapy is associated with a transient increase in antimicrobial resistance in fecal Escherichia coli of dairy calves. *J. Dairy Sci.* **2018**, *101*, 10126–10141. [CrossRef] [PubMed]
66. Ager, E.; Carvalho, T.; Silva, E.; Ricke, S.; Hite, J. Global trends in antimicrobial resistance on organic and conventional farms. *bioRxiv* **2023**, *13*, 22608. [CrossRef]
67. Lorenz, K.; Lal, R. Organic Agriculture and Greenhouse Gas Emissions. In *Organic Agriculture and Climate Change*; Springer: Berlin/Heidelberg, Germany, 2022; pp. 129–175.
68. Brito, A.F.; Silva, L.H.P. Symposium review: Comparisons of feed and milk nitrogen efficiency and carbon emissions in organic versus conventional dairy production systems. *J. Dairy Sci.* **2020**, *103*, 5726–5739. [CrossRef]
69. Średnicka-Tober, D.; Obiedzińska, A.; Kazimierczak, R.; Rembiałkowska, E. Environmental impact of organic vs. conventional agriculture-a review. *J. Res. Appl. Agric. Eng.* **2016**, *61*, 204–211.
70. Kilcawley, K.N.; Faulkner, H.; Clarke, H.J.; O'Sullivan, M.G.; Kerry, J.P. Factors influencing the flavour of bovine milk and cheese from grass based versus non-grass based milk production systems. *Foods* **2018**, *7*, 37. [CrossRef]
71. Król, J.; Brodziak, A.; Topyła, B. The nutritional value of the milk of Simmental cows in relation to the season and production system. *Anim. Prod. Rev.* **2016**, *6*, 20–24.
72. Sapbamrer, R.; Thammachai, A. A systematic review of factors influencing farmers' adoption of organic farming. *Sustainability* **2021**, *13*, 3842. [CrossRef]
73. Poudel, S.P.; Acharya, R.; Chetri, D.K. Somatic cell count: An indicator of intramammary infection in dairy animals. *Matrix Sci. Pharma* **2021**, *5*, 49–53. [CrossRef]
74. Stocco, G.; Summer, A.; Cipolat-Gotet, C.; Zanini, L.; Vairani, D.; Dadousis, C.; Zecconi, A. Differential somatic cell count as a novel indicator of milk quality in dairy cows. *Animals* **2020**, *10*, 753. [CrossRef]
75. Neculai-Valeanu, A.-S.; Ariton, A.-M. Udder health monitoring for prevention of bovine mastitis and improvement of milk quality. *Bioengineering* **2022**, *9*, 608. [CrossRef]
76. Moradi, M.; Omer, A.K.; Razavi, R.; Valipour, S.; Guimarães, J.T. The relationship between milk somatic cell count and cheese production, quality and safety: A review. *Int. Dairy J.* **2021**, *113*, 104884. [CrossRef]
77. Orjales, I.; Lopez-Alonso, M.; Miranda, M.; Rodríguez-Bermúdez, R.; Rey-Crespo, F.; Villar, A. The main factors affecting somatic cell count in organic dairy farming. *Span. J. Agric. Res.* **2017**, *15*, e06SC02. [CrossRef]
78. Orjales, I.; López-Alonso, M.; Rodríguez-Bermúdez, R.; Rey-Crespo, F.; Villar, A.; Miranda, M. Use of homeopathy in organic dairy farming in Spain. *Homeopathy* **2016**, *105*, 102–108. [CrossRef] [PubMed]
79. Orjales, I.; López-Alonso, M.; Rodríguez-Bermúdez, R.; Rey-Crespo, F.; Villar, A.; Miranda, M. Is lack of antibiotic usage affecting udder health status of organic dairy cattle? *J. Dairy Res.* **2016**, *83*, 464–467. [CrossRef] [PubMed]
80. FDA. *Grade "A" Pasteurized Milk Ordinance*; FDA: Silver Spring, MD, USA, 2017.
81. Martin, N.H.; Evanowski, R.L.; Wiedmann, M. Invited review: Redefining raw milk quality—Evaluation of raw milk microbiological parameters to ensure high-quality processed dairy products. *J. Dairy Sci.* **2023**, *106*, 1502–1517. [CrossRef]
82. Berg, G.; Rybakova, D.; Fischer, D.; Cernava, T.; Vergès, M.-C.C.; Charles, T.; Chen, X.; Cocolin, L.; Eversole, K.; Corral, G.H. Microbiome definition re-visited: Old concepts and new challenges. *Microbiome* **2020**, *8*, 103.
83. Dean, C.J.; Slizovskiy, I.B.; Crone, K.K.; Pfennig, A.X.; Heins, B.J.; Caixeta, L.S.; Noyes, N.R. Investigating the cow skin and teat canal microbiomes of the bovine udder using different sampling and sequencing approaches. *J. Dairy Sci.* **2021**, *104*, 644–661. [CrossRef] [PubMed]

84. Dahlberg, J.; Williams, J.E.; McGuire, M.A.; Peterson, H.K.; Östensson, K.; Agenäs, S.; Dicksved, J.; Waller, K.P. Microbiota of bovine milk, teat skin, and teat canal: Similarity and variation due to sampling technique and milk fraction. *J. Dairy Sci.* **2020**, *103*, 7322–7330. [CrossRef]
85. Dean, C.J.; Deng, Y.; Wehri, T.J.; Ray, T.; Peña-Mosca, F.; Crooker, B.A.; Godden, S.M.; Caixeta, L.S.; Noyes, N. The impact of kit, environment and sampling contamination on the observed microbiome of bovine milk. *bioRxiv* **2023**. [CrossRef]
86. Guo, W.; Liu, S.; Khan, M.Z.; Wang, J.; Chen, T.; Alugongo, G.M.; Li, S.; Cao, Z. Bovine milk microbiota: Key players, origins, and potential contributions to early-life gut development. *J. Adv. Res.* **2023**. [CrossRef] [PubMed]
87. Hagey, J.V.; Bhatnagar, S.; Heguy, J.M.; Karle, B.M.; Price, P.L.; Meyer, D.; Maga, E.A. Fecal microbial communities in a large representative cohort of California dairy cows. *Front. Microbiol.* **2019**, *10*, 1093. [CrossRef] [PubMed]
88. Loor, J.J.; Elolimy, A.A.; McCann, J.C. Dietary impacts on rumen microbiota in beef and dairy production. *Anim. Front.* **2016**, *6*, 22–29. [CrossRef]
89. Zhang, R.; Huo, W.; Zhu, W.; Mao, S. Characterization of bacterial community of raw milk from dairy cows during subacute ruminal acidosis challenge by high-throughput sequencing. *J. Sci. Food Agric.* **2015**, *95*, 1072–1079. [CrossRef] [PubMed]
90. Mbareche, H.; Veillette, M.; Bilodeau, G.J.; Duchaine, C. Fungal bioaerosols at five dairy farms: A novel approach to describe workers' exposure. *bioRxiv* **2018**, 308825. [CrossRef]
91. Guzzon, R.; Carafa, I.; Tuohy, K.; Cervantes, G.; Vernetti, L.; Barmaz, A.; Larcher, R.; Franciosi, E. Exploring the microbiota of the red-brown defect in smear-ripened cheese by 454-pyrosequencing and its prevention using different cleaning systems. *Food Microbiol.* **2017**, *62*, 160–168. [CrossRef] [PubMed]
92. Tommasoni, C.; Fiore, E.; Lisuzzo, A.; Gianesella, M. Mastitis in Dairy Cattle: On-Farm Diagnostics and Future Perspectives. *Animals* **2023**, *13*, 2538. [CrossRef] [PubMed]
93. Angelopoulou, A.; Holohan, R.; Rea, M.C.; Warda, A.K.; Hill, C.; Ross, R.P. Bovine mastitis is a polymicrobial disease requiring a polydiagnostic approach. *Int. Dairy J.* **2019**, *99*, 104539. [CrossRef]
94. Sharun, K.; Dhama, K.; Tiwari, R.; Gugjoo, M.B.; Iqbal Yatoo, M.; Patel, S.K.; Pathak, M.; Karthik, K.; Khurana, S.K.; Singh, R. Advances in therapeutic and managemental approaches of bovine mastitis: A comprehensive review. *Vet. Q.* **2021**, *41*, 107–136. [CrossRef]
95. Peña-Mosca, F.; Dean, C.; Machado, V.; Fernandes, L.; Pinedo, P.; Doster, E.; Heins, B.; Sharpe, K.; Ray, T.; Feijoo, V.; et al. Investigation of intramammary infections in primiparous cows during early lactation on organic dairy farms. *J. Dairy Sci.* **2023**, *106*, 9377–9392. [CrossRef] [PubMed]
96. Pol, M.; Ruegg, P.L. Relationship between antimicrobial drug usage and antimicrobial susceptibility of gram-positive mastitis pathogens. *J. Dairy Sci.* **2007**, *90*, 262–273. [CrossRef]
97. Cicconi-Hogan, K.M.; Gamroth, M.; Richert, R.; Ruegg, P.L.; Stiglbauer, K.E.; Schukken, Y.H. Risk factors associated with bulk tank standard plate count, bulk tank coliform count, and the presence of Staphylococcus aureus on organic and conventional dairy farms in the United States. *J. Dairy Sci.* **2013**, *96*, 7578–7590. [CrossRef]
98. Tikofsky, L.L.; Barlow, J.W.; Santisteban, C.; Schukken, Y.H. A comparison of antimicrobial susceptibility patterns for Staphylococcus aureus in organic and conventional dairy herds. *Microb. Drug Resist.* **2003**, *9*, 39–45. [CrossRef]
99. Hamilton, C.; Emanuelson, U.; Forslund, K.; Hansson, I.; Ekman, T. Mastitis and related management factors in certified organic dairy herds in Sweden. *Acta Vet. Scand.* **2006**, *48*, 1–7. [CrossRef]
100. Valle, P.S.; Lien, G.; Flaten, O.; Koesling, M.; Ebbesvik, M. Herd health and health management in organic versus conventional dairy herds in Norway. *Livest. Sci.* **2007**, *112*, 123–132. [CrossRef]
101. Richert, R.M.; Cicconi, K.M.; Gamroth, M.J.; Schukken, Y.H.; Stiglbauer, K.E.; Ruegg, P.L. Risk factors for clinical mastitis, ketosis, and pneumonia in dairy cattle on organic and small conventional farms in the United States. *J. Dairy Sci.* **2013**, *96*, 4269–4285. [CrossRef]
102. Hardeng, F.; Edge, V.L. Mastitis, ketosis, and milk fever in 31 organic and 93 conventional Norwegian dairy herds. *J. Dairy Sci.* **2001**, *84*, 2673–2679. [CrossRef]
103. Mullen, K.A.E.; Sparks, L.G.; Lyman, R.L.; Washburn, S.P.; Anderson, K.L. Comparisons of milk quality on North Carolina organic and conventional dairies. *J. Dairy Sci.* **2013**, *96*, 6753–6762. [CrossRef]
104. Ueda, Y.; Asakuma, S.; Miyaji, M.; Akiyama, F. Effect of time at pasture and herbage intake on profile of volatile organic compounds of dairy cow milk. *Anim. Sci. J.* **2016**, *87*, 117–125. [CrossRef]
105. Aprea, E.; Biasioli, F.; Carlin, S.; Endrizzi, I.; Gasperi, F. Investigation of volatile compounds in two raspberry cultivars by two headspace techniques: Solid-phase microextraction/gas chromatography—Mass spectrometry (SPME/GC−MS) and proton-transfer reaction−mass spectrometry (PTR−MS). *J. Agric. Food Chem.* **2009**, *57*, 4011–4018. [CrossRef]
106. Vazquez-Landaverde, P.A.; Torres, J.A.; Qian, M.C. Quantification of trace volatile sulfur compounds in milk by solid-phase microextraction and gas chromatography–pulsed flame photometric detection. *J. Dairy Sci.* **2006**, *89*, 2919–2927. [CrossRef]
107. Calvo, M.M.; de la Hoz, L. Flavour of heated milks. A review. *Int. Dairy J.* **1992**, *2*, 69–81. [CrossRef]
108. Urbach, G. Effect of feed on flavor in dairy foods. *J. Dairy Sci.* **1990**, *73*, 3639–3650. [CrossRef]
109. Smigic, N.; Djekic, I.; Tomasevic, I.; Stanisic, N.; Nedeljkovic, A.; Lukovic, V.; Miocinovic, J. Organic and conventional milk–insight on potential differences. *Br. Food J.* **2017**, *119*, 366–376. [CrossRef]

110. Schwendel, B.H.; Wester, T.J.; Morel, P.C.H.; Fong, B.; Tavendale, M.H.; Deadman, C.; Shadbolt, N.M.; Otter, D.E. Pasture feeding conventional cows removes differences between organic and conventionally produced milk. *Food Chem.* **2017**, *229*, 805–813. [CrossRef]
111. Liu, N.; Koot, A.; Hettinga, K.; De Jong, J.; van Ruth, S.M. Portraying and tracing the impact of different production systems on the volatile organic compound composition of milk by PTR-(Quad) MS and PTR-(ToF) MS. *Food Chem.* **2018**, *239*, 201–207. [CrossRef]
112. Clarke, H.J.; Griffin, C.; Rai, D.K.; O'Callaghan, T.F.; O'Sullivan, M.G.; Kerry, J.P.; Kilcawley, K.N. Dietary compounds influencing the sensorial, volatile and phytochemical properties of bovine milk. *Molecules* **2019**, *25*, 26. [CrossRef] [PubMed]
113. Alothman, M.; Hogan, S.A.; Hennessy, D.; Dillon, P.; Kilcawley, K.N.; O'Donovan, M.; Tobin, J.; Fenelon, M.A.; O'Callaghan, T.F. The "grass-fed" milk story: Understanding the impact of pasture feeding on the composition and quality of bovine milk. *Foods* **2019**, *8*, 350. [CrossRef]
114. Gallina Toschi, T.; Bendini, A.; Barbieri, S.; Valli, E.; Cezanne, M.L.; Buchecker, K.; Canavari, M. Organic and conventional nonflavored yogurts from the Italian market: Study on sensory profiles and consumer acceptability. *J. Sci. Food Agric.* **2012**, *92*, 2788–2795. [CrossRef]
115. Bloksma, J.; Adriaansen-Tennekes, R.; Huber, M.; van de Vijver, L.P.L.; Baars, T.; de Wit, J. Comparison of organic and conventional raw milk quality in the Netherlands. *Biol. Agric. Hortic.* **2008**, *26*, 69–83. [CrossRef]
116. Manzocchi, E.; Martin, B.; Bord, C.; Verdier-Metz, I.; Bouchon, M.; De Marchi, M.; Constant, I.; Giller, K.; Kreuzer, M.; Berard, J. Feeding cows with hay, silage, or fresh herbage on pasture or indoors affects sensory properties and chemical composition of milk and cheese. *J. Dairy Sci.* **2021**, *104*, 5285–5302. [CrossRef] [PubMed]
117. Brodziak, A.; Król, J.; Litwińczuk, Z.; Barłowska, J. Differences in bioactive protein and vitamin status of milk from certified organic and conventional farms. *Int. J. Dairy Technol.* **2018**, *71*, 321–332. [CrossRef]
118. Kiczorowska, B.; Samolińska, W.; Marczuk, J.; Winiarska-Mieczan, A.; Klebaniuk, R.; Kowalczuk-Vasilev, E.; Kiczorowski, P.; Zasadna, Z. Comparative effects of organic, traditional, and intensive production with probiotics on the fatty acid profile of cow's milk. *J. Food Compos. Anal.* **2017**, *63*, 157–163. [CrossRef]
119. Qin, N.; Faludi, G.; Beauclercq, S.; Pitt, J.; Desnica, N.; Pétursdóttir, Á.; Newton, E.E.; Angelidis, A.; Givens, I.; Juniper, D. Macromineral and trace element concentrations and their seasonal variation in milk from organic and conventional dairy herds. *Food Chem.* **2021**, *359*, 129865. [CrossRef] [PubMed]
120. Singh, A.; Duche, R.T.; Wandhare, A.G.; Sian, J.K.; Singh, B.P.; Sihag, M.K.; Singh, K.S.; Sangwan, V.; Talan, S.; Panwar, H. Milk-derived antimicrobial peptides: Overview, applications, and future perspectives. *Probiotics Antimicrob. Proteins* **2023**, *15*, 44–62. [CrossRef] [PubMed]
121. Brodziak, A.; Król, J.; Litwińczuk, Z.; Florek, M. Bioactive compound levels and sensory quality of partially skimmed organic yoghurts: Effects of the milk treatment, production season and starter culture. *Int. J. Dairy Technol.* **2021**, *74*, 139–147. [CrossRef]
122. Król, J.; Brodziak, A.; Zaborska, A.; Litwińczuk, Z. Comparison of whey proteins and lipophilic vitamins between four cow breeds maintained in intensive production system. *Mljekarstvo/Dairy* **2017**, *67*. [CrossRef]
123. Wójcik-Saganek, A. Nutritional Value and Technological Suitability of Farm Milk Together with a Technical and Economic Analysis of the Efficiency of Its Production. Doctoral Dissertation, The University of Life Sciences in Lublin, Lublin, Poland, 2019.
124. Zhao, X.; Xu, X.-X.; Liu, Y.; Xi, E.-Z.; An, J.-J.; Tabys, D.; Liu, N. The in vitro protective role of bovine lactoferrin on intestinal epithelial barrier. *Molecules* **2019**, *24*, 148. [CrossRef]
125. Anand, R.; Mohan, L.; Bharadvaja, N. Disease prevention and treatment using β-carotene: The ultimate provitamin A. *Rev. Bras. Farmacogn.* **2022**, *32*, 491–501. [CrossRef] [PubMed]
126. Fox, P.F.; Uniacke-Lowe, T.; McSweeney, P.L.H.; O'Mahony, J.A.; Fox, P.F.; Uniacke-Lowe, T.; McSweeney, P.L.H.; O'Mahony, J.A. Vitamins in milk and dairy products. In *Dairy Chemistry and Biochemistry*; Springer: Berlin/Heidelberg, Germany, 2015; pp. 271–297.
127. Liao, S.; Omage, S.O.; Börmel, L.; Kluge, S.; Schubert, M.; Wallert, M.; Lorkowski, S. Vitamin E and metabolic health: Relevance of interactions with other micronutrients. *Antioxidants* **2022**, *11*, 1785. [CrossRef] [PubMed]
128. Stobiecka, M.; Król, J.; Brodziak, A. Antioxidant activity of milk and dairy products. *Animals* **2022**, *12*, 245. [CrossRef] [PubMed]
129. Ponnampalam, E.N.; Kiani, A.; Santhiravel, S.; Holman, B.W.B.; Lauridsen, C.; Dunshea, F.R. The importance of dietary antioxidants on oxidative stress, meat and milk production, and their preservative aspects in farm animals: Antioxidant action, animal health, and product quality—Invited review. *Animals* **2022**, *12*, 3279. [CrossRef]
130. Kuczyńska, B.; Nałęcz-Tarwacka, T.; Puppel, K.; Gołębiewski, M.; Grodzki, H.; Slósarz, J. The content of bioactive components in milk depending on cow feeding model in certified ecological farms. *J. Res. Appl. Agric. Eng* **2011**, *56*, 7–13.
131. Fong, B.; Ma, K.; McJarrow, P. Quantification of bovine milk oligosaccharides using liquid chromatography–selected reaction monitoring–mass spectrometry. *J. Agric. Food Chem.* **2011**, *59*, 9788–9795. [CrossRef]
132. Kiely, L.J.; Busca, L.; Lane, J.A.; van Sinderen, D.; Hickey, R.M. Molecular strategies for the utilisation of human milk oligosaccharides by infant gut-associated bacteria. *FEMS Microbiol. Rev.* **2023**, *47*, fuad056. [CrossRef]
133. Taormina, V.M.; Unger, A.L.; Schiksnis, M.R.; Torres-Gonzalez, M.; Kraft, J. Branched-chain fatty acids—An underexplored class of dairy-derived fatty acids. *Nutrients* **2020**, *12*, 2875. [CrossRef]
134. O'Callaghan, T.F. The benefits of pasture-based dairy. *Ir. Food* **2019**, *1*, 34–35.

135. Hanuš, O.; Samková, E.; Křížová, L.; Hasoňová, L.; Kala, R. Role of fatty acids in milk fat and the influence of selected factors on their variability—A review. *Molecules* **2018**, *23*, 1636. [CrossRef]
136. Tzamaloukas, O.; Neofytou, M.C.; Simitzis, P.E.; Miltiadou, D. Effect of farming system (organic vs. conventional) and season on composition and fatty acid profile of bovine, caprine and ovine milk and retail Halloumi cheese produced in Cyprus. *Foods* **2021**, *10*, 1016. [CrossRef]
137. Vanbergue, E.; Peyraud, J.L.; Ferlay, A.; Miranda, G.; Martin, P.; Hurtaud, C. Effects of feeding level, type of forage and milking time on milk lipolytic system in dairy cows. *Livest. Sci.* **2018**, *217*, 116–126. [CrossRef]
138. Liu, S.; Zhang, R.; Kang, R.; Meng, J.; Ao, C. Milk fatty acids profiles and milk production from dairy cows fed different forage quality diets. *Anim. Nutr.* **2016**, *2*, 329–333. [CrossRef] [PubMed]
139. Benbrook, C.M.; Davis, D.R.; Heins, B.J.; Latif, M.A.; Leifert, C.; Peterman, L.; Butler, G.; Faergeman, O.; Abel-Caines, S.; Baranski, M. Enhancing the fatty acid profile of milk through forage-based rations, with nutrition modeling of diet outcomes. *Food Sci. Nutr.* **2018**, *6*, 681–700. [CrossRef] [PubMed]
140. Benbrook, C.M.; Butler, G.; Latif, M.A.; Leifert, C.; Davis, D.R. Organic production enhances milk nutritional quality by shifting fatty acid composition: A United States–wide, 18-month study. *PLoS ONE* **2013**, *8*, e82429. [CrossRef] [PubMed]
141. Tsiafoulis, C.G.; Papaemmanouil, C.; Alivertis, D.; Tzamaloukas, O.; Miltiadou, D.; Balayssac, S.; Malet-Martino, M.; Gerothanassis, I.P. NMR-based μetabolomics of the lipid fraction of organic and conventional bovine milk. *Molecules* **2019**, *24*, 1067. [CrossRef] [PubMed]
142. Ellis, K.A.; Innocent, G.; Grove-White, D.; Cripps, P.; McLean, W.G.; Howard, C.V.; Mihm, M. Comparing the fatty acid composition of organic and conventional milk. *J. Dairy Sci.* **2006**, *89*, 1938–1950. [CrossRef] [PubMed]
143. Collomb, M.; Bisig, W.; Bütikofer, U.; Sieber, R.; Bregy, M.; Etter, L. Fatty acid composition of mountain milk from Switzerland: Comparison of organic and integrated farming systems. *Int. Dairy J.* **2008**, *18*, 976–982. [CrossRef]
144. Butler, G.; Stergiadis, S.; Seal, C.; Eyre, M.; Leifert, C. Fat composition of organic and conventional retail milk in northeast England. *J. Dairy Sci.* **2011**, *94*, 24–36. [CrossRef]
145. Adler, S.A.; Jensen, S.K.; Govasmark, E.; Steinshamn, H. Effect of short-term versus long-term grassland management and seasonal variation in organic and conventional dairy farming on the composition of bulk tank milk. *J. Dairy Sci.* **2013**, *96*, 5793–5810. [CrossRef]
146. Miyamoto, J.; Igarashi, M.; Watanabe, K.; Karaki, S.-i.; Mukouyama, H.; Kishino, S.; Li, X.; Ichimura, A.; Irie, J.; Sugimoto, Y. Gut microbiota confers host resistance to obesity by metabolizing dietary polyunsaturated fatty acids. *Nat. Commun.* **2019**, *10*, 4007. [CrossRef]
147. Khiaosa-Ard, R.; Klevenhusen, F.; Soliva, C.R.; Kreuzer, M.; Leiber, F. Transfer of linoleic and linolenic acid from feed to milk in cows fed isoenergetic diets differing in proportion and origin of concentrates and roughages. *J. Dairy Res.* **2010**, *77*, 331–336. [CrossRef]
148. Ferreiro, T.; Gayoso, L.; Rodríguez-Otero, J.L. Milk phospholipids: Organic milk and milk rich in conjugated linoleic acid compared with conventional milk. *J. Dairy Sci.* **2015**, *98*, 9–14. [CrossRef] [PubMed]
149. Davis, H.; Chatzidimitriou, E.; Leifert, C.; Butler, G. Evidence that forage-fed cows can enhance milk quality. *Sustainability* **2020**, *12*, 3688. [CrossRef]
150. Bauman, D.E.; Lock, A.L.; Conboy Stephenson, R.; Linehan, K.; Ross, R.P.; Stanton, C. Conjugated linoleic acid: Biosynthesis and nutritional significance. In *Advanced Dairy Chemistry*; Springer: Cham, Switzerland, 2020; Volume 2, pp. 67–106.
151. Badawy, S.; Liu, Y.; Guo, M.; Liu, Z.; Xie, C.; Marawan, M.A.; Ares, I.; Lopez-Torres, B.; Martínez, M.; Maximiliano, J.-E. Conjugated linoleic acid (CLA) as a functional food: Is it beneficial or not? *Food Res. Int.* **2023**, *172*, 113158. [CrossRef] [PubMed]
152. Manuelian, C.L.; Penasa, M.; Visentin, G.; Zidi, A.; Cassandro, M.; De Marchi, M. Mineral composition of cow milk from multibreed herds. *Anim. Sci. J.* **2018**, *89*, 1622–1627. [CrossRef] [PubMed]
153. Litwińczuk, Z.; Koperska, N.; Chabuz, W.; Kędzierska-Matysek, M. Basic chemical composition and mineral content of the milk of cows of various breeds raised on organic farms and on traditional farms using intensive and traditional feeding systems. *Med. Weter.* **2018**, *74*, 309–313. [CrossRef]
154. Stergiadis, S.; Qin, N.; Faludi, G.; Beauclercq, S.; Pitt, J.; Desnica, N.; Pétursdóttir, Á.H.; Newton, E.E.; Angelidis, A.E.; Givens, I. Mineral concentrations in bovine milk from farms with contrasting grazing management. *Foods* **2021**, *10*, 2733. [CrossRef] [PubMed]
155. Kwiatkowski, C.A.; Harasim, E. Chemical properties of soil in four-field crop rotations under organic and conventional farming systems. *Agronomy* **2020**, *10*, 1045. [CrossRef]
156. Rodríguez-Bermúdez, R.; López-Alonso, M.; Miranda, M.; Fouz, R.; Orjales, I.; Herrero-Latorre, C. Chemometric authentication of the organic status of milk on the basis of trace element content. *Food Chem.* **2018**, *240*, 686–693. [CrossRef]
157. López-Alonso, M.; Rey-Crespo, F.; Herrero-Latorre, C.; Miranda, M. Identifying sources of metal exposure in organic and conventional dairy farming. *Chemosphere* **2017**, *185*, 1048–1055. [CrossRef]
158. Boudebbouz, A.; Boudalia, S.; Bousbia, A.; Habila, S.; Boussadia, M.I.; Gueroui, Y. Heavy metals levels in raw cow milk and health risk assessment across the globe: A systematic review. *Sci. Total Environ.* **2021**, *751*, 141830. [CrossRef]
159. Varol, M.; Sünbül, M.R. Macroelements and toxic trace elements in muscle and liver of fish species from the largest three reservoirs in Turkey and human risk assessment based on the worst-case scenarios. *Environ. Res.* **2020**, *184*, 109298. [CrossRef]

160. Ziarati, P.; Shirkhan, F.; Mostafidi, M.; Zahedi, M.T. An overview of the heavy metal contamination in milk and dairy products. *Acta Sci. Pharm. Sci.* **2018**, *2*, 1–14.
161. Zwierzchowski, G.; Ametaj, B.N. Minerals and heavy metals in the whole raw milk of dairy cows from different management systems and countries of origin: A meta-analytical study. *J. Agric. Food Chem.* **2018**, *66*, 6877–6888. [CrossRef] [PubMed]
162. Seğmenoğlu, M.S.; Baydan, E. Comparison of heavy metal levels of organic and conventional milk and milk products in Turkey. *Turk. J. Agric. -Food Sci. Technol.* **2021**, *9*, 696–700. [CrossRef]
163. Stergiadis, S.; Berlitz, C.B.; Hunt, B.; Garg, S.; Givens, D.I.; Kliem, K.E. An update to the fatty acid profiles of bovine retail milk in the United Kingdom: Implications for nutrition in different age and gender groups. *Food Chem.* **2019**, *276*, 218–230. [CrossRef] [PubMed]
164. Liu, N.; Pustjens, A.M.; Erasmus, S.W.; Yang, Y.; Hettinga, K.; van Ruth, S.M. Dairy farming system markers: The correlation of forage and milk fatty acid profiles from organic, pasture and conventional systems in the Netherlands. *Food Chem.* **2020**, *314*, 126153. [CrossRef] [PubMed]
165. Ormston, S.; Qin, N.; Faludi, G.; Pitt, J.; Gordon, A.W.; Theodoridou, K.; Yan, T.; Huws, S.A.; Stergiadis, S. Implications of organic dairy management on herd performance and milk fatty acid profiles and interactions with season. *Foods* **2023**, *12*, 1589. [CrossRef] [PubMed]
166. Soedamah-Muthu, S.S.; Guo, J. Dairy consumption and cardiometabolic diseases: Evidence from prospective studies. In *Milk and Dairy Foods*; Elsevier: Amsterdam, The Netherlands, 2020; pp. 1–28.
167. Wang, W.; Wu, Y.; Zhang, D. Association of dairy products consumption with risk of obesity in children and adults: A meta-analysis of mainly cross-sectional studies. *Ann. Epidemiol.* **2016**, *26*, 870–882.e872. [CrossRef]
168. Chen, M.; Pan, A.; Malik, V.S.; Hu, F.B. Effects of dairy intake on body weight and fat: A meta-analysis of randomized controlled trials. *Am. J. Clin. Nutr.* **2012**, *96*, 735–747. [CrossRef]
169. Booth, A.O.; Huggins, C.E.; Wattanapenpaiboon, N.; Nowson, C.A. Effect of increasing dietary calcium through supplements and dairy food on body weight and body composition: A meta-analysis of randomised controlled trials. *Br. J. Nutr.* **2015**, *114*, 1013–1025. [CrossRef]
170. Soedamah-Muthu, S.S.; De Goede, J. Dairy consumption and cardiometabolic diseases: Systematic review and updated meta-analyses of prospective cohort studies. *Curr. Nutr. Rep.* **2018**, *7*, 171–182. [CrossRef] [PubMed]
171. Tong, X.; Dong, J.Y.; Wu, Z.W.; Li, W.; Qin, L.Q. Dairy consumption and risk of type 2 diabetes mellitus: A meta-analysis of cohort studies. *Eur. J. Clin. Nutr.* **2011**, *65*, 1027–1031. [CrossRef] [PubMed]
172. O'Sullivan, T.A.; Hafekost, K.; Mitrou, F.; Lawrence, D. Food sources of saturated fat and the association with mortality: A meta-analysis. *Am. J. Public Health* **2013**, *103*, e31–e42. [CrossRef] [PubMed]
173. Drouin-Chartier, J.-P.; Li, Y.; Ardisson Korat, A.V.; Ding, M.; Lamarche, B.; Manson, J.E.; Rimm, E.B.; Willett, W.C.; Hu, F.B. Changes in dairy product consumption and risk of type 2 diabetes: Results from 3 large prospective cohorts of US men and women. *Am. J. Clin. Nutr.* **2019**, *110*, 1201–1212. [CrossRef] [PubMed]
174. Gijsbers, L.; Ding, E.L.; Malik, V.S.; De Goede, J.; Geleijnse, J.M.; Soedamah-Muthu, S.S. Consumption of dairy foods and diabetes incidence: A dose-response meta-analysis of observational studies. *Am. J. Clin. Nutr.* **2016**, *103*, 1111–1124. [CrossRef] [PubMed]
175. Guo, J.; Astrup, A.; Lovegrove, J.A.; Gijsbers, L.; Givens, D.I.; Soedamah-Muthu, S.S. Milk and dairy consumption and risk of cardiovascular diseases and all-cause mortality: Dose–response meta-analysis of prospective cohort studies. *Eur. J. Epidemiol.* **2017**, *32*, 269–287. [CrossRef]
176. Alexander, D.D.; Bylsma, L.C.; Vargas, A.J.; Cohen, S.S.; Doucette, A.; Mohamed, M.; Irvin, S.R.; Miller, P.E.; Watson, H.; Fryzek, J.P. Dairy consumption and CVD: A systematic review and meta-analysis. *Br. J. Nutr.* **2016**, *115*, 737–750. [CrossRef]
177. Chen, G.-C.; Wang, Y.; Tong, X.; Szeto, I.M.Y.; Smit, G.; Li, Z.-N.; Qin, L.-Q. Cheese consumption and risk of cardiovascular disease: A meta-analysis of prospective studies. *Eur. J. Nutr.* **2017**, *56*, 2565–2575. [CrossRef]
178. Power, M.L.; Heaney, R.P.; Kalkwarf, H.J.; Pitkin, R.M.; Repke, J.T.; Tsang, R.C.; Schulkin, J. The role of calcium in health and disease. *Am. J. Obstet. Gynecol.* **1999**, *181*, 1560–1569. [CrossRef]
179. Kalkwarf, H.J.; Khoury, J.C.; Lanphear, B.P. Milk intake during childhood and adolescence, adult bone density, and osteoporotic fractures in US women. *Am. J. Clin. Nutr.* **2003**, *77*, 257–265. [CrossRef] [PubMed]
180. Huncharek, M.; Muscat, J.; Kupelnick, B. Impact of dairy products and dietary calcium on bone-mineral content in children: Results of a meta-analysis. *Bone* **2008**, *43*, 312–321. [CrossRef] [PubMed]
181. Wallace, T.C.; Bailey, R.L.; Lappe, J.; O'Brien, K.O.; Wang, D.D.; Sahni, S.; Weaver, C.M. Dairy intake and bone health across the lifespan: A systematic review and expert narrative. *Crit. Rev. Food Sci. Nutr.* **2021**, *61*, 3661–3707. [CrossRef] [PubMed]
182. Rizzoli, R. Dairy products and bone health. *Aging Clin. Exp. Res.* **2022**, *99*, 1256S–1262S. [CrossRef] [PubMed]
183. Clinton, S.K.; Giovannucci, E.L.; Hursting, S.D. The world cancer research fund/American institute for cancer research third expert report on diet, nutrition, physical activity, and cancer: Impact and future directions. *J. Nutr.* **2020**, *150*, 663–671. [CrossRef] [PubMed]
184. Ralston, R.A.; Truby, H.; Palermo, C.E.; Walker, K.Z. Colorectal cancer and nonfermented milk, solid cheese, and fermented milk consumption: A systematic review and meta-analysis of prospective studies. *Crit. Rev. Food Sci. Nutr.* **2014**, *54*, 1167–1179. [CrossRef] [PubMed]
185. Zang, J.; Shen, M.; Du, S.; Chen, T.; Zou, S. The association between dairy intake and breast cancer in Western and Asian populations: A systematic review and meta-analysis. *J. Breast Cancer* **2015**, *18*, 313–322. [CrossRef] [PubMed]

186. Dong, J.-Y.; Zhang, L.; He, K.; Qin, L.-Q. Dairy consumption and risk of breast cancer: A meta-analysis of prospective cohort studies. *Breast Cancer Res. Treat.* **2011**, *127*, 23–31. [CrossRef]
187. Lumsden, A.L.; Mulugeta, A.; Hyppönen, E. Milk consumption and risk of twelve cancers: A large-scale observational and Mendelian randomisation study. *Clin. Nutr.* **2023**, *42*, 1–8. [CrossRef]
188. Palupi, E.; Jayanegara, A.; Ploeger, A.; Kahl, J. Comparison of nutritional quality between conventional and organic dairy products: A meta-analysis. *J. Sci. Food Agric.* **2012**, *92*, 2774–2781. [CrossRef]
189. Simopoulos, A.P. The omega-6/omega-3 fatty acid ratio: Health implications. *Oléagineux Corps Gras Lipides* **2010**, *17*, 267–275. [CrossRef]
190. Kummeling, I.; Thijs, C.; Huber, M.; van de Vijver, L.P.L.; Snijders, B.E.P.; Penders, J.; Stelma, F.; Van Ree, R.; van den Brandt, P.A.; Dagnelie, P.C. Consumption of organic foods and risk of atopic disease during the first 2 years of life in the Netherlands. *Br. J. Nutr.* **2008**, *99*, 598–605. [CrossRef]
191. Christensen, J.S.; Asklund, C.; Skakkebæk, N.E.; Jørgensen, N.; Andersen, H.R.; Jørgensen, T.M.; Olsen, L.H.; Høyer, A.P.; Moesgaard, J.; Thorup, J. Association between organic dietary choice during pregnancy and hypospadias in offspring: A study of mothers of 306 boys operated on for hypospadias. *J. Urol.* **2013**, *189*, 1077–1082. [CrossRef]
192. Murali, A.P.; Trząskowska, M.; Trafialek, J. Microorganisms in Organic Food-Issues to Be Addressed. *Microorganisms* **2023**, *11*, 1557. [CrossRef]
193. Sadiq, M.; Paul, J.; Bharti, K. Dispositional traits and organic food consumption. *J. Clean. Prod.* **2020**, *266*, 121961. [CrossRef]
194. Rana, J.; Paul, J. Consumer behavior and purchase intention for organic food: A review and research agenda. *J. Retail. Consum. Serv.* **2017**, *38*, 157–165. [CrossRef]
195. Precedence Research. Organic Dairy Market. 2022. Available online: https://www.precedenceresearch.com/organic-dairy-market#:~:text=The%20global%20organic%20dairy%20market,10.1%25%20from%202022%20to%202030.&text=The%20organic%20dairy%20products%20are,reared%20using%20organic%20farming%20techniques (accessed on 1 February 2024).
196. Barkema, H.W.; von Keyserlingk, M.A.G.; Kastelic, J.P.; Lam, T.J.G.M.; Luby, C.; Roy, J.P.; LeBlanc, S.J.; Keefe, G.P.; Kelton, D.F. Invited review: Changes in the dairy industry affecting dairy cattle health and welfare. *J. Dairy Sci.* **2015**, *98*, 7426–7445. [CrossRef] [PubMed]
197. Carlson, A.; Greene, C.; Raszap Skorbiansky, S.; Hitaj, C.; Ha, K.; Cavigelli, M.; Ferrier, P.; McBride, W. *US Organic Production, Markets, Consumers, and Policy, 2000–21*; United States Department of Agriculture (USDA): Washington, DC, USA, 2023.
198. Willer, H.; Schaack, D.; Lernoud, J. Organic farming and market development in Europe and the European Union. In *The World of Organic Agriculture. Statistics and Emerging Trends 2019*; Research Institute of Organic Agriculture FiBL: Frick, Switzerland; IFOAM-Organics International: Bonn, Germany, 2019; pp. 217–254.
199. Willer, H.; Sorensen, N.; Yussefi-Menzler, M. The world of organic agriculture 2008: Summary. In *the World of Organic Agriculture*; Routledge: London, UK, 2010; pp. 15–22.
200. EU Agricultural Outlook 2021-31: Lower Demand for Feed to Impact Arable Crops. December 2021. Available online: https://agriculture.ec.europa.eu/news/eu-agricultural-outlook-2021-31-lower-demand-feed-impact-arable-crops-2021-12-09_en (accessed on 1 February 2024).
201. Maji, S.; Meena, B.S.; Paul, P.; Rudroju, V. Prospect of organic dairy farming in India: A review. *Asian J. Dairy Food Res.* **2017**, *36*, 1–8. [CrossRef]
202. Mahesh, M.S. Integrated organic farming and organic milk production: Opportunities and challenges in India. *Indian Dairym.* **2013**, *65*, 56–60.
203. KPMG. Global Organic Milk Production Market Report. Available online: https://ciorganicos.com.br/wp-content/uploads/2020/09/global-organic-milk-production-market-report.pdf (accessed on 1 February 2024).
204. Managi, S.; Yamamoto, Y.; Iwamoto, H.; Masuda, K. Valuing the influence of underlying attitudes and the demand for organic milk in Japan. *Agric. Econ.* **2008**, *39*, 339–348. [CrossRef]
205. McBride, W.D. *Characteristics, Costs, and Issues for Organic Dairy Farming*; DIANE Publishing: Collingdale, PA, USA, 2010; Volume 82.
206. Sabunevica, S.; Zagorska, J. Organic Milk as Medium for Lactic Acid Bacteria Growth: A Review. *Rural Sustain. Res.* **2023**, *49*, 73–86. [CrossRef]

Disclaimer/Publisher's Note: The statements, opinions and data contained in all publications are solely those of the individual author(s) and contributor(s) and not of MDPI and/or the editor(s). MDPI and/or the editor(s) disclaim responsibility for any injury to people or property resulting from any ideas, methods, instructions or products referred to in the content.

MDPI AG
Grosspeteranlage 5
4052 Basel
Switzerland
Tel.: +41 61 683 77 34

Foods Editorial Office
E-mail: foods@mdpi.com
www.mdpi.com/journal/foods

Disclaimer/Publisher's Note: The title and front matter of this reprint are at the discretion of the Guest Editors. The publisher is not responsible for their content or any associated concerns. The statements, opinions and data contained in all individual articles are solely those of the individual Editors and contributors and not of MDPI. MDPI disclaims responsibility for any injury to people or property resulting from any ideas, methods, instructions or products referred to in the content.

www.ingramcontent.com/pod-product-compliance
Lightning Source LLC
LaVergne TN
LVHW072319090526
838202LV00019B/2309